Advances in Soil Science

SOILS AND GLOBAL CHANGE

Edited by
R. Lal
John Kimble
Elissa Levine
B. A. Stewart

Boca Raton   London   Tokyo

Library of Congress Cataloging-in-Publication Data

Catalog record available from the Library of Congress.

This book contains information obtained from authentic and highly regarded sources. Reprinted material is quoted with permission, and sources are indicated. A wide variety of references are listed. Reasonable efforts have been made to publish reliable data and information, but the author and the publisher cannot assume responsibility for the validity of all materials or for the consequences of their use.

Neither this book nor any part may be reproduced or transmitted in any form or by any means, electronic or mechanical, including photocopying, microfilming, and recording, or by any information storage or retrieval system, without prior permission in writing from the publisher.

All rights reserved. Authorization to photocopy items for internal or personal use, or the personal or internal use of specific clients, may be granted by CRC Press, Inc., provided that $.50 per page photocopied is paid directly to Copyright Clearance Center, 27 Congress Street, Salem, MA 01970 USA. The fee code for users of the Transactional Reporting Service is ISBN 1-56670-118-X/95/ $0.00+$.50. The fee is subject to change without notice. For organizations that have been granted a photocopy license by the CCC, a separate system of payment has been arranged.

CRC Press, Inc.'s consent does not extend to copying for general distribution, for promotion, for creating new works, or for resale. Specific permission must be obtained in writing from CRC Press for such copying.

Direct all inquiries to CRC Press, Inc., 2000 Corporate Blvd., N.W., Boca Raton, Florida 33431.

© 1995 by CRC Press, Inc.
Lewis Publishers is an imprint of CRC Press

No claim to original U.S. Government works
International Standard Book Number 1-56670-118-X
Printed in the United States of America 1 2 3 4 5 6 7 8 9 0
Printed on acid-free paper

Preface

Global change and its causes and effects in relation to natural and anthropogenic activity has been the recent focus of concern within the scientific community. An important component of this issue is the role management of soils plays in contributing as a source or sink of carbon in the environment. Soil management extends across agriculture, urban, and natural environments and is a critical player in controlling carbon dynamics. Management also must be considered in the context of policy and economics in order to ensure the well being of society over the long term.

The chapters in this book are presented to emphasize the importance of managing soils properly with an awareness of their effect on global change and, specifically, the greenhouse effect. We have chosen to publish these manuscripts in the *Advances in Soil Science* series because of its proven record of disseminating knowledge about soils to a large international community. This volume will provide the scientific community with valuable information about how soil management will affect carbon ecosystems. Issues which deal with policy options and their affects on soil management and decisions which need to be made with regards to the best utilizaiton of the pedosphere (soil resources) are also addressed.

This volume is an integrated effort funded by the United States Department of Agriculture's Global Change Office, Soil Conservation Service (SCS), and Forest Service (FS), the Environmental Protection Agency (EPA), the National Aeronautics and Space Administration (NASA), and the Ohio State University to bring together the ideas of many different scientists on the topic of *Soils and Global Change*. The information here is an attempt to address the gaps in our knowledge of the role of soil management and policy options in global change, and at the same time, to present a "state of the art" compendium of our present knowledge on these issues. There are still questions regarding the role that soils play as a sink or source of carbon in the global carbon cycle. This book adresses our current knowledge of this role and identifies gaps in our understanding.

The editors would like to thank the authors for their efforts in documenting what they know about *Soils and Global Change* which will greatly assist us in moving forward in soil science and environmental research. We have enjoyed working with the authors and have appreciated their stimulating contributions. The authors have described past changes and made predictions of what can be expected in the future. Their efforts will greatly help others appreciate the important role that soils play in potential global change. We would also like to thank the staff of Lewis Publishers and Kirsten Sturart of the SCS in Lincoln, Nebraska for all their efforts in completing this volume so that the scientific community would have timely information on this important topic.

The Editors

About the Editors:

Dr. R. Lal is a Professor of Soil Science in the School of Natural Resources at The Ohio State University, Columbus, Ohio. Prior to joining Ohio State in 1987, he served as a soil scientist for 18 years at The International Institute of Tropical Agriculture, Ibadan, Nigeria. Prof. Lal is a fellow of the Soil Science Society of America, American Society of Agronomy, and The Third World Academy of Sciences. He is recipient of both the International Soil Science Award, and the Soil Science Applied Research Award of the Soil Science Society of America.

Dr. John Kimble is a Research Soil Scientist at the USDA Soil Conservation Service National Soil Survey Laboratory in Lincoln, Nebraska. For the past 4 years, Dr. Kimble has managed the Global Change project of the Soil Conservation Service, and has worked for the last 13 years with US Agency for International Development projects dealing with soils-related problems in more than 40 developing countries. He is a member of the American Society of Agronomy, the Soil Science Society of America, the International Soil Science Society, and the International Humic Substances Society.

Dr. Elissa Levine is a Soil Scientist in the Biospheric Sciences Branch at the NASA Goddard Space Flight Center since September, 1986. Prior to her employment at NASA, Dr. Levine served as a Resident Research Associate sponsored by the National Academy of Sciences, National Research Council. She was also a Post-Doctoral Fellow in the Plant and Soils Department at the University of Connecticut. Dr. Levine is a member of the American Society of Agronomy, the Soil Science Society of America, Society of Soil Scientists of Southern New England, Gamma Sigma Delta (Agricultural Honor Society), and Sigma Delta Epsilon (Graduate Women in Science).

Dr. B.A. Stewart is a Distinguished Professor of Soil Science, and Director of the Dryland Agriculture Institute at West Texas A&M University, Canyon, Texas. Prior to joining West Texas A&M University in 1993, he was Director of the USDA Conservation and Production Research Laboratory, Bushland, Texas. Dr. Stewart is past president of the Soil Science Society of America, and was a member of the 1990-93 Committee of Long Range Soil and Water Policy, National Research Council, National Academy of Sciences. He is a Fellow of the Soil Science Society of America, American Society of Agronomy, Soil and Water Conservation Society, a recipient of the USDA Superior Service Award, and a recipient of the Hugh Hammond Bennett Award by the Soil and Water Conservation Society.

Contributors

I.P. Abrol, Indian Center for Agriculture, Krishi Bhawan, New Delhi, India.

Robert J. Ahrens, Lead Scientist Soil Taxonomy, Soil Conservation Service, Federal Building, Room 152, 100 Centennial Mall North, Lincoln, NE 68508-3866.

Darwin W. Anderson, University of Saskatchewan, Department of Soil Science, Saskatoon, Saskatchewan, Canada S7N 0W0.

R. Aravena, Department of Earth Sciences, University of Waterloo, Waterloo, Ontario, Canada N2L 3G1

Richard W. Arnold, Director, Soil Survey Division, Soil Conservation Service, P.O. Box 2890, Room 4238-D, Washington, D.C. 20013.

D. Bachelet, ManTech Environmental Technology Inc., 200 SW 35th Street, US EPA Environmental Research Laboratory, Corvallis, OR 97333.

D.P. Billesbach, Department of Agricultural Meterology, University of Nebraska, L.W. Chase Hall, Lincoln, NE 68583-0728.

R.L. Blevins, Department of Agronomy, University of Kentucky, Lexington, KY 40506.

N.B. Bliss, Applications Branch, EROS Data Center, Sioux Falls, SD 57198.

J. Bogner, Energy Systems Division, Argonne National Laboratory, 9700 South Cass Avenue, Argonne, IL 60439.

Ray B. Bryant, Department of Civil & Engineering, 220 Hinds Hall, Syracuse University, Syracuse, NY 13244-1190.

G.A. Buyanovsky, Department of Soil & Atmospheric Sciences, University of Missouri, 144 Mumford Hall, Columbia, MO 65211.

Oliver A. Chadwick, Jet Propulsion Laboratory, MS 183-501, California Institute of Technology, 4800 Oak Grove Drive, Pasadena, CA.

H.H. Cheng, Department of Soil Science, University of Minnesota, 1991 Upper Buford Circle, St. Paul, Minnesota 55108-6028.

R.V. Chinnaswamy, Aqua Terra Consultants, 2672 Bayshore Parkway, Suite 1001, Mountian View, California 94043.

R.J. Clement, Department of Agricultural Meterology, University of Nebraska, L.W. Chase Hall, Lincoln, NE 68583-0728.

C.R. Crozier, Wetland Biogeochemistry Institute, Louisiana State University, Baton Rouge, LA 70803-7507.

R.D. DeLaune, Wetland Biogeochemistry Institute, Louisiana State University, Baton Rouge, LA 70803-7507.

A.S. Donigian, Jr., Aqua Terra Consultants, 2672 Bayshore Parkway, Suite 1001, Mountian View, CA 94043.

K.J. Elliott, USDA Forest Service, Coweeta Hydrologic Laboratory, 999 Coweeta Lab Road, Otto, NC 28763.

H. Eswaran, National Leader, World Soil Resources, Soil Conservation Service, P.O Box 2890, Room 4838-S, Washington, D.C. 20013.

R. Follett, USDA Agricultural Research Service, Soil-Plant-Nutrient Research Unit, 301 S. Howes St., Ft. Collins, CO 80521.

W.W. Frye, Department of Regulatory Service, University of Kentucky, Lexington, KY 40506.

Leland H. Gile (Retired), Soil Conservation Service, Las Cruces, NM 88801.

R.W. Gillham, Department of Earth Sciences, University of Waterloo, Waterloo, Ontario, Canada N2L 3G1.

C.G.M. Klein Goldewyk, Global Change Department, RIVM-MTV, PO Box 1 3720 BA Bilthoven, The Netherlands.

R.B. Grossman, Research Soil Scientist, Soil Conservation Service, Federal Building, Room 152, 100 Centennial Mall North, Lincoln, NE 68508-3866.

D.W. Johnson, USDA Forest Service, Coweeta Hydrologic Laboratory, 999 Coweeta Lab Road, Otto, NC 28763.

D. Harris, Department of Crop & Soil Sciences, Michigan State University, East Lansing, MI 48824-1325.

J.L. Hatfield, National Soil Tilth Laboratory, USDA Agricultural Research Service, 2150 Pammel Drive, Ames, IA 50011.

Marcel R. Hoosbeek, Department of Soil Science and Geology, Wageningen Agricultural University, PO Box 37, 6700 AA Wageningen, The Netherlands.

W.R. Horwath, Oregon State University, Department of Bioresource Engineering, 116 Gilmore, Corvallis, OR 97331.

R.A. Houghton, National Aeronautics and Space Administration, Code YSE, Washington, D.C. 20546.

G.L. Hutchinson, USDA-ARS, 301 S. Howes Street, Room 435, Fort Collins, CO 80522.

R.C. Izarralde, Department of Soil Science, University of Alberta, 434 Earth Sciences Building, Edmonton, Alberta, Canada T6G 2E1,

R.B. Jackson IV, U.S. EPA, 960 College Station Road, Athens, GA 30605.

D.W. Johnson, Desert Research Institute, University of Nevada at Reno, Reno, NV 89506.

K. Kennedy, US. Geological Survey, 5293 Ward Road, MS 408/NRP, Arvada, CO 80002.

J.S. Kern, ManTech Environmental Technology Inc., 200 SW 35TH Street, US EPA Environmental Research Laboratory, Corvallis, OR 97333.

J. Kim, Department of Agricultural Meterology, University of Nebraska, L.W. Chase Hall, Lincoln, NE 68583-0728.

B.A. Kimball, USDA Agricultural Research Service, Water Conservation Laboratory, 4331 E. Broadway Road, Phoenix, AZ 85040.

J. Kimble, Research Soil Scientist, Soil Conservation Service, Federal Building, Room 152, 100 Centennial Mall North, Lincoln, NE 68508-3866.

R. Lal, School of Natural Resources, Ohio State University, 2021 Coffey Road, Columbus, OH 43210-1085.

J.W. Laidlaw, Department of Soil Science, University of Alberta, 434 Earth Sciences Building, Edmonton, Alberta, T6G 2E1, Canada.

S.W. Leavitt, Laboratory of Tree-Ring Research, University of Arizona, Tucson, AZ 85721.

M.J. Lindstrom, USDA-ARS-MWA, North Iowa Avenue, Morris, MN 56267.

Elissa R. Levine, Biospheric Sciences Branch, NASA Goddard Space Flight Center, Code 923, Greenbelt, MD 20771.

R.L. Malcolm, US. Geological Survey, 5293 Ward Road, MS 408/NRP, Arvada, CO 80002.

Kim G. Mattson, Department of Forest Sciences, Oregon State University, Corvallis, OR 97331.

C.J. Merry, Ohio State University-Department of Civil Engineering, 2070 Neil Avenue, 470 Hitchock Hall, Columbus, OH 43210.

A.K. Metherell, AgResearch, Canterbury Agriculture and Science Center, P.O. Box 60, Lincoln, New Zealand.

G.J. Michaelson, University of Alaska-Fairbanks, 533 E. Fireweed, Palmer, AK 99645.

J.A.E. Molina, Department of Soil Science, University of Minnesota, 1991 Upper Buford Circle, St. Paul, MN 55108-6028.

Clarence E. Montoya, R.C. and D. Coordinator, Technical Support Office, Soil Conservation Service, 1926 7th Street, Las Vegas, NM 87701.

T.R. Moore, McGill University, Geology Department, Burnside Hall, 805 Sherbrooke Street West, Montreal, Quebec, Canada H3A 2K6.

M. Nyborg, Department of Soil Science, University of Alberta, 434 Earth Sciences Building, Edmonton, Alberta, Canada T6G 2E1.

W.C. Oechel, Department of Biology, San Diego State University, San Diego, CA 92182-0057.

T.B. Parkin, National Soil Tilth Laboratory, USDA Agricultural Research Service, 2150 Pammel Drive, Ames, IA 50011.

W.H. Patrick, Jr., Wetland Biogeochemistry Institute, Louisiana State University, Baton Rouge, LA 70803-7507.

A.S. Patwardhan, Aqua Terra Consultants, 2672 Bayshore Parkway, Suite 1001, Mountian View, CA 94043.

E.A. Paul, Department of Crop & Soil Sciences, Michigan State University, East Lansing, MI 48824-1325.

K. Paustian, Natural Resource Ecology Laboratory, Colorado State University, Fort Collins, CO 80523.

Gary W. Petersen, Environmental Resources Research Institute and Agronomy Department, Pennsylvania State University, University Park, PA 16802.

C.L. Ping, University of Alaska-Fairbanks, 533 E. Fireweed, Palmer, AK 99645.

K. Pregitzer, Michigan State University, Department of Forestry, East Lansing, MI 48824-1325.

J.H. Prueger, National Soil Tilth Laboratory, USDA Agricultural Research Service, 2150 Pammel Drive, Ames, IA 50011.

M.C. Rabenhorst, Department of Agronomy, University of Maryland, College Park, MD 20742.

P. Reich, World Soil Resources, Soil Conservation Service, P.O Box 2890, Room 4838-S, Washington, D.C. 20013.

D. Reicosky, USDA-ARS-MWA, North Iowa Avenue, Morris, MN 56267.

N.T. Roulet, York University, Department of Geography, 4700 Keele St., North York, Ontario, Canada M3J 1P3.

A.L. Rowell, Computer Sciences Corporation, 960 College Station Road, Athens, GA 30605.

C. Ryan, Department of Earth Sciences, University of Waterloo, Waterloo, Ontario, Canada N2L 3G1.

W.H. Schlesinger, Department of Botany & Geology, Duke University, Durham, NC 27708.

N.J. Shurpali, Department of Agricultural Meterology, University of Nebraska, L.W. Chase Hall, Lincoln, NE 68583-0728.

K. Spokas, Energy Systems Division, Argonne National Laboratory, 9700 South Cass Avenue, Argonne, IL 60439.

M. Tölg, Fraunhofer Institute for Atmospheric Environmental Research, Garmisch-Partenkirchen, Germany.

E. Van den Berg, World Soil Resources, Soil Conservation Service, P.O Box 2890, Room 4838-S, Washington, D.C. 20013.

S.B. Verma, Department of Agricultural Meterology, University of Nebraska, L.W. Chase Hall, Lincoln, NE 68583-0728.

M. Vloedbeld, Global Change Department, RIVM-MTV, PO Box 1 3720 BA Bilthoven, The Netherlands.

J.M. Vose, USDA Forest Service, Coweeta Hydrologic Laboratory, 999 Coweeta Lab Road, Otto, NC 28763.

George L. Vourlitis, Department of Biology, San Diego State University, San Diego, CA 92182-0057.

G.H. Wagner, Department of Soil & Atmospheric Sciences, University of Missouri, 144 Mumford Hall, Columbia, MO 65211.

Sharon W. Waltman, Soil Scientist, Soil Conservation Service, Federal Building, Room 152, 100 Centennial Mall North, Lincoln, NE 68508-3866.

K.B. Weinrich, Department of Earth Sciences, University of Waterloo, Waterloo, Ontario, Canada N2L 3G1.

C. Whitman, Natural Resources and Environment, 217 East Administration Building, USDA, 14th and Independence, Washington, D.C.

Contents

Chapter 1. World Soils and Greenhouse Effect: An Overview 1
R. Lal, J. Kimble, E. Levine, and C. Whitman

A. Global Carbon and Nitrogen Reserves

Chapter 2. An Overview of the Carbon Cycle 9
William H. Schlesinger

Chapter 3. Global Soil Carbon Resources 27
H. Eswaran, E. Van den Berg, P. Reich, and J. Kimble

Chapter 4. Changes in the Storage of Terrestrial Carbon Since 1850 45
R.A. Houghton

Chapter 5. Carbon Storage in Landfills 67
J. Bogner and K. Spokas

Chapter 6. Areal Evaluation of Organic and Carbonate Carbon in a Desert Area of Southern New Mexico 81
Robert B. Grossman, Robert J. Ahrens, Leland H. Gile, Clarence E. Montoya, and Oliver A. Chadwick

Chapter 7. Carbon Storage in Tidal Marsh Soils 93
Martin C. Rabenhorst

Chapter 8. Spatial Modeling Using Partially Spatial Data 105
R.B. Jackson IV, A.L. Rowell, and K.B. Weinrich

B. Soil Processes and Gaseous Emissions

Chapter 9. Effect of Global Change on Carbon Storage in Cold Soils 117
Walter C. Oechel and George L. Vourlitis

Chapter 10. Global Soil Erosion by Water and Carbon Dynamics 131
R. Lal

Chapter 11. Gaseous Emissions from Agro-Ecosystems in India 143
I.P. Abrol

Chapter 12. Methane Emission from Canadian Peatlands 153
T.R. Moore and N.T. Roulet

Chapter 13. Decomposition of Organic Matter and Carbon Emissions from Soils ... 165
Darwin W. Anderson

C. Factors Affecting Gaseous Emissions

1. CO_2 and CO Flux

Chapter 14. Impact of Fall Tillage on Short-Term Carbon Dioxide Flux 177
D.C. Reicosky and M.J. Lindstrom

Chapter 15. Organic Matter Inputs and Methane Emissions from Soils in Major Rice Growing Regions of China 189
J.S. Kern, D. Bachelet, and M. Tölg

Chapter 16. Soil CO_2 Flux in Response to Elevated Atmospheric CO_2 and Nitrogen Fertilization: Patterns and Methods 199
J.M. Vose, K.J. Elliott, and D.W. Johnson

Chapter 17. Soil Respiration and Carbon Dynamics in Parallel Native and Cultivated Ecosystems 209
G.A. Buyanovsky and G.H. Wagner

2. NO_x Flux

Chapter 18. Biosphere - Atmosphere Exchange of Gaseous N Oxides 219
G.L. Hutchinson

Chapter 19. Nitrous Oxide Flux from Thawing Soils in Alberta 237
J.W. Laidlaw, M. Nyborg, and R.C. Izaurralde

3. CH_4 Flux

Chapter 20. Methane Production in Mississisppi Deltaic Plain Wetland Soils As a Function of Soil Redox Species 247
C.R. Crozier, R.D. DeLaune, and W.H. Patrick, Jr.

D. Monitoring and Assessment

1. Soil Survey and GIS

Chapter 21. Role of Soil Survey in Obtaining a Global Carbon Budget 257
Richard W. Arnold

Chapter 22. Methods to Assess Soil Carbon Using Remote Sensing Techniques 265
Carolyn J. Merry and Elissa R. Levine

Chapter 23. Preparing a Soil Carbon Inventory for the United States using Geographic Information Systems 275
Norman B. Bliss, Sharon W. Waltman, and Gary W. Peterson

2. Analytical Techniques

Chapter 24. Establishing the Pool Sizes and Fluxes in CO_2 Emissions from Soil Organic Matter Turnover ... 297
E.A. Paul, W.R. Horwath, D. Harris, R. Follett, S.W. Leavitt, B.A. Kimball, and K. Pregitzer

Chapter 25. Fractionation and Carbon Balance of Soil Organic Matter in Selected Cryic Soils in Alaska ... 307
C.L. Ping, C.J. Michaelson, and R.L. Malcolm

Chapter 26. Fractionation, Characterization, and Comparison of Bulk Soil Organic Substances and Water-Soluble Soil Interstitial Organic Constituents in Selected Cryosols of Alaska ... 315
R.L. Malcolm, K. Kennedy, C.L. Ping, and G.T. Michaelson

Chapter 27. CO_2 Efflux from Coniferous Forest Soils: Comparison of Measurement Methods and Effects of Added Nitrogen ... 329
Kim G. Mattson

Chapter 28. In Search of Bioreactive Soil Organic Carbon: The Fractionation Approaches ... 343
H.H. Cheng and J.A.E. Molina

Chapter 29. The Use of ^{13}C Natural Abundance to Investigate the Turnover of the Microbial Biomass and Active Fractions of Soil Organic Matter under Two Tillage Treatments ... 351
M.C. Ryan, R. Aravena, and R.W. Gillham

3. Climatic Approach

Chapter 30. Trace Gas and Energy Fluxes: Micrometeorological Perspectives ... 361
S.B. Verma, J. Kim, R.J. Clement, N.J. Shurpali, and D.P. Billesbach

Chapter 31. A Micrometeorological Technique for Methane Flux Determination from A Field Treated with Swine Manure ... 377
J.H. Prueger, T.B. Parkin, and J.L. Hatfield

4. Modeling

Chapter 32. Application of the CENTURY Soil Organic Matter Model to a Field Site in Lexington, KY ... 385
A.S. Patwardhan, R.V. Chinnaswamy, A.S. Donigian, Jr., A.K. Metherell, R.L. Blevins, W.W. Frye, and K. Paustian

Chapter 33. The Exchange of Carbon Dioxide between the Atmosphere and the Terrestrial Biosphere in Latin America ... 395
C.G.M. Klein Goldewijk and M. Vloedbed

Chapter 34. Modeling the Dynamics of Organic Carbon in a Typic Haplorthod ... 415
Marcel R. Hoosbeek and Ray B. Bryant

E. Research and Development Priorities

Chapter 35. Towards Improving the Global Data Base on Soil Carbon 433
R. Lal, J. Kimble, E. Levine, and C. Whitman

CHAPTER 1

World Soils and Greenhouse Effect: An Overview

R. Lal, J. Kimble, E. Levine, and C. Whitman

I. Introduction

The average temperature of the earth's surface, currently at about 15°C, is controlled by the gaseous composition of the atmosphere. Radiatively-active or greenhouse gases in the atmosphere trap outgoing solar radiation which warms the earth. Important greenhouse gases in the atmosphere include water vapor (H_2O), carbon dioxide (CO_2), methane (CH_4), nitrous oxide (N_2O), oxides of nitrogen (NO_x), tropospheric ozone (O_3), carbon monoxide (CO) and chloroflourocarbon (CFC). Per molecule, CH_4 is 32 times more effective in trapping longwave radiation than CO_2, and N_2O is approximately 150 times more effective (Bouwman, 1990). Major natural sources of these gases are terrestrial ecosystems, including world soils, biota, wetlands, and volcanic eruptions. Emission and reabsorption of these gases from natural ecosystems have been in equilibrium for millions of years. However, this balance has recently been disturbed by human activities. Consequently, atmospheric concentrations of several of these gases (e.g. CO_2, CH_4, and N_2O) have been increasing since the onset of the industrial revolution and more rapidly since the 1950's (Table 1).

II. Global Carbon Pools and Fluxes

There are four principal pools of global carbon, e.g. oceans, atmosphere, terrestrial ecosystems, and geological formations containing fossil and mineral carbon (Figure 1). Disturbance of any of these pools has a direct effect on others because of interlinkages among the pools. An increase in the atmospheric pool at the expense of soils, biotic, and oceanic pools can lead to global warming and climate change. Global warming can also shift the delicate balance among various pools.

The global budget of organic carbon is shown in Figure 2. The amount of carbon fixed annually by world biota through photosynthesis is balanced by the release of carbon by plant respiration and decomposition of organic residues from biomass and soil organic carbon. Principal sources of carbon are burning fossil fuel, and deforestation and land use. Burning fossil fuel releases about 5.4 Pg carbon annually, and deforestation of tropical rainforest and other land uses release about 1.6 Pg C annually.

There are two known sinks for carbon. Increase in atmospheric CO_2 concentration at 0.5%/yr accounts for about 3.2 Pg of carbon. An additional 2.0 Pg/yr is absorbed by the ocean. The remainder of 1.8 Pg C/yr is unaccounted for, and is believed to be absorbed by terrestrial ecosystems. A big unknown in this global C balance is the role of the world soils, especially with regards to processes involved in carbon emission and sequestration.

Table 1. Composition and changes in concentration of greenhouse gases in the atmosphere

Gas	Concentration in 1985	Annual increase since 1985 to present (%)	Contribution to global warming (%)
CO_2	345 ppm	0.5	50
CH_4	90 ppb	0.8	19
N_2O	1.65 ppm	1.0	5
CFC	0.24 ppb	3.0	15
Others	----	----	11

Adapted from Bouwman, 1990; USEPA, 1990.

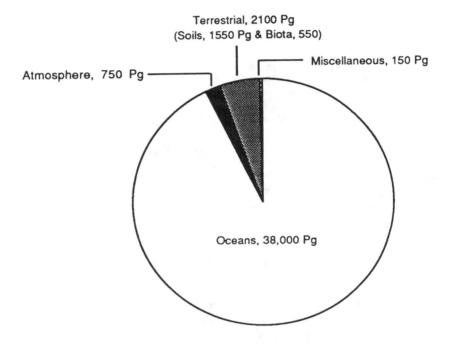

Figure 1. Global carbon pools (Pg = petagrams = 1×10^{15} grams). (Adapted from Post et al., 1990; Houghton and Skole, 1990; Eswaran et al., 1993; and Schlesinger, 1993.)

III. Global Carbon Balance and World Soils

A. Soil Carbon and Nitrogen Pools

World soils are an important pool of active carbon and play a major role in the global carbon cycle. There are two types of carbon pools in world soils: organic (SOC) and inorganic (SIC) pools. The SOC pool is concentrated near the soil surface within the top 1 m depth, and is estimated at about 1550 Pg. The SIC pool is contained in the deeper layer (below 1 m depth), and is composed of inorganic carbon in the form of calcium carbonate ($CaCO_3$) or caliche. The SIC pool contains as much as 1700 Pg of carbon (Post et al., 1982; Schlesinger, 1991; Eswaran et al., 1993; Monger, 1993).

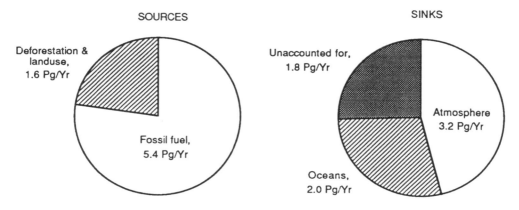

Figure 2. Global CO_2 budget for 1980-1989. (Adapted from Sarmiento and Sundquist, 1992; Watson et al., 1992; and Sundquist, 1993.)

World soils also contain about 95 Tg (Tg = Teragram = 10^{12} grams) of nitrogen. Along with fertilizers applied to cultivated soils, this nitrogen pool is the principal source of N_2O and other gases (NO_x) released into the atmosphere. About 80 percent of nitrogen stored in world soils is contained in the top 10 cm of the soil surface.

The atmospheric pools of carbon and other gases are increasing to a large extent at the cost of soil pools. Small changes in the pools of carbon and nitrogen in world soils could have large effects on atmospheric concentrations of CO_2, CH_4, and N_2O. Important agricultural activities that lead to change in pools of carbon and nitrogen in world soils include deforestation and afforestation, biomass burning, cultivation, including rice paddies residue management, fertilizer application, farming systems, etc. During the past two centuries alone, soils have been a net source of greenhouse gases, with the cumulative loss of carbon from vegetation and soils for the period from 1850 to 1980 ranging from 90 to 120 Pg (Houghton and Skole, 1990). The current annual net loss of carbon from plant and soils is estimated to be 0.2 Pg from temperate regions and about 2 Pg from tropical regions.

B. Cold Soils

Northern ecosystems (comprising Arctic, boreal forest, and northern bogs) contain an estimated 350 to 455 Pg of carbon in the permafrost and soil active layer. This carbon pool represents 22.5 to 29.4% of the total world soil carbon pool (Billings, 1987; Post, et al., 1990; Oechel and Vourlitis, 1993). The Arctic tundra ecosystem alone contains about 192 Pg of soil carbon, or approximately 12.4% of the global soil pool.

In the event of global warming, northern ecosystems are anticipated to undergo the most significant increases in surface temperature. Surface temperature increases of 4 °C in summer and as much as 17 °C in winter have been predicted (Schlesinger and Mitchell, 1987). Drastic climate change could increase the depth of the soil's active layer, resulting in warmer soil temperatures, more soil aeration, and higher rates of SOC decomposition and CO_2 efflux. In the warming scenario, northern ecosystems could cease to be a carbon sink and become a net carbon source.

The loss of continuous and discontinuous permafrost in the northern ecosystems may also increase the land area of more productive and warmer soils that may lead to more carbon sequestration. Therefore, net result of the perceived climate change on C balance of cold soils is difficult to predict.

C. Forest Ecosystems

Forest ecosystems and forest soils are estimated to account for 60% of the terrestrial carbon pool. Forest soils represent a large, stable pool of SOC and nutrients, and act as a buffer from the effects of natural and human disturbances to forest systems. Forest floor decomposition and CO_2 efflux are very sensitive to temperature and moisture conditions. Accumulation of SOC is generally inversely correlated to the primary production of the forest system and to the forest floor decomposition rate. Accumulation is highest at high latitudes and higher elevations and least in equatorial climate and low latitudes of tropical rain forests. The net primary productivity follows the reverse trend. Carbon storage in forest soils are also affected by forest type (species, deciduous, evergreen) and site quality.

D. Wetlands and Rice Paddies

Wetlands include the swamps, bogs, marshes, mires, fens and several other wet ecosystems found throughout the world. These are highly diverse ecosystems, and are found in every continent, except Antarctica, and in every climate from the tropics to the tundra. It is estimated that wetlands occupy over 6% of the land surface of the world (about 856 million hectares). About 56% of the world wetlands occur in tropical and sub-tropical regions, where a large proportion has been developed and is cultivated for rice production. Boreal regions account for 30% of world's wetlands and temperate regions about 12%. Coastal wetlands have high primary productivity with carbon storage of 20-200 kg/m^2 with C sequestration rate of 0.05-0.5 kg/m^2/yr.

Carbon transformation in wetlands mostly occurs under reducing or anaerobic conditions. Some important anaerobic processes include fermentation, methanogenesis, and sulfur reduction. Peat deposition in wetlands leads to carbon storage in these ecosystems, and is estimated at about 29 g C/m^2/yr. Currently, the world's wetlands are estimated to be a sink of about 0.08 Pg C/yr and a source of 0.055 Pg C/yr. Net global retention of carbon in wetland peats is about 0.057 to 0.083 Pg/yr. About one-fourth of the annual carbon stored in wetlands is released by methane emissions. Total methane emission from wetlands is estimated at 0.09 Pg C/yr.

Agricultural conversion of peatlands also have an important impact on the global carbon cycle. By 1980, the total carbon shift due to agricultural drainage of peatlands in North America was 0.063-0.085 Pg C/yr, with an additional release of 0.032-0.039 Pg C/yr from peat combustion.

Rice is cultivated on about 146 million ha, and rice paddies are a major source of methane emission into the atmosphere. Methane flux from rice paddies is estimated at 0.08 Pg C/yr, out of a total methane release of about 0.4 Pg C/yr from all sources.

E. Landfills and Waste Disposal

Worldwide landfills are a major waste disposal method compared with alternative techniques of open dumping and burning. Worldwide approximately 340 Tg of municipal solid waste were landfilled in the mid-1980's. In the United States, 178 Tg of solid waste was transferred to more than 9,000 municipal solid waste landfills during the mid-1980's. Similar to wetlands and rice paddies, anaerobic decomposition is the predominant process of biomass breakdown in landfills. Consequently, landfills emit methane and carbon dioxide to the atmosphere. Worldwide estimates of methane emission from landfills range from 9-70 Tg/yr (Bogner and Spokes, 1993).

IV. Processes and Practices

Important soil processes that affect the global carbon balance include global soil-erosion, biomass burning, and soil fertility depletion.

A. Global Soil Erosion

The total amount of soil displaced annually by water erosion from world soils is estimated at 190 Pg (Lal, 1993a), of which 19 Pg is transported by world rivers to the oceans (Walling, 1987). Assuming a mean carbon content of 3%, total carbon displaced from world soils by water erosion is 5.7 Pg/yr. Assuming that one-fifth of the carbon displaced by erosion is easily decomposed and mineralized, it will lead to carbon efflux of 1.14 Pg/yr into the atmosphere. Remainder of the carbon displaced, about 4 Pg/yr, is translocated within the landscape. The fate of the carbon translocated within the landscape is not known.

B. Biomass Burning

Fire is an important factor in global carbon balance and is a natural occurrence in all fire-dependent ecosystems, e.g. savannas and grasslands. In addition, it is an important tool in soil management for agricultural ecosystems. Fire affects the global carbon pool in two principal ways. First, it directly releases carbon from the biomass during combustion. Second, it indirectly accentuates carbon release from soil from which vegetation has been burnt. Indirect effects of fire may lead to biogenic emissions of nitrous and nitric oxides, and methane. Furthermore, soil exposed by burning is prone to accelerated erosion, and possibly an enhanced rate of SOC mineralization.

Fire is an important management tool in regions of shifting cultivation and tropical grasslands. Shifting cultivation is practiced on about 25 million ha of freshly cut forest annually. The loss of carbon from soil by shifting cultivation may be as much as 6.25 Tg/yr with a range of 3.75 Tg to 9.18 Tg/yr (Lal, 1993b). In comparison, tropical grasslands and savannas occupy 1.5 billion ha. The loss of soil carbon due to natural fire in tropical savannas may be as much as 187.5 Tg/yr with a range of 112.5 Tg to 275.6 Tg/yr (Lal, 1993b).

C. Soil Fertility Depletion

Inappropriate land use and soil mismanagement can render world soils as a major source of greenhouse gases. Soil degradation, caused by land misuse, ecologically incompatible farming systems, and inappropriate soil management practices, can be a major cause of fertility depletion and gaseous emissions from soil. Soil degradation and desertification are serious problems in several tropical ecosystems, especially in dry and hot climates. Soil degradation in the tropics is responsible for total emission of about 130 Tg C/yr. Tropical deforestation may cause an additional loss of 100 to 200 Tg C/yr (Lal, 1993a). Total emission from soils of the tropics may be 0.5 Pg C/yr. In addition to increasing emissions of greenhouse gases from soils, soil degradation reduces the net primary productivity of land that is the rate of carbon uptake by plants/biota from the atmospheric pool of carbon.

Consequently, the current rate of loss of carbon from plant and soils in the tropics is about 10 times that of temperate regions e.g. 2 Pg C/yr vs. 0.2 Pg C/yr (Houghton and Skole, 1990). Agricultural practices that exacerbate emission include: mechanized methods of deforestation, plow-till farming, continuous cropping on marginal lands and ecologically-sensitive ecoregions, low-input and resource-based shifting agriculture and related bush fallow systems, subsistence farming leading to fertility depletion and soil degradation, and overstocking and overgrazing. In contrast, agricultural practices that replenish SOC and restore soils capacity as a carbon sink include afforestation, conservation tillage and mulch farming techniques, planted fallows and cover crops, science-based agriculture with judicious chemical inputs, managed pastures with low stocking rate, agroforestry, etc.

Science-based, economically profitable, and ecologically-sustainable agricultural systems are soil-restorative and likely to sequester carbon in world soils. Low input systems of subsistence farming and shifting cultivation can be soil-degradative leading to net carbon emission from soil. Judicious use of chemical fertilizers and organic manures can replenish soil fertility, improve net primary productivity, enhance biomass addition into the soil, restore soil structure, improve SOC content, and sequester carbon from atmosphere to the soil. Science-based soil management is crucial to reversing the greenhouse effect.

V. Uncertainties

An accurate assessment of global carbon balance is faced with many uncertainties. These uncertainties may lead to erroneous data and misinterpretations, and arise from a variety of sources including unstandardized methods, data quality and reliability, and missing and incomplete data.

A. Unstandardized Methods

The principal source of uncertainty lies in procedures and methodologies used for assessment of various pools and fluxes. There continues to be a lack of standard methods to measure and monitor SOC, and gaseous fluxes between soil and the atmosphere. Commonly used methods for carbon determination may lead to substantial errors. Progress is being made in understanding processes that govern gas fluxes, but methods employed to measure them differ and often give variable results. Technologically elegant methods are expensive, require a substantial technical know-how and may not work as well for forested ecosystems as they do in agricultural fields or open peat lands. Static chamber and diffusion gradient methods are much less expensive, but consider only a small column of soil, sample only a short time period, and require many replications. The size and design of the chamber itself may significantly influence results. It is important to standardize and compare methods, integrate point measurements in time and space, and improve quality. The objective is to take valid measurements using a well-documented methodology, and to understand the processes within the soil ecosystem that contribute to emissions, and the controlling environmental factors.

B. Data Quality and Reliability

There is a great deal of uncertainty in the data on global soil carbon pool and fluxes. Information generated can only be as good as the quality of input data. At present, the quality of available data is extremely uneven. Users unaware of data quality can be easily misled. The data quality is also related to the problems of scaling and extrapolation from point source to landscape and ecosystem or region. There is also a tendency of over reliance on modeling. While models are a powerful tool to examine the feasibility and impact of options, they do not capture exceptions and anecdotal information relevant to policy analysis.

C. Missing and Incomplete Data

There is also a lack of data for specific ecosystems, including detailed measurements of fluxes. Most problematic are missing and incomplete data. The cause and effect relationships of management practices are uncertain in many instances and need to be evaluated for different land use systems.

D. Knowledge Gaps

Several processes related to gaseous emissions from soils are not well understood. Processes that are often overlooked in field experiments include soil erosion and effects of termites, earthworms, and other soil fauna. There is also a possibility that the different carbon pools in soils differ not only in chemistry but in physical entrapment in soil aggregates. Understanding processes and controls is necessary for developing and validating models required to extrapolate findings, both in time, and into areas where measurements are neither available nor feasible. The magnitude of CO_2 fluxes from soil and root respiration processes under elevated levels of CO_2 and interaction with nitrogen in soil is not known and processes involved are poorly understood.

VI. Policy Issues

Efforts to sequester carbon in soil through management are hindered by biophysical and socio-economic factors. While biophysical factors can be managed, policy and economic issues may be limiting and require careful consideration for sequestering carbon. Forest soils are a major pool of carbon in most ecoregions. Impact factors with regard to C sequestration in forest soils include forest management, age, location and forest species. Other options for maintaining or increasing carbon stocks include creating markets for biomass feedstocks, the promotion of short rotation woody crops, the use of trees for windbreaks and shelterbelts, and urban tree planting programs which reduce energy usage in buildings.

Policy considerations and economics are inter-related issues. While economics plays a major role in policy making, there are also social and cultural factors which have equally significant roles, and must be addressed.

References

Billings, W.D., 1987. Carbon balance of Alaskan tundra and taiga ecosystems: past, present and future. *Quart. Sci. Rev.* 6:165-177.

Bogner, J.E. and K. Spokas. 1993. Landfill CH_4: Rates, fate and role in global carbon cycle. *Chemosphere* 26: 369-386.

Bouwman, A.F. (ed.)., 1990. *Soils and the greenhouse effect*. John Wiley & Sons, United Kingdom. 575 pp.

Eswaran, H., E. VandenBerg, P. Reich and J. Kimble, 1993. Global soil carbon resources. Proc. Int'l Symp. on "Soil Processes and Management Systems: Greenhouse Gas Emissions and Carbon sequestration", Columbus, Ohio 4-9 April, 1993.

Houghton, R.A. and D.L. Skole, 1990. The long term flux of carbon between terrestrial ecosystems and the atmosphere as a result of changes in land use. Research Project of the March-July. Carbon Dioxide Research Program, Office of Health and Environmental Research, US Department of Energy, Washington, D.C.

Lal, R., 1989. Soil as a potential source or sink of carbon in relation to greenhouse effect. USEPA Workshop "Greenhouse Gas Emissions From Agricultural Systems. Vol. 2. 12-14 Dec., 1989.

Lal, R. and T.J. Logan. 1993a. Agricultural activities and carbon emissions from soils of the topics. Proc. Int'l. Symp. on "Soil Processes and Management Systems: Greenhouse Gas Emissions and Carbon Sequestration", Columbus, Ohio 4-9 April, 1993.

Lal, R., 1993b. Global soil erosion by water and carbon dynamics. Proc. Int'l Symp. on "Soil Processes and Management Systems: Greenhouse Gas Emissions and Carbon Sequestration", Columbus, Ohio. 4-9 April, 1993.

Monger, H.C., 1993. Inorganic carbon in desert soils of New Mexico: Microbial precipitation and isotopic significance. Proc. Int'l Symp. on "Soil Processes and Management Systems: Greenhouse Gas Emissions and Carbon Sequestration", Columbus, Ohio 4-9 April, 1993.

Oechel, W.C. and G.L. Courlitis, 1993. Effects of global change on carbon storage in cold soils. Proc. Int'l. Symp. on "Soil Processes and Management Systems: Greenhouse Gas Emissions and carbon Sequestration", Columbus, Ohio. 4-9 April, 1993.

Post, W.M., W.R. Emmanuel, P.J. Zinke, and A.G. Stangenberger, 1982. Soil carbon pool in world life zones. *Nature* 298:156-159.

Post, W.M., T.H. Peng, W.R. Emmanuel, A.W. King, V.H. Dale and D.L. DeAngelis, 1990. The global carbon cycle. *Amer. Sci.* 78:310-326.

Sarmiento, J.L. and E.T. Sundquist, 1992. Revised budget for the oceanic uptake of anthropogenic carbon dioxide. *Nature* 356:589-593.

Schlesinger, W.H., 1991. *Biogeochemistry: analysis of global change*. Academic Press. San Diego.

Schlesinger, W.H., 1993. An overview of the global carbon cycle. Proc. Int'l. Symp. on "Soil Processes and Management Systems: Greenhouse Gas Emissions and Carbon Sequestration", Columbus, Ohio. 4-9 April, 1993.

Schlesinger, W.H. and J.F.B. Mitchell, 1987. Climate model simulations of the equilibrium climatic response to increased carbon dioxide. *Rev. Geophys.* 25:760-798.

Sundquist, E.T., 1993. The global carbon dioxide budget. *Science*. 259:934-941.

USEPA, 1990. Greenhouse gas emissions from agricultural systems. Workshop. Vol. I. 12-14 Dec., 1989. Washington, D.C.

Walling, D.E., 1987. Rainfall, runoff, and erosion of the land: a global view. p. 89-117. In: K.J. Gregory (ed.), *Energetics of Physical Environment*. John Wiley & Sons, United Kingdom.

Watson, R.T., L.G. Meira Filho, E. Sanhueza, and A. Janetos, 1992. p. 29-46. In: J.T. Houghton, B.A. Callander, and S.K. Varney (eds.), *Climate Change*. The Supplementary Report to the IPCC Scientific Assessment. Cambridge Univ. Press, Cambridge.

CHAPTER 2

An Overview of the Carbon Cycle

William H. Schlesinger

I. Introduction

A. The Geochemical Cycle of Carbon

The Earth contains about 10^{23} g of carbon, which it obtained early during its formation as a planet. Most of the accretion probably took place in the interval from 5.0 to 3.5 bya--a period of great meteor bombardment. The present-day receipt of extra-terrestrial materials would be about 1000x too low to yield the current inventory of carbon even if it had continued for all of Earth's history (Anders, 1989). Although a number of theories can explain the formation and differentiation of the planet, the source of the Earth's carbon is often linked to the receipt of carbonaceous chondrites, a type of meteor that typically contains from 0.4 to 3.6% carbon (Anders and Owen, 1977). Carbonaceous chondrites are thought to have been particularly abundant during the late phases of the Earth's accretion, providing a layer of carbon-rich materials to the Earth's developing crust. The receipt of comets was also a potential source of both water and carbon to the Earth's surface (Chyba, 1987; Delsemme, 1992), including some organic molecules that may have been precursors for biochemistry (Anders, 1989).

In its early history, the Earth experienced massive heating, melting, and differentiation of its materials, resulting in the separation of the crust from the underlying mantle. During this process, most of the planet's carbon was released to the atmosphere as carbon dioxide. Even today, a portion of the Earth's volcanic emissions represents primordial degassing of the mantle, which continues at a relatively low rate (Lupton and Craig, 1981). Higher rates of degassing in the past must have resulted in high contents of carbon dioxide and water vapor in the atmosphere, especially before the planet cooled to the condensation point of water. Walker (1985) suggests that a high concentration of CO_2 in the Earth's primitive atmosphere may have been important in maintaining the Earth's temperature above the freezing point of water during periods when the Sun's luminosity was as much as 30% lower than today.

With the condensation of water vapor to form the oceans, atmospheric carbon dioxide dissolved in seawater, according to Henry's Law for the proportionation of gases between gaseous and dissolved phases in a closed system:

$$CO_2 + H_2O \rightleftarrows H_2CO_3.$$

The solubility of CO_2 in water is 3.2 g/l under conditions of standard temperature and pressure, and the present-day oceans contain about 56× more dissolved CO_2 than the atmosphere. The dissolution of CO_2 in water creates a weak acid, viz:

$$H_2CO_3 \rightleftarrows H^+ + HCO_3^-,$$

but from the time of the earliest oceans, this acid must have been consumed by reaction with crustal minerals in the process of carbonation weathering. For example, the reaction of CO_2 with albite consumes protons, and delivers Na^+ to the sea:

$$2NaAlSi_3O_8 + 2H_2CO_3 + 9H_2O \rightarrow$$
$$2Na^+ + 2HCO_3^- + 4H_4SiO_4 + Al_2Si_2O_5(OH)_4.$$

The removal of CO_2 from the Earth's atmosphere by rock weathering is succinctly described by the following equation (Siever, 1974):

Acid Volatiles + Igneous Rocks = Sedimentary Rocks + Salty Oceans,

which recognizes that the removal of trace gases containing carbon, nitrogen, or sulfur from the atmosphere involves the formation of "acid" anions (HCO_3^-, NO_3^-, or SO_4^{2-}) and weathering of the exposed crust.

As early as 1918, Arrhenius recognized that the interaction of the Earth's atmosphere with the crust consumed CO_2, which, if not restored, would eventually lead to a cooling of the planet through a loss of its natural "greenhouse effect":

> As the crust grew thicker, the supply of this gas [CO_2] diminished and was further used up in the process of disintegration [weathering]. As a consequence the temperature slowly decreased, although decided fluctuations occurred with changing volcanic activity during different periods. Supply and consumption of carbon dioxide fairly balanced as disintegration ran parallel with the proportion of this gas in the air.
>
> From Svante Arrhenius, *The Destinies of the Stars*

Carbon dioxide removed from the atmosphere and transferred to the ocean by rock weathering is eventually deposited on the seafloor in carbonate rocks, adding to the Earth's crust. Fortunately, CO_2 is later released from the crust as a result of the Earth's tectonic activity. Kasting et al. (1988) have outlined the complete geochemical cycle for carbon (Figure 1), in which subduction of the oceanic crust carries carbonate minerals deposited on the sea floor to the interior of the Earth, where CO_2 and other volatile elements are once again released by hydrothermal and volcanic emissions. Presumably most of the CO_2 being vented by volcanoes today has made at least one previous trip through this cycle.

Through the "greenhouse effect," the content of H_2O and CO_2 in the Earth's modern atmosphere adds 33 °C to the Earth's mean temperature (16 °C), preventing our planet from becoming a frozen ball of ice. In the past, Mars also enjoyed a period of substantial tectonic activity and crustal degassing and a surface temperature that allowed the presence of liquid water (Carr, 1987). Now, however, Mars shows no tectonic activity, the total pressure of CO_2 in its atmosphere is very low, and the planet is very cold (-53 °C). On Venus, where the temperature is too hot to allow the formation of carbonate minerals on the surface, most of the carbon dioxide is in the atmosphere, where it contributes to a strong "greenhouse" effect and a surface temperature of 474 °C (Nozette and Lewis, 1982). As a warning of our own future, Wally Broecker has said that "God put Venus [in the heavens] as a symbol of what can go wrong if a planet is mismanaged" (quoted in Nisbet, 1991).

Despite their long-term significance, the annual transfers of carbon in the geochemical cycle are relatively small. The massive quantities of CO_2 that are now tied up in the carbonate minerals of the Earth's crust (7.7 $\times 10^{22}$ g C) are the result of the slow accumulation of these materials over long periods of the Earth's history. Today, rivers carry about 480×10^{12} g of calcium to the sea. For seawater to maintain fairly constant concentrations of Ca, an equivalent amount of Ca must be deposited as $CaCO_3$ in ocean sediments, carrying 0.14×10^{15} g of carbon to the oceanic crust. Dividing the mass of carbonate rocks by the rate of annual formation, we find that each atom of carbon sequestered in marine carbonate spends about 500,000,000 years in that reservoir.

Each year, between 0.018 and 0.13×10^{15} g C are released as CO_2 from volcanoes around the world (Allard et al., 1991; Williams et al., 1992), roughly balancing the burial of carbonate-carbon. During periods of great tectonic activity, for example in the Eocene, the release of CO_2 from hydrothermal activity may have exceeded the rate of CO_2 consumption in rock weathering, yielding high concentrations of CO_2

in the atmosphere and a temporary greenhouse-induced warming of the planet (Owen and Rea, 1985). Luckily this process seems never to have exceeded the capacity for the geochemical cycle of carbon to buffer the concentrations of atmospheric CO_2 during a subsequent quiescent period of crustal activity.

II. The Biogeochemical Cycle of Carbon

The earliest evidence of life on Earth is found in 3.5-billion-year-old rocks from western Australia, where in 1983, Stan Awramik and his colleagues found microfossils that resemble primitive bacteria of today. Evidence of photosynthesis dates to 3.5 bya, the age of the earliest sediments that contain organic carbon that is depleted in ^{13}C to the extent seen in modern photosynthetic organisms (Schidlowski, 1983). At times when the production of organic carbon by photosynthesis has exceeded its destruction by respiration, organic carbon has accumulated in geologic sediments. Significant organic carbon is thought to have been present by 2.5 bya, with the pool increasing to 1.56×10^{22} g by about 540 mya (Des Marais et al., 1992). During that interval, between 10 and 20% of all the carbon buried in marine sediments was organic--similar to the ratio in modern marine sediments (Li, 1972). Presently about 0.13×10^{15} g C is buried in organic forms in marine sediments each year (Berner, 1982).

Despite the early evolution of photosynthesis, between 3.5 and 2.0 bya the O_2 content of the atmosphere was negligible. Until recently, most researchers attributed the lack of oxygen solely to the oxidation of reduced iron (Fe^{2+}) in seawater and the deposition of Fe_2O_3 in Banded Iron Formations (Cloud, 1973; Francois and Gerrard, 1986). Now, it seems that the quantity of Fe^{2+} may not have been sufficient to remove all the O_2 (Towe, 1990; Kump and Holland, 1992); the oxidation of other reduced species, perhaps sulfide (S^{2-}), may have also played a role. The relatively late appearance of microbial sulfate reduction-- about 2.4 bya--may indicate the time needed to accumulate sufficient sulfate in seawater to allow the evolution of this metabolic pathway (Cameron, 1982). The termination of the Banded Iron Formation marks the time when oxygen began to accumulate in the atmosphere--about 2.0 bya--which subsequently allowed the evolution of land plants and the appearance of a significant terrestrial component to biogeochemistry.

The presence of life on Earth means that an understanding of the dynamics of carbon on the planet moves from the realm of geochemistry to that of *bio*geochemistry. Models of the carbon cycle must focus on the large annual transfer of CO_2 from the atmosphere to plants during the process of photosynthesis, and the large return of CO_2 to the atmosphere during the process of decomposition. On the present-day Earth, these fluxes dwarf the movements of CO_2 in the geochemical cycle by a factor of 1000.

However, the appearance of life on land and sea also stimulated the rate of some of the reactions in the underlying geochemical cycle of carbon (Figure 1). Various marine organisms enhance the deposition of calcareous sediments, which now cover more than half of the oceans' seafloor (Kennett, 1982). Land plants, by maintaining high concentrations of CO_2 in the soil pore space, raise the rate of carbonation weathering, speeding the reaction of atmospheric CO_2 with the Earth's crust. Various models developed and summarized by Berner (1992) suggest that the atmospheric concentration of CO_2 declined precipitously as land plants gained dominance about 350 mya (Figure 2).

The flux of organic carbon to sediments has varied by a factor of about 7 during the last 500 million years (Figure 3). During the Carboniferous, large deposits of organic carbon were stored in freshwater environments, leading to modern economic deposits of coal; during the Tertiary, the precursors of most of the large modern petroleum deposits were added to marine sediments. Together, all conventional fossil fuel reserves contain about 4×10^{18} gC globally, but the total storage of organic carbon in the Earth's crust is about $10,000\times$ greater--10^{22} g or 10% of the total global inventory of carbon (Schidlowski, 1983). The majority of the difference is found in dispersed deposits of organic carbon, known as kerogen, which are contained in most sedimentary rocks.

Garrels and Lerman (1981) provide a simple model that unifies various aspects of the biogeochemical cycle of carbon and its interactions with the atmosphere, the hydrosphere, and the crust of Earth (Figure 4). The model assumes that the atmosphere and oceans have not changed in their composition during geologic time. With this constraint, the model couples reactions in the atmosphere and oceans to seven compartments that represent major crustal minerals, including gypsum, pyrite, and calcium carbonate. If weathering of limestone transfers 8 units of Ca to the sea, and the Ca content of seawater does not change, then 8 units of Ca must be deposited as an alternative Ca mineral--in this case, gypsum. The biosphere comprises all life, which in Figure 4 appears in the compartment labeled CH_2O, representing the approximate stoichiometric composition of living tissues. Changes in the mass of organic material through

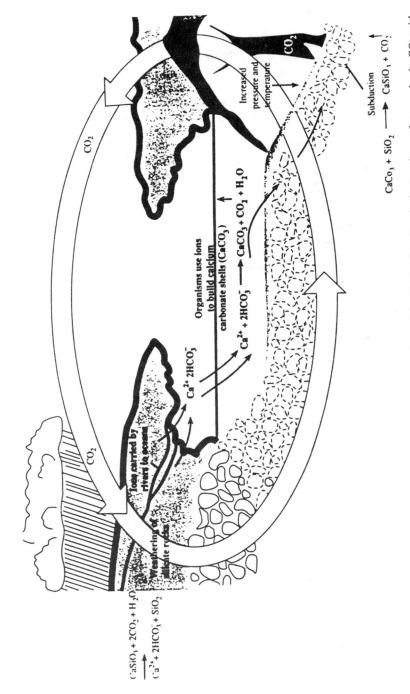

Figure 1. The long-term geochemical cycle of carbon at the surface of the Earth, showing the interactions of atmospheric CO_2 with the exposed crust and the deposition of carbonate minerals on the sea floor. When the ocean sediments are eventually returned to the Earth's interior, CO_2 is returned to the atmosphere by volcanoes. (Modified from Kasting et al., 1988.)

An Overview of the Carbon Cycle

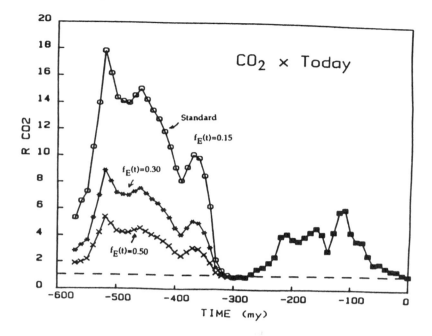

Figure 2. Various assumptions used in a model that predicts the historical concentration of atmospheric CO_2 as a fraction of the modern level (R) all yield a dramatic decline in the CO_2 concentration about 350 mya--close to the proliferation of the earliest land plants. F_E indicates the weathering rate before the advent of vascular land plants as a fraction of the modern rate. (From Berner, 1992.)

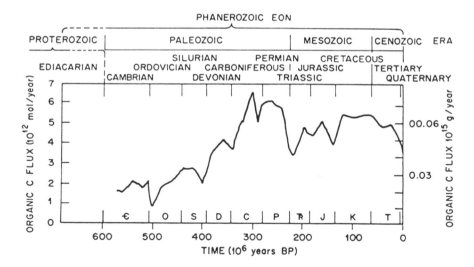

Figure 3. Calculated changes in the burial of organic carbon in geologic sediments over the last 600 million years. (From Olson et al., 1985.)

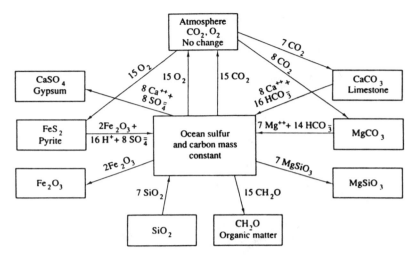

Figure 4. A model linking the sedimentary reservoirs near the Earth's surface, showing the adjustments in each that accompany the storage of 15 moles of carbon in the biosphere. (From Garrels and Lerman; 1981.)

geologic time are modeled by changes in the size of that compartment. Consider the increase in the total mass of organic matter that must have occurred during the Carboniferous when large areas of land were covered by coal-producing swamps. The storage of organic carbon in coal represents an increase in the mass of the biosphere--in this case the storage of its dead materials. With no change in the CO_2 content of the atmosphere or the CO_2 dissolved in seawater as HCO_3^-, this carbon must have been derived from the weathering of carbonate minerals, transferring Ca and Mg to the sea. With no change in the Ca or Mg content of seawater, the Ca must have been deposited as gypsum and the Mg as a Mg-silicate mineral. To deposit $CaSO_4$ with no change in the SO_4^{2-} content of the world's oceans, the sulfur in gypsum must be derived from another pool. Oxidative weathering of pyrite would supply SO_4^{2-} to the oceans, and consume some of the O_2 that would have been added to the atmosphere by photosynthesis. The remaining O_2 would be consumed in the deposition of Fe_2O_3, so the atmospheric content of O_2 would not change. The total moles of O_2 available for reaction exactly balance the moles of carbon stored as organic matter by photosynthesis.

Although this model is a simplistic representation of the global carbon cycle, it shows how interactions among the near-surface compartments on Earth could buffer changes in the concentration of atmospheric CO_2 over geologic time. For example, a release of CO_2 from a widespread loss of global vegetation, such as during a glaciation, should lead to an increase in the deposition of carbonate minerals in marine sediments. As evidence of the effectiveness of this buffering, Holland (1965) points out that during the last several million years, neither gypsum nor dolomite has been an important constituent of marine sedimentary rocks. This sets the limits of atmospheric CO_2 between 200 and 1300 ppm, since concentrations greater than 1300 would lead to the precipitation of dolomite and concentrations less than 200 ppm would lead to the deposition of gypsum. Evidently, the interactions between atmospheric and surface compartments have maintained the chemical, and thus the climatic, conditions on the planet within relatively narrow limits for much of geologic time.

The historical record of atmospheric CO_2, extending to 160,000 ybp, is obtained from an analysis of gas bubbles trapped in Antarctic ice (Figure 5). Until the last century, concentrations varied only between 200 and 280 ppm, with the lowest values found in the layers of ice that were deposited during the last period of continental glaciation. The lower limit is remarkably close to that predicted by Holland (1965) based on the buffering that would be derived by the deposition of gypsum in marine sediments. It is impossible to tell whether changes in CO_2 led to changes in climate or vice versa, but the association of high CO_2 and warm climates is undeniable. Most alarming, of course, is the recent increase in atmospheric CO_2 to the present-day value of about 350 ppm. If and when the temperature of the planet "catches up" with the rise in CO_2, we are destined for a significant global warming.

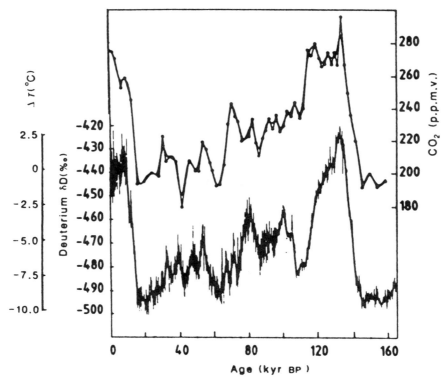

Figure 5. Atmospheric CO_2 concentrations (upper line) derived from analysis of trapped bubbles in the Vostok ice core, Antarctica, and accompanying estimates of temperature, relative to today, derived from changes in the deuterium composition of the ice. (From Barnola et al., 1987.)

III. The Modern Carbon Cycle

Figure 6 shows a simple model of the global carbon cycle on Earth. The largest exchange occurs between the atmosphere and land plants, although the exchange with seawater is not far behind. The mean residence time for a molecule of CO_2 in the atmosphere before it is removed to another reservoir is about 3 years. About half of the carbon fixed by land plants (gross primary production) is respired by the plants themselves, so net primary production (NPP) on land is only about 60 Pg C/yr. Estimates of the current terrestrial biomass, 560 Pg C, yield a mean residence time of 9 years for carbon in live biomass. Globally, about two-thirds of the terrestrial vegetation occurs in regions with seasonal periods of growth, and the well-known annual oscillations in the atmospheric CO_2 concentration reflect changes in the seasonal storage of carbon in vegetation. The oscillation is most pronounced in the northern hemisphere, which contains most of the world's continental area and temperate vegetation. At high, northern latitudes, vegetation accounts for about 50% of the annual variation in atmospheric CO_2 (D'Arrigo et al., 1987). In the southern hemisphere, the smaller fluctuations in atmospheric CO_2 appear to be driven mainly by exchange with seawater (Keeling et al., 1984).

The pool of organic carbon in soils (ca. 1500 Pg) represents a steady-state between the input of dead plant debris from NPP and losses due to the activity of decomposers. The production of plant debris is about 55 Pg C/yr--i.e., NPP minus the consumption by herbivores (ca. 3 PgC/yr, Whittaker and Likens, 1973) and losses by fire (2-5 Pg C/yr, Crutzen and Andreae, 1990). Soil respiration, about 68 Pg C/yr (Raich and Schlesinger, 1992), is slightly greater than the inputs of dead organic materials to the soil, due to the portion of plant respiration that occurs belowground. As we shall see in a later section, estimates of the long-term

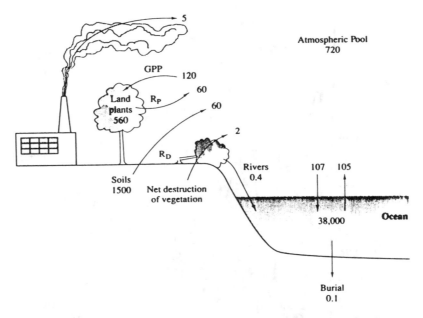

Figure 6. The global carbon cycle, showing the major annual transfers between land, sea and atmosphere, expressed as 10^{15} g C/yr. (Modified from Schlesinger, 1991a.)

accumulation of organic carbon in soils are small, ca. 0.4 Pg C/yr (Schlesinger, 1990), reflecting the efficiency of aerobic decomposers on land.

The release of CO_2 in fossil fuels, now nearly 6 Pg C/yr, is one of the best-known values in the global carbon cycle. If all this CO_2 accumulated in the atmosphere, the annual increment would be about 0.7%/yr. In fact, the atmospheric increase is about 0.4%/yr or 1.5 ppm; only 58% of the fossil fuel release remains in the atmosphere as the "airborne fraction."

Quay et al. (1992) and Sarmiento and Sunquist (1992) suggest that net ocean uptake of CO_2 is between 1.7 and 2.8 Pg C/yr—about 40% of the fossil fuel release. Thus, Figure 6 shows an uptake by the sea (92 Pg C/yr) that is slightly greater than its return of CO_2 to the atmosphere (90 Pg C/yr). The net flux into the ocean is driven by the death of marine phytoplankton, which sink through the thermocline carrying carbon to the deep ocean. In addition to replenishing the CO_2 removed from the surface waters by this "biotic pump," CO_2 enters the ocean as a result of the increasing concentration of CO_2 in the atmosphere, which leads to dissolution in seawater following Henry's law. Seawater acidity is presumably buffered by the dissolution of marine carbonates. If the release of CO_2 were curtailed, nearly all the CO_2 that has accumulated in the atmosphere would eventually dissolve in the oceans, and the global carbon cycle would return to a steady-state. It is the rate of the current release relative to the rate at which the oceans can buffer the global cycle that accounts for the current increase in the atmosphere.

Taken alone, this budget suggests that we have a fairly complete understanding of the global carbon cycle. Many terrestrial ecologists believe, however, that there are substantial additional releases of CO_2 to the atmosphere, caused by the destruction of forest vegetation in favor of agricultural land, especially in the tropics. Indeed, the net biotic flux appears to have been larger than the fossil fuel flux until about 1960 (Houghton et al., 1983). A net release of carbon from land is also suggested by measurements of the $\delta^{13}C$ in tree rings and ice cores (Leavitt and Long, 1988; Siegenthaler and Oeschger, 1987), which both show a decline in atmospheric ^{13}C content that is consistent with a reduction in the pool of organic carbon on land. Globally the net release from land may be as large as 1.8 Pg C/yr (Houghton et al., 1987). If these estimates are accurate, then the atmospheric budget is misbalanced, and a large amount of carbon that ought to be in the atmosphere is missing.

Assuming that the estimates of ocean uptake are correct, various attempts to balance the carbon dioxide budget of the atmosphere have failed, to a greater or lesser degree, unless they included a substantial increase in the carbon storage on land (Table 1). The recent provocative paper by Tans et al. (1990), which presents substantially lower estimates of CO_2 uptake by the oceans, shows that we are no closer to understanding the budget for atmospheric CO_2 today than we were in the early 1970s, despite an enormous scientific effort directed toward this problem.

A variety of possibilities involving the terrestrial biosphere could reconcile the discrepancy in the atmospheric CO_2 budget. First, estimates of release of CO_2 from the destruction of tropical forests may be too high. Indeed, Lugo and Brown (1992) suggest that rapid regrowth of vegetation on deforested areas is a carbon sink that nearly balances the release from cleared areas, yielding no net source from the tropics. Their results are consistent with the observed changes in atmospheric O_2, which show no net decrease due to the destruction and oxidation of land vegetation (Keeling and Shertz, 1992). A similar result could derive from the regrowth of forests on abandoned agricultural land in the temperate zone. A regional net sink for CO_2 in the northern mid-latitudes is consistent with the modeling results of Tans et al. (1990). Kauppi et al. (1992) suggest that the biomass of forests in Europe has increased by about 0.1 Pg C/yr since 1971, mostly in stands that were in place by 1971. It is worth noting, however, that the global inventory of Houghton et al. (1987), showing a large net release from land vegetation, also included an accumulation of 0.08 Pg C/yr in trees regrowing on abandoned lands in Europe. For both of these explanations, I worry that any regional carbon sink today may disappear rapidly in the face of increasing human population and pressures for higher food production in the future. A carbon sink due to changes in land use is not a direct negative feedback to rising concentrations of CO_2 in the atmosphere.

Numerous authors have postulated that the size of the terrestrial biosphere has increased due to CO_2-fertilization, providing an explanation for the "missing sink" of CO_2 that ought to be in the atmosphere. There is ample evidence that the growth of crops in high CO_2 atmospheres, with abundant water and fertilizer, is greater than in ambient CO_2 (Strain and Cure, 1985; Allen, 1990; Bazzaz, 1990), and there is some evidence that enhanced growth of crops increases the organic matter in agricultural soils (Leavitt et al., in press). We do not know if the higher production of plants at elevated CO_2 translates into greater long-term storage of carbon on land; it is possible that it could lead to greater rates of carbon turnover and little net sequestration (e.g., Korner and Arnone, 1992).

Although there is much interest in a CO_2-fertilization effect, there is little or no empirical evidence that it exists in natural ecosystems. Working in montane forests of the Sierra Nevada, Graumlich (1991) found no statistically significant trend in tree-ring growth that could be linked to a CO_2-induced stimulation of plant growth since 1900. Field experiments in the tundra of Alaska showed that rapid acclimation of native plants to high CO_2 reduced their potential for enhanced carbon acquisition (Tissue and Oechel, 1987), and Grulke et al. (1990) concluded that there was "little if any long-term stimulation of ecosystem carbon acquisition by increases in atmospheric CO_2." Billings et al. (1984) found that the storage of carbon in tundra soils increased at high CO_2 only if additional nitrogen was provided as a plant nutrient. Similarly, Norby et al. (1992) found that seedlings of *Liriodendron tulipifera* showed no net increase in size when grown at high CO_2 with ambient levels of water and nutrients. In the latter experiment, the turnover of fine roots increased, potentially increasing the input of plant residues to the pool of soil organic matter.

IV. Soils and the Global Carbon Cycle

The pool of carbon in soils is one of the largest near-surface stores of carbon on Earth. Although estimates vary widely, most workers now agree that between 1400 and 1500 Pg C are contained in soil organic matter, which includes undecomposed litter on the soil surface and humic materials dispersed throughout the soil profile (Schlesinger, 1977; Post et al., 1982; Eswaran et al., 1993). A large portion of this pool is held in the organic soils of tundra and boreal forest ecosystems, where dramatic changes in climate are expected during the next century.

In addition to organic forms of carbon, soils contain about 800 to 1000 Pg C in inorganic forms, largely $CaCO_3$, in arid and semi-arid regions, where it is deposited in pedogenic calcites, known variously as caliche or calcic horizons (Schlesinger, 1982; 1985). Although the turnover of this pool is slow, accumulations of soil carbonate in non-carbonate terrain represent a net sink of CO_2 from the atmosphere.

Even small changes in such large pools of carbon would be expected to have dramatic feedbacks in the global climate system. Losses of soil organic matter by oxidation could contribute to atmospheric CO_2 and exacerbate global warming. On the other hand, increases in soil organic matter could slow the rise of atmospheric CO_2 and provide a negative feedback to global warming.

Table 1. A history of attempts to balance the atmospheric CO_2 budget; all data are given in 10^{15} g C/yr

Inputs			Fates			Reference
Fossil fuel combustion	Net biomass destruction		Increase in atmospheric pool	Oceanic uptake	Unknown sinks	
3.6		=	1.8	0.5–0.8	1.0–1.3	Reiners (1973) The Broookhaven Symposium
5.2	3.3	=	2.5	2.0	4.0	Woodwell et al. (1983)
5.0	1.3	=	2.9	2.4	1.0	Trabalka (1985) U.S. Department of Energy State-of-the-Art Report
5.4	1.6	=	3.4	2.0	1.6	Houghton et al. (1990) Intergovernmental Panel on Climate Change
5.3	1.8	=	3.0	1.0–1.6	2.5–3.1	Tans et al. (1990)

(From Schlesinger, 1993; references given therein.)

A significant portion of the carbon in soils is relatively labile--subject to return to the atmosphere as CO_2 as a result of decomposition and other soil processes. The annual flux of CO_2 to the atmosphere, known as soil respiration, amounts to about 68 Pg C globally (Raich and Schlesinger, 1992). Adjusting for the contributions of roots, the flux of CO_2 from soils indicates a mean residence time of 32 years for the global pool of carbon in soil organic matter. Most plant residues decompose rapidly at the soil surface, but a small fraction of the material is very old; radiocarbon ages of soil humic materials often exceed 500 yr (Campbell et al., 1967).

Rates of CO_2 efflux vary as a function of soil temperature (Schleser, 1982; Raich and Schlesinger, 1992; Peterjohn et al., 1993), and there is good reason to believe that rates of soil respiration will increase with global warming. Even a 1% increase in the rate of CO_2 evolution from soils globally (i.e., increasing to 68.7 Pg C/yr) would be equivalent to approximately 14% of the annual flux of CO_2 to the atmosphere from the burning of fossil fuels. Alternatively, if a slightly higher percentage of the annual net primary productivity of Earth were to escape decomposition and accumulate in soils, a substantial sink for fossil fuel CO_2 might be found in the terrestrial biosphere.

Ten thousand years ago, 29.5×10^6 km^2 of the Earth's present land area was covered with ice and presumably contained little or no soil organic matter (Flint, 1971; Bell and Laine, 1985). Much of this area now supports tundra and boreal forest ecosystems, with substantial pools of soil carbon. Adams et al. (1990) calculated that the global pool of soil organic matter increased 490 Pg C from the last glacial maximum (18,000 ybp) to the present. Similarly, Prentice and Fung (1990) estimate that the soil carbon pool increased between 13 and 148 Pg during this interval. The average rate of accumulation on glaciated lands has been 0.04 Pg C/yr during the last 10,000 years (Schlesinger, 1990), with current rates ranging from 0.075 to 0.18 Pg C/yr (Harden et al., 1992). Accumulations in peatlands account for much of the total (Armentano and Menges, 1986; Gorham, 1991). Despite the large range in these estimates, there is no doubt that glaciated soils have served as a net sink for carbon as the climate has warmed from glacial conditions to today.

Over most of the rest of the Earth, the geomorphic surfaces are much older than 10,000 years, and the accretion of soil organic matter is minimal. Schlesinger (1990) calculated an upper-limit for the global formation of soil humic substances of 0.4 Pg C/yr, or about 0.7% of terrestrial NPP. Significantly, this value is about equivalent to the riverine transport of organic carbon to the sea (Schlesinger and Melack, 1981; Meybeck, 1982), suggesting a steady-state condition in soil organic matter globally. Thus, the formation of humic substances on land is roughly balanced by their net transport to the sea, where they may ultimately be added to ocean sediments (Berner, 1982; Lugo and Brown, 1986). Despite accumulations of organic matter in some glaciated soils, it is not likely that, globally, soils were a significant net source or sink for atmospheric CO_2 at the beginning of the industrial revolution.

The advent of mechanized agriculture in the late 1800s led to dramatic losses of organic matter from cultivated soils. Typically 20 to 40% of the native soil organic matter is lost when virgin lands are converted to agriculture (Schlesinger, 1986; Mann, 1986; Detwiler, 1986). The losses are greatest during the first few years of land conversion, and slow after about 20 years of cultivation. These percentage losses are consistent with independent estimates of the relative portion of soil organic matter that exists in labile versus refractory pools (Spycher et al., 1983). Schlesinger (1984) calculated that 36 Pg C were lost from soils between 1860 and 1960, with a current rate of loss of about 0.8 Pg C/yr. Thus, the loss from soils is a significant component of the net biotic flux. At present, this loss is mostly confined to the tropics, where the rates of new land conversion are greatest (Houghton et al., 1987).

Vegetation is one of the major "state factors" determining soil development and the amount of carbon that a soil will ultimately contain (Jenny, 1980). Thus, changes in the distribution of vegetation with changes in climate are likely to alter the input of plant residues to the soil and the accumulation of soil organic matter (e.g., Tate, 1992). From one of the first models predicting the future distribution of vegetation in response to climate (Emanuel et al., 1985), Schlesinger (1991b) estimated that soils will lose 45 Pg C as the climate warms from today's conditions to those anticipated with a doubling of atmospheric CO_2. Smith et al. (1992) derive a similar prediction from each of four more recent climate models--finding that potential changes in soil carbon range from a loss of 19.5 Pg C to a gain of 57.3 Pg C under climatic shifts that include both temperature and precipitation (Table 2). Relative to soils, in each case, the vegetation offers a substantial sink for atmospheric CO_2 when it has fully equilibrated with the anticipated new climate.

These models provide a picture of the carbon pool in the biosphere at two points in time; they do not include potential transient states that may yield dramatic fluxes of CO_2 from the biosphere to the atmosphere (Smith and Shugart, 1993). For example, while the climate may change rapidly, changes in the distribution of vegetation are likely to take 100s of years (Overpeck et al., 1991; MacDonald et al., 1993). At the same

Table 2. Changes in carbon storage (Pg) in above-ground biomass and soil for four climate change scenarios. Valus in parentheses are percentage change from current

Scenario	Aboveground biomass	Soil	Total	Change
Current	737.2	1158.5	1895.7	
OSU	860.4 (16.7)	1215.8 (4.9)	2076.2	180.5 (9.5)
GFDL	782.3 (6.1)	1151.3 (-0.6)	1933.6	37.9 (2.0)
GISS	829.6 (12.5)	1213.0 (4.7)	2042.6	146.9 (7.7)
UKMO	765.2 (3.8)	1139.0 (-1.7)	1904.2	8.5 (0.4)

(From Smith et al., 1992.)

time, soil microbial communities are likely to show an immediate response to higher soil temperature, increasing the rate of soil respiration. Jenkinson et al. (1991) suggested that during the next 60 years, as much as 61 Pg C may be lost from the global pool in soils and released to the atmosphere as CO_2. Recent data from the Arctic tundra of Alaska suggest that a net release of soil carbon may already be in progress (Oechel et al., 1993). These losses of soil organic matter may provide one indirect long-term feedback leading to a greater storage of carbon in the terrestrial biosphere. In a soil-warming experiment in a black spruce forest of Alaska, Van Cleve et al. (1990) noted higher nutrient concentrations in the soil solution and in the foliage of spruce growing on heated plots. Presumably, these nutrients were made available as a result of higher rates of decomposition when the soil was maintained at a higher temperature. A greater mineralization of nutrients in globally warmer soils may help alleviate nutrient deficiencies and allow vegetation to respond to high CO_2. Because the C/N ratio of soil (ca. 12) is much lower than the C/N ratio of wood (ca. 160), a small increment in the rate of nitrogen mineralization in soil could support a large increase in vegetation production and net carbon sequestration by the terrestrial biosphere (Rastetter et al., 1991; McGuire et al., 1992; Bonan and Van Cleve, 1992). We know little about whether this sink is reasonable; it is also possible that losses of nitrogen from soils could increase with greater rates of mineralization. Although it would not be invoked until substantial climatic warming is already in progress, this mechanism is a potential negative feedback of the biosphere to higher atmospheric CO_2.

V. Soil Management

Several recent reports of the Environmental Protection Agency assess the potential for soil management to maintain, restore, and enlarge the pool of carbon in soils (Kern and Johnson, 1993). These reports focus on agricultural lands, where substantial losses of soil organic matter contribute CO_2 to the atmosphere (Schlesinger, 1984; 1986). When natural land is converted to agriculture using "no-tillage" techniques, there are often smaller losses of soil carbon compared to those seen when traditional cultivation is practiced (Blevins et al., 1977; Dick, 1983). Small increases in soil organic matter are sometimes seen when existing cultivated fields are converted to no-till agriculture. For example, Wood et al. (1991) found that intensive no-till management resulted in accumulations of 7 to 16 g $C/m^2/yr$ in the 0-10 cm depth of prairie soils that were previously cultivated. Nevertheless, the most optimistic scenario for the widespread adoption of no-till agriculture in the United States offsets only 0.7 to 1.1% of the projected fossil fuel use by the U.S. during the next 30 years (Kern and Johnson, 1993).

Maintaining or improving soil fertility by the application of fertilizer may also maintain or enlarge the pool of carbon in soils, provided the "carbon costs" of fertilizer production do not exceed the gain in soil organic matter. When a pasture soil containing 5.2 kg C/m^2 was converted to a heavily fertilized (336 kgN/ha/yr) no-till cornfield, soil organic matter increased to 5.5 kg C/m^2 in the 0-30 cm layer over five years (Blevins et al., 1977). Assuming 100% industrial efficiency, the release of CO_2 during the production of the fertilizer applied over this interval was equivalent to 54 g C/m^2, or 18% of the apparent carbon gain. The *net* gain of soil organic matter, 49.2 g $C/m^2/yr$, would be further reduced by subtracting the carbon costs of producing herbicides and pesticides, and the fossil fuel emissions of CO_2 during the application of all of these chemicals. Again, the potential for intensive management of agricultural lands to yield a substantial sequestration of carbon seems minimal.

Thornley et al. (1991) suggested that substantial carbon storage has occurred in British grasslands in response to growth at higher concentrations of atmospheric CO_2 and at greater levels of available nitrogen from fertilizer and atmospheric deposition. However, their values for net carbon sequestration are actually very low when expressed on an annual basis and compared to annual values of CO_2 emission by fossil fuels. Peterson and Melillo (1985) suggested that the inadvertent fertilization of the biosphere by the combustion of fossil fuels might sequester only 0.2 Pg C/yr in vegetation and soils (cf. Schlesinger, 1993).

I suggest that the potential for enhanced carbon storage in the terrestrial biosphere is much greater in vegetation, especially forest vegetation, than in soils. After all, if <1% of net primary production escapes decomposition and accumulates in soils (Schlesinger, 1990), terrestrial net primary production would have to increase by >300% (60 to 260 Pg C/yr) to account for the missing sink (2 Pg C/yr) of atmospheric CO_2 in soils. Reforestation and forest fertilization are attractive short-term practices to increase the sequestration of atmospheric CO_2 on land. Fertilization of *Pinus radiata* resulted in a 97% increase in tree growth, versus a 21% increase in soil organic matter (Neilsen et al. 1992). The shorter mean residence time of carbon in vegetation means that the pool of carbon in vegetation is more responsive than that in soils to changes in atmospheric CO_2 and fertility (cf. Harrison et al., 1993). For the same reason, however, the period of net carbon uptake by vegetation is short-lived, because forest regrowth is nearly complete within a few decades (Schiffman and Johnson, 1989; Vitousek, 1991). The potential for global reforestation to help attenuate the rise in atmospheric CO_2 is attractive, but its implementation will strongly depend on our ability to control human population growth and the increasing conversion of land to agriculture.

Acknowledgements

I thank Lisa Dellwo Schlesinger and Pat Megonigal for their critical comments on an earlier draft of this paper.

References

Adams, J.M., H. Faure, L. Faure-Denard, J.M. McGlade, and F.I. Woodward. 1990. Increases in terrestrial carbon storage from the last glacial maximum to the present. *Nature* 348:711-714.
Allard, P., J. Carbonnell, D. Dajlevic, J. Le Bronec, P. Morel, M.C. Robe, J.M. Maurenas, R. Faivre-Pierret, D. Martin, J.C. Sabroux, and P. Zettwoog. 1991. Eruptive and diffuse emissions of CO_2 from Mount Etna. *Nature* 351:387-391.
Allen, L.H. 1990. Plant responses to rising carbon dioxide and potential interactions with air pollutants. *J. Environ. Qual.* 19:15-34.
Anders, E. 1989. Pre-biotic organic matter from comets and asteroids. *Nature* 342:255-257.
Anders, E. and T. Owen. 1977. Mars and Earth: Origin and abundance of volatiles. *Science* 198:453-465.
Armentano, T.V. and E.S. Menges. 1986. Patterns of change in the carbon balance of organic soils of the temperate zone. *J. Ecol.* 74:755-774.
Arrhenius, S. 1918. *The Destinies of the Stars*. G.P. Putnam's Sons, New York.
Awramik, S.M., J.W. Schopf, and M.R. Walter. 1983. Filamentous fossil bacteria from the Archean of western Australia. *Precambrian Res.* 20:357-374.
Barnola, J.M., D. Raynaud, Y.S. Korotkevich, and C. Lorius. 1987. Vostok ice core provides 160,000-year record of atmospheric CO_2. *Nature* 329:408-414.
Bazzaz, F.A. 1990. The response of natural ecosystems to the rising global CO_2 levels. *Annu. Rev. Ecol. Syst.* 21:167-196.
Bell, M. and E.P. Laine. 1985. Erosion of the Laurentide region of North America by glacial and glaciofluvial processes. *Quatern. Res.* 23:154-174.
Berner, R.A. 1982. Burial of organic carbon and pyrite sulfur in the modern ocean: Its geochemical and environmental significance. *Am. J. Sci.* 282:451-473.
Berner, R.A. 1992. Weathering, plants, and the long-term carbon cycle. *Geochim. Cosmochim. Acta* 56:3225-3231.
Billings, W.D., K.M. Peterson, J.O. Luken, and D.A. Mortensen. 1984. Interaction of increasing atmospheric carbon dioxide and soil nitrogen on the carbon balance of tundra ecosystems. *Oecol.* 65:26-29.

Blevins, R.L., G.W. Thomas, and P.L. Cornelius. 1977. Influence of no-tillage and nitrogen fertilization on certain soil properties after 5 years of continuous corn. *Agron. J.* 69:383-386.

Bonan, G.B. and K. Van Cleve. 1992. Soil temperature, nitrogen mineralization, and carbon source-sink relationships in boreal forests. *Can. J. Forest Res.* 22:629-639.

Cameron, E.M. 1982. Sulphate and sulphate reduction in early Precambrian oceans. *Nature* 296:145-148.

Campbell, C.A., E.A. Paul, D.A. Rennie, and K.J. McCallum. 1967. Factors affecting the accuracy of the carbon-dating method in soil humus studies. *Soil Sci.* 104:81-85.

Carr, M.H. 1987. Water on Mars. *Nature* 326:30-35.

Chyba, C.F. 1987. The cometary contribution to the oceans of primitive Earth. *Nature* 330:632-635.

Cloud, P.E. 1973. Paleoecological significance of the banded iron formation. *Econ. Geol.* 68:1135-1143.

Crutzen, P.J. and M.O. Andreae. 1990. Biomass burning in the tropics: Impact on atmospheric chemistry and biogeochemical cycles. *Science* 250:1669-1678.

D'Arrigo, R., G.C. Jacoby, and I.Y. Fung. 1987. Boreal forests and atmosphere-biosphere exchange of carbon dioxide. *Nature* 329:321-323.

Delsemme, A.H. 1992. Cometary origin of carbon, nitrogen, and water on Earth. *Origins Life* 21:279-298.

Des Marais, D.J., H. Strauss, R.E. Summons, and J.M. Hayes. 1992. Carbon isotope evidence for the stepwise oxidation of the Proterozoic environment. *Nature* 359:605-609.

Detwiler, R.P. 1986. Land use change and the global carbon cycle: The role of tropical soils. *Biogeochem.* 2:67-93.

Dick, W.A. 1983. Organic carbon, nitrogen, and phosphorus concentrations and pH in soil profiles as affected by tillage intensity. *Soil Sci. Soc. Am. J.* 47:102-107.

Emanuel, W.R., H.H. Shugart, and M.P. Stevenson. 1985. Climatic change and the broad-scale distribution of terrestrial ecosystem complexes. *Climatic Change* 7:29-43.

Eswaran, H., E. Van den Berg, and P. Reich. 1993. Organic carbon in soils of the world. *Soil Sci. Soc. Amer. J.* 57: 192-194.

Flint, R.F. 1971. *Glacial and Quaternary Geology.* John Wiley and Sons, New York.

Francois, L.M. and J.-C. Gerard. 1986. Reducing power of ferrous iron in the Archean ocean. 1. Contribution of photosynthetic oxygen. *Paleoceanogr.* 1:355-368.

Garrels, R.M. and A. Lerman. 1981. Phanerozoic cycles of sedimentary carbon and sulfur. *Proc. Nat. Acad. Sci. U.S.* 78:4652-4656.

Gorham, E. 1991. Northern peatlands: Role in the carbon cycle and probable responses to climatic warming. *Ecol. Applications* 1:182-195.

Graumlich, L.J. 1991. Subalpine tree growth, climate, and increasing CO_2: An assessment of recent growth trends. *Ecol.* 72:1-11.

Grulke, N.E., G.H. Riechers, W.C. Oechel, U. Hjelm, and C. Jaeger. 1990. Carbon balance in tussock tundra under ambient and elevated atmospheric CO_2. *Oecol.* 83:485-494.

Harden, J.W., E.T. Sundquist, R.F. Stallard, and R.K. Mark. 1992. Dynamics of soil carbon during deglaciation of the Laurentide ice sheet. *Science* 258:1921-1924.

Harrison, K., W. Broecker, and G. Bonani. 1993. A strategy for estimating the impact of CO_2 fertilization on soil carbon storage. *Global Biogeochem. Cycles* 7:69-80.

Holland, H.D. 1965. The history of ocean water and its effect on the chemistry of the atmosphere. *Proc. Nat. Acad. Sci. U.S.* 53:1173-1183.

Houghton, R.A., J.E. Hobbie, J.M. Melillo, B. Moore, B.J. Peterson, G.R. Shaver, and G.M. Woodwell. 1983. Changes in the carbon content of terrestrial biota and soils between 1860 and 1980: A net release of CO_2 to the atmosphere. *Ecol. Monogr.* 53:235-262.

Houghton, R.A., R.D. Boone, J.R. Fruci, J.E. Hobbie, J.M. Melillo, C.A. Palm, B.J. Peterson, G.R. Shaver, G.M. Woodwell, B. Moore, D.L. Skole, and N. Myers. 1987. The flux of carbon from terrestrial ecosystems to the atmosphere in 1980 due to changes in land use: Geographic distribution of the global flux. *Tellus* 39B:122-139.

Jenkinson, D.S., D.E. Adams, and A. Wild. 1991. Model estimates of CO_2 emissions from soil in response to global warming. *Nature* 351:304-306.

Jenny, H. 1980. *The Soil Resource.* Springer-Verlag, New York.

Kasting, J.F., O.B. Toon, and J.B. Pollack. 1988. How climate evolved on the terrestrial planets. *Scientific Am.* 258(2):90-97.

Kauppi, P.E., K. Mielikainen, and K. Kuusela. 1992. Biomass and carbon budget of European forests, 1971 to 1990. *Science* 256:70-74.

Keeling, C.D., A.F. Carter, and W.G. Mook. 1984. Seasonal, latitudinal, and secular variations in the abundance and isotopic ratios of atmospheric CO_2. 2. Results from oceanographic cruises in the tropical Pacific ocean. *J. Geophys. Res.* 89:4615-4628.

Keeling, R.F. and S.R. Shertz. 1992. Seasonal and interannual variations in atmospheric oxygen and implications for the global carbon cycle. *Nature* 358:723-727.

Kennett, J.P. 1982. *Marine Geology.* Prentice-Hall, Englewood Cliffs, New Jersey.

Kern, J.S. and M.G. Johnson. 1993. Conservation tillage impacts on national soil and atmospheric carbon levels. *Soil Sci. Soc. Am. J.* 57:200-210.

Korner, C. and J.A. Arnone. 1992. Responses to elevated carbon dioxide in artificial tropical ecosystems. *Science* 257:1672-1675.

Kump, L.R. and H.D. Holland. 1992. Iron in Precambrian rocks: Implications for the global oxygen budget of the ancient Earth. *Geochim. Cosmochim. Acta* 56:3217-3223.

Leavitt, S.W. and A. Long. 1988. Stable carbon isotopic chronologies from trees in the southwestern United States. *Global Biogeochem. Cycles* 2:189-198.

Leavitt, S.W., E.A. Paul, B.A. Kimball, P.J. Pinter, G.F. Hendrey, K.F. Lewin, J. Nagy, J.R. Mauney, R. Rauschkolb, H. Rogers, and H.B. Johnson. 1994. Carbon isotopes in soils indicate rapid input of carbon under free-air CO_2 enrichment. *Agric. Forest Meteorol.*, In press.

Li, Y.-H. 1972. Geochemical mass balance among lithosphere, hydrosphere, and atmosphere. *Am. J. Sci.* 272:119-137.

Lugo, A.E. and S. Brown. 1986. Steady state terrestrial ecosystems and the global carbon cycle. *Vegetation* 68:83-90.

Lugo, A.E. and S. Brown. 1992. Tropical forests as sinks of atmospheric carbon. *Forest Ecol. Manag.* 54:239-255.

Lupton, J.E. and H. Craig. 1981. A major helium-3 source at 15 °S on the east Pacific Rise. *Science* 214:13-18.

MacDonald, G.M., T.W.D. Edwards, K.A. Moser, R. Pienitz, and J.P. Smol. 1993. Rapid response of treeline vegetation and lakes to past climate warming. *Nature* 361:243-246.

Mann, L.K. 1986. Changes in soil carbon storage after cultivation. *Soil Sci.* 142:279-288.

McGuire, A.D., J.M. Melillo, L.A. Joyce, D.W. Kicklighter, A.L. Grace, B. Moore, and C.J. Vorosmarty. 1992. Interactions between carbon and nitrogen dynamics in estimating net primary productivity for potential vegetation in North America. *Global Biogeochem. Cycles* 6:101-124.

Meybeck, M. 1982. Carbon, nitrogen, and phosphorus transport by world rivers. *Am. J. Sci.* 282:401-450.

Neilsen, W.A., W. Pataczek, T. Lynch, and R. Ryrke. 1992. Growth response of *Pinus radiata* to multiple applications of nitrogen fertilizer and evaluation of the quantity of added nitrogen remaining in the forest system. *Plant Soil* 144:207-217.

Nisbet, E.G. 1991. *Leaving Eden.* Cambridge University Press, Cambridge.

Norby, R.J., C.A. Gunderson, S.D. Wullschleger, E.G. O'Neill, and M.K. McCracken. 1992. Productivity and compensatory responses of yellow-poplar trees in elevated CO_2. *Nature* 357:322-324.

Nozette, S. and J.S. Lewis. 1982. Venus: Chemical weathering of igneous rocks and buffering of atmospheric composition. *Science* 216:181-183.

Oechel, W.C., S.J. Hastings, G. Vourlitis, M. Jenkins, G. Riechers, and N. Grulke. 1993. Recent change of Arctic tundra ecosystems from a net carbon dioxide sink to a source. *Nature* 361:520-523.

Olson, J.S., R.M. Garrels, R.A. Berner, T.V. Armentano, M.I. Dyer, and D.H. Yaalon. 1985. The natural carbon cycle. p. 175-213. In: J.R. Trabalka (ed.). *Atmospheric Carbon Dioxide and the Global Carbon Cycle.* DOE/ER-0239, National Technical Information Service, Washington, D.C.

Overpeck, J.T., P.J. Bartlein, and T. Webb. 1991. Potential magnitude of future vegetation change in eastern North America: Comparisons with the past. *Science* 254:692-695.

Owen, R.M. and D.K. Rea. 1985. Sea-floor hydrothermal activity links climate to tectonics; The Eocene carbon dioxide greenhouse. *Science* 227:166-169.

Peterson, B.J. and J.M. Melillo. 1985. The potential storage of carbon caused by eutrophication of the biosphere. *Tellus* 37B:117-127.

Peterjohn, W.T., J.M. Melillo, F.P. Bowles, and P.A. Steudler. 1993. Soil warming and trace gas fluxes: Experimental design and preliminary flux results. *Oecol.* 93:18-24.

Post, W.M., W.R. Emanuel, P.J. Zinke, and A.G. Stangenberger. 1982. Soil carbon pools and world life zones. *Nature* 298:156-159.

Prentice, K.C. and I.Y. Fung. 1990. The sensitivity of terrestrial carbon storage to climate change. *Nature* 346:48-51.

Quay, P.D., B. Tilbrook, and C.S. Wong. 1992. Oceanic uptake of fossil fuel CO_2: Carbon-13 evidence. *Science* 256:74-79.

Raich, J.W. and W.H. Schlesinger. 1992. The global carbon dioxide flux in soil respiration and its relationship to vegetation and climate. *Tellus* 44B:81-99.

Rastetter, E.B., M.G. Ryan, G.R. Shaver, J.M. Melillo, K.J. Nadelhoffer, J.E. Hobbie, and J.D. Aber. 1991. A general biogeochemical model describing the responses of the C and N cycles in terrestrial ecosystems to changes in CO_2, climate and N deposition. *Tree Physiol.* 9:101-126.

Sarmiento, J.L. and E.T. Sundquist. 1992. Revised budget for the oceanic uptake of anthropogenic carbon dioxide. *Nature* 356:589-593.

Schidlowski, M. 1983. Evolution of photoautotrophy and early atmospheric oxygen levels. *Precambrian Res.* 20:319-335.

Schiffman, P.M. and W.C. Johnson. 1989. Phytomass and detrital carbon storage during forest regrowth in the southeastern United States piedmont. *Can. J. Forest Res.* 19:69-78.

Schleser, G.H. 1982. The response of CO_2 evolution from soils to global temperature changes. *Z. Naturforsch.* 37a:287-291.

Schlesinger, W.H. 1977. Carbon balance in terrestrial detritus. *Ann. Rev. Ecol. Syst.* 8:51-81.

Schlesinger, W.H. 1982. Carbon storage in the caliche of arid soils: A case study from Arizona. *Soil Sci.* 133:247-255.

Schlesinger, W.H. 1984. Soil organic matter: A source of atmospheric CO_2. p.111-127. In: G.M. Woodwell (ed.). *The Role of Terrestrial Vegetation in the Global Carbon Cycle: Measurement by Remote Sensing.* John Wiley and Sons, New York.

Schlesinger, W.H. 1985. The formation of caliche in soils of the Mojave desert, California. *Geochim. Cosmochim. Acta* 49:57-66.

Schlesinger, W.H. 1986. Changes in soil carbon storage and associated properties with disturbance and recovery. p.194-220. In: J.R. Trabalka and D.E. Reichle (eds.). *The Changing Carbon Cycle: A Global Analysis.* Springer-Verlag, New York.

Schlesinger, W.H. 1990. Evidence from chronosequence studies for a low carbon-storage potential of soils. *Nature* 348:232-234.

Schlesinger, W.H. 1991a. *Biogeochemistry: An Analysis of Global Change.* Academic Press, San Diego.

Schlesinger, W.H. 1991b. Climate, environment and ecology. p.371-378. In: J. Jager and H.L. Ferguson (eds.). *Climate Change: Science, Impacts and Policy. Proceedings of the Second World Climate Conference.* Cambridge University Press, Cambridge.

Schlesinger, W.H. 1993. Responses of the terrestrial biosphere to global climate change and human perturbation. *Vegetatio* 104:295-305.

Schlesinger, W.H. and J.M. Melack. 1981. Transport of organic carbon in the world's rivers. *Tellus* 33:172-187.

Siegenthaler, U. and H. Oeschger. 1987. Biospheric CO_2 emissions during the past 200 years reconstructed by deconvolution of ice core data. *Tellus* 39B:140-154.

Siever, R. 1974. The steady state of the Earth's crust, atmosphere and oceans. *Scientific Amer.* 230(6):72-79.

Smith, T.M., R. Leemans, and H.H. Shugart. 1992. Sensitivity of terrestrial carbon storage to CO_2-induced climatic change: Comparison of four scenarios based on general circulation models. *Climatic Change* 21:367-384.

Smith, T.M. and H.H. Shugart. 1993. The transient response of terrestrial carbon storage to a perturbed climate. *Nature* 361:523-526.

Spycher, G., P. Sollins, and S. Rose. 1983. Carbon and nitrogen in the light fraction of a forest soil: Vertical distribution and seasonal patterns. *Soil Sci.* 135:79-87.

Strain, B.R. and J.D. Cure (eds.). 1985. *Direct Effects of Increasing Carbon Dioxide on Vegetation.* DOE/ER-0238, National Technical Information Service, Washington, D.C.

Tans, P.P., I.Y. Fung, and T. Takahashi. 1990. Observational constraints on the global atmospheric CO_2 budget. *Science* 247:1431-1438.

Tate, K.R. 1992. Assessment, based on a climosequence of soils in tussock grasslands, of soil carbon storage and release in response to global warming. *J. Soil Sci.* 43:697-707.

Tissue, D.T. and W.C. Oechel. 1987. Response of *Eriophorum vaginatum* to elevated CO_2 and temperature in Alaskan tussock tundra. *Ecol.* 68:401-410.

Thornley, J.H.M., D. Fowler, and M.G.R. Cannell. 1991. Terrestrial carbon storage resulting from CO_2 and nitrogen fertilization in temperate grasslands. *Plant, Cell, Environ.* 14:1007-1011.

Towe, K.M. 1990. Aerobic respiration in the Archaean? *Nature* 348:54-56.

Vitousek, P.M. 1991. Can planted forests counteract increasing atmospheric carbon dioxide? *J. Environ. Qual.* 20:348-354.

Van Cleve, K., W.C. Oechel, and J.L. Hom. 1990. Response of black spruce (*Picea mariana*) ecosystems to soil temperature modification in interior Alaska. *Can. J. Forest Res.* 20:1530-1535.

Walker, J.C.G. 1985. Carbon dioxide on the early Earth. *Origins Life* 16:117-127.

Whittaker, R.H. and G.E. Likens. 1973. Carbon in the biota. p.281-302. In: G.M. Woodwell and E.V. Pecan (eds.). *Carbon and the Biosphere.* CONF 720510, National Technical Information Service, Washington, D.C.

Williams, S.N., S.J. Schaefer, M.L. Calvache, and D. Lopez. 1992. Global carbon dioxide emission to the atmosphere by volcanoes. *Geochim. Cosmochim. Acta* 56:1765-1770.

Wood, C.W., D.G. Westfall, and G.A. Peterson. 1991. Soil carbon and nitrogen changes on initiation of no-till cropping systems. *Soil Sci. Soc. Am. J.* 55:470-476.

CHAPTER 3

Global Soil Carbon Resources

H. Eswaran, E. Van den Berg, P. Reich, and J. Kimble

1. Introduction

In recent years the emission of greenhouse gases, specifically through anthropogenic activities, has come under intense scrutiny. This has led to a more concerted effort to understand the "carbon cycle". Concerns regarding sources of atmospheric C stems from human activities such as utilization of fossil fuels which releases about 6 Pg yr^{-1} of which about 3.4 Pg yr^{-1} accumulates in the atmosphere. Houghton et al., 1990, considers this accumulation to be sufficient to result in global warming. Though sources and sinks of carbon are well documented, the global carbon budget has proved to be more elusive. In their study, Houghton et al., 1991, indicate that the C flux in terrestrial ecosystems are a net source of an additional 0.3 to 2.5 Pg yr^{1}. As a result Lugo (1992) and others have called for a better identification and quantification of global carbon sinks.

Figure 1 from Downing and Cataldo (1992) represents a typical schematic of the carbon cycle. They indicate that most analyses of the carbon cycle have focused on the atmospheric and oceanic systems while treating the terrestrial systems as a neutral component with respect to carbon. They emphasize that more detailed studies are needed for the components of the soil/plant ecosystem.

Soils are an important source and sink of carbon. The purpose of this paper is largely to provide global estimates of carbon in soils. In addition, aspects of carbon in soils are considered to aid in the modelling of carbon fluxes and to enhance the carbon storage process through land management.

Recent concerns of greenhouse gases and damages to the ozone layer have resulted in more concerted studies on the quantities, kinds, distributions, and behavior of carbon in different systems. Although the purpose of many of these studies is related to impacts of potential global climate change, the contributions would have applications in all fields ranging from energy to agriculture. A very important product of such studies would be in the area of mitigating global climate change which has a direct relationship to agriculture and specifically to organic matter management in soils (Johnson and Kerns, 1991). Thus a better understanding of the terrestrial reservoir has benefits far beyond the current objectives of carbon sequestration in soils and the detrimental effects of greenhouse gases.

II. Forms of Carbon in Soils

Carbon is sequestered in soils as organic carbon and in carbonates. Carbonates may be present in the parent rock as in limestones or calcareous sediments, or may be pedogenically precipitated in soils. Sources of carbonate are bicarbonate and carbonates in ground or laterally moving water in soils, aerosolic dusts, or recycled from carbonatic substratum. Soils with calcic, hypercalcic, or petrocalcic horizons are evidences of secondary accumulations in soils. Schlesinger (1985) estimates that in the eastern Mojave Desert of California, the rate of addition has averaged 1.0 to 3.5 g $CaCO_3/m^2$/yr over the last 2,000 years. In an earlier study of soils of Arizona, Schlesinger (1982) provided a global estimate of carbonate carbon in soils

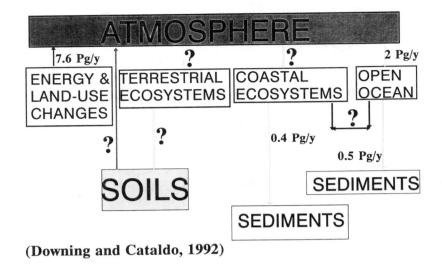

(Downing and Cataldo, 1992)

Figure 1. Schematic diagram of carbon equilibrium. (From Downing and Cataldo, 1992.)

Table 1. Organic and carbonate carbon on soils of Arizona, USA

Soil Order	Organic	Carbonate
	Carbon as kg/m^2	
Aridisols	3.4	33.9
Entisols	4.9	11.4
Mollisols	8.8	16.0
Alfisols	6.2	17.7

(From Schlesinger, 1982.)

as being about 800 Pg. Table 1, provides estimates of organic and carbonate carbon in different kinds of soils in Arizona. As seen in Table 1, carbonate carbon contributes significant amount of carbon in these calcareous soils. However, this carbon does not participate in the flux to other carbon systems as rapidly as organic carbon, except if the soil is irrigated or acidified.

Organic carbon in soils is present in many forms. The litter layer on the soil surface consists of undecomposed to completely humified organic matter. Few data are available on the carbon content of this layer in different soils and there are also few data on bulk density of this layer. The contribution of this layer to total organic carbon in soils is underestimated or ignored in most global estimates. Using simulation models, Meentemyer et al. (1982) have estimated an annual litter production of 54.8×10^9t (55 Pg). Most of the litter layer accumulates under forests and when deforestation takes place, this carbon is lost quickly from terrestrial ecosystems.

In soils, the carbon is present in different forms which have been allocated as pools. Based on carbon dynamics in the soil, Duxbury (1991) suggested four pools:

Active or labile pool -- readily oxidizable;
- controlling factors include residue inputs and climate;
- agronomic factors (management) affect the size of this pool;

Slowly oxidized pool -- macro-aggregates
- controlling factors are soil aggregation and mineralogy;
- agronomic factors, particularly tillage affect the size of this pool;

Very slowly oxidized pool -- micro-aggregates
- controlling factor is water stable micro-aggregates;
- agronomic factors have little impact on this pool;

Passive or recalcitrant pool
- controlling factors include clay mineralogy (as complexes of clay minerals); microbial decomposition may have reduced carbon to elemental form;
- agronomic factors do not influence this pool.

Table 2. Estimates of relative amounts of carbon pools in different soils (*=detectable; **=low; ***=moderate; ****=high; *****=dominance)

Soil	Carbon pools in subsurface horizons			
	Active	Slowly oxidizable	Very slowly oxidizable	Passive
Histosols	*****	***	**	*
Andisols	***	***	**	*
Spodosols	**	**	***	****
Oxisols	*	**	***	****
Vertisols	****	***	**	*
Aridisols	**	**	*	*
Mollisols	*****	***	**	*
Ultisols	***	**	***	****
Inceptisols	*****	*	*	*
Entisols	*****	*	*	*

There are insufficient detailed studies to proportion carbon into each of these pools and thereby to estimate the rates of change of these pools. Table 2, is a first approximation of the relative amounts of these pools in soils. From the point of view of losses from the system, the active pool is the most important and consequently is the most transient. This pool dominates the surface horizons of soils where they are not only most easily oxidized but also most easily lost through erosion. The very slowly and passive pools would be present in significant amounts in Andisols, and the highly weathered Oxisols and Ultisols. In the latter two soils, stable micro-aggregate formation with entrapped organic carbon is frequent.

In most soils there is an exponential decrease of carbon with depth (Figure 2). The highest concentration is in the surface horizons, usually the top 20 cm and this is also the zone with the most active carbon. Under poor management, this layer may be lost through erosion and the loss may be 50 to 80% of the total organic carbon in the one meter of the soil.

There are soils where the pattern of Figure 2, is not followed (as shown in Figure 3) due to specific reasons. Soils on recent sediments (Fluvents) are stratified and so there is a marked irregular distribution of carbon with depth. In Vertisols and Andisols, organic carbon complexes with the mineral colloid. The soils are black but the blackness has no relation to quantity of carbon. In Vertisols, due to cracking, fragments of plant residues and soil aggregates from surface horizons fall into cracks and are assimilated into the matrix of the subsoil. Carbon thus shows an irregular distribution with depth.

Organic matter is solubilized in soils with very high pH (Halaquepts and natric horizons) or soils with very low pH (Sulfaquepts). The solubilized material is translocated downwards and penetrates peds in the subsurface horizons. For all practical purposes, this carbon is very slowly oxidizable or even passive. In Spodosols, soluble organic matter as complexes with iron and aluminum is translocated and accumulates in the spodic horizon. The C/N ratio in spodic horizons is generally more than 20 and may be as high as 60, suggesting an active form of carbon. This carbon is, however, very stable and generally is not readily decomposed by microbial action. The carbon in spodic horizons is present as complexes with iron and aluminum and is not available for microbial decomposition. This is one of the sources of error if emphasis is placed on C/N ratios to differentiate active and passive forms of carbon in carbon models such as that of the grassland model of Parton et al. (1987).

In highly weathered soils, as in Oxisols and Ultisols, micro-aggregate formation (Eswaran, 1979) is a common process. Carbon is entrapped in the micro-aggregates and remains in a passive pool. In a study of

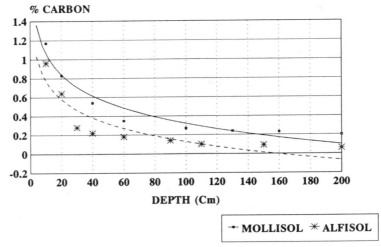

USDA-SCS/WSR

Figure 2. Organic carbon variation with depth.

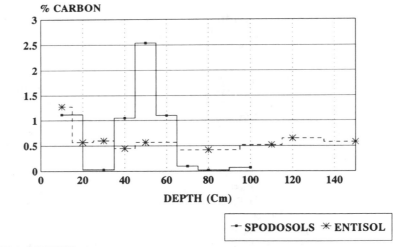

USDA-SCS/WSR

Figure 3. Organic carbon variation with depth.

a deeply weathered Oxisol from Zaire the senior author (unpublished, 1975) detected elemental carbon in the micro-aggregates. This is an unique form of microbial decomposition of organic matter where the final product is elemental carbon and not carbon dioxide. The extent of this pathway in soils is not known but the low cation exchange properties of many highly weathered soils of the tropics, despite a relatively high measured carbon content, suggests that the process may be significant. The Nipe Series (Typic Acrudox) of Puerto Rico has more than 4% organic carbon in the oxic horizon but has almost no permanent charge and a very low pH dependent charge.

Microbial decomposition of organic matter in wet soils is a function of many factors. Under anaerobic conditions, the processes are specific and slow. In general the organic matter is preserved as recognizable plant tissues. Decomposition is initiated with drainage and concomitant aerobic conditions when the rates are rapid. Wet soils are also the major soil contributors of greenhouse gases. It is from this point of view that a more reliable estimate of the spatial distribution of wet soils is urgently needed.

Maximum accumulation of organic matter occurs in Histosols. The accumulation is due to anaerobic conditions or low temperatures or both. Decomposition processes are significantly lower than accumulation processes. The amount of organic matter in Histosols is not easy to estimate. Histosols formed in cold environments are derived mainly from sphagnum species and depth of accumulation may exceed 50 m. In most estimates, only the top meter is considered and results in gross underestimates of sequestered organic carbon. In woody peats and particularly in the tropics, where the peat deposit is composed of partially decomposed trees, few estimates are available of the total organic carbon. In the estimates given later, an arbitrary factor of ten is used to obtain a more realistic estimate.

There is a desire to develop algorithms to estimate soil carbon. Burke et al. (1989, 1990) indicated a good positive relationship between organic carbon and percent clay for the Great Plains region of the U.S. In an area dominated by Spodosols as in the Study of Davidson and Lefebvre (1993) a similar relationship does not hold. This points to the danger of haphazard use of algorithms.

III. Microvariability -- Spatial Distribution

There is considerable variability in the distribution of carbon in soils. This includes variability as a depth function and spatially as a function of position of soils in a landscape. The variability complicates the task of making reliable estimates of carbon content in soils and in studies relating carbon content to soil processes.

Variability may be observed at the microscopic scale. Thin-sections of soils show zones of organic staining or accumulation. Roots decaying in place or faunal accumulations give rise to this point of enrichment. Soluble organic matter translocated down in the voids diffuses from void walls into the matrix of peds. The rims of peds are consequently enriched with organic carbon while the cores may be free of carbon. In sandy Spodosols, the organic matter precipitates forming a coating on sand grains.

Many soils, particularly the highly weathered Oxisols and Ultisols, form micro-aggregates and the stability of the aggregates is a function of the cementing agents, which is usually iron. Finely comminuted organic matter is trapped in these micro-aggregates and persists. In such soils, if the cation exchange capacity (CEC) of sieved samples is compared with finely ground samples, the latter has a significantly higher CEC and exchangeable bases. This increase is the contribution of the entrapped carbon belonging to the 'passive pool' of Duxbury (1991).

Short distance horizontal variability is observed in some soils. This is most evident in some Vertisols (Dudal and Eswaran, 1988). Due to the seasonal shearing forces experienced by the soil, bowl shaped three dimensional entities are formed. On the soil surface, the micro-low of the gilgai corresponds to the center of the bowl while the rims correspond to micro-highs. The formation of such features is explained by Dudal and Eswaran (1988).

Microvariability was also studied by analysis of Vertisols subject to different farming system practices at the experimental station of the International Crops Research Institute for the Semi-Arid Tropics (ICRISAT) at Hyderabad, India. Table 3, shows the total carbon to 1m depth in five pedons sampled at this station. The station was established in 1976. The two uncultivated sites have about the same content of organic carbon; the site under grass appears to have slightly higher amounts than the clean weeded site of the meteorological station. During the 14 year period, the two cultivated sites have lost 1 and 1.7 Kg/m/m^2 of organic carbon respectively. On Field BW4, ICRISAT has mimicked the traditional farming practice wherein the land is fallowed during the rainy season and one crop is grown on the stored moisture at the end of the rains. There is a very high soil loss of about 8 tons/Ha/year while the soil loss on the neighboring

Table 3. Total organic carbon on Vertisols under different land use at ICRISAT Experimental Station, Hyderabad, India

Pedon number	Year sampled	Field number	Land use	Total O.C. Kg/m/m^2
003	82	BW5	Uncultivated/ Met. Station	5.59
003	90	BW8	Uncultivated/grass and shrubs	6.07
006	90	BW1	Improved technology	4.99
008	90	BW4	Traditional farmer's technology	4.30

field with improved technology is less than 1 ton/Ha (Miranda et al. 1982). The marked difference in organic carbon content is primarily the result of management.

The geographic distribution of carbon is equally complicated. Losses of carbon on conversion of forest soils to arable lands introduces marked changes (Brown and Lugo, 1984) in carbon content of soils. Accepting the errors in geographical variations of carbon in soils, patterns of geographic distributions can be established. An example of a regional study is the work of Franzmeier et al. (1985) for North Central United States. Their organic carbon map shows that the regions with the lowest carbon content, are in the Badlands of North Dakota (1.9 Kg m^{-2}), the sandy outwash plains of Wisconsin and Minnesota (2.2 Kg m^{-2}), and the sand hills of Nebraska (2.5 Kg m^{-2}). The highest amounts are in the fine-textured and wet soils of eastern North Dakota and Minnesota (19.4 to 20.7 Kg m^{-2}). The Histosols, which do not form sufficiently large extent to be portrayed on the map, have about 75 Kg m^{-2}. Franzmeier et al. (1985) related this sequestration pattern to soils and environmental conditions, with water-saturation contributing significantly to the process.

These regional patterns do not provide information regarding the factors controlling the sequestration process. Considering a more homogeneous environment such as the Great Plains Grasslands area in the U.S. and employing a model to evaluate the process, Parton et al. (1987) analyzed the factors controlling soil organic matter levels on a regional scale. For their specific region, they determined that organic matter is predicted using four site-specific variables -- temperature, moisture, soil texture, and plant lignin content. Land use is also an important variable. The "Century model" which they employed assumes a loamy soil and so it overestimates the carbon content for clayey soils and underestimates it for sands. The model, however, appears to provide reliable estimates of biomass production. The study of Parton et al. (1987) was followed by Burke et al. (1989) who attempted to evaluate the major controls of soil organic carbon sequestration and to predict regional patterns of carbon in cultivated soils. Carbon losses due to cultivation increased with precipitation and the least losses were in clayey soils. Such regional studies are very useful for validating ecosystem models.

Another example of a regional study is that of Alexander et al. (1989) who evaluated the soils of S.E. Alaska. They showed that in the cold environment, more organic carbon is stored in soils than in vegetation. The soils contain an average of 28 Kg m^{-2} of organic carbon and they postulate that this is increasing because many glaciers in the region are receding and also because of the continuing accumulation of organic carbon in Histosols which occupy large areas.

Few studies have attempted national or continental level evaluations of carbon distribution. One assessment of global evaluation was by Wilson and Henderson-Sellers (1985). They archived soil type and land cover data in 1° × 1° cells for use in general circulation models. This database holds great potential for global carbon assessments. With international collaboration, a reliable database can be incorporated into this archive.

IV. Rates of Sequestration

Unlike mineral soils, sequestration rates of organic carbon in peat (Histosols) deposits have been well studied. A recent Canadian study (National Wetlands Working Group, 1988) has evaluated the radiocarbon dates performed by several workers. Radiocarbon dates were available for several depths at each of four sites in the subarctic region of Canada. Long-term average peat accumulation rates of between 2 to 5 Cm/100 years were obtained. Most of the subarctic was glaciated during the Wisconsin glaciation and the retreat of the glaciers took place about 13,000 years BP. The Wetlands study estimates that peat accumulation commenced about 1,000-2,000 years after retreat of the glaciers.

The Canadian study also evaluated rates of accumulation. They show a negative trend with lower rates for recent times. The lower rates may be largely influenced by the height of the peat surface from the water-saturated basal area and consequently a function of the physiography of the area. It is also due to elevation of surface by permafrost. Other studies have shown that the accumulation rates in non-frozen peatlands are higher than those of frozen peatlands. However, all four sections show at least one period of high rate of accumulation occurring in their lower half.

There are no similar assessments for rates of carbon accumulation in other parts of the world, particularly for the warmer parts of the world. Linking carbon content of soils to geomorphic surfaces or studying carbon content of geologic deposits such as loess will provide better understanding of the time factor in sequestration process.

Radiocarbon dating has been done in a few soils. The studies of Scharpenseel and Schiffman (1977) show that much of the carbon in the top 20 cm of the soil is recent. Below this depth, the radiocarbon age increases abruptly to about 2,000 years and reaches levels of about 8,000 years or more. This suggests that subsoil carbon is mostly "passive" and in these well drained soils do not play a role in the carbon equilibrium. This pattern of radiocarbon age with depth is intriguing and requires much more studies to appreciate its significance.

VI. Global Carbon Estimates

The most reliable estimate of global soil distribution is the FAO-UNESCO (1971-1981) Soil Map of the World. Unfortunately this map is not accompanied by appropriate attribute files providing properties for the map units. One of the earliest estimates of global carbon stocks was made by Buringh (1984) using the soils data base of the USDA Soil Conservation Service. The studies of Bohn (1976, 1982) were made using estimates of soils based on this map and carbon content of soils from a sparsely distributed set of pedons. The studies of Post et al. (1982,1990) are widely cited and they have used a database of about 2,700 pedons. A significant number of the pedon data comes from the Soil Conservation Service while the remaining is from publications. Although they employ soils data to compute the soil organic carbon, the carbon content is not related to distribution of soils. Instead, each of the pedons is designated to a life-zone according to Holdridge life zone groups (Holdridge, 1947). The global amounts are calculated by multiplying the carbon estimates by the land area of the life zone groups. This approach is different from most other workers and serves as a valid comparison. A good review of the different approaches is provided by Bouwman (1990).

A. Materials and Methods

During the last two years, the staff of World Soil Resources of the Soil Conservation Service (WSR-SCS) has attempted to collate data from national sources. A report on this was made in 1992 (Eswaran et al., 1993b). The compilation is a continuing process, and the current data set has about 1,000 pedons from 45 countries (mostly in the tropics) and an additional 15,000 pedons from the United States. WSR-SCS, in collaboration with other agencies, is in the process of developing a map showing "Major Soil Regions of the World" (Eswaran et al., 1993a). This map has been digitized, and initial estimates for the different map units of soils are available. WSR-SCS has also developed a data base on carbon in soils of the world, using the WSR-SCS data base and published information. Based on this global data base on organic carbon, each map unit on the Major Soil Regions of the World map is assigned a value for the organic carbon content to a meter depth. If the map unit has representative pedons, the average value is employed for the map unit. If the map unit has no pedons, a "best value" is assigned based on the soil classification and the soil moisture and temperature regime of the area. This assigned value is multiplied by the area of the map unit to obtain the total carbon content for the unit. Because the data for carbonate carbon is incomplete, these estimates are less reliable in the current study.

Table 4. Estimates of carbon in different soils and error associated with estimate

Author/soils	Mean O.C. kg/m^2	Number of pedons	CV %
Alexander et al., 1989 -- Alaska			
Shallow Entisols	16.9	7	28
Deep Entisols	32.4	7	34
Shallow Spodosols	17.4	26	42
Deep Spodosols	29.8	95	35
Cryofolists	14.2	4	30
Kimble et al., 1990 -- Global			
Tropical Oxisols	9.7	71	42
Tropical Ultisols	8.3	53	70
Temperate Mollisols	9.1	522	46
Temperate Alfisols	5.5	354	62
Aridisols	4.2	98	60

B. Error and Reliability

Estimates of global carbon content in terrestrial ecosystems have been made by several persons and the most recent is that of Bohn (1976, 1982) and of Kimble et al. (1990). The problems of making accurate global estimates results from:

    a. very high variability in carbon content of soils;
    b. absence of reliable estimates of area occupied by kinds of soils;
    c. availability of reliable data;
    d. the confounding effect of vegetation and land use.

Every study has shown considerable variability in the range of organic carbon in classes of soils. If the coefficient of variation (CV) is taken to express the variability, Table 4 shows the unreliability of most generalizations. In order to obtain global estimates, this variability must be accepted and appreciated.

Digitized soil maps are the most reliable base information for estimating the amount and spatial distribution of carbon in soils. The accuracy of such estimates is a function of scale. Studies of small geographic areas are just commencing and a good example is that of Davidson and Lefebvre (1993) for the State of Maine. They employed a digitized State soil map at scale of 1:250,000 for their evaluation. They compared estimates of stocks of soil carbon in Maine using three scales of resolution: (1) multiplying a published mean of soil carbon for temperate forests or for Spodosols by the area of the State (assuming that all of Maine is composed of Spodosols); (2) calculating soil C for each soil series and map unit in the 1:250,000 State soil map; (3) calculating soil C for a 7.5 minute quadrangle using the 1:20,000 scale soil maps. Scale dependent errors and biases are significant when the first two approaches are compared with the third. A major source of errors when coarser scales are used result from high C inclusions which are omitted in the small scale maps and which can significantly increase the estimates of terrestrial carbon.

VI. Results and Discussion

The organic carbon content for each of the suborders of soils as defined in Soil Taxonomy (Soil Survey Staff, 1975) is provided in Table 5. The total mass of organic carbon stored in soils of the world is 1,555 petagrams (Pg). In the previous study (Eswaran et al., 1993b) 1,576 Pg was estimated and the difference with the most recent estimate is due to the availability of data from many more countries. The global estimate is close to the estimate made by Buringh (1984) and deviates from those of Kimble et al. (1990), and Bohn (1982). Histosols were not included in the study of Kimble et al. (1990) and additionally, they made their estimates by using unpublished area data of soils estimated by the Soil Conservation Service in the fifties. The amount of carbonate carbon is estimated to be about 1,738 Pg, giving a total of 3,293 Pg stored in soils. The estimates for carbonate carbon are still not very reliable due to the lack of representative pedons from different parts of the world. In addition, data for soils with petrocalcic horizons are usually not available in most data bases. In our study, if the profile description indicated the presence of a petrocalcic horizon, the carbonate carbon was estimated arbitrarily. Error on the positive side results from fine fragments of limestone rocks in the fine-earth fraction which are measured together with pedogenic carbonates in the total carbonate analysis.

Maps 1 and 2 show the distribution of global organic carbon and global carbonate carbon. These maps are generated using the spatial and tabular data base using a Geographic Information System. They show the geographic pattern of sequestered carbon in soils. Desert soils have very low levels of carbon while the soils of the higher latitudes have significantly higher amounts. Maximum amount of carbonate carbon is in desert soils followed by the soils of the semi-arid regions.

Lack of suitable data, in terms of measured organic carbon and bulk density, is still a problem with respect to arriving at more reliable estimates. This is particularly the case with Histosols for which bulk density measurements are frequently lacking. The amount of carbon in Histosols is a gross under-estimation because only a meter depth is considered; in many cases, the actual depth of the organic soil is much greater than one meter. In addition, undecomposed timber is a characteristic feature of many tropical Histosols and this is not measured. Published data for tropical forest aboveground biomass was used to correct for the tropical Histosols.

Litter layer, a significant component in the cooler regions of the world and under forest canopy is also not considered in the current estimate. Finally, reliable data is also not available for the Russian territories, Mongolia and China. U.S. information was employed to estimate both organic and carbonate carbon for similar soils in these countries.

Post et al. (1982) compiled using 3,100 pedons and the Holdridge's life zone approach, the global stock of nitrogen in soils. Understanding of the content and distribution of nitrogen in soils, in addition to organic carbon, is important for evaluating the dynamics of carbon and for evaluating greenhouse gas fluxes. Post et al. (1985) estimated that the nitrogen pool in the surface meter of soils is about 95 Tg. In the current study, table 6, an almost identical amount was estimated using a completely different database. The highest amount of soil nitrogen is contributed by the Inceptisols, partly because of their wide extent. Although Histosols contribute about 390 Pg of carbon, they only contribute about 2.7 Tg of N reflecting the poorly decomposed nature of the fibrous materials in such soils. In all the mineral soils, more than 80% of the nitrogen is stored in the top 10 cm of the soil.

Table 6, also illustrates the differential amounts of carbon stored at different depths. The amount of organic carbon is calculated for 0-25 cm and 0-50 cm, and these amounts are expressed as percentage of the total for the 0-100 cm depth. In the Ultisols, more than 95% of the total carbon is in the upper 50 Cm of the soil with more than 70% in the upper 25 cm of the soil. In the Andisols, Aridisols, Oxisols, and Inceptisols, about 50% or more of the organic carbon is stored in the top 25 cm while the top 50 cm holds as much as 80 to 90%. In the Vertisols and Entisols, less than 50% is stored in the upper 50 cm. These are soils which generally have an irregular distribution of carbon with depth due to the mixing process in the Vertisols or the differential deposition in the alluvial soils. Finally, in the Histosols, less than 7% of the total is present in the top 25 cm due to admixtures with mineral soil materials. Like some of the alluvial Entisols, the Histosols are also layered and this is reflected in the organic carbon depth function.

As indicated previously, from the point of view of emission of green house gases, wetlands are the major contributors. There is no reliable estimate of the global area of wetlands. Information on both the spatial distribution and amount of carbon held, is required for wetlands as they are perhaps the largest sinks of carbon among the soil ecosystems. In addition, the wetlands are being drained for agricultural and other purposes in many parts of the world and as a result are becoming a significant source of atmospheric carbon. Finally, the wetlands are also points for emission of other green house gases such as methane.

Table 7, gives an estimate of area of organic soils (Histosols) and wet soils in general. The wet mineral soils occupy about 6% of the land mass, of which about 5.8% is present in the tropics. About a third of the organic soils which occupy about 1.9% of the global land mass is present in the tropics and dominantly in S.E. Asia. Armentano et al., (1984, 1986), estimated that the non-tropical organic soils occupy about 88% of the global organic soils. Our estimate is similar with tropical Histosols occupying 16% of the global Histosols. For pragmatic purposes, the term wetlands as used here includes the Histosols and the wet mineral soils. The stock of carbon in the wetlands is estimated to be about 498 Pg of which about 12% is

Table 5. Organic and carbonate carbon mass in soils of the world

Suborder/Order	Eswaran et al., 1993b Organic carbon	Recent revision (1994)		
		Organic carbon	Carbonate carbon	Total carbon
	------------------------------Peta gram------------------------			
Folists	1	1	0	1
Fibrists	207	250	0	250
Hemists	72	68	0	68
Saprists	77	71	0	71
Histosols	357	390	0	390
Aquands	1	1	0	1
Cryands	18	18	0	18
Torrands	1	1	1	2
Xerands	3	2	0	2
Vitrands	1	1	0	1
Ustands	16	13	0	13
Udands	38	33	0	33
Andisols	78	69	1	70
Aquods	5	4	0	4
Ferrods	0	0	0	0
Humods	49	41	0	41
Orthods	17	53	0	53
Spodosols	71	98	0	98
Aquox	1	1	0	1
Torrox	0	0	0	0
Ustox	47	41	0	41
Perox	7	16	0	16
Udox	64	92	0	92
Oxisols	119	150	0	150
Aquerts	0	1	1	1
Xererts	2	5	1	6
Torrerts	4	12	14	26
Uderts	3	5	0	5
Usterts	10	15	9	24
Vertisols	19	38	25	63
Salids	5	5	113	118
Gypsids	3	3	12	15
Calcids	17	17	407	424
Durids	0	0	1	1

Table 5. -- continued

Argids	38	38	112	150
Cambids	47	47	399	446
Aridisols	110	110	1,044	1,154
Aquults	7	5	0	5
Humults	3	4	0	4
Udults	43	50	0	50
Ustults	50	40	0	40
Xerults	2	2	0	2
Ultisols	105	101	0	101
Albolls	2	2	0	2
Aquolls	1	1	1	2
Rendolls	0	0	1	1
Xerolls	14	13	23	36
Borolls	22	15	29	54
Ustolls	12	17	32	49
Udolls	21	24	53	77
Mollisols	72	72	139	139
Aqualfs	7	6	0	6
Boralfs	37	35	0	35
Ustalfs	31	45	71	116
Xeralfs	10	14	0	14
Udalfs	42	36	56	92
Alfisols	127	136	127	236
Aquepts	54	67	12	79
Plaggepts	0	0	0	0
Tropepts	36	20	26	46
Ochrepts	252	135	247	382
Umbrepts	10	45	0	45
Inceptisols	352	267	285	552
Aquents	8	20	0	20
Arents	0	0	0	0
Psamments	15	21	30	51
Fluvents	16	3	6	9
Orthents	109	62	81	143
Entisols	148	106	117	223
Rocky land	13	13	0	13
Shifting sand	5	5	0	5
Misc. land	18	18	0	18
TOTAL	1,576	1,555	1,738	3,293

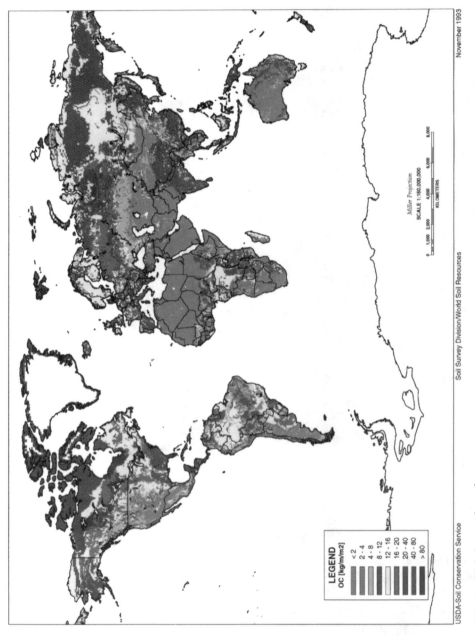

Map 1. Global organic carbon.

Global Soil Carbon Resources

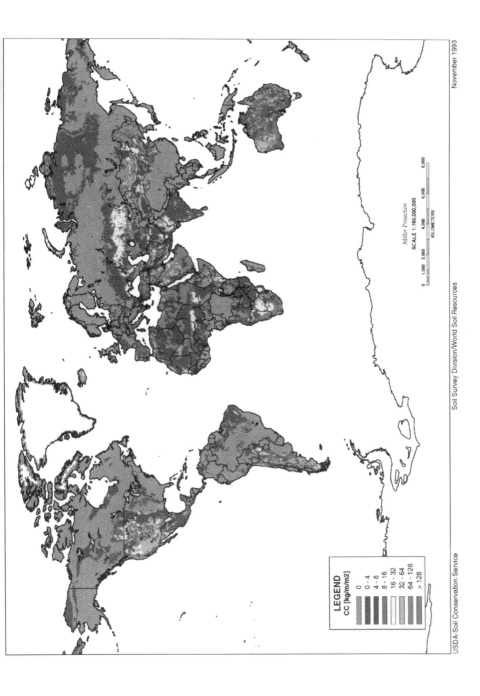

Map 2. Global carbonate carbon.

Table 6. Organic carbon, its amount for different soil depths, and nitrogen mass in soils of the world (values in parentheses are percentages of total carbon in the 0-100 cm depth)

Order	Organic carbon 0-100 cm	Organic carbon 0-25 cm		Organic carbon 0-50 cm		Nitrogen 0-100 cm
	----------Peta gram----------					Tera gram
Ultisols	101	74	(73)	96	(95)	4.19
Andisols	69	38	(55)	61	(88)	4.44
Aridisols	110	57	(52)	95	(86)	11.74
Oxisols	150	88	(59)	128	(85)	11.42
Inceptisols	267	162	(61)	215	(81)	30.00
Alfisols	136	73	(54)	100	(74)	15.36
Mollisols	72	41	(57)	52	(72)	3.56
Vertisols	38	17	(45)	21	(55)	2.37
Spodosols	98	39	(40)	53	(54)	3.66
Entisols	106	37	(35)	52	(49)	6.27
Histosols	390	26	(7)	54	(14)	2.68
Misc. Land	18					
TOTAL	1,555	652	(42)	927	(60)	95.69

Table 7. Initial estimates of organic carbon mass in wet soils of the world (values in parentheses are percentages of the total land mass)

Order	Area[a] Global		Area[a] Tropical		Organic C Global
	----------------10³ km²----------------				
Wet mineral soils					
Aquands	-		-		
Aquods	210	(0.16)	-		4
Aquerts	84	(0.06)	84	(0.17)	1
Salids	1,130	(0.84)	134	(0.27)	5
Aquults	563	(0.42)	501	(1.01)	5
Aquolls	-		-		
Aqualfs	701	(0.52)	69	(0.14)	6
Aquepts	4,644	(3.44)	2,050	(4.13)	67
Aquents	1,362	(1.01)	30	(0.06)	20
Total	8,808	(6.53)	2,868	(5.78)	108
Organic soils					
Histosols	1,745	(1.9)	286	(0.57)	390
TOTAL	10,553	(7.82)	3,154	(6.35)	498

[a] Most recent estimates by USDA, Soil Conservation Service. (From Eswaran et al., 1993.)

in the wet mineral soils. The remaining 88% in the Histosols is slowly being reduced due to drainage of the land for agriculture. Armentano et al. (1986) estimate that about 5 to 6 Gt would be lost over the next thirty year period due to burning of peat and utilization of peatlands for agriculture.

VII. Conclusions

The limitations of a database in terms of number of samples and their representation is still a major problem which will be resolved as the database is enhanced through the collaborative efforts under the International Geosphere Biosphere Program. Information on soil microvariability reduces the confidence in assigning values to soils. This becomes more significant when small areas of the landscape are modelled based on the current data base. Another source of error results from changes in land use. Currently, it is assumed that all soils have the same land use as at time of sampling.

The quantification of soil carbon losses and gains resulting from land use/land management changes is a prerequisite to the understanding of greenhouse gases fluxes in different ecosystems. Modelling the fluxes of greenhouse gases, specifically due to land use changes, requires detailed knowledge of the factors regulating carbon and nitrogen cycles and the major soil controls. Providing the spatial and temporal databases to backstop these models is a challenge which requires detailed soil resource assessments. A second major challenge deals with characterization of the forms of carbon in soils. As quantity and form of carbon varies within and between soils, it is necessary prior to developing databases, to establish methods to characterize the different forms.

Finally, current soil carbon databases are from soil assessment studies where evaluation of soil carbon was not the basic objective. Consequently, in most databases critical information needed for modelling purposes is still not available. This implies that a concerted effort is needed to initiate sampling activities specifically to provide those kind of information required by the models. Initially this must be in well-defined ecosystems, such as the Tundra, tropical swamps, defined agricultural-ecosystems, etc. so that not only the amount and forms of soil carbon are quantified but also the relationship with respect to the major controls are understood. This research endeavor will provide a better understanding of how ecosystems will respond to global environmental changes in the near future.

VIII. Summary

In recent years the emission of greenhouse gases, specifically through anthropogenic activities, has come under intense scrutiny. This has led to a more concerted effort to understand the "carbon cycle". Concerns regarding sources of atmospheric C stems from human activities such as utilization of fossil fuels which releases about 6 Pg yr^{-1} of which about 3.4 Pg yr^{-1} accumulates in the atmosphere. The picture is further complicated by the fact that although some sources and sinks of carbon are well documented, the global carbon budget has proved to be more elusive. There is thus a need for a better identification and quantification of global carbon sinks. This is the basic objective of this study.

The World Soil Resources (WSR) of the Soil Conservation Service maintains a database of soils of the world. WSR in collaboration with other Agencies has also prepared a draft map of "Major Soil Regions of the World". Based on the database, normative values for total carbon to a meter depth is allocated to each Great Group of soil. This is employed to evaluate carbon contents of different soils and also permits global estimates. The data base on carbonate carbon is less representative. Though some initial estimates are presented, it is indicated that a significantly larger data base is required before more confident estimates are provided. The initial estimates show that there is 1,555 Pg of organic carbon and 1,738 Pg of carbonate carbon with a total of 3,293 Pg stored in global soils. There is also about 95 Tg of nitrogen stored in global soil resources.

References

Alexander, E.B., E. Kissinger, R.H. Huecker, and P. Cullen. 1989. Soils of Southeast Alaska as sinks for organic carbon from atmospheric carbon dioxide. Publ. Conf. on the stewardship of soil, air, and water resources. USDA Forest Service, Juneau, Alaska.

Armentano, T.V., E.S. Menges, J. Molofsky, and D.J. Lawler (1984). Carbon exchange of organic soils ecosystems of the world. Report to the Carbon Dioxide Research Division. U.S. Department of Energy, Holcomb Research Institute, Butler University, Indianapolis, Indiana.

Armentano, T.V. and E.S. Menges. 1986. Patterns of change in the carbon balance of organic soil-wetlands of the temperate zone. *J. Ecology.* 74:755-774.

Bohn, H.L. 1976. Estimate of organic carbon in world soils. *Soil Sci. Soc. Amer. J.* 40:468-470.

Bohn, H.L. 1982. Estimate of organic carbon in world soils II. *Soil Sci. Soc. Amer. J.* 46:1118-1119.

Bouwman, A.F. 1990. Global distribution of the major soils and land cover types. p. 31-59. In: A.F. Bouwman (ed.), *Soils and the greenhouse effect: Proceedings of the international conference on soils and the greenhouse effect.* John Wiley and Sons, New York.

Brown, S. and A.E. Lugo. 1984. Biomass of tropical forests. A new estimate based on forest volumes. *Science* 223:1290-1293.

Buringh, P. 1984. Organic carbon in soils of the world. p 91-109. In: Woodwell, G.M. (ed.), The role of terrestrial vegetation in the global carbon cycle. Measurement by remote sensing. *SCOPE* Vol. 23. Wiley and Sons, New York.

Burke, J.C., D.S. Schimel, C.M. Yonker, W.J. Parton, L.A. Joyce, and W.K. Lauenroth. 1990. Regional modeling of grassland biogeochemistry using GIS. *Landscape Ecology* 4: 45-54.

Burke, J.C., C.M. Yonker, W.J. Parton, K. Flach, and D.S. Schimel. 1989. Texture, climate, and cultivation effects on soil organic matter content in the U.S. Grassland soils. *Soil Sci. Soc. Amer. J.* 53:800-805.

Davidson, E.A. and P.A. Lefebvre. 1993. Soil carbon stocks estimated from Geographic Information Systems at three scales. *Biogeochemistry.* In press.

Downing, J.P and D.A. Cataldo. 1992. Natural sinks of CO_2: Technical synthesis from the Palmas Del Mar Workshop. p. 439-453. In: J. Wisniewski and A.E. Lugo (eds.), *Natural sinks of CO_2*. Kluwer Academic Publ. Holland.

Dudal, R and H. Eswaran. 1988. Distribution, properties and classification of Vertisols. In: L.P. Wilding and R. Puentes (eds.), Vertisols: Their distribution, properties, classification and management. Publ. Soil Management Support Services, Washington D.C. 1-22.

Duxbury, J. 1991. In: M.G. Johnson and J.S. Kerns (eds.), Sequestering carbon in soils: a workshop to explore the potential for mitigating global climate change. Publ. EPA Corvallis, Oregon. 85pp.

Eswaran, H. 1979. Micromorphology of Oxisols. Proc. 2nd Inter. Workshop Soil Classification. Kuala Lumpur, Malaysia.

Eswaran, H., N. Bliss, D. Lytle, and D. Lammers. 1993a. Major soil regions of the world. In preparation. USDA Soil Conservation Service.

Eswaran, H, E. Van Den Berg, and P.F. Reich. 1993b. Organic carbon in soils of the world. *J. Soil Sci. Soc. Amer.*

FAO/UNESCO. 1971-1981. Soil Map of the World 1:5,000,000. Vols. I-X., Food and Agricultural Organization, Rome.

Franzemeier, D.P., G.D. Lemme, and R.J. Miles. 1985. Organic carbon in soils of North Central United States. *Soil Sci. Soc. Amer. J.* 49:702-708.

Holdridge, L.R. 1947. Determination of world plant formations from simple climatic data. *Science* 105: 367-368.

Houghton J.T., G.J. Jenkins, and J.J. Ephraums (eds.). 1990. *Climate change, the IPCC scientific assessment.* Cambridge University Press, Cambridge. 365 pp.

Houghton, R.A., D.L. Skole, and D.S. Lefkowitz. 1991. Changes in the landscape of Latin America between 1850 and 1985. II. Net release of CO_2 to the atmosphere. *For. Ecol. Manage.* 38:173-199.

Johnson, M.G. and J.S. Kerns. 1991. Sequestering carbon in soils: A workshop to explore the potential for mitigating global climate change. EPA/600/3-91/031. U.S. Environmental Protection Agency, Environmental Research Laboratory, Corvallis, OR. 85 pp.

Kimble, J, T. Cook, and H. Eswaran. 1990. Organic matter in soils of the tropics. Proc. Symposium on Characterization and role of organic matter in different soils. 14th. Int. Congress Soil Science, Kyoto, Japan. V:258-250.

Lugo, A.E. 1992. The search for carbon sinks in the tropics. p. 3-9. In: J. Wisniewski and A.E. Lugo (eds.), *Natural sinks of CO_2*. Kluwer Academic Publ. Holland.

Meentemeyer, V., E.O. Box, and R. Thompson. 1982. World patterns and amounts of terrestrial plant litter production. *Bioscience* 32:125-128.

Miranda, S.M., P. Pathak, and K.L. Srivastava. 1982. Runoff management on small agricultural watersheds -- the ICRISAT experience. In: National Seminar on "Decade of Dryland Research in India and the thrust in the eighties". Publ. CRIDA, Hyderabad, India.

National Wetlands Working Group. 1988. *Wetlands of Canada*. Ecological Land Classification Series No. 24. Environment Canada, Ontario, Canada. 452 pp.

Parton, W.J., D.S. Schimel, C.V. Cole, and D.S. Ojima. 1987. Analysis of factors controlling soil organic matter levels in Great Plains Grasslands. *Soil Sci. Soc. Amer. J.* 51: 1173-1179.

Post, W.M., W.R. Emmanuel, P.J. Zinke, and A.G. Stangenberger. 1982. Soil carbon pools and world life zones. *Nature* 298:156-159.

Post, W.M., J. Pastor, P.J. Zinke, and A.G. Stangenberger. 1985. Global patterns of soil nitrogen. *Nature* 317:613-616.

Post, W.M., T.H. Peng, W.R. Emmanuel, A.W. King, V.H. Dale, and D.L. De Angelis. 1990. The global carbon cycle. *American Scientist* 78:310-326.

Scharpenseel, H.W and H. Schiffman. 1977. Radiocarbon dating of soils, a review. *Zeitschrift fur Pflanzenesnachrung und Bodenkunde* 140:159-174.

Schlesinger, W.H. 1982. Carbon storage in the caliche of arid soils: A case study of Arizona. *Soil Sci.* 133: 247-255.

Schlesinger, W.H. 1985. The formation of caliche in soils of the Mojave Desert, California. *Geochimica et Cosmochimica Acta* 49:57-66.

Soil Survey Staff. 1975. Soil Taxonomy: A basic system of soil classification for making and interpreting soil surveys. USDA-SCS Agric. Handb. 436 U.S. Gov. Print. Office. Washington D.C.

Wilson, M.F. and A. Henderson-Sellers. 1985. A Global archive of land cover and soils data for use in general circulation climate models. *J. Climatology* 5:119-143.

CHAPTER **4**

Changes in the Storage of Terrestrial Carbon Since 1850

R.A. Houghton

I. Introduction

Are terrestrial ecosystems releasing carbon to the atmosphere, or are they withdrawing carbon from the atmosphere and accumulating it in vegetation and soils? The question was first asked more than 20 years ago when early models of the global carbon cycle found that the carbon emitted annually from combustion of fossil fuels could not be accounted for by the sum of the observed increase in atmospheric carbon and the modeled uptake of carbon by the world's oceans (Bacastow and Keeling, 1973). In order to balance the global carbon budget, geophysicists assumed that terrestrial ecosystems were accumulating carbon.

$$\begin{matrix}\text{Fossil fuel} \\ \text{emissions}\end{matrix} = \begin{matrix}\text{Atmospheric} \\ \text{increase}\end{matrix} + \begin{matrix}\text{Oceanic} \\ \text{uptake}\end{matrix} + \begin{matrix}\text{Terrestrial} \\ \text{uptake}\end{matrix} \qquad (1)$$

The required terrestrial uptake was achieved in models by assuming a "biotic growth factor" (Bacastow and Keeling, 1973) that increased global net primary production and carbon storage in proportion to the atmospheric concentration of CO_2. The atmospheric increase in CO_2 had been observed for a little more than ten years when these first results were obtained. Continuous measurements of the concentrations of CO_2 in the atmosphere began in 1958 at Mauna Loa, Hawaii, and at the South Pole (Keeling et al., 1976a,b). Ocean uptake was calculated with a two-box model that included a shallow surface ocean and a deep ocean.

The question of whether terrestrial ecosystems were storing or releasing carbon was addressed by terrestrial ecologists at about the same time (Woodwell and Pecan, 1973), but the first quantitative estimates, based on the historical increase in agricultural lands at the expense of forest area, showed terrestrial ecosystems to be releasing carbon (Bolin, 1977; Woodwell and Houghton, 1977; Woodwell et al., 1978). The finding upset contemporary understanding of the global carbon cycle which required for a balanced budget that terrestrial ecosystems be accumulating carbon, not releasing it. The additional release calculated on the basis of changes in land use indicated a poorer understanding of the carbon cycle than seemed possible given the data constraining the first three terms in (1), above (Broecker et al., 1979).

Today, more than 20 years later, the same question can be asked: Are terrestrial ecosystems releasing carbon to the atmosphere, or are they withdrawing it from the atmosphere and accumulating it in vegetation and soils? And the answer is almost as controversial. Direct analyses based on changes in land use have consistently shown terrestrial ecosystems to be releasing carbon to the atmosphere (Houghton et al., 1983, 1985, 1987; Detwiler and Hall, 1988; Houghton and Skole, 1990; Hall and Uhlig, 1991; Houghton, 1991b; Flint and Richards, 1994). Indirect estimates based on geophysical constraints from the atmosphere and oceans indicate that terrestrial ecosystems must be accumulating carbon, albeit a small amount (Keeling et al., 1989; Oeschger et al., 1975; Sarmiento et al., 1992; Tans et al., 1990). Both geophysicists and terrestrial ecologists recognize that ecosystems have released carbon over the last century or so. The controversy pertains only to the last two or three decades. For 1990, the following "equation" applies (units are Pg carbon; 1 Pg = 10^{15} g):

$$\begin{array}{llll}
\text{Fossil fuel} = \text{Atmospheric} + \text{Oceanic} + \text{Terrestrial} \\
\text{emissions} \quad\;\; \text{increase} \quad\;\;\;\; \text{uptake} \quad\;\;\; \text{uptake} \\
\\
6\ (\pm 0.6) \quad\quad 3\ (\pm 0.1) \quad\;\; 2\ (\pm 0.5) \quad\;\; ??
\end{array} \quad (2)$$

If terrestrial uptake is calculated by difference, terrestrial ecosystems are accumulating about 1 Pg C/yr, and the carbon budget is balanced. On the other hand, if the terrestrial term is determined on the basis of changes in forest area, terrestrial ecosystems are releasing about 1.5 Pg C/yr to the atmosphere, and the terms of the global carbon equation are not balanced.

The purpose of this review is, first, to describe the basis for the estimate of changes in terrestrial carbon, or flux of carbon from terrestrial ecosystems to the atmosphere. What are the data and calculations on which it is based? Second, an argument will be presented for reconciling the two conflicting estimates of the flux, those based on changes in land use, and those based on analyses using geophysical data and models.

II. Changes in Terrestrial Carbon Resulting from Changes in Land Use

Clearing forests for new agricultural land causes a release of carbon to the atmosphere. The carbon initially held in trees and other vegetation is released through burning or through decomposition of above- and belowground plant material left in the soil at the time of clearing. Even if the productivity of the new agricultural land is as high as it was in the forest, less of the crop production accumulates as litter; most of it is harvested and subsequently consumed or respired. This reduction in litter input is not initially balanced by a reduction in soil respiration. In fact, the respiratory release is often enhanced from the cultivation itself, which exposes more of the organic matter to oxygen, and from increased soil moisture and temperature associated with the microclimate of cleared land. As a result, some of the carbon originally held in forest soil is released to the atmosphere after clearing.

The reverse is true when agricultural lands are abandoned and allowed to return to forests. With forest growth, carbon is withdrawn from the atmosphere and accumulated in biomass and soil. Thus, the net contribution of terrestrial ecosystems to atmospheric carbon dioxide will depend on changes in the areas of carbon rich and carbon poor ecosystems. Generally, the carbon rich systems are forests, and carbon poor systems are cleared lands. When one considers soils, however, temperate zone grasslands are analogous to forests: they hold large amounts of carbon in their undisturbed state and hold less after cultivation. Changes in the areas of undisturbed and cultivated lands are, in turn, related to changes in land use. Determining the net flux of carbon over any interval of time is, thus, dependent upon two types of data: (1) rates of land-use change and (2) changes in the carbon stored in vegetation and soil per unit area in those ecosystems affected by the land-use change.

These two types of data (change in area and change in carbon per unit area) have been assembled for the major ecosystems of the world and have been used in a number of analyses to compute regional and global changes in terrestrial carbon storage. The specific data used in these analyses are published elsewhere and will not be given here. The analyses have included the entire globe (Moore et al., 1981; Houghton et al., 1983, 1987; Woodwell et al., 1983); the tropics (Houghton et al., 1985; Detwiler and Hall, 1988; Hall and Uhlig, 1991; Houghton, 1991b); the former Soviet Union (Melillo et al., 1988); Latin America (Houghton et al., 1991a,b); Southeast Asia (Palm et al., 1986); and tropical Asia (Houghton, 1991a; Flint and Richards, 1994; Richards and Flint, 1994; Houghton and Hackler, 1994). These studies should be consulted for the specific data and assumptions used to calculate changes in the carbon stocks of terrestrial ecosystems. Estimates of change, reviewed below, are based on the most recent of these analyses. The next section reviews the data and models used in these analyses.

A. Changes in Land Use

Because forests contain so much more carbon than the lands that replace them, analyses of terrestrial carbon have emphasized forests and rates of deforestation and reforestation. Before 1980, however, there were no estimates of deforestation for much of the earth, and analyses of long-term trends in carbon storage relied on changes in the area of agricultural lands, rates of logging, and other changes in land use to determine changes in the area of forests. Six of the major types of land-use change are described below. The emphasis

1. Cultivated Areas or Croplands

The area in cultivated land is generally much better documented historically than the area of forest (Richards, 1990). Thus, changes in the area of cultivated lands, together with maps or other descriptions of where the cultivated lands are located, have been used to infer the types of ecosystems converted to cultivated lands. The global area in croplands was somewhat more than 1400×10^6 ha in 1980. In 1850, it was about 500×10^6 ha (Tables 1 and 2). About 60% of the 900×10^6 ha increase is estimated to have been in tropical regions.

2. Pastures

Changes in the area of pastures is more difficult to document historically because the distinction between grasslands and pastures is a fuzzy one. The number of cattle can be used to infer changes in the area of pasture, but use of livestock requires independent data on stocking densities and, as with cultivated areas, maps of where pastures are located (Houghton et al., 1991a). Most of the worldwide increase in grasslands, both grazed and ungrazed, seems to have occurred in the temperate zone (Table 1). The large areas of South America now grazed by cattle were once natural grasslands supporting herds of indigenous animals. The change has probably affected carbon stocks little. Large scale conversion of tropical forests to pastures is a relatively recent phenomenon, most of it occurring in the last two decades.

3. Degradation of Land

Despite the fact that most deforestation in the tropics is for new agricultural land, the annual net expansion of agricultural lands is considerably less than the annual net reduction in forest area. For the entire tropics, for example, the expansion of croplands accounted for only 27% of total deforestation between 1980 and 1985 (FAO, 1990b). Adding the increase in pasture area accounted for an additional 18% of deforestation. Fully 55% of the deforestation between 1980 and 1985 was explained by an increase in a category defined by FAO as "other land". Although some of this "other land" is urban land, roads, and other settled lands, these uses are unlikely to have accounted for more than a few per cent of the area deforested. Most of the "other land" seems likely to be abandoned, degraded croplands and pastures --- lands that no longer support crop or livestock production, but that do not revert readily to forest, either.

Forests are not converted directly to degraded areas, of course. The transformation of land is from forest to agriculture and, subsequently, to degraded land. The important point is that only about one half of the area of tropical forest lost each year actually expands the area in agriculture. The other half is only temporarily useful. After a few years it is lost and is neither agriculturally productive nor forested.

The fraction of deforestation used to expand the area in agriculture, as opposed to replace worn-out land, varies among tropical regions. In Africa agricultural expansion accounted for only about 12% of the net area deforested. Eighty-eight percent of the decrease in forest area was matched by the expansion of "other land". In tropical Asia only 40% of the net reduction in forests appeared as an expansion of agricultural lands. In Latin America about 2/3 of the reduction in forests could be accounted for by the expansion of croplands and pastures. If agriculture could be made sustainable throughout the tropics, rates of deforestation might be reduced by about 50% without reducing the expansion of agricultural lands. Large areas of marginal or degraded lands might become available for reforestation or plantations.

4. Shifting Cultivation

Shifting cultivation is a rotational form of agriculture, predominantly in the tropics, in which the period of cropping is short relative to the period of fallow. During the fallow period, forests begin to grow.

Table 1. Area, total carbon, and mean carbon content of vegetation and soils in major ecosystems of the earth in 1980

	Area (10^6 ha)	Carbon in vegetation (10^{15} g)	Carbon in soil (10^{15} g)	Mean carbon content of vegetation (Mg C/ha)	Mean carbon content of soil (Mg C/ha)
Tropical evergreen forest	602	107	62	177	104
Tropical seasonal forest	1,459	169	125	116	86
Temperate evergreen forest	508	81	68	161	134
Temperate deciduous forest	368	48	49	131	134
Boreal forest	1,168	105	241	90	206
Tropical fallows (shifting cultivation)	227	8	19	36	83
Tropical open forest/woodland	307	15	20	50	64
Tropical grassland and pasture	1,021	17	49	16	48
Temperate woodland	264	7	18	27	69
Temperate grassland and pasture	1,235	9	233	7	189
Tundra and alpine meadow[1]	800	2	163	3	204
Desert scrub[1]	1,800	5	104	3	58
Rock, ice, and sand[1]	2,400	0.2	4	0.1	2
Cultivated, temperate zone	751	3	96	4	128
Cultivated, tropical zone	655	4	35	7	53
Swamp and marsh[1]	200	14	145	68	725
TOTAL	13,767	594	1,431		

[1] From Whittaker and Likens (1973) and Schlesinger (1984).

Table 2. Areas and carbon content of major terrestrial ecosystems in 1850 and 1980, and changes over the 130-year period. Negative values indicate an increase in area or total carbon.

	Area (10^6 ha)			Carbon (10^{15} g)						
				Vegetation			Soil		
	1850	1980	Change	1850	1980	Change	1850	1980	Change
Tropical evergreen forest	674	602	72	126	107	19	75	62	13
Tropical seasonal forest	1,757	1,459	298	212	169	43	158	125	33
Temperate evergreen forest	537	508	29	86	81	5	72	68	4
Temperate deciduous forest	429	368	61	56	48	8	57	49	8
Boreal forest	1,171	1,168	3	105	105	0	241	241	0
Tropical fallows (shifting cultivation)	203	227	−24	7	8	−1	17	19	−2
Tropical open forest/woodland	456	307	149	24	15	9	30	20	10
Tropical grassland and pasture	1,029	1,021	8	16	17	−1	44	49	−5
Temperate woodland	274	264	10	7	7	0	19	18	1
Temperate grassland and pasture	1,507	1,235	272	11	9	2	285	233	52
Tundra and alpine meadow[1]	800	800	0	2	2	0	163	163	0
Desert scrub[1]	1,800	1,800	0	5	5	0	104	104	0
Rock, ice, and sand[1]	2,400	2,400	0	0.2	0.2	0	4	4	0
Cultivated, temperate zone	384	751	−367	1	3	−1	47	96	−49
Cultivated, tropical zone	153	655	−502	0	4	−3	9	35	−26
Swamp and marsh[1]	200	200	0	14	14	0	145	145	0
TOTAL	13,767	13,767	0	672	592	80	1471	1431	40

[1]From Whittaker and Likens (1973) and Schlesinger (1984).

Depending on the length of fallow, considerable biomass may accumulate. The increasing need for agricultural land in much of the tropics has led to a shortening of the fallow period. Often the result is a degradation of land (see above).

Changes in the area occupied by shifting cultivation and fallow are difficult to obtain, historically and currently. Current estimates by FAO (FAO/UNEP, 1981) and by Myers (1980) do not agree. According to the FAO study (FAO/UNEP, 1981), the area of fallow increased between 1980 and 1985, and shifting cultivation was responsible for 35, 70, and 50% of the deforestation of closed forests in tropical America, Africa, and Asia, respectively [closed forests have dense canopies that restrict the penetration of light and, hence, grasses are absent. Open forests have grasses growing between trees or between patches of trees (FAO/UNEP 1981).]. Myers (1980), on the other hand, argues that shifting cultivation is largely being replaced by permanently cleared land, and that the area of fallow is not increasing, but decreasing. He estimated that fallow forests were being cleared at a rate of about 10×10^6 ha yr^{-1} (Houghton et al., 1985). This annual rate is almost as large as the rate of deforestation of non-fallow tropical forests. The discrepancy concerning the fate of forest fallow has not been resolved, although satellite data have the potential to resolve it.

5. Logging and Degradation of Forests

Logging is defined here as harvest followed by regrowth of the forest. Degradation refers to a gradual reduction in the biomass of forests. It occurs when harvests of wood exceed rates of growth.

6. Afforestation and Reforestation

These terms refer to the establishment or re-establishment of forests or tree plantations. Sometimes the regeneration of forests occurs naturally following logging or abandonment of agriculture. In other cases, the establishment of plantations requires intensive management.

7. Total Changes in Land Use

In 1980 and again in 1990, Myers (1980, 1991) and the FAO (FAO/UNEP, 1981; FAO, 1990a, 1991) reported rates of deforestation for most of the tropics. These estimates of deforestation have been used to calculate losses of carbon from tropical regions. The rates vary, however. For the years around 1980, Myers' estimates were 1% higher in Africa and 28% higher in Asia; the FAO/UNEP estimate was 11% higher in Latin America. In the ten years since 1980, estimates of the rates of deforestation in the tropics have increased sharply. According to Myers (1991) the annual loss of closed tropical forests increased by 90% from 7.34×10^6 ha in 1979 to 13.86 in 1989. FAO's estimate (FAO, 1990a, 1991) (including both closed and open forests) was also higher than their earlier one (about a 50% increase), but the FAO maintains that some of the apparent increase is due to their underestimate of deforestation in the earlier period. They acknowledge that the rate of deforestation has generally increased in the moist tropics, although it may have declined in some Asian countries.

A major difference between the recent estimates of deforestation was the rate for Brazil. Myers (1991) based his estimate for Brazil on a study by Setzer and Pereira (1991), who used AVHRR data from the NOAA-7 satellite to determine the number of fires burning during the dry season (mid-July through September). They accounted for the fact that some fires burned for more than one day (and should not be counted twice), and that a small, hot fire would saturate the entire pixel (1km square) and overestimate the area actually burned. With these adjustments, Setzer and Pereira (1991) estimated that about 20×10^6 ha of fires burned in the Brazilian Amazon in 1987, about 60% of which, or 12×10^6 ha, were on lands that had already been deforested. Their estimate of deforestation was 8×10^6 ha.

Myers (1991) reduced their estimate to 5×10^6 ha, to account for other factors. Nevertheless, even this reduced rate seems high according to other recent studies. Using data from Landsat (80m resolution rather than 1km resolution), the National Institute for Space Research (INPE) in Brazil found the rate of deforestation of closed forests in Brazil's Legal Amazonia to have averaged about 2.1×10^6 ha yr^{-1} between 1978 and 1989 (Fearnside et al., 1990), about one fourth the rate initially determined by Setzer and

Pereira (1990). The actual rate probably increased between 1978 and 1987, but fell substantially after 1987 to 1.8×10^6 ha in 1988/89, to 1.38 in 1989/90, and to 1.11 in 1991. Recent work by Skole and Tucker (1993) shows rates similar to those reported by INPE.

When these more recent data from INPE are substituted for those reported by Myers (1991), the estimated rate of deforestation for all Latin America is revised from 7.7 to 4.5×10^6 ha/yr. In contrast to this downward revision, the FAO (1990a, 1991) raised its estimated rate of deforestation for tropical Asia. For the entire tropics, both Myers (1991) (revised for Brazil) and the FAO (1990a, 1991) estimate that the rate of deforestation increased by about 40% over the 10-year period 1980 to 1990.

B. Stocks of Carbon Per Unit Area

1. Vegetation

Deforestation of high biomass forests releases more carbon to the atmosphere than deforestation of low biomass forests. The biomass of tropical forests is variable, however, and not well known. Estimates of tropical forest biomass based on surveys of wood volumes are substantially lower than estimates based on direct measurements of biomass (Brown and Lugo, 1984, Brown et al., 1989, 1991). Because the lower estimates are based on a much larger sampling of area, they are often assumed to be more accurate than the higher estimates and more appropriate for calculation of carbon flux (Brown and Lugo, 1984; Brown et al., 1989, 1991; Detwiler and Hall, 1988). Recent evidence suggests, however, that the use of low estimates of biomass to calculate flux underestimates both the current and the long-term (>10 yr) flux. The evidence comes from the observed, widespread reduction in biomass of tropical forests over time (FAO, 1990a, 1991). A gradual reduction in biomass affects calculation of the long-term flux of carbon because current measures of biomass underestimate past biomass. Thus, for the same area deforested, then and now, more carbon will have been released in the past.

Use of current (low) estimates of biomass also underestimates the current release of carbon if the calculations are based on deforestation alone, and do not include the carbon lost as a result of reduction of biomass *within* forests (degradation). The total net flux includes that due to deforestation and reforestation (changes in area) as well as that due to degradation and regrowth within forests. Most analyses have not considered the loss of carbon from within forests, although some have approached the issue of degradation by consideration of logging (Houghton et al., 1983, Palm et al., 1986, Hall and Detwiler, 1988), which may reduce biomass if the rates of logging remove biomass more rapidly than it regrows. The net flux of carbon calculated for this logging has generally been small, however (Houghton et al., 1985; Palm et al., 1986), while the flux from degradation has been large (Brown et al., 1994; Flint and Richards, 1991, 1994; Houghton, 1991a; Houghton and Hackler, 1994).

2. Soil

Estimates of the organic carbon in the top meter of soil have been summarized by Sanchez et al. (1982), Schlesinger (1984), and Zinke et al. (1986) for the major types of vegetation by Eswaran et al. (this volume) for the major soil groups, and by Brown and Lugo (1982), Post et al. (1982), and Zinke et al. (1986) for life zones. Life zones are classes of potential vegetation determined from climatic variables. On average, the soils of the world contain 1400-1500 Pg C, about three times more organic carbon than is contained in vegetation, 500-600 Pg C. Tropical forests differ from forests of temperate and boreal zones in that more of the carbon of tropical forests is contained in vegetation than in soils (Table 1). When inventories of soil carbon are not limited to the top 1m, this difference between regions may no longer apply. New work in the northeast region of the Brazilian Amazon has shown that roots and soil organic matter extend to 8m or more (Nepstad et al., 1991). Although the carbon content of soils greater than 1m in depth is low, the great depths to which carbon is found indicates that the total organic carbon of soils may be considerably greater than usually thought. If this carbon is old and refractory, its mass is largely irrelevant in exchanges with the atmosphere.

C. Changes in the Stocks of Carbon for Different Kinds of Land Use

The net flux of carbon to the atmosphere from deforestation depends not only on rates of deforestation and stocks of carbon in forests but on uses of deforested lands. The changes in land use considered in analyses of carbon flux include conversion of natural ecosystems to permanent croplands, to shifting cultivation, and to pasture; abandonment of croplands and pastures; degradation of croplands and pastures; harvest of timber; and establishment of tree plantations. The kinds of data used to calculate the rates of release and accumulation of carbon on land following changes in land use are discussed below. Detailed information can be found in Palm et al. (1986), and Houghton et al. (1987; 1991a,b).

1. Cultivated Land

When forests and woodlands are cleared for cultivated land, much of the aboveground biomass is burned and released immediately to the atmosphere as CO_2. Some of the wood may be harvested for products and oxidized more slowly. The rest of the above-ground and below-ground material decays, as does the organic matter of newly cultivated soil (Figure 1). Rates of decay vary with climate. A small fraction of the organic matter burned is converted to elemental carbon or charcoal that resists decay. When croplands are abandoned, the lands may return to forests at rates determined by the intensity of disturbance and climatic factors (Brown and Lugo, 1982; Uhl et al., 1988).

Disturbance, and particularly cultivation, of forest soils generally results in a loss of organic carbon (Detwiler, 1986; Schlesinger, 1986; Davidson and Ackerman, 1993). The variability is large, but, on average, 25-30% of the carbon in the surface meter seems to be lost to the atmosphere when forest soils are cleared of vegetation and cultivated. Most of the loss occurs rapidly within the first 5 years of clearing. However, if some of the large pool of organic carbon below 1m (Nepstad et al., 1991) is not refractory but actively cycling, the net release of carbon associated with removing the forest may be considerably greater than previously thought. Recent work with isotopes in Amazonian soils shows that more than half of the carbon released after conversion of forest to pasture is from depths greater than 1m (Davidson et al., 1993). More research is needed to determine the importance of this loss globally.

2. Pastures

Because pastures are generally not cultivated, the loss of carbon from pasture soils is usually less than 25% of the initial carbon contained in the top 1m. The findings are extremely variable. Most studies show a loss of carbon, sometimes as much as 40% of the carbon originally in the forest soil (Falesi, 1976; Hecht, 1982). On the other hand, under some conditions there appears to be no loss of soil carbon (Buschbacher, 1984; Cerri et al., 1991), and there may even be an increase (Lugo et al., 1986).

3. Degradation of Land

Lands abandoned from agriculture are generally assumed to return to the ecosystem from which they were originally cleared, at rates that vary with climate and land-use history. If abandoned tropical lands do not return readily to forest, as may be the case for degraded lands, rates of accumulation are much reduced. Analyses have attempted to use appropriate rates of regrowth.

4. Shifting Cultivation

Deforestation for shifting cultivation releases less carbon to the atmosphere than deforestation for permanently cleared land because of the partial recovery of the forests. The length of fallow varies considerably due to both ecological and cultural differences (Turner et al., 1977). Decay rates for dead plant material and accumulation rates for regrowing vegetation during the fallow periods have been determined for a variety of ecosystems (Brown and Lugo, 1982; Uhl et al., 1982; Uhl, 1987; Saldariagga et al., 1988).

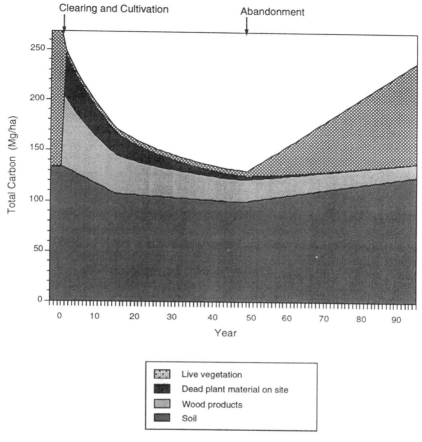

Figure 1. Changes in the initial stocks of carbon (Mg C/ha) following clearing and cultivation of a temperate deciduous forest (year 0) and following abandoment of agriculture (year 50).

Less soil organic matter is oxidized during the shifting cultivation cycle than during continuous cultivation (Detwiler, 1986; Schlesinger, 1986).

5. Logging

Although logging is not generally considered deforestation, it causes a release of carbon to the atmosphere from the mortality and decay of trees damaged in the harvest operations, from the decay of logging debris, and from the oxidation of the wood products. It also causes a net withdrawal of carbon from the atmosphere if logged forests are allowed to regrow. If there is no other use of land following harvest, the net flux of carbon from logging is probably close to zero in the long term, but in the short term logging may either release or accumulate carbon, depending on whether rates of logging are increasing or decreasing, respectively.

Rates of harvest are reported annually by the Yearbook of Forest Products (FAO 1946-1987). Logging in the tropics is selective; extraction rates in different regions average between 8.4 and 56.9 m^3/ha out of total growing stocks of 100 to 250 m^3/ha (Lanly, 1982). One third or more of the original biomass may be damaged or killed in the process of harvest. The dead material decays exponentially. The undamaged, live vegetation accumulates carbon again at rates that vary with the type of forest and intensity of logging (Brown and Lugo, 1982; Horne and Gwalter, 1982). The harvested products decay at rates dependent upon their end use (FAO, 1946-1987) (for example, fuelwood in 1 year, paper in 10 years, construction material in 100 years).

6. Afforestation and Reforestation

Most tree plantations are established on forest lands and replace existing forests. The conversion of natural forests to plantations may cause a net loss of carbon from the soil (Holt and Spain, 1986). The rate of accumulation of carbon in tropical plantations averages about 6.8 t ha^{-1} yr^{-1} (Brown et al., 1986); rotation lengths are as long as 40 years.

D. A Model to Calculate Regional and Global Changes in Terrestrial Carbon

Deforestation and other changes in land use initiate changes in vegetation and soil. With deforestation, a large amount of carbon may be released through burning. Afterwards, decay of soil organic matter, logging residue, and wood products continue to release carbon to the atmosphere, but at lower rates. If croplands are abandoned, regrowth of live vegetation and redevelopment of soil organic matter withdraw carbon from the atmosphere and accumulate it again on land. Such changes have been defined for different types of land use and different types of ecosystems (see above). Annual changes in the different reservoirs of carbon (live vegetation, soils, slash, and wood products) determine the annual net flux of carbon between the land and atmosphere. Because of the variety of ecosystems and land uses, and because the calculations require accounting for cohorts of different ages, bookkeeping models have been used for the calculations (Houghton et al., 1983, 1985, 1987; Detwiler and Hall, 1988).

E. The Net Flux of Carbon Between Terrestrial Ecosystems and the Atmosphere

1. Long-term flux of carbon from changes in land use.

The long-term (1850-1990) flux of carbon to the atmosphere from global changes in land use is estimated to have been 120 Pg C (Figure 2). The annual flux increased from about 0.4 Pg C/yr in 1850 to about 1.7 Pg C/yr in 1990. The rate of the release has increased. It took 100 years for the first increase of 0.6 Pg C/yr ((from 0.4 to 1.0 Pg C/yr) and only 35 years (1950 to 1985) for the next increase of 0.6 Pg C/yr. Until about 1940, the region with the greatest flux was the temperate zone. Since 1950, the tropics have been increasingly important. Current emissions from northern temperate and boreal zones are thought to be close to zero, with releases from decaying wood products approximately balanced by annual accumulations of carbon in regrowing forests (Houghton et al., 1987; Melillo et al., 1988; Houghton, 1993).

About 2/3 of the total long-term flux, or 80 Pg C, was from oxidation of plant material, either burned or decayed (Table 2). About 1/3, or 40 Pg C, was from oxidation of soil carbon, largely from cultivation. Relative to the stocks of carbon in 1850, carbon in vegetation was reduced by about 12% over this 140-year interval, and organic carbon in soil was reduced by about 4%, worldwide.

The greatest loss of carbon from soils occurred in the grasslands of the temperate zone (52 Pg C). About 12 Pg C were lost from temperate forests, and about 56 Pg C were lost from all types of tropical forests and woodlands, combined. These estimates of loss, however, are overestimates. They include not only the losses from oxidation of soil carbon but the bookkeeping losses resulting from the "loss" of natural ecosystems. The latter kind of loss is balanced by a "gain" in the area (and hence carbon) of cultivated lands. Accounting for changes in area, the net loss of carbon from soils was 40 Pg C, globally. About 23 Pg C were lost from tropical soils, and about 17 Pg C were lost from soils in the temperate zone.

The ratio of soil loss to biomass loss (1:2 in units of carbon) is somewhat surprising given the ratio of total stocks (1450:600, or approximately 2.4:1). The difference is explained by the fact that, on average, only 25-30% of the soil carbon is lost with cultivation, and a smaller fraction is usually lost when forests are converted to pastures or to shifting cultivation. In contrast, about 100% of the biomass is often oxidized and released as CO_2 to the atmosphere. Assuming that, on average, 25% of soil carbon is lost, the effective ratio of soil carbon to biomass carbon becomes about 350:600, or approximately 1:2, in agreement with the ratio observed above. The ratio varies with ecosystem as well as with land use. For example, as long as the

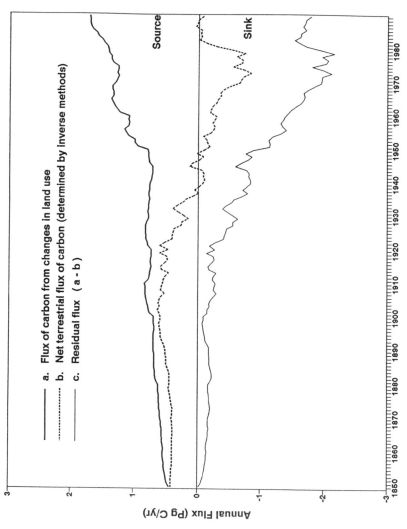

Figure 2. Different components of the net annual flux of carbon between terrestrial ecosystems and the atmosphere. The heavy solid curve is the flux of carbon from changes in land use. The heavy dashed curve is the net terrestrial flux, determined from inversion of an ocean model. (From Sarmiento et al., 1992.). The thin solid curve is the residual flux of carbon, the difference between the net flux and the flux from land-use change. The negative residual flux is assumed to indicate an accumulation of carbon in undisturbed as a result of environment changes.

soil carbon from deep in the soil column of tropical forests is refractory, soils may contribute less than 20% of the total carbon released from tropical deforestation. In contrast, they may contribute more than 50% of the total carbon lost from cultivation of temperate zone grasslands.

If some of the carbon deep in the soil of tropical forests is actively cycled, the net release associated with removing the forest may be considerably greater than previously thought, and the proportion lost by soils considerably greater, as well. Recent work with isotopes in Amazonian soils suggests that this may be true. About one-third of the carbon released after conversion of forest to pasture was from depths greater than 1m (Davidson et al., 1993).

The estimate of 120 Pg C is about 15% lower than an earlier one (Houghton, in press), with most of the difference occurring between 1985 and 1990 as a result of revisions in the rate of deforestation. For South and Southeast Asia, the estimate is about 30% lower than an independent study (Flint and Richards, 1994). The error associated with this value of 120 Pg C is estimated here to be 30% or less.

2. Current Flux of Carbon from Changes in Land Use

Recent published estimates of the flux of carbon from changes in land use vary between 0.6 and 2.5 for 1980 (Houghton et al., 1987; Hall and Uhlig, 1991; Houghton, 1991b) and between 1.1 and 3.6 PgC for 1990 (Houghton 1991b), almost entirely from the tropics. The high end of this range now seems almost certainly too high, as discussed in Section II.7. If the revised rates of deforestation for Brazil are used to recalculate emissions of carbon, the estimate for Latin America is reduced from about 0.7 Pg C/yr (Houghton et al., 1991b) to about 0.5 Pg C/yr in 1980, and from about 0.9 to 0.7 Pg C/yr in 1990.

Recent estimates of biomass and deforestation in tropical Asia and Africa require new estimates of carbon flux for those regions as well. Rates of deforestation in Southeast Asia were revised upward recently by FAO (1990a, 1991), and historical rates of degradation in the region were reassessed by Flint and Richards (1994). These revisions give an estimate of flux for South and Southeast Asia that is about 0.7 PgC in 1990 (Houghton and Hackler, 1994). A reanalysis of land-use change in Africa, using improved estimates of biomass (Brown et al., 1989), gives an estimate of flux there of 0.35 Pg C/yr in 1990.

Globally, the annual flux of carbon from changes in land use is now estimated to have been about 1.4 Pg C in 1980 (1.3 Pg from the tropics and 0.1 Pg from outside the tropics) and 1.7 Pg C in 1990 (essentially all of it from the tropics). The estimate for all the tropics in 1980 is about 45% higher than the recent estimate by Hall and Uhlig (1991), which did not explicitly consider reduction of biomass within forests. The estimate near zero for the world's temperate and boreal zones includes the accumulation of carbon in growing forests (Kauppi et al., 1992; Sedjo, 1992). This accumulation is approximately balanced by releases of carbon from the decay of logging debris and wood products (Houghton et al., 1987; Melillo et al., 1988; Houghton, 1993).

III. Changes in Terrestrial Metabolism

A. An Indirect Estimate of Change in Terrestrial Carbon

The above estimate of 120 Pg C lost from terrestrial ecosystems since 1850 is in contrast to the estimate obtained independently from analyses based on atmospheric CO_2 data and models of oceanic uptake. Geophysical analyses suggest a net release of 25 to 50 Pg C from terrestrial ecosystems since 1850 (Siegenthaler and Oeschger, 1987; Keeling et al., 1989; Sarmiento et al., 1992). The approach is based on an inverse technique, sometimes referred to as deconvolution. Normally, ocean models of carbon uptake are used to calculate atmospheric concentrations of CO_2 from specified annual emissions of carbon from fossil fuels and deforestation. The models can be inverted, however. If the historic pattern of atmospheric concentrations of CO_2 is known, the same models can be used to calculate the annual emissions and accumulations required to generate that pattern. Subtraction of the annual emissions of fossil fuel carbon from the total emissions calculated from these inversions yields a residual flux, a non-fossil flux of carbon. The flux need not be a terrestrial flux, but it is often assumed to be, because the other terms in the budget are specified. Furthermore, short-term evidence from $^{13}C/^{12}C$ ratios suggests that the changes are of terrestrial rather than oceanic origin (Siegenthaler and Oeschger, 1987; Keeling et al., 1989).

Table 3. Factors included and not included in analyses of the flux of carbon from changes in land use

Included	Not included

Change in the area of forests
  Deforestation
    Croplands
    Pastures
  Degradation of land
  Reforestation
  Abandoment of agriculture
  Afforestation
	Unmanaged, undisturbed ecosystems are assumed to be in steady state with respect to carbon
	
	Forest decline or enhancement
	
	Europhication (including CO_2 fertilization)

Change in biomass within forests
  Logging and regrowth
  Shifting cultivation
  Degradation of forests
    (reduction of biomass)
	Frequency of fires
	
	Climatic change
	
	Desertification

A recent inversion of atmospheric CO_2 data with an ocean model (Sarmiento et al., 1992) shows the residual (presumably terrestrial) flux to have been a cumulative net release of about 25 Pg C over the period 1850 to 1990. In 1800 the net release was about 0.4 Pg C/yr, according to this analysis, rising gradually to 0.6 Pg C/yr by 1900, remaining at 0.6 Pg C/yr until about 1920, and then falling to zero by the late 1930's (Figure 2). The flux continued to fall, indicating a net accumulation of carbon in terrestrial ecosystems after 1930. A maximum accumulation rate of 0.8 Pg C/yr occurred in the 1970's. Thereafter, the flux returned to zero and has remained near zero since 1982 or so. The estimate of 25 Pg C for the cumulative net flux between 1850 and 1990 is about half that calculated by Siegenthaler and Oeschger (1987) with their box-diffusion model, but the temporal patterns of the two estimates are almost identical.

B. The Difference between the Two Estimates

The latest estimate of flux obtained indirectly from geophysical data (25 Pg C) is considerably less than the revised estimate of flux from changes in land use (120 Pg C over the period 1850-1990), and the temporal patterns of the two estimates do not bear much resemblance (Figure 2). The disagreement, however, may not be the result of errors in one or the other of the approaches. The difference between these two estimates of flux may represent an actual flux of carbon between terrestrial ecosystems and the atmosphere, a flux that is not related to changes in land use but to changes in the global environment (Table 3). Analyses of flux based on changes in land use assume that ecosystems not directly affected by land-use change are in steady state with respect to carbon. This assumption is probably not valid. The amount of carbon stored in vegetation and soils is changed not only by deliberate human activity, but from inadvertent changes in climate, CO_2 concentrations, nutrient deposition, or pollution.

It is useful to distinguish between these two causes or types of change: deliberate and inadvertent. Deliberate changes result directly from management, for example, cultivation. Inadvertent changes occur through changes in the metabolism of ecosystems, more specifically through a change in the ratio of the rates of primary production and respiration (respiration includes autotrophic and heterotrophic respiration, or plant and microbial respiration, or decomposition). Clearly, management practices affect metabolism, and the distinction between deliberate and inadvertent effects is blurred. Nevertheless, the distinction is a useful one because changes in carbon stocks that result from deliberate human activity are more easily documented than changes resulting from metabolic changes alone. Deliberate changes usually involve a well defined area, for example, the area in agriculture or the area reforested; and the changes in the density of carbon (Mg C/ha) associated with land-use change are generally large: forests hold 20 to 50 times more carbon in their vegetation than the lands that replace them following deforestation. In contrast to these large and well defined changes associated with management, changes that occur as a result of metabolic changes are generally subtle. They occur slowly over poorly defined areas. They are difficult to measure against the large background levels of carbon in vegetation and soils and against the natural variability of diurnal, seasonal, and annual metabolic rates.

Assuming that the entire flux obtained through inversion of ocean models is terrestrial, the inverse approach provides an estimate of the *net* terrestrial flux, including both a flux from land-use change and a flux due to other changes in terrestrial ecosystems. The flux of carbon from land-use change, on the other hand, includes only those lands directly and deliberately managed. Thus, *the difference between the net flux obtained through inversion and the flux obtained through analysis of land-use change may define the flux of carbon from terrestrial ecosystems not directly modified by humans.*

There has never been a direct measurement of even the direction of inadvertent change in carbon storage, globally. Neither is the estimate proposed here direct; it is, once again, by difference. The errors involved in calculating a flux of carbon from the difference between the inversion technique and the land-use approach are potentially large, but the gross trends may indicate which environmental factors were important in the past century and which, perhaps, may be important over the next century. Any clues as to the direction and magnitude of terrestrial feedbacks between global carbon and global climate systems are important. Arguments are made, on the one hand, that terrestrial ecosystems will amplify a global warming (positive feedback) by releasing additional carbon to the atmosphere as a result of a warming-enhanced increase in respiration rates; and, on the other hand, that they will reduce the warming (negative feedback) by accumulating carbon in response to increasing concentrations of CO_2 in the atmosphere. The historical variation in the residual flux, calculated here, allows us to ask whether the variation is related to temperature, atmospheric CO_2, or some other environmental factor. Correlations do not demonstrate a cause-effect relationship, of course, but they might reveal which environmental factor had been most important in the past, and they might suggest the direction of response of terrestrial ecosystems to global change in the future. Will terrestrial ecosystems act as a net source or sink for carbon?

The residual (inadvertent) flux, or the difference between the net biotic flux (from inversion) and the land-use flux, is shown in Figure 2 (thin solid line). The flux has always been negative and is interpreted here to mean that some terrestrial ecosystems were accumulating carbon independent of land-use change. Positive differences, if they occurred, would indicate that some terrestrial ecosystems were releasing carbon to the atmosphere in addition to that released from changes in land use. The pattern of this residual flux over time exhibits three features: first, a period before 1920 showing a small terrestrial sink (not different from zero) with little variation; second, a period between about 1920 and 1975, or so, when some of the world's terrestrial ecosystems were apparently accumulating carbon at an increasing rate; and, finally, a period since the mid-1970's in which the rate of accumulation has decreased.

C. Possible Feedback Mechanisms

The amount of carbon held in terrestrial ecosystems is determined by the balance between photosynthesis and respiration, the latter including decay of soil organic matter. Both of these metabolic processes are affected by many environmental factors, but the progressive increase in the residual flux of carbon after 1920 (Figure 2) parallels the trend in industrial activity, and one cause of an increasing accumulation of carbon on land might be increasing concentrations of CO_2 in the atmosphere. Elevated concentrations of CO_2 have been shown to enhance photosynthesis and growth in many plants (Strain and Cure, 1985; Allen et al., 1987), and may increase the storage of carbon in terrestrial ecosystems if the CO_2-enhanced growth is not respired. Annual concentrations of CO_2 are highly correlated with variation in the residual flux over the period 1850-1990 (Houghton 1992). About 0.035 Pg C/yr are stored per ppmv (parts per million by volume) increase in CO_2. The apparent increase in carbon storage between 1850 and 1990 may not have been driven by increased concentrations of CO_2, of course, but by some other factor associated with increased industrial activity. One such factor is the increased availability of nitrogen, often the mineral element thought to limit productivity. A previous analysis of increased loading of nitrogen and phosphorus on land and in coastal waters (Peterson and Melillo, 1985) estimated the enhanced storage of carbon to be no more than 0.2 Pg C annually, considerably less than the rate of storage suggested here (2.3 Pg C/yr). However, nitrogen may also have become more available through a warming-caused increase in the rate of mineralization (decomposition) (see below).

The role of temperature in affecting metabolism and, hence, in affecting the storage of carbon in plants and soils is also important. The net long-term effect on ecosystems, however, is unclear. On the one hand, rates of respiration (including decomposition) are temperature sensitive, and higher temperatures should increase these rates (Woodwell, 1983, 1989; Raich and Schlesinger, 1992). On the other hand, respiration/mineralization not only releases CO_2; it also releases nitrogen and other nutrients. Because the ratio of carbon:nitrogen is so much higher in woody tissue than in soil organic matter, the same amount of nitrogen can sequester much more carbon in wood than in soil (Rastetter et al., 1992; Shaver et al., 1992). For such a storage to occur, the nitrogen released during decomposition must be taken up by the plant and used in the production of additional wood. If the processes are decoupled in time, as they might be in winter and spring, nitrogen would leak from the system and not permit more plant growth.

If the net effect of warmer temperatures is to release carbon, the pattern of the residual flux should parallel the trend in temperature deviations (Figure 3): a decreasing sink (increasing source) of carbon between 1880 and 1940, an increasing sink between 1940 and 1970, and then an abruptly decreasing sink again after 1970. If the net effect of warming is to store carbon (changing its form from soil to wood), the expected flux would be reversed. Comparison of the temperature trend with the missing flux is ambiguous (Figure 3). The flux shows no response to the early warming between 1880 and 1940. After 1940, however, the increasing sink of carbon matches the decrease in global temperature until the 1970's; and the decreased sink after the 1970's matches the recent warming. The results before 1940 show no effect; but the results after 1940 suggest a warming-caused release of carbon, a positive feedback. There is no support for the hypothesis that warming would cause enhanced mineralization of soil nitrogen and concomitant increases in aboveground storage of carbon and nitrogen.

Regression coefficients determined from an analysis of the residual flux against global mean temperature (Houghton, 1992) suggest that the 0.5 °C warming between 1880 and 1990 may have been responsible for an additional release that is currently 2.4 to 3.7 Pg C/yr. If temperature is assumed to have affected only the rate of heterotrophic (largely soil) respiration, then the Q_{10} is between 2.5 and 3.2, somewhat higher than the value of 2 commonly measured in the laboratory and field but within the range observed (Raich and Schlesinger, 1992). If the change in temperature is assumed to affect also the rate of plant respiration (autotrophic + heterotrophic = total respiration of the ecosystem, Rs_e), the values of Q_{10} range between 1.6 to 2.1. These values of Q_{10} assume that only respiration is affected by temperature. If rates of photosynthesis have also been increased by warmer temperatures, the Q_{10} for respiration alone is higher.

If increasing concentrations of CO_2 in the atmosphere and variation in surface temperature have been the "causes" of variation in the annual flux of carbon to undisturbed ecosystems, their effects have largely canceled each other thus far. They may not continue to do so in the future, however. Assuming the linearity of the relationships persists into the future, and it almost certainly will not, one can calculate the effect of these feedbacks on an earth with doubled CO_2. The scenario for a doubled CO_2 concentration (about 700 ppmv) is based on projection IS92a from the IPCC Supplement (Houghton et al., 1992; pp. 173-175). The doubled CO_2 would include increases in most other greenhouse gases, as well, would occur around 2080, and would carry with it an increase in global temperature of about 2.5 °C (range 1.5 to 4.5 °C). If only CO_2 affects the residual or inadvertent flux, the increased concentration of CO_2 would sequester 15 Pg C/yr in the terrestrial ecosystems of the earth in 2080. Considering the effects of both temperature and CO_2, feedbacks could range from a net annual storage of 10 Pg C to a net release of 13 Pg C/yr (Houghton, 1992). The estimates are unrealsistic. The results demonstrate the ambiguity thus far contained in these results. Even the net direction of the feedbacks is uncertain, to say nothing of the magnitudes possible under environmental conditions very different from those of the present day.

The abrupt decline (about 0.5 Pg C/yr) in the global (terrestrial?) sink for carbon since the end of the 1970's is, perhaps, important. It has been observed by others (Keeling et al., 1989; Sarmiento et al., 1992), but the cause remains uncertain. Clearly, there has been an increased release of carbon from tropical deforestation over the last decade. It appears, if the interpretation given here is correct, that there has also been an increased release (or reduced accumulation) elsewhere, perhaps caused by a warming-enhanced increase in respiration and decomposition.

IV. Summary and Conclusions

The net flux of carbon between terrestrial ecosystems and the atmosphere since 1850 is estimated by inverse methods (Sarmiento et al., 1992) to have been a net release to the atmosphere of about 25 Pg carbon. In contrast, changes in land use are estimated to have released about 120 Pg C over the same period. By difference, terrestrial ecosystems not directly disturbed by changes in land use seem to have accumulated about 95 Pg C, presumably as a result of changing environmental conditions.

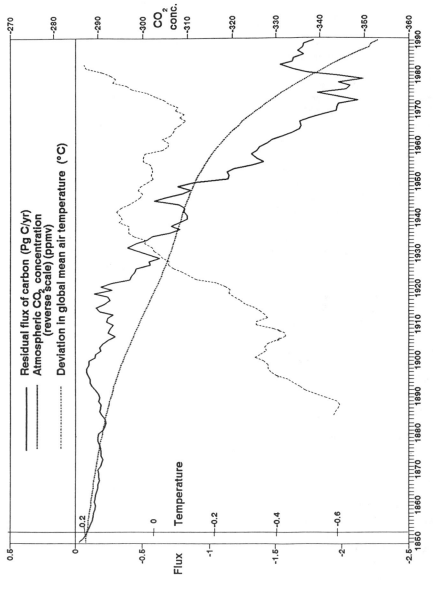

Figure 3. The residual flux of terrestrial carbon (heavy solid line); the concentration of atmospheric CO_2 (dotted line plotted with the ordinate reversed so the overall trend in concentration parallels the trend in the residual flux of carbon)(from Neftel et al., 1985; Friedli et al., 1986; and Keeling et al., 1989; and mean global land surface temperature deviations (thin dashed line) from the 1951-1980 baseline (from Hansen and Lebedeff, 1987).

$$\begin{array}{cccc} \text{Fossil fuel} & = \text{Atmospheric} & + \text{Oceanic} & - \text{Terrestrial} \\ \text{emissions} & \text{increase} & \text{uptake} & \text{net release} \\ 220\ (\pm 22) & 145\ (\pm 10) & 100\ (\pm 25) & 25\ (\pm 25) \end{array} \quad (3)$$

$$\begin{array}{ccc} \text{Terrestrial} & = \text{Release from} & - \text{Accumulation in} \\ \text{net release} & \text{land-use change} & \text{undisturbed ecosystems} \\ 25\ (\pm 25) & 120\ (\pm 30) & 95\ (\pm 55) \end{array} \quad (4)$$

Assuming for the moment that all 95 Pg C accumulated in vegetation, the increased storage of carbon in plants should have been about 15% (95/600). In fact, the relative increase would have been larger than 15% because 10% of the world's forests were lost between 1850 and 1990, and 90% of the earth's vegetation is in forests. If the remaining trees of the world now hold 15-20% more carbon than they did 140 years ago, the change should be detectable in tree rings. Local and regional studies of tree rings, do not show consistent trends of increased growth, but a systematic investigation of trees from high latitudes and high elevations in both hemispheres indicates increased rates of growth, apparently related to the warming over the last century (Jacoby and D'Arrigo, 1992). The authors caution, however, that the trees selected were not necessarily representative of the earth's forested regions. Indeed, studies in mid-latitude regions indicate that rates of forest growth there are consistent with the logging histories in those regions (Houghton, 1993); there is no indication of a widespread enhanced growth. Nevertheless, a systematic survey relating global changes in tree rings to carbon storage has not been carried out.

The possibility that carbon is accumulating in soil organic matter, other than in reforested regions, also seems remote because observed rates of accumulation are too low (Schlesinger, 1990). Thus, the inability to measure directly an accumulation of terrestrial carbon large enough to account for the 1.5 to 2 Pg C/yr imbalance in the global carbon equation, together with uncertainties in the ocean (Siegenthaler and Sarmiento, 1993), leaves open the possibility that at least some of the missing carbon is not on land, and that the interpretations given here are incorrect.

To the extent the trends in missing carbon are of terrestrial origin, however, the net flux of carbon from the world's terrestrial ecosystems over the last decade was about zero (Sarmiento et al., 1992). The balance may have been the result of three offsetting fluxes: a flux from changes in land use, releasing 1.4 to 1.7 Pg C/yr to the atmosphere; a CO_2-, nutrient-, or cooling-enhanced storage, annually storing 1.4 to 2.8 Pg C/yr on land; and a warming-enhanced respiration, releasing on the order of 0.7 to 0.9 Pg C/yr (Houghton, 1992). These calculated sources and sinks of carbon and the factors responsible are, of course, suggestive only. The analysis did not consider changes in precipitation, which could well be the dominating influence on the storage of terrestrial carbon.

The net flux of carbon between terrestrial ecosystems and the atmosphere is comprised of one flux resulting from deliberate changes in land use and another flux, inadvertent, resulting from global changes in the environment (atmospheric CO_2, surface temperature, moisture, nitrogen, UVb, or other factor or combination of factors). Over the period 1850 to 1990, the net flux has been a release of carbon from land. Most of this release occurred early in this century, as a result of clearing lands for agriculture outside the tropics. Over the last decades, the net flux seems to have been close to zero, but the release from changes in land use has been increasing over this period and seems likely to remain high in the near future. If inadvertent changes have caused an accumulation of carbon in terrestrial ecosystems over the last 70 years, the implications for the future are ambiguous: global warming seems likely to cause a release of carbon from soil, but increased mineralization of soil nitrogen or elevated concentrations of CO_2 in the atmosphere may enhance processes that counter this release. The net direction of the feedbacks is difficult to predict.

On the other hand, the future of land-use change is less ambiguous. Deforestation in the tropics is about 15×10^6 ha (almost 1%) yr^{-1} at present and has increased substantially over the last ten years. The net effect of this conversion will be a reduction in the capacity of ecosystems to accumulate carbon in the future even if they have in the past.

Acknowledgements

I thank Eric Davidson and William Schlesinger for helpful comments on the manuscript. The research was supported by the U.S. Department of Energy, Office of Energy Research, Office of Health and Environmental Research, Carbon Dioxide Research Program (Grant Number DE-FG02-90ER61079).

References

Allen, L.H., K.J. Boote, J.W. Jones, P.H. Jones, R.R. Valle, B. Acock, H.H. Rogers, and R.D. Dahlman. 1987. The response of vegetation to rising carbon dioxide: Photosynthesis, biomass, and seed yield of soybean. *Global Biogeochemical Cycles* 1:1-14.

Bacastow, R. and C. D. Keeling. 1973. Atmospheric carbon dioxide and radio-carbon in the natural carbon cycle. II. Changes from A. D. 1700 to 2070 as deduced from a geochemical model. p. 86-135. In: G.M. Woodwell and E. V. Pecan (eds.). Carbon and the Biosphere. U.S. Atomic Energy Commission, Symposium Series 30, National Technical Information Service, Springfield, Virginia.

Bolin, B. 1977. Changes in land biota and their importance for the carbon cycle. *Science* 196:613-615.

Broecker, W. S., T. Takahashi, H. H. Simpson, and T.-H. Peng. 1979. Fate of fossil fuel carbon dioxide and the global carbon budget. *Science* 206:409-418.

Brown, S., and A. E. Lugo. 1982. The storage and production of organic matter in tropical forests and their role in the global carbon cycle. *Biotropica* 14(3):161-187.

Brown, S. and A. E. Lugo. 1984. Biomass of tropical forests: A new estimate based on volumes. Science 223:1290-1293.

Brown, S. A.E. Lugo, and J. Chapman. 1986. Biomass of tropical tree plantations and its implications for the global carbon budget. *Can. J. For. Res.* 16:390-394.

Brown, S., A.J.R. Gillespie, and A.E. Lugo. 1989. Biomass estimation methods for tropical forests with applications to forest inventory data. *For. Sci.* 35:881-902.

Brown, S., A.J.R. Gillespie, and A.E. Lugo. 1991. Biomass of tropical forests of south and southeast Asia. *Can. J. For. Res.* 21:111-117.

Brown, S., L. Iverson, and A.E. Lugo. 1994. Land-use and biomass changes of forests in Peninsular Malaysia from 1972 to 1982: A GIS approach. p. 117-143. In: V.H. Dale (ed.). *Effects of Land Use Change on Atmospheric CO_2 Concentrations: South and Southeast Asia as a Case Study*. Springer-Verlag, New York.

Buschbacher, B. 1984. Changes in productivity and nutrient cycling following conversion of Amazon rainforest to pasture. Dissertation, University of Georgia, Athens.

Cerri, C.C., B. Volkoff, and F. Andreaux. 1991. Nature and behaviour of organic matter in soils under natural forest, and after deforestation, burning and cultivation near Manaus. *For. Ecol. Manag.* 38:247-257.

Davidson, E.A. and I.L. Ackerman. 1993. Changes in soil carbon inventories following cultivation of previously untilled soils. *Biogeochemistry*. 20:161-193.

Davidson, E.A., D.C. Nepstad, and S.E. Trumbore. 1993. Soil carbon dynamics in pastures and forests of the eastern Amazon. *Bull. Ecol. Soc. Amer.* 74(suppl.):208.

Detwiler, R. P. 1986. Land use change and the global carbon cycle: The role of tropical soils. *Biogeochemistry* 2:67-93.

Detwiler, R.P. and C.A.S. Hall. 1988. Tropical forests and the global carbon cycle. Science 239:42-47.

Eswaran, H., E. Van den Berg, P. Reich, and J. Kimble. 1994. Global Soil Carbon Resources. (This volume).

Falesi, I.C. 1976. Ecossistema de Pastagem Cultivada na Amazonia Brasiliera. Belem: Centro de Pesquisa Agropecuario do Tropico Umido.

FAO. 1946-1987. Yearbook of Forest Products. FAO, Rome.

FAO. 1990a. Interim Report on Forest Resources Assessment 1990 Project. FAO COFO-90/8(a). FAO, Rome, Italy.

FAO. 1990b. 1989 Production Yearbook. Rome, Italy.

FAO. 1991. Second interim report on the state of tropical forests. 10th World Forestry Congress, Paris, France (September, 1991).

FAO/UNEP. 1981. Tropical Forest Resources Assessment Project. FAO, Rome.
Fearnside, P.M., A.T. Tardin, and L.G.M. Filho. 1990. Deforestation rate in Brazilian Amazonia. National Secretariat of Science and Technology, Brazilia. 8 pp.
Flint, E.P., and J.F. Richards. 1991. Historical analysis of changes in land use and carbon stock of vegetation in south and southeast Asia. *Can. Jour. For. Res.* 21:91-110.
Flint, E.P. and J.F. Richards. 1994. Trends in carbon content of vegetation in south and southeast Asia associated with changes in land use. p. 201-299. In: V.H. Dale (ed.). *Effects of Land Use Change on Atmospheric CO_2 Concentrations: South and Southeast Asia as a Case Study*. Springer-Verlag, New York.
Friedli, H. H. Lotscher, H. Oeschger, U. Siegenthaler, and B. Stauffer. 1986. Ice core record of the $^{13}C/^{12}C$ ratio of atmospheric CO_2 in the past two centuries. *Nature* 324:237-238.
Hall, C.A.S. and J. Uhlig. 1991. Refining estimates of carbon released from tropical land-use change. Candian Journal of Forest Research 21:118-131.
Hansen, J., and S. Lebedeff. 1987. Global trends of measured surface air temperature. *J. Geophys. Res.* 92:13345-13372.
Hecht, S.B. 1982. Cattle ranching in the Brazilian Amazon: Evaluation of a development strategy. Dissertation, University of California, Berkeley.
Horne, R. and J. Gwalter. 1982. The recovery of rainforest overstorey following logging. I. Subtropical rainforest. *Aust. For. Res.* 13:29-44.
Holt, J.A. and A.V. Spain. 1986. Some biological and chemical changes in a North Queensland soil following replacement of rainforest with *Araucaria cunninghammii* (*Coniferae:Araucariaceae*). *J. Appl. Ecol.* 23:227-237.
Houghton, J.T., B.A. Callander, and S.K. Varney. 1992. Climate Change 1992. *The Supplementary Report to the IPCC Scientific Assessment*. Cambridge University Press, Cambridge, U.K.
Houghton, R.A. 1991a. Releases of carbon to the atmosphere from degradation of forests in tropical Asia. Canadian Journal of Forest Research 21:132-142.
Houghton, R.A. 1991b. Tropical deforestation and atmospheric carbon dioxide. *Climatic Change* 19:99-118.
Houghton, R.A. 1992. Effects of land-use change, surface temperature, and CO_2 concentration on terrestrial stores of carbon. Paper presented at an IPCC-sponsored International Workshop on Biotic Feedbacks in the Global Climatic System. Woods Hole, Massachusetts. October 26-29, 1992.
Houghton, R.A. 1994. Changes in terrestrial carbon over the last 135 years. 1992. In: M. Heimann (ed.). *The Global Carbon Cycle*. Springer-Verlag, Berlin. (In press.)
Houghton, R.A. 1993. Is carbon accumulating in the northern temperate zone? *Global Biogeochemical Cycles.* 7:611-617.
Houghton, R.A. and J.L. Hackler. 1994. The net flux of carbon from deforestation and degradation in South and Southeast Asia. p. 301-327. In: V.H. Dale (ed.). *Effects of Land Use Change on Atmospheric CO_2 Concentrations: South and Southeast Asia as a Case Study*. Springer-Verlag, New York.
Houghton, R.A. and D.L. Skole. 1990. Carbon. p. 393-408. In: B.L. Turner, W.C. Clark, R.W. Kates, J.F. Richards, J.T. Mathews, and W.B. Meyer (eds.). *The Earth As Transformed by Human Action*. Cambridge University Press, Cambridge, U.K.
Houghton, R. A., J. E. Hobbie, J. M. Melillo, B. Moore, B. J. Peterson, G. R. Shaver, and G. M. Woodwell. 1983. Changes in the carbon content of terrestrial biota and soils between 1860 and 1980: A net release of CO_2 to the atmosphere. *Ecological Monographs* 53:235-262.
Houghton, R. A., R. D. Boone, J. M. Melillo, C. A. Palm, G. M. Woodwell, N. Myers, B. Moore and D. L. Skole. 1985. Net flux of CO_2 from tropical forests in 1980. *Nature* 316:617-620.
Houghton, R.A., R.D. Boone, J.R. Fruci, J.E. Hobbie, J.M. Melillo, C.A. Palm, B.J. Peterson, G.R. Shaver, G.M. Woodwell, B. Moore, D.L. Skole and N. Myers. 1987. The flux of carbon from terrestrial ecosystems to the atmosphere in 1980 due to changes in land use: geographic distribution of the global flux. Tellus 39B:122-139.
Houghton, R. A., D.S. Lefkowitz, and D.L. Skole. 1991a. Changes in the landscape of Latin America between 1850 and 1980. I. A progressive loss of forests. *For. Ecol. and Manag.* 38:143-172.
Houghton, R. A., D. L. Skole, and D.S. Lefkowitz. 1991b. Changes in the landscape of Latin America between 1850 and 1980. II. A net release of CO_2 to the atmosphere. *For. Ecol. and Manag.* 38:173-199.
Jacoby, G.C. and R.D. D'Arrigo. 1992. Indicators of climatic and biospheric change: Evidence from tree-rings. Paper presented at an IPCC-sponsored International Workshop on Biotic Feedbacks in the Global Climatic System. Woods Hole, Massachusetts. October 26-29, 1992.

Kauppi, P.E., K. Mielikainen, and K. Kuusela. 1992. Biomass and carbon budget of European forests, 1971-1990. *Science* 256:70-74.

Keeling, C.D., R.B. Bacastow, A.E. Bainbridge, C.A. Ekdahl, Jr., P.R. Guenther, L.S. Waterman, and J.F.S. Chin. 1976a. Atmospheric carbon dioxide variations at Mauna Loa Observatory, Hawaii. *Tellus* 28:538-551.

Keeling, C.D., J.A. Adams, Jr., C.A. Ekdahl, Jr., and P.R. Guenther. 1976b. Atmospheric carbon dioxide variations at the South Pole. *Tellus* 28:552-564.

Keeling, C.D., R.B. Bacastow, A.F. Carter, S.C. Piper, T.P. Whorf, M. Heimann, W.G. Mook, and H. Roeloffzen. 1989. A three-dimensional model of atmospheric CO_2 transport based on observed winds: 1. Analysis of observational data. p. 165-236. In: D.H. Peterson (ed.). Aspects of Climate Variability in the Pacific and the Western Americas. Geophysical Monograph 55, American Geophysical Union, Washington, D.C.

Lanly, J.-P. 1982. Tropical forest resources. FAO Forestry Paper 30. FAO, Rome.

Lugo, A.E., M.J. Sanchez, and S. Brown. 1986. Land use and organic carbon content of some subtropical soils. *Plant and Soil* 96:185-196.

Melillo, J.M., J.R. Fruci, R.A. Houghton, B. Moore, and D.L. Skole. 1988. Land-use change in the Soviet Union between 1850 and 1980: causes of a net release of CO_2 to the atmosphere. *Tellus* 40B:116-128.

Moore, B., R.D. Boone, J.E. Hobbie, R.A. Houghton, J.M. Melillo, B.J. Peterson, G.R. Shaver, C.J. Vorosmarty, and G.M. Woodwell. 1981. A simple model for analysis of the role of terrestrial ecosystems in the global carbon budget. p. 365-385. In: B. Bolin (ed.). *Carbon Cycle Modelling. Scope 16.* John Wiley & Sons, New York.

Myers, N. 1980. *Conversion of Tropical Moist Forests.* National Academy of Sciences Press, Washington, D.C.

Myers, N. 1991. Tropical forests: present status and future outlook. *Climatic Change* 19:3-32.

Neftel, A., E. Moor, H. Oeschger, and B. Stauffer. 1985. Evidence from polar ice cores for the increase in atmospheric CO_2 in the past two centuries. *Nature* 315:45-47.

Nepstad, D.C., C. Uhl, and E.A.S. Serrao. 1991. Recuperation of a degraded Amazonian landscape: Forest recovery and agricultural restoration. *Ambio* 20:248-255.

Oeschger, H., U. Siegenthaler, U. Schotterer, and A. Gugelmann. 1975. A box diffusion model to study the carbon dioxide exchange in nature. *Tellus* 27:168-192.

Palm, C.A., R.A. Houghton, J.M. Melillo, and D.L. Skole. 1986. Atmospheric carbon dioxide from deforestation in Southeast Asia. *Biotropica* 18:177-188.

Peterson, B.J. and J.M. Melillo. 1985. The potential storage of carbon by eutrophication of the biosphere. *Tellus* 37B:117-127.

Post, W.M., W.R. Emanuel, P.J. Zinke, and A.G. Stangenberger. 1982. Soil carbon pools and world life zones. *Nature* 298:156-159.

Raich, J.W. and W.H. Schlesinger. 1992. The global carbon dioxide flux in soil respiration and its relationship to vegetation and climate. *Tellus* 44B:81-99.

Rastetter, E.B., R.B. McKane, G.R. Shaver, and J.M. Melillo. 1992. Changes in C storage by terrestrial ecosystems: How C-N interactions restrict responses to CO_2 and temperature. *Water, Air, and Soil Pollution* 64:327-344.

Richards, J.F. 1990. Land transformation. In: B.L. Turner, W.C. Clark, R.W. Kates, J.F. Richards, J.T. Mathews, and W.B. Meyer (eds.). p. 163-178 In: *The Earth As Transformed by Human Action.* Cambridge University Press, Cambridge, U.K.

Richards, J.F. and E.L. Flint. 1994. A century of land use change in South and Southeast Asia. p. 15-66. In: V.H. Dale (ed.). *Effects of Land Use Change on Atmospheric CO_2 Concentrations: South and Southeast Asia as a Case Study.* Springer-Verlag, New York.

Saldarriaga, J.G., D.C. West, M.L. Tharp, and C. Uhl. 1988. Long-term chronosequence of forest succession in the Upper Rio Negro of Colombia and Venezuela. *Jour. Ecol.* 76:938-958.

Sanchez, P.A., M.P. Gichuru, and L.B. Katz. 1982. Organic matter in major soils of the tropics and temperate regions. p. 99-114. In: Non-Symbiotic Nitrogen Fixation and Organic Matter in the Tropics. 12th International Congress of Soil Science, New Delhi.

Sarmiento, J.L, J.C. Orr, and U. Siegenthaler. 1992. A perturbation simulation of CO_2 uptake in an ocean general circulation model. *Jour. Geophys. Res.* 97:3621-3645.

Schlesinger, W. H. 1984. The world carbon pool in soil organic matter: A source of atmospheric CO_2. p 111-124. In: G.M. Woodwell (ed.). *The Role of Terrestrial Vegetation in the Global Carbon Cycle: Measurement by Remote Sensing SCOPE 23.* J. Wiley & Sons, New York.

Schlesinger, W.H. 1986. Changes in soil carbon storage and associated properties with disturbance and recovery. p 194-220. In: J.R. Trabalka and D.E. Reichle, (eds.). *The Changing Carbon Cycle: A Global Analysis.* Springer-Verlag, New York.

Schlesinger, W.H. 1990. Evidence from chronosequence studies for a low carbon-storage potential of soils. *Nature* 348:232-234.

Sedjo, R.A. 1992. Temperate forest ecosystems in the global carbon cycle. *Ambio* 21:274-277.

Setzer, A.W. and Pereira, M.C. 1991. Amazonia biomass burnings in 1987 and an estimate of their tropospheric emissions. *Ambio* 20:19-22.

Shaver, G.S., W.D. Billings, F.S. Chapin, A.E. Giblin, K.J. Nadelhoffer, W.C. Oechel, and E.B. Rastetter. 1992. Global change and the carbon balance of arctic ecosystems. *BioScience* 42:433-441.

Skole, D.L. and C.J. Tucker. 1993. *Tropical deforestation and habitat fragmentation in the Amazon: Satellite data from 1978 to 1988.* Science 260:1905-1910.

Siegenthaler, U. and J.L. Sarmiento. 1993. Atmospheric carbon dioxide and the ocean. *Nature* 365:119-125.

Siegenthaler, U. and H. Oeschger. 1987. Biospheric CO_2 emissions during the past 200 years reconstructed by deconvolution of ice core data. *Tellus* 39B:140-154.

Strain, B.R. and J.D. Cure (eds.). 1985. Direct Effects on Increasing Carbon Dioxide on Vegetation. DOE/ER-0238, U.S. Department of Energy, Washington, D.C.

Tans, P.P., I.Y. Fung, and T. Takahashi. 1990. Observational constraints on the global atmospheric CO_2 budget. *Science* 247:1431-1438.

Turner, B.L., R.Q. Hanham, and A.V. Portararo. 1977. Population pressure and agricultural intensity. Ann. Assoc. Amer. Geogr. 67:384-396.

Uhl, C. 1987. Factors controlling succession following slash-and-burn agriculture in Amazonia. *Jour. Ecol.* 75:377-407.

Uhl, C., H. Clark, K. Clark, and P. Maquirino. 1982. Successional pattern associated with slash-and-burn agriculture in the Upper Rio Negro region of the Amazon Basin. *Biotropica* 14:249-254.

Uhl, C., R. Buschbacher, and E.A.S. Serrao. 1988. Abandoned pastures in eastern Amazonia. I. Patterns of plant succession. *J. Ecol.* 76:663-681.

Whittaker, R.H. and G.E. Likens. 1973. Carbon in the biota. p. 281-302. In: G.M. Woodwell and E.V. Pecan (eds.). Carbon and the Biosphere. U.S. Atomic Energy Commission, Symposium Series 30, National Technical Information Service, Springfield, Virginia.

Woodwell, G.M. 1983. Biotic effects on the concentration of atmospheric carbon dioxide: a review and projection. p. 216-241. In: *Changing Climate*. National Academy Press, Washington, D.C.

Woodwell, G.M. 1989. The warming of the industrialized middle latitudes 1985-2050: Causes and consequences. *Climatic Change* 15:31-50.

Woodwell, G.M. and R.A. Houghton. 1977. Biotic influences on the world carbon budget. Pages 61-72. In: W. Stumm (ed.). Global Chemical Cycles and their Alterations by Man. Dahlem Konferenzen, Berlin, Germany.

Woodwell, G.M., and E.V. Pecan (eds.). 1973. Carbon and the Biosphere. U.S. Atomic Energy Commission, Symposium Series 30, National Technical Information Service, Springfield, Virginia.

Woodwell, G.M., R.H. Whittaker, W.A. Reiners, G.E. Likens, C.C. Delwiche, and D.B. Botkin. 1978. The biota and the world carbon budget. *Science* 199:141-146.

Woodwell, G. M., J. E. Hobbie, R. A. Houghton, J. M. Melillo, B. Moore, B. J. Peterson, and G. R. Shaver. 1983. Global deforestation and the atmospheric carbon dioxide problem. Science 222:1081-1086.

Zinke, P.J., A.G. Stangenberger, W.M. Post, W.R. Emanuel, J.S. Olson. 1986. Worldwide organic soil carbon and nitrogen data. ORNL/CDIC-18. Oak Ridge National Laboratory, Oak Ridge, Tennessee.

CHAPTER 5

Carbon Storage in Landfills

J. Bogner and K. Spokas

I. Introduction

Recently, global carbon balance calculations have suggested that the sum of current estimates for carbon sinks fails to account for over a billion tonnes of carbon per year (Tans et al., 1990). Concurrently, conventional wisdom regarding high rates of carbon storage in anoxic marine sediments has been challenged (Calvert et al., 1991), and terrestrial carbon sinks such as soils and forest biomass have been given increased attention with respect to carbon sequestration. For example, Kauppi et al. (1992) suggested that 85-120 millions tonnes of carbon per year may be sequestered in European forests while Harden et al. (1992) have suggested high rates of carbon storage in northern hemisphere peats. This paper will focus on calculated rates of carbon storage in landfill sediments, contrast landfills with other anoxic sedimentary environments, and speculate on long-term implications of landfill burial of carbon.

In 1990, approximately 65% of the 178 million tonnes of refuse generated in the United States was buried in engineered sanitary landfills (U.S. EPA, 1992). In 1985, more than 9,000 municipal solid waste landfills existed in the U.S. (Brown, Fallah, and Thompson, 1986); in the same year, worldwide, approximately 340 million tonnes of municipal solid waste were landfilled (Bingemer and Crutzen, 1987). For the last 25-30 years, landfilling has been a major waste disposal method in the U.S. and abroad--a preferred alternative to open dumping and burning with respect to rodent control, spread of pathogens, and aesthetic considerations. Often, the lower cost of landfilling compared with controlled incineration and other waste disposal options contributes to its common use by local waste disposal authorities. As a minimum, landfilling is one of a mix of several waste disposal practices in an integrated waste management plan. Even with increased recycling during recent years, landfilling remains widespread and will continue to do so into the next century.

By definition and by design, anaerobic decomposition of refuse predominates after landfill burial. A landfill functions largely as a moisture-deficient, unstirred anaerobic batch digester in the ground. Under anaerobic conditions, limited conversion of the organic carbon contained in degradable organics (food, yard waste, and paper products) occurs. The terminal reaction over short time frames is the production of methane by methanogenic bacteria present in refuse and interbedded soils.

A valid comparison can be made with anaerobic burial of organic matter in more traditional sedimentary environments. Moreover, the importance of landfill burial with respect to long-term preservation of organic carbon warrants closer scrutiny. It is reasonable to suggest that the bulk mass of organic carbon buried in many thousands of landfills worldwide will not be exhumed over short time scales (decades). Landfills thus function as dispersed anthropogenic anoxic basins. These basins may be relatively large--in the U.S., particularly, average landfill size is increasing so that depth may exceed 100 m. and land surface may exceed several hundred hectares. This paper will explore: (1) landfill burial of refuse in the larger context of long-term sedimentary burial of organic matter; (2) preliminary conservative estimates for optimum conversion of landfill carbon to biogas carbon based on the literature and our own work; and (3) annual estimates for landfill burial of organic carbon compared to estimates for other environments. In part,

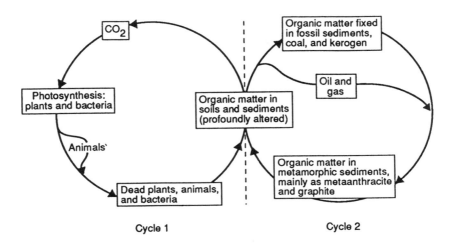

Figure 1. The two main cycles of organic carbon. Most carbon is recycled in Cycle 1. The cross-over from Cycle 1 to long-term sedimentary storage in Cycle 2 amounts of 0.1% or less of primary organic productivity. (Adapted from Tissot and Welte, 1978; after Welte, D.H., 1970, Organischer Kohlenstoff und die Entwicklung der Photosynthese auf der Erde, Naturwissenschaffen 57:17.)

previous work (Bogner, 1992a; Bogner, 1992b; and Bogner and Spokas, 1993) was consulted for the preparation of this paper.

II. Anaerobic Burial of Organic Carbon

Organic matter is preserved in geologic environments through rapid burial with exclusion of air, often by fine-grained sediments in an aqueous environment. Figure 1 illustrates a temporal division of the global organic carbon cycle. The first part is a short-term cycle in which carbon is rapidly recycled to atmospheric carbon dioxide; the second part is a long-term cycle where organic carbon enters sedimentary storage to be chemically and physically transformed by increased heat and pressure through geologic time into future fossil fuels. In general, less than 0.1% of primary organic productivity of carbon enters long-term sedimentary storage via kerogen precursors of fossil fuels (Tissot and Welte, 1978).

Microbial production of methane, as occurs in landfills and other anoxic sedimentary environments, lies at the cross-over point of the two parts of the organic carbon cycle shown in Figure 1. After anaerobic burial, methane is produced from some portion of the organic matter present. Geologically, decomposition of organic materials in a landfill constitutes change that would be considered "early diagenetic"--shallow burial, small increases of pressure and temperature over ambient, and elevated microbial activity aiding decomposition. During later diagenesis, after the cross-over point of Figure 1, organic matter evolves into various kerogen pathways, with the particular pathway dependent on C, H, O content and origin.

Major differences and similarities are apparent between landfilled refuse and more traditional pathways for organic matter entering sedimentary storage. First, landfills typically contain a wide diversity of carbon-containing materials, including natural and processed lignocellulosic substrates (paper and packaging, garden wastes, construction materials), food waste, plastics, and small amounts of household pesticides, herbicides, fertilizers, and cleaning agents. Some organic materials in landfills, such as petroleum-based polymers, have undergone extensive processing to make them particularly resistant to biodegradation. The major sources of organic material in the geologic record, although derived from multiple plant and animal sources, can be conveniently grouped into an evolutionary series based on dominant source as follows: pre-Devonian--bacteria, marine algae, and zooplankton; Devonian to Jurassic-- bacteria, algae, and zooplankton plus remnants of higher plants strongly degraded by microorganisms; and Devonian to Holocene --remnants of higher plants, little to moderately altered with fewer planktonic organisms and bacteria. Most organic matter derived from continental higher plants evolves to precursors for natural gas or coal (Type III Kerogen of Tissot and Welte, 1978).

Secondly, one must consider the duration of observed biodegradation and diagenesis. The historically-observed microbial decomposition of refuse in landfills has proceeded over several decades, with rates depending on moisture circulation and other factors such as nutrients, toxins, pH, and temperature, which influence microbial population development. In the geologic record, transformation of organic materials to fossil fuels occurs over millennia. A minimum time for hydrocarbon evolution of several thousand years is observed in active tectonic environments with elevated geothermal gradients and rapid burial to thousands of meters (Peter et al., 1991).

Third, it is important to consider pressure-temperature relationships. Landfills have a temperature regime that is most frequently mesophilic (25-40 °C) with pressures at or slightly elevated above ambient. As mentioned above, such temperatures and pressures represent early diagenesis with regard to organic matter in the geologic record. There, later diagenesis and catagenesis occur with higher temperatures and pressures as kerogen is formed and subsequently transformed into petroleum crudes, natural gas, and coal (ranks below anthracite). Temperatures are generally in the range of 50-150 °C with pressures of 30,000 to 100,000 kPa or more. Metagenesis at even more elevated temperatures and pressures produces methane and carbon residues in which crystalline ordering begins to develop, as in anthracite and meta-anthracite coals (Tissot and Welte, 1978).

Finally, with respect to the biochemistry of decomposition, one might argue that landfill processes are typical of early burial stages in anoxic basins. Anaerobic digestion of biodegradable refuse produces intermediate carboxylic acids, particularly acetic, which are converted to a biogas consisting of approximately half methane and half carbon dioxide. Figure 2a gives a schematic for the chemistry of refuse decomposition while Figure 2b presents actual results from a laboratory study of refuse decomposition in 2 l. containers (Barlaz et al., 1989a). As concluded by the laboratory study, three major groups of bacteria are involved in waste decomposition in a landfill: (1) hydrolytic and fermentative bacteria which convert biologic polymers to sugars which are then fermented to carboxylic acids, alcohols, carbon dioxide, and hydrogen; (2) acetogenic bacteria which convert carboxylic acids and alcohols to acetate and hydrogen; and (3) methanogenic bacteria which produce methane from either acetate or carbon dioxide/hydrogen. For contrast, Figure 3 presents a geologic perspective on the evolution of organic matter through kerogen formation to fossil fuels. There, the time may be extended to millions of years with biochemical degradation occurring at a very early stage. Thus, an argument can be made that landfill processes are reasonably typical of early stage decomposition in anoxic basins. The question of carbon conversion will be addressed in the next section.

III. Landfill Carbon Balance and Carbon Conversion Rates

A simple carbon balance for a landfill can be presented as:
$C_{in} - C_{out} = \Delta C$ stored where
C_{in} = input carbon = $(C_r + C_s + C_{aq})$
and C_r = refuse C
C_s = soil C
C_{aq} = aqueous C (groundwater inputs)
C_{out} = carbon removed = $(C_g + C_{aq})$
and C_g = biogas C (methane and carbon dioxide)
C_{aq} = aqueous C (carboxylic acids, bicarbonate)

Figure 4 presents a schematic for change in carbon storage (dC/dt) over long time frames. Initially, C_{in} is high and C_{out} negligible (Stage 1). When active biogas production begins concurrent with continued filling operations, there is still a net increase in C storage, but at a reduced rate (Stage 2). After filling is completed and biogas production continues over future decades, carbon is lost from storage (Stage 3). Finally, over the longer time frames shown in the figure, carbon storage is stagnant, except for small perturbations due to groundwater exchange (Stage 4). Over geologic time (Stage 5), further declines in carbon storage are postulated due to tectonic or eustatic processes influencing refuse exhumation and the groundwater carbon balance. Tissot and Welte (1978) suggested a low ultimate preservation rate for organic carbon (mainly marine) of less than 0.1%; it is unclear whether this number has any application to terrestrial settings. A simple calculation for relative daily volumes of refuse C and soil C during active landfilling

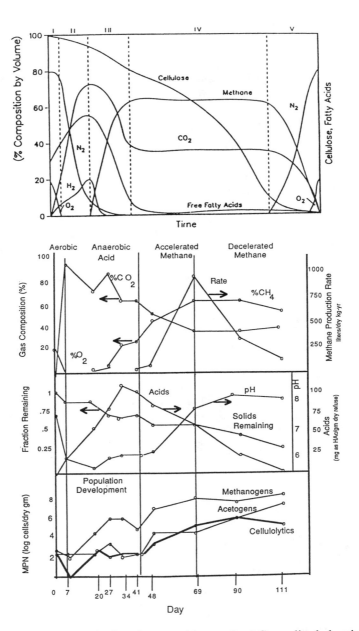

Figure 2. Chemistry of refuse decomposition. a. (top) Generalized chemistry of refuse decomposition in a landfill. I, II, III, IV, and V refer to landfill gas production phases. (Modified from Rees, 1980.) b. (bottom) Observed trends in refuse decomposition with leachate recycle in 2 L laboratory containers. The total carboxylic acids are expressed as acetic acid equivalents. Methanogen MPN (most probable number) data are the log of the average of the acetate and hydrogen/carbon dioxide-utilizing populations. "Solids remaining" equals the ratio of (cellulose + hemicellulose) removed from a container divided by the weight of (cellulose + hemicellulose) added to the container initially. (From Barlaz et al., 1989, used by permission.)

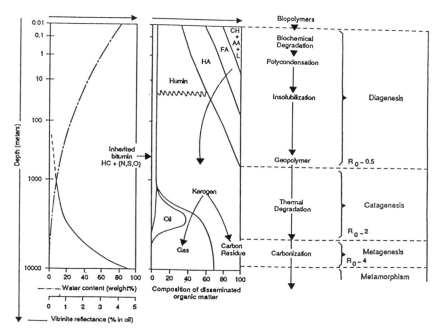

Figure 3. General scheme of geologic evolution of buried organic matter, from the freshly deposited sediment to the metamorphic zone. CH = carbohydrates; AA = amino acids; FA = fulvic acids; HA = humic acids; L = lipids; HC = hydrocarbons; N,S,O = N,S,O compounds (nonhydrocarbons); Ro = vitrinite reflectance (%). (Adapted from Tissot and Welte, 1978.)

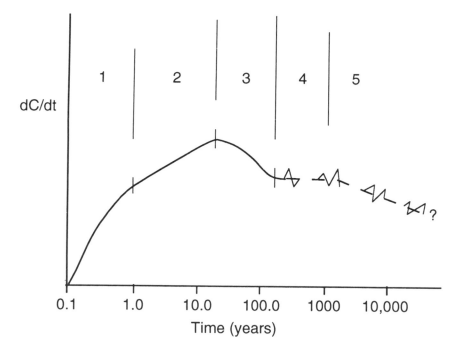

Figure 4. Schematic for changing carbon storage in landfills over long time frames. See text for discussion of Stages 1-5.

suggests that daily input of soil C would be less than 5% of refuse C, based on the following assumptions: 1,000 tonnes of refuse with 20% organic carbon (based on Bingemer and Crutzen, 1987, averages) put into an area of 75 x 75 m with 0.15 m of soil cover; soil has wet density of 1.2 g cm^{-2} and organic carbon content of 1%.

Let us specifically consider the long-term fate of lignin and cellulose following anaerobic burial. Barlaz et al. (1989a, 1989b) showed that over 90% of the methane potential of typical Madison, Wisconsin refuse was from cellulose. Moreover, in the laboratory study summarized in Figure 2b, they showed that almost 80% of the cellulose plus hemicellulose but negligible lignin were decomposed. Lignin is recalcitrant to anaerobic microbial deposition. Similarly, the limited preservation of cellulose and more abundant preservation of lignin have been documented in peats, with the amount of cellulose relative to lignin generally declining with depth (Given et al., 1984). It has been suggested that, through geologic time from the Paleozoic to the Tertiary, dominant land plants have become more sparing in their use of lignin as the diversity and abundance of aerobic lignin-degrading organisms have risen; the end result is a decline in the fraction of terrestrial primary production preserved in coals and kerogens (Robinson, 1990). The widespread practice of engineered anaerobic burial of refuse would seem to alter this trend, with higher than normal rates of lignin preservation.

A practical approach to assessing the state of landfill decomposition was proposed by Bookter and Ham, 1982; they suggested that declining cellulose:lignin ratios could be correlated with increasing decomposition. In our own laboratory studies, we observed decreasing cellulose content in landfill samples after controlled anaerobic degradation in the laboratory with added aqueous nutrient media (methods detailed in Bogner, 1990; nutrient media described in Shelton and Tiedje, 1984). Figure 5a compares cellulose and lignin for a series of landfill samples; Figure 5b gives cellulose and lignin for the same samples following anaerobic degradation in the laboratory for more than 500 days with added aqueous nutrient media, resulting in a water content of 200% (wt/dry wt). In Figure 5b, an obvious decrease in the cellulose relative to the lignin occurs. Since the standard method for cellulose and lignin analysis relies on an acid hydrolysis with the residue quantified as lignin, one must be aware that, for refuse samples, the "lignin" includes everything not specifically quantified in cellulose and hemicellulose fractions; thus, sample to sample variations in this residue fraction may cloud cellulose:lignin ratios and preclude application of universal ratios meaningful to assess state of decomposition.

Figure 6 compares total carbon content to the sum of (cellulose + hemicellulose) for a variety of landfill samples from sites in Illinois and Wisconsin. The figure includes data for samples of landfilled refuse dated at 1973-1988 as well as for the same samples subjected to controlled anaerobic decomposition in the laboratory. Samples were incubated "as is" (controls) or with added water or aqueous nutrient media as discussed above. A simple linear relationship describes the carbon to (cellulose + hemicellulose) relationship, regardless of whether the samples were analyzed as obtained from the landfill or following extensive laboratory decomposition. Thus, even though limited cellulose degradation was documented in Figure 5, the limited cellulose conversion did not substantially alter carbon:cellulose ratios. It should be pointed out that the conditions of incubation for these studies (25°C, no leachate recycle) were designed to mimic conditions observed in most landfills and were less optimized than those employed by Barlaz et al., 1989a (41°C, with leachate recycle).

Figure 7 shows a whole landfill carbon balance with the landfill portrayed schematically as an excavation in geologic materials with compacted clay final cover. It is recognized that state-of-the-art landfills have a more complex design with side and bottom liners, composite covers, leachate collection systems, and gas recovery and control systems. In the figure, carbon partitioning to recalcitrant solid carbon, aqueous intermediates, and gaseous products occurs. Aqueous intermediates may be removed from the landfill environment; however, the trend today is toward increased recycle of landfill liquids where permitted by existing regulations. In the U.S., a lined landfill with a leachate collection system is necessary for approval of leachate recycle under RCRA (Resource Conservation and Recovery Act) Subtitle D regulations. In a landfill, partitioning of carbon dioxide to gaseous or aqueous phases according to Henry's Law also occurs. Ideally, the recalcitrant C is dominated by lignin with a minor contribution from plastic wastes (at most, a few % by weight of the total refuse). Practically, as can be seen by excavating old landfills, substantial amounts of cellulose will also be nondegraded, even after more than 20 years of landfill burial. As will be addressed in the following paragraphs, moisture is a critical variable influencing decomposition rates at landfills.

Laboratory studies of refuse degradation under controlled anaerobic conditions can be used to indicate optimum conversion rates of refuse carbon (C_r) to biogas products (C_g methane and carbon dioxide). Table

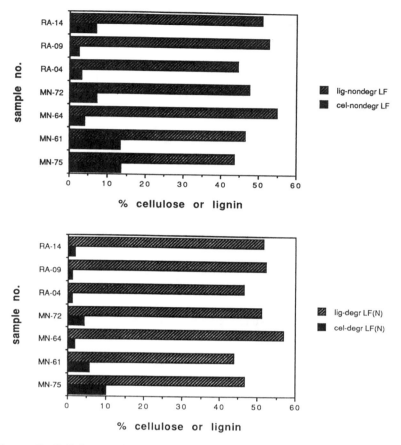

Figure 5. Cellulose and lignin content of landfill samples before and after controlled anaerobic degradation in the laboratory at 25° C. All values % dry weight basis. a. (top) Landfill samples before incubation. b. (bottom) Same landfill samples after more than 500 days of incubation under controlled anaerobic conditions with added aqueous nutrient media to equal moisture content of 200% (wt/dry wt). (Methods from Pettersen and Schwandt, 1988.)

1 summarizes results from several studies. Although the results vary from negligible to a high approaching 50%, the maximum observed conversion rates across all studies are in the range of 25-40%. In our own studies (Bogner and Spokas, 1993), we observed substantial improvement in biogas yields by addition of moisture or moisture plus nutrients (roughly a doubling of yields observed in controls consisting of landfill samples alone). The highest conversion rates among the controls were correlated with landfill samples exhibiting high natural water contents and substantial percentages of interbedded calcareous soils with high indigenous nutrients (N,P).

Controls on landfill decomposition and carbon conversion in actual field settings are directly related to waste composition, site construction practices, and site hydrogeology. Waste composition varies worldwide with regard to the overall percent of degradable organics as well as the relative proportions of paper, food waste, and garden waste. Water content is related to the original refuse moisture content, the pre-filling depth to the top of zone of saturation, and construction practices. With respect to the latter, the timing of precipitation events relative to placement of compacted daily cover may drastically influence site water content and thus decomposition rates. Historically, older landfills, which include many closed sites with low levels of maintenance, were designed according to less stringent regulations regarding both cover materials and compaction; hence, they receive higher rates of infiltration. The amount and characteristics of

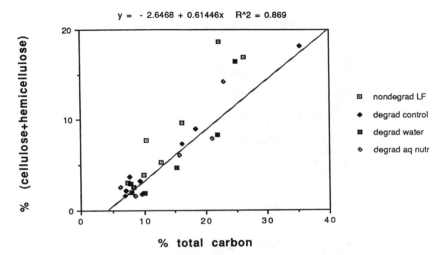

Figure 6. Comparison of total carbon content to the sum of (cellulose + hemicellulose) for landfill samples from Illinois and Wisconsin. Data include landfill samples and samples subjected to laboratory degradation under controlled anaerobic conditions (controls, water added, aqueous nutrient media added). Plot includes same samples shown in Figure 5. Total carbon method (1000°C) from LECO, 1991.

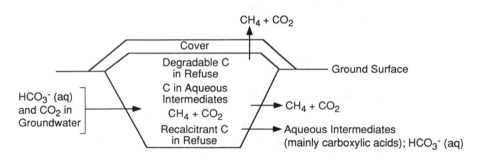

Figure 7. Landfill carbon balance.

interbedded soil materials used for daily cover or final cover influence decomposition rates by variable supply of nutrients, indigenous microorganisms, and carbonates. Due largely to poor circulation of liquids within many landfill sites, gas production rates may vary by two orders of magnitude within a single site; thus, within-site variability may be as great as between-site variability (Bogner, 1990).

Observed total gas production rates (methane plus carbon dioxide) vary from 0.007 to more than 1.0 m^3 dry kg^{-1} yr^{-1} (Ramaswami, 1970; Rovers and Farquhar, 1973; Augenstein et al., 1976; DeWalle et al., 1978; Buivid, 1980; Pohland, 1980; EMCON Associates, 1981; Klink and Ham, 1982; Jenkins and Pettus, 1985; Pacey and Dietz, 1986; Emberton, 1986; Barlaz, Milke, and Ham, 1987; Barlaz et al., 1989a). The highest rates are from laboratory studies with high percentages of degradable organics under optimized conditions; the low rates are from field pumping tests where the rate of biogas production is equated to the sustainable pumping rate for wells drawing from a given mass of refuse. The duration of gas generation remains problematical. Biogas production as well as substantial quantities of undecomposed organics are typically observed in old landfills (pre-1970's). For commercial landfill methane recovery, currently practiced at more than 100 sites in the U.S. and more than 250 world wide (Berenyi and Gould, 1991; Richards, 1989), the conservative assumption of 20-30 years duration is often used. Commercial gas recovery schemes will

Table 1. Carbon conversion to methane and carbon dioxide during anaerobic refuse decomposition. Fraction of total carbon converted to biogas carbon

Source	Fraction of C converted to [$CH_4 + CO_2$] Carbon		
Barlaz et al., 1989a, calculated from CH_4 production data and solids analysis	0.275		
Bogner and Spokas, 1993, ranges for:	Controls	Added water-- 200% wt/dry wt	Added aqueous nutrient media-- 200% wt/dry wt
Wisconsin, landfill	0.0056-0.249	0.104-0.354	0.107-0.397
Colorado, landfill	0-0.0165	0.02-0.419	----
Illinois, landfill	0-0.249	0.03-0.273	0.0478-0.281
Cossu et al., 1991	Composted and anaerobically digested refuse mixed w/acetic acid and/or olive oil waste water in laboratory columns 0.059-0.416		
Ehrig, 1991	0.283 (mean) 0.47 (max)		

Modified from Bogner and Spokas, 1993.

frequently rely on models based on a first-order decomposition reaction where the results of field pumping tests are matched to an assumed gas generation rate function (EMCON, 1981; Findikakis et al., 1988). If refuse composition is known, the refuse mass may be partitioned according to assigned percentages of readily, moderately, and slowly degradable materials; then individual production curves may be developed for the various fractions over the assumed period of gas generation.

IV. Worldwide Landfill Carbon Storage and Greenhouse Gas Emissions

For purposes of long-term, worldwide climatic considerations, organic carbon deposited in sanitary landfills is partitioned to one of three major pathways: (1) eventual atmospheric emission as methane, (2) eventual atmospheric emission as carbon dioxide following microbial oxidation or combustion, and (3) long-term sedimentary storage in landfill reservoirs. Taking Bingemer and Crutzen's (1987) estimates of worldwide quantities of landfilled municipal waste (but neglecting their estimated quantities of industrial waste) and the laboratory work on refuse decomposition by Barlaz et al. (1989a), one can develop a preliminary estimate of long-term organic carbon partitioning to sedimentary storage. This is a highly conservative estimate, based on idealized laboratory decomposition studies using well-mixed refuse at elevated moisture contents and temperatures. In field settings, particularly at post-1980 sites with high rates of compaction and moisture limitation, the quantities of organic materials recalcitrant to anaerobic digestion and entering sedimentary storage would be expected to be much greater.

Table 2 summarizes assumptions, calculations, and results; it also provides some estimates of current primary productivity of organic carbon (mainly oceanic phytoplankton) and annual flux of organic carbon to sedimentary storage. Barlaz et al. (1989a), under idealized laboratory decomposition conditions at 41°C, observed 72% decomposition of cellulose and hemicellulose but negligible decomposition of lignin. On a dry-weight basis, the raw refuse had 40-50% cellulose, 12% hemicellulose, 10-15% lignin, and 4% protein --a total of 64-81% potentially degradable organic materials. If these values are applied to Bingemer and Crutzen's (1987) worldwide estimates of landfilled municipal refuse, excluding the uncertain contribution from developing countries, one can develop conservative estimates for landfill organic carbon that enters

Table 2. Assumptions and calculated results for sedimentary storage of organic carbon from landfilled refuse (developed countries only, annual basis)

Location	Waste carbon from municipal refuse[a,b] 10^6 T/yr	Degradable organic C in waste (%)	Total waste 10^6 T/yr
U.S., Canada, Australia	37	22	168.2
Other OECD	19	19	100.0
USSR and Eastern Europe	13	17.5	74.3
Total			342.5[c]

			Recalcitrant to landfill decomposition assumed to go to sedimentary storage[f]			
Fraction	(%)[d]	10^6 T/yr[e]	%	10^6 T/yr	% C in fraction	C in fraction 10^6 T/yr
Cellulose[h]	45	115.6	18	20.8	45	9.4
Hemicellulose[h]	12	30.8	18	5.5	40	2.2
Lignin[h]	13	33.4		33.4	60	20.0
Total		256.9[g]				31.6

Source/location	Total organic C productivity 10^9 T/yr	Organic C to long-term sedi-mentary storage 10^6 T/yr
Tissot and Welte, 1978; Krey, 1970 (Marine)	20	20[i]
Fleming, 1957 (Marine)	20	
Koblents-Mishke et al., 1968 (Marine)	23	
Berger et al., 1987 (Marine)		
Sundquist, 1985 (Terrestrial)	27	
Garrels et al., 1976; Lerman, 1979[j]	45-60	
		30

[a]Bingemer and Crutzen, 1987; [b]excludes industrial waste from developing countries; [c]approximately 20% degradable organic carbon content; [d]amount in fresh U.S. refuse, Barlaz et al., 1989a; [e]estimated annual quantities in refuse from developed countries, derived in part from data presented above; [f]Barlaz et al., 1989a; [g]based on assumption of 25% water; [h]dry-weight basis; [i]0.1% of primary productivity; [j]using worldwide organic carbon model.
(Modified from Bogner, 1992b.)

sedimentary storage. Results shown suggest that the annual amount of recalcitrant organic carbon entering sedimentary storage from landfill deposition, about 30 million metric tonnes of carbon per year, is approximately equal to previously published estimates of worldwide organic carbon entering sedimentary storage from other sources (mainly marine)--a total of about 20-30 million metric tonnes of carbon per year. For comparison, if one considers the base estimates by Bingemer and Crutzen (1987) of annual organic carbon landfilled (top of Table 2) with about 18% recalcitrant, one still derives about 12.5 million metric tonnes of carbon to sedimentary storage annually. Although refuse composition varies greatly worldwide, the sum of degradables (food waste, yard/garden waste, paper) in municipal solid waste is fairly constant at 55-75% (wt/wt) for the developed countries when data for the European Economic Community (Trunick, 1991), Canada (Municipal Solid Waste News, 1991), and the United States (U.S. EPA, 1992) are compared.

Landfills emit methane and carbon dioxide to the atmosphere. Worldwide estimates for methane, based on quantities of landfilled refuse, assumed methane generation rates, and various methane emission rates, range from 9-70 Tg/yr (Richards, 1989; Orlich, 1990; Bingemer and Crutzen, 1987). Most strategies for mitigating methane emissions rely on conversion of methane to carbon dioxide; these include flaring and various gas recovery schemes to fuel boilers, gas turbines, or internal combustion engines. Rates for landfill methane emissions, based on closed chamber measurements range from 10^{-6} to 10^{-12} g cm^{-2} s^{-1} (Lytwynyshyn et al., 1982; Kunz and Lu, 1980; Bogner and Spokas, 1993). One study of landfill methane emissions using micrometeorological techniques yielded preliminary results in the middle of this range, on the order of 10^{-8} g cm^{-2} s^{-1} (Meyers et al., 1992). At the present time, we are studying methane emissions around and between gas recovery wells at a northern Illinois landfill. Previously, we obtained limited measurements of both methane and carbon dioxide emissions at a site in southern California; these results should be considered maxima due to dry porous cover soils and, at the time of measurement, no commercial gas recovery. Results from static chamber measurements over a 2 day period (n = 8) indicated mean methane emissions of 1.3×10^{-6} g cm^{-2} s^{-1} and mean carbon dioxide emissions of 2.5×10^{-5} g cm^{-2} s^{-1} (Bogner et al., 1989b).

V. Conclusions and Uncertainties

Laboratory studies of refuse decomposition under controlled anaerobic conditions suggest that, at best, 25-40% of landfill carbon is converted to biogas carbon (methane and carbon dioxide). The carbon remaining is proportioned to aqueous intermediates (mainly carboxylic acids and bicarbonate) and to recalcitrant carbon in solids (mainly lignin). Under actual field conditions, the fraction of carbon converted would be less than in laboratory studies, suggesting that considerable quantities of carbon are annually shifted to landfill storage. Preliminary conservation estimates, applying high laboratory conversion rates to worldwide estimates of landfilled refuse in developed countries, indicate that, at present, the annual storage of 30 million tons of carbon may be occurring in landfills.

Landfilling has been a major waste disposal method throughout much of the developed world over the last 25-30 years; in most landfills, limited conversion of organic carbon extends over several decades. It is problematical whether landfill carbon will be recycled over longer time frames by anthropogenic activities or geologic processes. This paper has addressed carbon storage in landfills in the larger framework of anoxic burial of organic carbon in sedimentary environments. Emphasis has been on two time frames--shorter time frames (decades) characterized by limited conversion of organic carbon to biogas and very long (geologic) time frames in which buried organic carbon evolves to precursors of fossil fuels. Intermediate time scales have largely been omitted because very few conclusions can be made at this time with respect to burial of landfilled refuse for hundreds or thousands of years. Recovery of undecomposed organics from archaeological sites suggests that, if landfills are not extensively excavated in future decades or centuries, longer term storage of carbon is possible. From our current perspective, in spite of some limited pilot studies examining the feasibility of landfill "mining", widespread exhumation of refuse in various stages of decomposition at thousands of sites appears unlikely. For the future, landfills in the U.S. and elsewhere are becoming larger and deeper. A provocative question remains--what further chemical transformations will occur in landfill carbon over the next thousand years?

Acknowledgments

Project funding was provided by the U.S. Dept. of Energy (DOE), Assistant Secretary for Conservation and Renewable Energy, through the Municipal Solid Waste Program at NREL (National Renewable Energy Laboratory, Golden, CO). Special thanks to B. Gupta and P. Shepherd at NREL and S. Friedrich and D. Walter at DOE. In addition, we are grateful to the Forest Preserve District of DuPage County, IL; Browning-Ferris, Inc.; Waste Management of North America, Inc.; and Orange County, CA for access to field sites and site assistance. Finally, we are grateful to Irene Fox of Argonne for carbon analysis as well as to Roger Petterson and Virgil Schwandt at the USDA Forest Products Laboratory, Madison, Wisconsin for cellulose and lignin analysis.

References

Augenstein, D.C., C.L. Cooney, D.L. Wise, and R.L. Wentworth. 1976. Fuel gas recovery from controlled landfilling of municipal wastes. *Resource Recovery and Conservation* 2:103-107.

Barlaz, M.A., M.W. Milke, and R.K. Ham. 1987. Gas production parameters in sanitary landfill simulators. *Waste Management & Research* 5:27-39.

Barlaz, M.A., D.M. Schaefer, and R.K. Ham. 1989a. Bacterial population development and chemical characteristics of refuse decomposition in a simulated sanitary landfill. *Applied and Environmental Microbiology* 55:55-65.

Barlaz, M.A., R.K. Ham, and D.M. Schaefer. 1989b. Mass-balance analysis of anaerobically decomposed refuse. *J. Environmental Engineering Division ASCE* 115:1088-1102.

Berenyi, E. and R. Gould. 1991-1992. Methane recovery from landfill yearbook. Governmental Advisory Associates, Inc. New York. 508 pp.

Berger, W.H., K. Fischer, C. Lai, and G. Wu. 1987. Ocean productivity and organic carbon flux. Part I. Overview and maps of primary production and export production. University of California, San Diego. SIO Reference 87-30. Cited in Berger, W.H. 1989. Global Maps of Ocean Productivity. In: *Productivity of the Oceans: Present and Past*. W. Berger, V. Smetacek, and G. Wefer (eds.). Report of the Dahlem workshop held in Berlin. April 1988. Wiley-Interscience, New York. p. 429-445.

Bingemer, H.G. and P.J. Crutzen. 1987. The production of methane from solid wastes. *J. Geophysical Research* (D2)92:2181-2187.

Bogner, J.E. 1990. Controlled study of landfill biodegradation rates using modified BMP assays. *Waste Management and Research* 8:329-352.

Bogner, J.E. 1992a. Garbage as a sedimentary deposit: Challenges in interpretation of landfilled refuse. *Earth Interpreters: F.M. Fryxell, Geology, and Augustana*. D.A. Schroeder and R.C. Anderson (eds.) *Augustana College Library Publication* 36. 207pp.

Bogner, J.E. 1992b. Anaerobic burial of refuse in landfills: Increased atmospheric methane and implications for increased carbon storage. *Ecol. Bull.* 42:98-108.

Bogner, J.E. and K. Spokas. 1993. Landfill CH_4: Rates, fates, and role in global carbon cycle. *Chemosphere* 26(1-4):369-386.

Bogner, J.E. M. Vogt, and R. Piorkowski. 1989 Landfill gas generation and migration, review of current research II. In: *Proc. Anaerobic Digestion Review Meeting*. Solar Energy Research Institute. Golden, Colorado.

Bookter, T.J. and R.K. Ham. 1982. Stabilization of solid waste in landfills. *J. Environmental Engineering Division ASCE* 108(EE6):1089-1096.

Brown, G., S. Fallah, and C. Thompson. 1986. Census of state and territorial subtitle D non-hazardous waste programs. U.S. Environmental Protection Agency. EPA/530-SW-86-039.

Buivid, M.G. 1980. *Laboratory simulation of fuel gas production enhancement from municipal solid waste landfills*. Cambridge, Mass: Dynatech R&D Company.

Calvert, S.E., R.E. Karlin, L.J. Toolin, D.J. Donahue, J.R. Southon, and J.S. Vogel. 1991. Low organic carbon accumulation rates in Black Sea sediments. *Nature* 350:692-695.

Cossu, R., R. Stegmann, and C. Acaia. 1991. Treatment of vegetation water on layers of MSW organic fractions. In: *Sardinia '91, Third International Landfill Symposium*. p. 1461-1476. Published by CISA. Environmental Sanitary Engineering Center. University of Cagliari, Sardinia.

DeWalle, F.B., E.S.K. Chian, and E. Hammerberg. 1978. Gas production from solid waste in landfills. *Journal of the Environmental Engineering Division ASCE*. 104(EE3):415-432.

Ehrig, H.J. 1991. Prediction of gas production from laboratory scale tests. In: *Sardinia '91 Third International Landfill Symposium*. p. 87-1214. Published by CISA. Environmental Sanitary Engineering Center. University of Cagliari. Sardinia.

Emberton, J.R. 1986. The biological and chemical characterization of landfills. p. 150-163. In: *Energy from landfill gas*. Proceedings of a conference jointly sponsored by U.S. and U.K. Departments of Energy. Solihull, West Midlands, U.K. October 1986. J.R. Eberton & R.F. Emberton (eds.). Oxfordshire, U.K. Harwell Laboratory.

EMCON Associates. 1981. State of the art of methane gas enhancement in landfills. *Report ANL-CNSV-23*. Argonne, Ill. Argonne National Laboratory.

Findikakis, A.N., C. Papelis, C.P. Halvadakis, and J.O. Leckie. 1988. Modelling gas production of managed sanitary landfills. *Waste Management & Research* 6:115-123.

Fleming, R.H. 1957. General features of the ocean. In: *Treatise on marine ecology and paleoecology.* J.W. Hedpeth (ed.). Geological Society of America Memoir 67:87-107. Cited in Berger, W.H. 1989. Global maps of ocean productivity. In: *Productivity of the oceans: present and past.* W. Berger, V. Smetacek, and G. Wefer (eds.). p. 429-455. Report of the Dahlem workshop held in Berlin. April 1988. Wiley-Interscience. N.J.

Garrels, R.M., A. Lerman, and F.T. MacKenzie. 1976. Controls of atmospheric O_2 and CO_2: past, present, and future. *American Scientist* 64:306-314.

Given, P.H., W. Spackman, P.C. Painter, C.A. Rhoads, and N.J. Ryan. 1984. The fate of cellulose and lignin in peats: An exploratory study of the input to coalification. *Organic Geochemistry.* 6:399-407.

Harden, J.W., E.T. Sundquist, R.F. Stallard, and R.K. Mark. 1992. Dynamics of soil carbon during deglaciation of the Laurentide ice sheet. *Science* 258:1921-1924.

Jenkins, R.L. and J.A. Pettus. 1985. The use of in-vitro anaerobic landfill samples for estimating landfill gas generation rates. In: *Proceedings of the first symposium on biotechnological advances in processing municipal wastes for fuels and chemicals.* Aug. 1984. Minneapolis, Minn. Report ANL/CNSV-TN-167. A. Antonopoulos (ed.). Argonne, Ill. Argonne National Laboratory.

Kauppi, P.E., K. Mielikäinen, and K. Kuusela. 1992. Biomass and carbon budget of European forests, 1971 to 1990. *Science* 256:70-79.

Klink, R.E. and R.K. Ham. 1982. Effects of moisture movement on methane production in solid waste landfill samples. *Resources and Conservation.* 8:29-41.

Koblents-Mishke, O.I., V. Volkovinshiy, and Y. Kabanova. 1968. Noviie danniie or velichine pervichnoi producktsii mirovogo okeana. Doklady Akad. Nuak SSSR 183. p. 1189-1192. Cited in Berger, W.H. 1989. Global maps of ocean productivity. In: *Productivity of the oceans: present and past.* W. Berger, V. Smetacek, and G. Wefer (eds.). p. 429-455. Report of the Dahlem workshop held in Berlin. April 1988. Wiley-Interscience. New York.

Krey, J. 1970. Die urproduktion des meeres. p. 183. In: *Erforschung des Meeres.* G. Dietrich, (ed.). Umschau. Frankfurt, Germany.

Kunz, C. and A.H. Lu. 1980. Methane production rate studies and gas flow modeling for the fresh kills landfill. *New York State Energy Research and Development Administration Report No. 80-21.* Albany, New York.

LECO. 1991. Manual for CHN-90 carbon, hydrogen, and nitrogen analyzer. Version 2.0. Form No. 200-512-051.

Lerman, A. 1979. *Geochemical processes -- water and sediment environments.* Wiley-Interscience. New York.

Lytwynyshyn, G., E. Zimmerman, and R. Wigender. 1984. Landfill methane recovery -- Part II: gas characterization. *Argonne National Laboratory Report* ANL/CNSV-TM-118.

Meyers, T.P., D.C. Hovde, A.C. Stanton, D.R. Matt. 1992. Micrometeorological measurements of methane emission rates from a sanitary landfill. National Oceanic and Atmospheric Administration. ATDL Contribution No. 92/2. Oak Ridge, Tennessee.

Municipal Solid Waste News. 1991. Canada's green plan. March, 1991.

Orlich, J. 1990. Methane emissions from landfill sites and waste water lagoons. In: *Proc. International Workshop on Methane Emissions from Natural Gas Systems, Coal Mining, and Waste Management Systems.* p 465-472. Held Washington, D.C. April 9-13, 1990. Published by U.S. Environmental Protection Agency. Washington, D.C.

Pacey, J.G., and A.M. Dietz. 1986. Gas production enhancement techniques. In: *Energy from landfill gas.* Proceedings of a conference jointly sponsored by U.S. and U.K. Departments of Energy, Solihull, West Midlands. U.K. October 1986. J.R. Emberton and R.F. Emberton (eds.). Oxfordshire, U.K. Harwell Laboratory

Peter, J.M., P. Peltonen, S.D. Scott, B.R.T. Simoneit, O.E. Kawka. 1991. ^{14}C ages of hydrothermal petroleum and carbonate in Guaymas Basin, Gulf of California: implication for oil generation, expulsion, and migration. *Geology* 19:253-256.

Pettersen, R.C. and V.H. Schwandt. 1988. Wood sugar analysis by anion chromatography. *J. of Wood Chemistry and Technology* 11:495-502.

Pohland, F.G. 1980. Leachate recycle as landfill management option. *Journal of the Environmental Engineering Division ASCE* 106(EE6):1057-1069.

Ramaswami, J.N. 1970. Nutritional effects on acid and gas production in sanitary landfills. Ph.D. thesis. West Virginia University. Morgantown, W.Va.

Rees, J. 1980. The fate of carbon compounds in the landfill disposal of organic matter. *J. Chemical Technology and Biotechnology* 30:458-465.

Richards, K. 1989. Landfill gas: Working with gaia. *Biodeterioration Abstracts* 3:525-539.

Robinson, J.M. 1990. Lignin, land plants, and fungi: Biological evolution affecting phanerozoic oxygen balance. *Geology.* 15:607-610.

Rovers, F.A. and G.J. Farquhar. 1973. Infiltration and landfill behavior. *Journal of the Environmental Engineering Division ASCE.* 9(EE5):671-690.

Shelton, D. and J. Tiedje. 1984. General method for determining anaerobic biodegradation potential. *Applied and Environmental Microbiology* 47:850-857.

Sundquist, E.T. 1985. Geological perspectives on carbon dioxide and the carbon cycle. p. 5-59. In: Sundquist, E.T. and W.S. Broecker (ed.). The carbon cycle and atmospheric CO_2: Natural Variations. Archaean to present. *Geophysical Monograph* 32. American geophysical union. Washington, D.C.

Tans, P.P., I.Y. Fung, and T. Takahashi. 1990. Observational constraints on the global atmospheric CO_2 budget. *Science* 247:1431-1438.

Tissot, B.P. and D.H. Welte. 1978. Petroleum formation and occurrence. Springer-Verlag. New York.

Trunick. P. 1991. Sharing MSW planning in the European community. MSW Management. 1:46-49.

U.S. Environmental Protection Agency. 1992. Characterization of municipal solid waste in the United States. 1992 Update. Executive Summary. EPA/530-S-92-019.

CHAPTER **6**

Areal Evaluation of Organic and Carbonate Carbon in a Desert Area of Southern New Mexico

Robert B. Grossman, Robert J. Ahrens, Leland H. Gile,
Clarence E. Montoya, and Oliver A. Chadwick

I. Introduction

An understanding of global climate change requires estimates of terrestrial carbon that includes soil carbon, which consists of both organic carbon and carbonate carbon. Globally, the two sources in soils are in the ratio of about 2:1 (Schlesinger, 1991a; Houghton and Skole, 1990). Understandably, more emphasis has been placed on organic carbon than carbonate carbon. In fact, global estimates of carbon storage in soils often ignore carbonate carbon accumulations in arid and semi-arid environments (Schlesinger, 1990; Post et al., 1982; Schlesinger, 1977; Bohn, 1976). The magnitude of carbonate carbon storage in these environments is similar to organic carbon storage in humid environments (Schlesinger, 1982).

Areal estimates of both organic carbon and carbonate carbon are presented for a 90,000 ha area in southern New Mexico astride the Rio Grande (Figure 1), referred to as the Desert Soil-Geomorphology Project, that has been the object of intensive geomorphic and pedologic study and for which both parent materials and soil age range widely. Some 50 professional years have been devoted to the project, which is unique in terms of detail and sophistication of the information.

The area receives 150 to 300 mm annual precipitation increasing with elevation. Annual pan evaporation at lower elevations exceeds precipitation 15-fold (Gile and Grossman, 1979). At the time of selection of the study area, the genesis of carbonate horizons was a matter of controversy. The area selected has rock sources that range from limestone to Soledad rhyolite relatively low in calcium. Dunham (1935) reports 0.16 percent CaO in Soledad rhyolite. A criterion for selection of the area was that it encompass parent material sources low in calcium in order to provide a setting to explore the source of calcium for horizons of strong carbonate accumulation. The low calcium rocks are in the southern part of the Organ Mountains shown in Figure 1. The northern part of the Organ Mountains is formed from monzonite, relatively high in calcium. Limestone occurs in the San Andres Mountains.

The approach rests heavily on approaches developed by the National Cooperative Soil Survey. The appendix is a brief discussion of aspects of soil survey for people in other fields.

II. Literature Review

Organic matter has extreme importance for agriculture and a wealth of data exists. Compilations on a broad scale have been made using the soil survey data base (Kimble et al., 1990, Bryant et al., 1991, Sanchez et al., 1982). These studies organize the data by taxa of the US taxonomy system (Soil Survey Staff, 1975). Areal estimates in these studies, if made, employ small scale maps and a high categorical level of soil classification. Strategies that employ detailed soil maps and assign analyzed pedons to the map unit

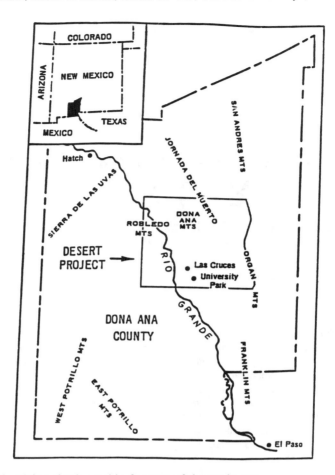

Figure 1. Major physiographic features of the study area.

components such as has been done in this study and in Grossman et al. (1992) are rare. Agricultural data bases have been employed to evaluate models for organic carbon prediction. A notable example of the application of the soil survey data base is Parton et al. (1987).

Studies of carbonate carbon have followed a somewhat different path. They may not employ the soil survey data base and the thrust has been mainly by geologists, not soil scientists. Three approaches have been employed to document carbonate carbon accumulation in arid and semi-arid soils: 1) chronosequence sampling to estimate rates of carbon accumulation, but with no areal extrapolation (Harden et al., 1991, 1992; Machette, 1985); 2) soil mapping and taxonomic database analysis to determine carbon storage, but with little indication of rates (Schlesinger, 1982); and 3) modelling the effect of climate and its change to determine the depth distribution and quantity of carbonate in soil profiles, but with no areal extrapolation (McFadden et al., 1991; Mayer et al., 1988; McFadden and Tinsley, 1985; Marion et al., 1985; Arkley, 1963). Machette (1985) has incorporated the information in Gile and Grossman (1979) into a general chronosequence. Gile and Grossman (1979) used approach 1 and implicitly approach 2, but quantitative areal values of carbonate carbon (and also organic carbon) for mapped areas were not obtained. In this paper quantitative areal estimates for mapped areas are obtained using the soil map and the laboratory measurements in Gile and Grossman (1979) as enlarged and modified by subsequent largely unpublished studies.

Table 1. Specifications for the project

Project	
Total area	90,000 ha[a]
Measurement area (total minus rock land)	75,000 ha[b]
Mapping	
Scale	1:15,840
Number of map units	66
Number of map unit components	
Total	509
80% map unit coverage[c]	3
Laboratory sampling[d]	
Pedon totals	
Available	113
Employed	93
Pedon assignments	
Total	744
80% map unit coverage[e]	391
Areally weighted area per assigned pedon[f]	
All map units	244 ha
>5,000 ha map units	2050 ha

[a] Some minor components were grouped, assigned an aggregate areal percentage, and the components were assumed to have the same composition.
[b] Information presented subsequently is for the measurement area. Streamwash was included as a map unit component and assigned zero carbonate and organic carbon.
[c] The most frequent number of map unit components to encompass 80 percent of the map unit area.
[d] Data are drawn from Gile and Grossman (1979), Gile (1987), Gile (in press), Monger et al. (1991b), Tatarko (1980), and unpublished data, Soil Survey Laboratory, SCS.
[e] The minimum number of pedon assignments to encompass 80 percent of the area of each map unit was determined and the sum obtained.
[f] The sum for all map units of the product of the average area per pedon and the areal fraction of the map unit.

III. Methods

Table 1 presents a numerical overview of the project. The area of each map unit (see appendix) was determined by planimeter. The photography was not rectified. Field observations of section corners, however, indicate that for all but a very few photographs the error is small. For the portion that is not rockland (referred to as the measurement area), areal proportions of the map unit components were assigned. An initial estimate of the proportions of the map unit components was obtained from information in Gile and Grossman (1979). In recent years on the basis of field studies, Gile has made unpublished revisions in these proportions as well as in the areas of some of the map unit delineations. The revisions are ongoing and a final report on the work will be presented later. Sites have been sampled for laboratory analyses since 1959. Most of the sites were selected to be representative of map unit components. Some were selected as part of transects to study soil changes and may not be highly representative of the map unit component in which they occur. Some map unit components were assigned several pedons, others only one pedon. The pedon assignments were made on the basis of detailed field experience gained not only over the course of the formal project but to the present time. Assignments were made for all map unit components using whatever laboratory data were at hand. This is an important point. There were no prior standards for

Table 2. Composition of the Bucklebar complex map unit

Map unit component	Classification	% of map unit	Soil carbon Carbonate	Organic
			kg m^{-2}	kg m^{-2}
Bucklebar	Typic Haplargid, fine-loamy	35	6	3.9
Bucklebar, Ustollic[a]	Ustollic Haplargid, fine-loamy	35	7	4.8
Onite	Typie Haplargid, coarse-loamy	25	13	3.7
Berino; Berino, Ustollic[a]	Ustollic Haplargid, fine-loamy	5	26	1.8
All components		100	9	4.1

[a] The soil is similar in terms of behavior to Berino or Bucklebar but would be classified differently -- in an Ustollic subgroup instead of Typic.

applicability of the laboratory data. In some instances, data were assigned which were identified with a different soil series. The exercise has substance because the density of laboratory pedons to draw from is quite high and the aggregate area of the map unit components taken individually is relatively small. As a consequence of both circumstances together, over 700 pedon assignments were made to map unit components.

For each assigned laboratory pedon, the kilograms of organic carbon and/or carbonate carbon (0.12 times the CaCO$_3$ equivalent percentage) per square meter to a variable depth were calculated. The organic carbon was determined by the Walkley-Black method, which is 6A1 in Soil Survey Laboratory Staff (1992). The carbon values are sums for the various layers (horizons) of the quantities obtained by the relationship to follow:

$$\frac{L \times Db \times Xc \times (1 - Vr)}{10}$$

where L is the thickness in centimeters, Db is the bulk density of the <2 mm in Mg m^{-3}, Xc is the <2 mm carbon percentage, and Vr is the volume of rock fragments expressed in fractional form. For carbonate-cemented horizons developed in low-carbonate parent materials, the weight of >2 mm after removal of the carbonate was calculated to a carbonate-containing volume basis. Organic carbon was summed to the depth measured, which usually was to where the values were below 0.1 percent. Carbonate carbon was summed to the depth encompassed by the concept of the map unit component. Commonly carbonate in buried soils was excluded unless it was part of the concept of the map unit component. This will be discussed subsequently.

Determination of the average soil carbon for a map unit is illustrated in Table 2. Bucklebar soils predominate in the area labeled Bucklebar complex map unit in Figure 2. Each component was assigned soil carbon values based on the average for the laboratory pedons assigned to the component. For Onite two laboratory pedons were assigned for both organic and carbonate carbon. A single pedon was assigned for the other components. Using the soil carbon values for the map unit components and the areal proportion of the components in the map unit, a weighted average for the map unit area was computed. The weighted average for a map unit is then multiplied by the areal fraction that the map unit is of the measurement area to obtain the contribution to the measurement area.

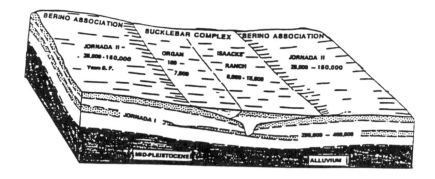

Figure 2. Generalized diagram showing the physiographic landscape position of the Bucklebar complex map unit. The stippled zones on the face of the diagram depict argillic horizons. Horizons of carbonate accumulation occur in and/or beneath the argillic horizons.

IV. Results

As shown in Table 3, organic carbon mean for the whole measurement area is 3.1 kg m^{-2} with a higher value for the semiarid portion of the measurement area. Lower organic carbon is associated with abundance of >2 mm rock fragments. Soils that contain ≥35 percent by volume rock fragments within depth limits that depend on the classification of the soil are referred to as skeletal. Skeletal map units have less organic carbon than nonskeletal units. The map units strongly influenced by limestone that are nonskeletal have the highest organic carbon.

For the soils developed in parent materials high in limestone, the carbonate values are misleading for the purposes here. The arid area exclusive of the map units with strong limestone influence has 25 kg m^{-2} of carbonate carbon. This value is applicable to 74 percent or the majority of the measurement area. For the arid area exclusive of strong limestone influence, nonskeletal map units have substantially more carbonate carbon than do skeletal units. As a generality, carbonate carbon exceeds organic carbon about 10-fold.

V. Discussion

A. Areal Soil Carbon Quantities

The areal weighted average organic carbon for a generally cultivated county in southwest Iowa is 12 kg m^{-2} (Grossman et al., 1992), about 4 times that for the measurement area of the project discussed here. Some pedons developed in fine-silty sediments high in limestone reach about 10 kg m^{-2} organic carbon, approximately that for average of the county in southwest Iowa. A reason for the high organic carbon is under study (Wilding, 1992). The weighted average organic carbon for the arid part of the study area of 3.0 kg m^{-2} is similar to the 3.3 kg m^{-2} for Desert Shrub given in Schlesinger (1991a) and exceeds the value by Post et al. (1982) of 1.2 kg m^{-2} for warm deserts.

B. Population of STATSGO Map Units

A pertinent question is how to generalize from the carbon values for the measurement area to other areas with similar soils. STATSGO is a statewide digital soil geographic database which was developed by generalizing existing detailed soil survey maps to a scale of 1:250,000. Jessie Rossbach, Soil Scientist, New Mexico State Office, prepared the STATSGO map for the project area (Figure 3). The part of the state-wide STATSGO map for the measurement area includes about 10 map units, compared to the 66 for the detailed

Table 3. Organic and carbonate carbon expressed areally for parts of the measurement area

Part of measurement area	Portion of measurement area	Carbon[c]	
		Organic	Carbonate
	%	kg m^{-2}	kg m^{-2}
Total area	100	3.1	---
Exclude limestone	79	2.7	24
Arid part	94	3.0	29
Exclude limestone	74	2.6	25
Skeletal[a]	16	2.2	19
Nonskeletal	58	2.7	27
Limestone	20	4.3	43
Skeletal	7	2.4	40
Nonskeletal	13	5.3	44
Semiarid part[b]	6	4.3	13
Exclude limestone	5	4.3	8
Skeletal	3	2.9	13
Nonskeletal	2	6.5	tr
Limestone	1	4.3	37

[a] >50% of the map unit has >35% volume rock fragments. The depth zone considered may change depending on the classification of the soil.
[b] >1500 m elevation.
[c] The measurement area contains 2.3 Tg (10^{12} g) of organic carbon. The measurement area minus the limestone influenced area contains 14 Tg of carbonate carbon.

map on which this study is based. A portion of a delineation of map unit NM 877 encompasses the basin floor of the project. The portion of the delineation not included in the map has a similar distribution of soils to the area shown in Figure 3. The similarity of the excluded portion permits considerations of the part within the project as representative of the whole. For illustrative purposes, this STATSGO map unit has been populated with areally weighted quantities of organic and carbonate carbon for included map units from the detailed soil map. Excluding a single map unit strongly influenced by limestone, the organic carbon is 3.7 kg m^{-2} and the carbonate carbon is 66 kg m^{-2}. Organic carbon exceeds the average for the arid area generally because some of the soils are fine textured and of those a portion are subject to overflow. (The relationship of organic carbon to clay is discussed in the next section.) The carbonate carbon is high relative to the measurement area as a whole because the soils are old.

C. Prediction of Soil Carbon

Taxonomic placement is quite predictive for organic carbon. Soils in the arid part that are high in rock fragments (skeletal) and have coarse or medium textures, are low (<2 kg) in organic carbon. The presence of petrocalcic horizons within 50 cm of the ground surface coupled with the aforementioned coarse particle size leads to low organic carbon even in some instances if the soil is semiarid. High relative amounts of organic carbon (>6 kg) are the usual situation in semiarid areas with the exceptions previously given. As mentioned previously, soils in fine-silty families derived from limestone that are calcareous throughout have high organic carbon. Haplargids in fine families also usually have relatively high organic carbon. Some coarse soils in coppice dunes commonly have high organic carbon, in part related to the strong biological activity.

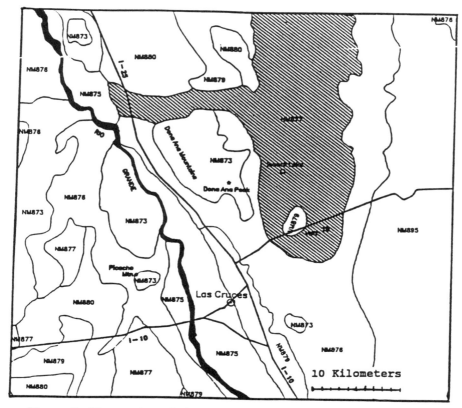

Figure 3. The portion of the statewide digital soil geographic database (STATSGO) map of New Mexico that covers the study area.

To follow is a summerization of some relationships between organic carbon and soil taxa:

<u>Low Relative Amounts <2 kg m^{-2}</u>

Amount predominant if:
  Typic and sandy, sandy-skeletal, or loamy-skeletal soils. (Two sandy-skeletal pedons have <1 kg.).
  Petrocalcic subgroups or Paleorthids whether Typic or Ustollic, and with previous particle sizes.

Amount common if:
  Typic or coarse-loamy and not Petrocalcic subgroups or Paleorthids.

<u>High Relative Amounts >6 kg m^{-2}</u>

Amount predominant if:
  Ustollic subgroups or Haplustolls, both irrespective of particle size.
  Fine-silty families with a calcic horizon or in a calcareous family (Ustollic subgroup reaches 10 kg m^{-2}).

Amount common if:
  Fine Haplargids (Ustollic subgroups reach 10 kg m^{-2}).
  Torripsamments formed by eolian accretion.

The reason for the increase in organic carbon as the particle size becomes finer may have several contributing explanations and ramifications. First, there is a tendency for soils with finer particle size to be lower on the landscape where run-on can provide water in excess of the precipitation. Secondly, there is the possibility that grasses instead of shrubs are favored by a nonskeletal particle size (Hallmark and Allen, 1975). Third, the water retention of the finer particle size materials should be higher, although there is the counter argument that precipitation moves deeper in the coarser textures and can be used more effectively by plants. Fourth, there is the possibility, as has been discussed, that fine grained carbonate acts to enhance organic carbon accumulation (Wilding, 1992). Fine-silty soils of the measurement area tend to be high in fine carbonate throughout the rooting depth and have the highest organic carbon in the study. Fifth, there is the common observation that organic carbon tends to increase with noncarbonate clay (Nichols, 1984; Parton et al., 1987). For the measurement area, the following relationship for argillic horizons was found: $OC(kg\ m^{-2}) = 0.06 + 0.15$ (clay %); n = 28, r = 0.86. Organic carbon was computed to the base of the horizon nearest to 1 m. The noncarbonate clay was calculated on a carbonate-containing weighted average basis for 0-50 cm. Rock fragments were <10% by volume. The aforementioned tendency for finer soils to occur lower on the landscape where they are more commonly subject to run-on may be a confounding factor. Finally, as indicated, skeletal soils in the arid portion tend to have lower organic carbon. Soils that are skeletal tend to have less clay throughout the zone of highest organic carbon accumulation. This may be a contributing reason why the organic carbon is lower.

If soils high in limestone are excluded, taxonomic placement is also predictive of carbonate carbon except for the Haplargids. Petrocalcic Paleargids, Paleorthids and Calciorthids averaged 70 kg m^{-2} of carbonate carbon while Haplustolls and Torriorthents had <1 kg m^{-2}. Haplargids that formed in low limestone materials have 1 to 100 kg m^{-2} depending on soil age, climate, and rock fragment volume. The main reason for this wide range for Haplargids is that the taxonomic criteria permit from zero to very prominent carbonate accumulation. In this regard several Haplargids have petrocalcic horizons that are too deep (below 1 m depth) to be diagnostic for the Paleargids. The Haplargid studied by Monger (1991b) is illustrative.

D. Deep Carbonate Accumulation

The carbonate carbon reported pertains only to the map unit concepts. Commonly buried soils high in carbonate are excluded. Thus the reported carbonate carbon underestimates the total storage by an unknown and possibly large amount. Figure 2 shows a coalescent alluvial fan with two buried soils having associated carbonate. For the Bucklebar complex map unit (Table 2), neither buried soil is part of the soil concept for mapping purposes. For the Berino association the upper buried soil is part of the mapping concept. Neither mapping concept includes the lower buried soils. The Bucklebar complex map unit contains 9 kg m^{-2} of carbonate carbon compared to 17 kg m^{-2} for the Berino association. Both values markedly underestimate carbonate carbon to the base of the two buried soils. Furthermore, additional buried soils probably occur at greater depth. This problem of carbonate below the depth of the map unit concept has not been addressed.

For questions of global warming, an issue is the residence time of carbonate carbon. The residence time of the excluded carbonate in general would be longer than for the carbonate included in the evaluation. However, there is a very wide range in the residence time of carbonate included in the study. In instances, carbonate included has a longer residence time than some of that excluded.

E. Rate of Carbonate Accumulation

The accumulation rate of carbonate carbon in soils of late Holocene age (1 to 4 yrs) has been evaluated based on seven pedons from the aridic part of the measurement area (Gile and Grossman, 1979). These pedons developed in nominally noncalcareous parent materials and are in sandy, coarse-loamy or fine-loamy families (Soil Survey Staff, 1975). An initial assumed carbonate carbon content of 1 percent for the <2 mm was subtracted. The range in the rate of accumulation is 0.1 to 1.4 g m^{-2} yr, which is below the 3 g m^{-2} yr given by Schlesinger (1991a). The rates of accumulation for 24 pedons of late Pleistocene to mid-Pleistocene age are about the same as for the Holocene soils. The similarity suggests that net carbonate carbon accumulation over periods of time measured in tens of thousands of years in the Pleistocene (including pluvials) did not differ too much from that in the late Holocene. Machette (1985) estimated accumulation rates for the soils discussed here plus soils of latest Pleistocene. He concluded that in the Holocene carbonate accumulated at roughly twice the rate as during pluvial episodes.

The rate of carbonate carbon accumulation in the late Holocene is markedly less than the 20-30 g m^{-2} yr of organic carbon (2-3 g m^{-2} yr of nitrogen) for soil materials about 100 years old or less (Grossman, 1983;

Schlesinger, 1991b) and relatively similar to the value of 2 g m^{-2} yr for 10,000 year old forest given by Schlesinger (op cit). No information was available to the authors on the rate of accumulation of organic carbon for very young parent materials (<100 years) under a warm arid environment to which the rate of carbonate carbon could be more directly compared.

F. Carbonate as a CO$_2$ Sink

For carbonate carbon to be a sink for atmospheric CO$_2$, at least a portion of the calcium must not come from carbonate. If the authigenic soil carbonate originated by the dissolving of carbonate in dust deposited on the ground surface followed by precipitation of the carbonate within the soil, there would not be a net transfer of atmospheric CO$_2$ to the soil. Similarly, if the calcium in the authigenic carbonate came from calcium in the precipitation, to the extent that this calcium originated by dissolving carbonate dust, there would not be a net transfer of CO$_2$ to the soil. On the other hand, if the calcium originated by weathering of the host non-carbonate minerals in the soil or from noncalcareous atmospheric sources, then the authigenic carbonate would be a sink for atmospheric CO$_2$. It has been postulated that much of the calcium did originate from carbonate in dry dust or from calcium dissolved in the precipitation that came from carbonate dust (Gile and Grossman, 1979). Monger et al. (1991a) presents evidence that microorganisms are involved in the precipitation of carbonate, which may be a further complication. Chadwick and Capo (1992), report based on ^{87}Sr/^{86}Sr (Graustein, 1989) that at least 95 percent of the calcium for samples from the project area is derived from dust and precipitation. The implication is that little of the soil carbonate represents a sink for atmospheric CO$_2$.

VI. Summary

We present the organic and carbonate carbon on an areal basis for an intensively studied 75,000 ha area in south-central New Mexico. Based on a very detailed knowledge of the soils, areal amounts were obtained by estimating the percentage of each component in every map unit and assigning laboratory pedons with measured organic carbon and carbonate to each map unit component. The arid part of the area has a weighted average organic carbon of 3.1 kg m^{-2}. The arid part, excluding that strongly influenced by limestone, contains 25 kg m^{-2} of carbonate carbon. The results may be extended through use of 1:250,000 STATSGO state soil maps which are coordinated nationally by the Soil Conservation Service. Taxonomic placement of major soils is quite predictive of organic carbon and also of carbonate carbon except for the Haplargids. The wide range in carbonate carbon of Haplargids is primarily because carbonate accumulations in Haplargids may range widely and horizons of strong carbonate accumulation may occur below the depth relevant to the taxonomic placement. A rate of accumulation of carbonate carbon for soils 1-4 yrs old of 0.1-1.4 kg m^{-2} yr has been computed. One problem not explored is how to include in the estimates of carbonate carbon buried soils horizons high in carbonate that are below the depth of the map unit component concept. Another problem is the extent to which the carbonate carbon actually is a sink for atmospheric CO$_2$.

Acknowledgments

The Desert Soil-Geomorphology Project was begun in 1957 under Dr. R.V. Ruhe and reflects his philosophy. Dr. Ruhe died February 10, 1993. This paper is dedicated to him.

References

Arkley, R.J. 1963. Calculation of carbonate and water movement in soil from climatic data. *Soil Sci.* 96:239-248.

Bohn, H.L. 1976. Estimates of organic carbon in world soils. *Soil Sci. Soc. Am. J.* 40:468-470.

Bryant, R.B., P.S. Puglia, J.M. Duxbury, J.M. Galbraith, and F.J. Ramos. 1991. Carbon sequestration in cool humid environments: a case study of New York. Report. Department of Agronomy, Cornell Univ., Ithaca, NY.

Chadwick, O.A. and R.C. Capo. 1993. Partitioning allogenic and authigenic sources of calcium in New Mexico calcretes. Agron. Abst. 1993 Annual Meetings. Am. Soc. Agron., Madison, WI.

Dunham, K.C. 1935. The geology of the Organ Mountains with an account of the geology and mineral resources of Dona Ana County, New Mexico. New Mexico Bureau of Mines and Mineral Resources Bulletin 11

Gile, L.H. and R.B. Grossman. 1979. The desert project soil monograph. USDA, SCS. Washington, D.C.

Gile, L.H. 1987. Late Holocene displacement along the Organ Mountains Fault in southern New Mexico. Circular 196, N.M. Bureau of Mines & Mineral Resources, Socorro, NM.

Gile, L.H. (In press). Soils, geomorphology, and multiple displacements along the Organ Mountains Fault in southern New Mexico. N.M. Bureau of Mines & Mineral Resources, Socorro, NM.

Graustein, W.C. 1989. Stable isotopes in ecological research. p. 491-512. In: P.W. Rundel, J.R. Ehleringer, and K.A. Nagy (eds.). *Stable isotopes in ecological research.* Springer-Verlag, NY.

Grossman, R.B. 1983. Entisols. p. 55-90. In: L.P. Wilding, N.E. Smeck, and G.F. Hall (eds.). *Pedogenesis and soil taxonomy. II. The soil orders.* Elsevier, NY.

Grossman, R.B., E.C. Benham, J.R. Fortner, S.W. Waltman, J.M. Kimble, and C.E. Branham. 1992. A demonstration of the use of soil survey information to obtain areal estimates of organic carbon. ASPRS/ACSM/RT Tech. Papers, Vol. 4. Am. Soc. Photogrammetry and Remote Sensing and Am. Cong. Surveying and Mapping.

Hallmark, C.T. and B.L. Allen. 1975. The distribution of Creosotebush in west Texas and eastern New Mexico as affected by selected soil properties. *Soil Sci. Soc. Am. J.* 39:120-124.

Harden, J.W., F.M. Taylor, C. Hill, R.K. Mark, L.D. McFadden, M.C. Reheis, J.M. Sowers, and S.G. Wells. 1991. Rates of soil development from four soil chronosequences in the southern Great Basin. *Quaternary Research* 35:383-399.

Harden, J.W., E.M. Taylor, L.D. McFadden, and M.C. Reheis. 1992. Calcic, gypsic, and siliceous soil chronosequences in arid and semi-arid environments. p. 1-16. In: W.D. Nettleton (ed.). *Occurrence, characteristics, and genesis of carbonate, gypsum, and silica accumulations in soils.* Soil Sci. Soc. Am. Special Pub. 26:1-16.

Houghton, R.A. and D.L. Skole. 1990. The long term flux of carbon between terrestrial ecosystems and the atmosphere as a result of changes in land use. Carbon Dioxide Res. Progress. Office of Health and Environ. Res. U.S. Dept. of Energy.

Kimble, J.M., H. Eswaran, and T. Cook. 1990. Organic carbon on a volume basis in tropical and temperate soils. Proc. 14th International Cong. Soil Sci. Vol. V:248-253.

Machette, M.N. 1985. Calcic soils of the southwestern United States. In: D.L. Weide (ed.) *Soils and Quaternary geomorphology of the southwestern United States.* Geological Society of America Special Paper, 203:1-21.

Marion, G.M., W.H. Schlesinger, and P.J. Fonteyn. 1985. CALDEP: a regional model for soil $CaCO_3$ (caliche) deposition in southwestern deserts. *Soil Sci.* 139:468-481.

Mayer, L., L.D. McFadden, and J.W. Harden. 1988. The distribution of calcium carbonate: a model. *Geology* 16:303-306.

McFadden, L.D. and J.C. Tinsley. 1985. Rate and depth of pedogenic-carbonate accumulation in soils: formulation and testing of a compartment model. In: D.L. Weide (ed.). *Soils and Quaternary geomorphology of the southwestern United States.* Geological Society of America Special Paper. 203:23-41.

McFadden, L.D., R.G. Amundson, and O.A. Chadwick. 1991. Numerical modeling, chemical, and isotopic studies of carbonate accumulation in soils of arid regions. p. 17-35. In: W.D. Nettleton (ed.). *Occurrence, characteristics, and genesis of carbonate, gypsum, and silica accumulations in soils.* Soil Sci. Soc. Am. Special Pub. No. 26, Madison, WI.

Monger, H.C., L.A. Daugherty, W.C. Lindemann, and C.M. Lidell. 1991a. Microbial precipitation of pedogenic calcite. *Geology* 19:997-1000.

Monger, H.C., L.A. Dougherty, and L.H. Gile. 1991b. A microscopic examination of pedogenic calcite in an Aridisol of southern New Mexico. p. 37-60. In: W.D. Nettleton (ed.). Occurrence, characteristics, and genesis of carbonate, gypsum, and silica accumulations in soils. *Soil Sci. Soc. Am.* Special Pub. 26, Madison, WI.

Nichols, J.D. 1984. Relationship of organic carbon to soil properties and climate in the southern Great Plains. *Soil Sci. Soc. Am. J.* 48:1382-1384.

Parton, W.J., D.S. Schimel, C.V. Cole, and D.S. Ojima. 1987. Analysis of factors controlling soil organic matter levels in Great Plains grasslands. *Soil Sci. Soc. Am. J.* 51:1173-1179.

Post, W.M., W.R. Emanuel, P.J. Zinke, and A.G. Stangenberger. 1982. Soil carbon pools and world life zones. *Nature* 298:156-159.

Sanchez, P.A., M.P. Gichuru, and L.B. Katz. 1982. Organic matter in major soils of the tropic and temperate regions. p. 99-114. In: Non-symbiotic nitrogen fixation and organic matter in the tropics. Symposia Papers. 12th International Cong. Soil Sci.

Schlesinger, W.H. 1977. Carbon balance in terrestrial detritus. *Ann. Rev. Ecol. Syst.* 8:51-81.

Schlesinger, W.H. 1982. Carbon storage in the caliche of arid soils: case study from Arizona. *Soil Sci.* 133:247-255.

Schlesinger, W.H. 1990. Evidence from chronosequence studies for a low carbon-storage potential of soils. *Nature.* 348:232-234.

Schlesinger, W.H. 1991a. *Biogeochemistry - an analysis of global change.* p. 135. Academic Press, N.Y.

Schlesinger, W.H. 1991b. Presentation Summary. p. 26-27. In: M.G. Johnson and J.S. Kern. (eds.). Sequestering carbon in soils: a workshop to explore the potential for mitigating global climate change. EPA. Environmental Res. Lab., EPA/600/3-91/031. Corvallis, OR.

Soil Survey Staff. 1975. Soil Taxonomy. U.S. Dept. Agric. Hdbk 436. p. 754. USDA Washington, D.C.

Soil Survey Laboratory Staff. 1992. Soil survey laboratory methods manual. Soil Survey Invest. Rept. No. 42. USDA, SCS. Washington, D.C.

Tatarko, J. 1980. Effect of calcium carbonate on the distribution of creosotebush (*Larrea tridentata* (DC) Coville) in west Texas and southern New Mexico. M.S. thesis. Texas Tech University, Lubbock, TX.

Wilding, L.P. 1992. Personal communication.

Appendix: The US Soil Survey

The overarching organization for the US soil survey is the National Cooperative Soil Survey, which is an informal association of the Soil Conservation Service (the lead agency), the various state experiment stations, and certain other Federal agencies (U.S. Forest Service, Bureau of Land Management, Bureau of Indian Affairs). The Soil Conservation Service in the various states and in Puerto Rico conduct the soil survey of private land. Quality control and technical support are provided by state, regional, and national offices of the Soil Conservation Service. The National Cooperative Soil Survey developed Soil Taxonomy (Soil Survey Staff, 1975), the US system for classifying soils. On public lands Federal agencies other than the Soil Conservation Service are responsible for the operation of soil surveys; the role of the Soil Conservation Service is limited to the consistent naming, classification, and interpretation of the soils.

The map unit is central to an understanding of soil survey. Usually the kind of soils taxonomically are specified by the map unit definition as well as the slope, erosion, and possibly other properties important to management. The definition of the map unit should be applicable to each delineation identified as part of the map unit. Usually there are a considerable number of delineations of a given map unit in a standard soil survey. Map units usually consist of more than one map unit component. These map unit components are separable conceptually, but are not separated cartographically because of the scale of the map, the time devoted to mapping per unit area of land, or both. Each map unit component is assigned a set of interpretive properties. These sets of interpretive properties include a number that are derived from laboratory measurements. The interpretive data are available from the lead Federal Agency for the survey. For many components pedological laboratory data may be available (as is true for the soils of the project discussed in this paper).

Geographic point sites are referred to as pedons. Most pedons have only descriptive information. Some have also been sampled for laboratory analysis. The National Cooperative Soil Survey has sampled some 30,000 pedons for laboratory analysis with a wide range in completeness in terms of the layers sampled and the analyses performed. The data base of laboratory measurements on these pedons may be accessed through the National Soil Survey Center, Lincoln, Nebraska, as well as through State Soil Conservation Service offices, State Agricultural Experiment Stations, and for public lands through other Federal agencies.

CHAPTER 7

Carbon Storage in Tidal Marsh Soils

Martin C. Rabenhorst

I. Introduction

Concern over greenhouse gasses and global carbon budgeting has recently led researchers to focus on terrestrial systems. Estimates of the carbon stored in soils (1500 Petagrams) is roughly three times that stored in the vegetation (Post et al., 1990). Thus the storage of carbon in soils and the dynamics of soil carbon have become of paramount importance to understanding the global carbon budget. It is obvious that the organic soils commonly associated with wetlands must contain a large amount of stored C. As such it is equally clear that these C-rich soils have functioned as carbon sinks in the past. It is not, however, immediately apparent whether these soils will continue to serve as carbon sinks, or if they might become sources of carbon to the atmosphere. The soils of coastal marshes represent an important subset of the larger group of carbon enriched soils in wetlands. This paper is an attempt to evaluate the relative importance of coastal marsh soils with respect to carbon storage.

II. Areal Extent of Coastal Marshes

On a worldwide basis, wetlands have been estimated to occupy roughly 2.8×10^6 km^2 or 2.2% of the global land surface (Post et al., 1982). Estimates by Mitsch and Gosselink (1993) and Mitsch and Wu (this volume) place the areal extent of wetlands even higher at 5.3-8.6×10^6 km^2 or ~4-6% of the earth's surface. In either case, wetlands represent an important component in the global ecosystem. Coastal wetlands comprise a significant portion of the total wetlands, and they are extensive on every continent but Antarctica.

Following U.S. Fish and Wildlife accounting, it has been estimated that approximately 16% of the world's wetlands occur in the USA (4.0×10^5 km^2), and that nearly 30% of the nation's total wetlands are coastal (Field et al., 1991). Roughly one third of the nation's 1.1×10^5 km^2 of coastal wetlands are coastal marshes with the rest being forested and scrub-shrub wetlands. Most of the coastal marshes in the USA, including both salt marsh and fresh marsh, are found along the Gulf of Mexico (60%), the South Atlantic (25%), and Middle Atlantic (9%) regions (Figure1). After the Mississippi Delta drainage area, which by far has the greatest areal extent of salt marshes in the country, the Chesapeake Bay drainage area ranks next with approximately 150,000 ha.

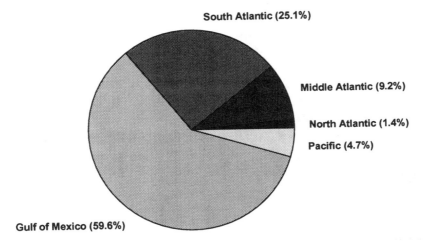

Figure 1. Areal extent of coastal marshes in the United States, distributed by region; total area 3.6 M ha. (Data from Fields et al., 1991.)

III. Tidal Marsh Soils

A. Formation of Marsh Soils

Toward the end of the Pleistocene Epoch, sea level was approximately 100 meters below the present level as vast amounts of water were tied up in the continental glaciers. As the glacial ice began to melt, and as global warming brought thermal expansion of the oceans, sea level began rise. It is generally held that the rates of sea level rise from the onset of glacial retreat up until approximately 3,000 years ago, were quite rapid (Redfield and Rubin, 1962; Stuiver and Daddario, 1963). During this period incised river valleys became drowned and low-lying areas became flooded and there was significant sedimentation in deeper water habitats. The rate of sea level rise was, however, too rapid to permit the establishment of emergent vegetation in the shallow water environments. Around three thousand years ago, the rate of sea level rise began to slow to a more modest rate (Redfield and Rubin, 1962), and during this period, marsh vegetation began to colonize areas of tidal flats. The vegetation served both to further entrap sediment and also to contribute an organic component to the accumulating marsh soils. While sea level continued to rise at a modest pace, the marsh soils accrued mineral and organic materials and underwent vertical accretion or growth at rates approximately equal to those of sea level rise. Thus, so long at the marshes remained healthy and protected against severe degradation such as storm driven erosion, the vertical growth of coastal marshes was able more or less to keep up with the sea level rise (Redfield, 1972).

What is commonly called sea level rise, is in fact a combination of both sea level rise and coastal submergence, and should probably be called *apparent* sea level rise. The worldwide average rate of sea level rise has been estimated at approximately 1.2 mm yr^{-1} (Gornitz et al, 1982). Rates of apparent sea level rise vary considerably, even around the Atlantic and Gulf coasts, with some estimates being as much as an order of magnitude higher than the worldwide estimates. The average rate of sea level rise along the North Atlantic coast during the last century has been estimated to range from 2.2 to 4.2 mm yr^{-1} (Hicks et al, 1983). These variations are due mainly to differences in the rates of coastal subsidence, which may be driven by sedimentary accumulations on the coastal shelf, and moderated by isostatic adjustments or even by subsidence related to the pumping of water from aquifers. Even higher rates of apparent sea level rise are reported along the gulf coast in Louisiana due to rapid subsidence of deltaic sediments of the Mississippi river (Delaune et al, 1983).

The soils of coastal marshes can be grouped into three principal types based on the physiography of the upland-marsh landscapes (Darmody and Foss, 1979). Those that are termed *Estuarine Marshes* have formed in stream channels and estuarine meanders filled with mineral and organic sediments. These soils typically have high *n-values*, indicating that the sediments were deposited under water and have never dried in place. Consequently these soils have a very low bearing capacity. The *Coastal Marsh* soils are found in marshes which ring lagoons or the sound area behind barrier islands. These marsh soils tend to be highly stratified with layers of organic and mineral materials. When located adjacent to the barrier islands and dunes, the mineral component is usually sandy. On the mainland side of the lagoon, silts and clays dominate the mineral fraction. The third main group of marsh soils has been termed *Submerged Upland Marshes* which have developed where slowly rising seas have inundated previously-formed soils occurring on gently sloping uplands. These marsh soils are comprised of low density or high n-value organic-rich sediments overlying a dense, consolidated soil profile, previously formed in an upland condition. Submerged upland marsh soils are the most extensive type occurring in the upper Chesapeake Bay estuary (Darmody and Foss, 1979).

B. Classification of Marsh Soils

Most of the coastal marsh soils are classified as either Histosols or Entisols. Histosols are formed predominantly from *organic soil materials*, which are defined as containing more than 12 to 20% organic carbon (Soil Survey Staff, 1992). If organic materials comprise as much as 40 cm of the upper 80 cm of the soil, then the soil will be classified in the Histosol order. Histosols may be further divided into suborders based principally upon the degree of organic matter decomposition in the subsurface tier (generally between 30-90cm). Fibrists, Hemists, and Saprists represent soils where the organic materials are relatively undecomposed, partially decomposed and highly decomposed, respectively. The classes of *Terric* and *Typic* are used in Histosols to describe the thickness of the organic materials. If less than 130 cm of the upper 160 cm of soil is organic, then the soils would be classified in *Terric* subgroups; otherwise they would be considered *Typic*.

Histosols, then, represent soils where the accumulation of organic materials has been high and where there has been a relatively low input of mineral sediment. Where the accumulation of organic materials has been lower, or where there has been a relatively high and prolonged input of mineral sediment within the marsh, the content of organic matter is generally too low for the soils to be Histosols. In most cases, these soils are classified as Entisols indicating their relatively youthful character (Soil Survey Staff, 1975). Because they are very wet, with essentially all having peraquic moisture regimes, these Entisols are inevitably classified in the suborder of Aquents (Soil Survey Staff, 1992). In coastal areas where the salinity is moderately high, many of the tidal marsh soils will contain sulfides produced by sulfate-reducing microorganisms (Pons et al., 1982; Rickard, 1973; Rabenhorst and James, 1992). Soils which contain sulfides would be classified in *Sulfi-* subgroups such as Sulfihemists or Sulfaquents. In areas less affected by saline water, soils would be classified into a variety of classes, the most common of which in Atlantic and Gulf coastal marshes include Medisaprists, Hydraquents, and Psammaquents.

IV. Carbon Storage in Marsh Soils

A. Magnitude of Carbon Storage

Wetland ecosystems taken as a whole comprise only about 2 to 6 % of the world's land area, but they account for a disproportionately high percentage (14.5%) of the stored soil carbon (Figure 2). This is largely due to the great amount of carbon sequestered in the organic soils common to wetlands. Post et al. (1982), for example, estimate that carbon storage in temperate forests is in the range of 7 to 13 kg C m^{-2}, while that in wetlands is nearly an order of magnitude greater.

Quantities of soil carbon stored in three cultivated coastal plain soils in Maryland are presented in Table 1. These soils are fine-silty Hapludults or Ochraquults and contain between 4.6 and 6.3 kg C m^{-2}. Estimates of carbon in soils presumed to be principally Mollisols and Alfisols in Potawattamie Co., Iowa by Grossman et al. (1992) averaged 12.7 kg C m^{-2}. World wide averages of carbon in cultivated soils as estimated by Post et al. (1982) to be 7.9 kg C m^{-2}, with carbon in most other non-wetland soils also occurring within the same basic range.

In comparison to carbon storage in non-wetland soils, the quantity of carbon stored in coastal marsh soils is very high. Table 2 shows the amount of carbon present in a variety of coastal marsh soils which have been sampled, described, or mapped along the Atlantic and Gulf coasts of the USA. The frequency distribution of carbon density within the various marsh soils from Table 2 are presented in Figure 3. It is immediately clear that the range of carbon densities in tidal marsh soils can be more than an order of

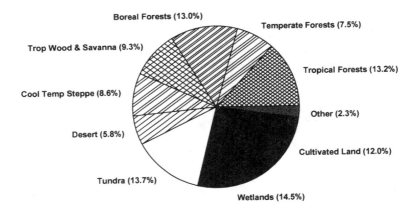

Figure 2. Estimates of carbon stored in soils of the world as distributed among various ecological zones; total carbon = 1395.3 Pg (Pg=10^{15}g). (Data from Post et al., 1982.)

magnitude greater than for non-wetland soils. Some of the extremely high values (>75 kg C m^{-2}) are related to deep deposits of organic materials, some of which approach 3 meters or more. But even when only the upper meters of the marsh soils is considered (Figure 4; the depth to which carbon content is commonly calculated for other soils), the carbon densities in marsh soils still range between three and ten times the levels in other soils, with values mostly depending on the presence and thickness of the C-rich (O and A) soil horizons.

B. Dynamics of Marsh Soil Carbon

The accumulation of organic carbon in tidal marsh soils is related in part to biomass production. Estimates of Reimold (1977) show above ground biomass production in Eastern marshes to range between 0.5 and 3 kg C m^{-2}yr^{-1}. Below ground production is thought even to exceed that formed above ground (Adam, 1990), and is likely to be more significant to carbon accumulation in soils. Long term estimates (author's unpublished data) of carbon accumulation in Atlantic Coast marshes range only between 0.05 and 0.5 kg C m^{-2}yr^{-1}. Most of the plant carbon produced must therefore be utilized at higher trophic levels, exported from the marsh, or decomposed by microorganisms. The efficiency of organic matter decomposition in marsh soils is, however, much lower than in soils which are not saturated for such extended periods.

It is the combination of inefficient decomposition with high primary production, which causes organic carbon to accumulate to high levels in marshes and other soils with peraquic moisture regimes. The contrast in decomposition efficiency between aerobic and anaerobic systems is seen most dramatically in instances where once anaerobic organic soils have been drained and become aerobic. The subsidence associated with the aerobic decomposition of drained organic soils can be as high as 3 cm/year. Thus, a meter or more of peat, which may have taken many centuries to accumulate, can be lost in a matter of decades.

Most soils reach some steady state condition where organic matter additions are approximately equal to losses. As described in the previous section, steady state carbon levels in a variety of soils range between 5 and 15 kg C m^{-2}. The precise quantity of soil carbon maintained under steady state soil conditions is of course dependent on a variety of soil properties in addition to climate, vegetation, and management. Perturbations in these controlling factors have been shown to significantly alter the quantity of carbon stored in the soil.

Figure 5 (curve 1) illustrates the impacts of various natural or man-induced disturbances on carbon levels in soils. Events such as fire (point F) may cause a sudden decrease in the soil C, particularly if the litter layer is consumed. But so long as the forest ecosystem is reestablished, the system should move back toward that level of soil carbon which can be maintained by the steady state condition of the system. Although climatic effects per se are not discussed here, any long term changes in precipitation or temperature which

Table 1. Organic carbon content of five selected soils constituting a drainage sequence, formed from silty Pleistocene and Holocene sediments on the Delmarva Peninsula (Note the dramatic increase in carbon density when the soils aquire a peraquic moisture regime)

Series	Natural soil drainage class	Classification	Moisture regime	Organic C (kg m^{-2} in upper meter)
Matapeake	Well	Typic Hapludult	Udic	5.6
Mattapex	Moderately well	Aquic Hapludult	Udic	4.6
Othello	Poor	Typic Ochraquult	Aquic	6.3
Sunken	Very poor	Typic Ochraqualf	Peraquic	18.1
Honga	Very poor	Terric Sulfihemist	Peraquic	45.9

Table 2. Estimates of carbon stored in selected tidal marsh soils from the Atlantic and Gulf coasts of the USA

Site	kg C m^{-2} in soil	Soil depth (cm)	kg C m^{-2} in upper 1 m	Soil series or other identifier	Soil classification	Source[a]
GA01	31.9	165	25.1	Bohicket	Typic Sulfaquents	1
GA02	49.9	152	45.8	Capers	Typic Sulfaquents	1
LA01	47.5	183	25.4	Salt water peat		2
LA02	30.2	86	34.0	Salt water peat		2
LA03	156.8	274	62.8	Brackish marsh		3
LA04	127.0	244	52.1	Brackish marsh		3
LA05	63.9	188	37.8	Bellpass	Terric Medisaprists	4
LA06	40.2	213	28.3	Scatlake	Typic Hydraquents	4
LA07	68.1	213	32.1	Timbalier	Typic Medisaprists	4
MD01	24.2	70	31.9	ST1	Typic Sulfaquents	5
MD02	52.9	70	86.9	ST3	Terric Sulfihemists	5
MD03	21.1	60	36.9	ST4	Histic Hydraquents	5
MD04	18.3	50	27.2	ST5	Histic Sulfaquents	5
MD05	166.0	350	43.5	ST9	Typic Sulfihemists	5
MD06	42.1	70	49.7	ST12	Terric Sulfihemists	5
MD07	47.4	165	44.2	Honga	Terric Sulfihemists	6
MD08	90.1	203	44.5	Puckum	Typic Medisaprists	6
MD09	89.3	203	42.1	Puckum	Typic Medisaprists	6
MD10	55.1	188	38.1	Bestpitch	Terric Sulfihemists	6
MD11	69.5	183	50.7	Bestpitch	Terric Sulfihemists	6
MD12	28.2	152	26.7	Honga	Terric Sulfihemists	6
NC01	11.1	203	6.9	Carteret	Typic Psammaquents	7
NC02	92.0	165	77.1	Currituck	Terric Medisaprists	7
NC03	190.9	183	85.3	Hobbonny	Typic Medisaprists	7

NH01	71.6	160	TM-12	Typic Sulfihemists	8
NH02	47.3	130	TM-36	Terric Sulfihemists	8
NH03	37.8	130	TM-41	Typic Sulfaquents	8
NH04	56.8	142	TM-42	Terric Sulfihemists	8
NH05	61.5	142	TM-45	Typic Sulfihemists	8
NH06	62.5	155	TM-46	Terric Sulfihemists	8
NH07	62.4	142	TM-51	Typic Sulfihemists	8
NH08	71.6	160	TM-55	Terric Sulfihemists	8
NH09	49.6	132	TM-59	Typic Sulfihemists	8
NH10	49.8	132	TM-69	Terric Sulfihemists	8
NH11	49.7	130	TM-77	Terric Sulfihemists	8
NH12	53.0	119	TM-78	Lithic Sulfihemist	8
NH13	52.3	152	TM-92	Typic Sulfihemists	8
NH14	126.9	343	S70NH-10-1	Typic Sulfihemists	8
NH15	130.9	305	S70NH-10-2	Typic Sulfihemists	8
NH16	59.1	183	S70NH-10-3	Terric Sulfihemists	8
VA01	9.0	102	Magotha	Typic Ochraqualfs	9
VA02	25.3	102	Chincoteague	Typic Sulfaquents	9
VA03	47.9	152	Backbay	Histic Humaquepts	10

[a] 1-Camden and Glynn counties, Georgia, Soil Survey Report; 2-Terrabonne Parish, Louisiana, Soil Survey Report; 3-St. Mary Parish, Louisiana, Soil Survey Report; 4-Lafourche Parish, Louisiana, Soil Survey Report; 5-T.M. Griffin MS thesis, Univ. Maryland, College Park; 6-Dorchester County, Maryland, Soil Characterization Data, National Soil Survey Laboratory; 7-Dare County, North Carolina, Soil Survey Report; 8-Soil Survey of New Hampshire Tidal Marshes, New Hampshire Agric. Exp. Sta. Res. Rept. No.40; 9-Soil Classifications and floral relationships of seaside salt marsh soils, Virginia Agric. Exp. Sta. Bull 85-8; 10-City of Virginia Beach, Virginia, Soil Survey Report.

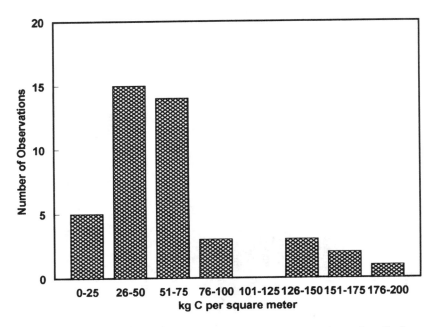

Figure 3. Frequency distribution of carbon stored in selected marsh soils from the Atlantic and Gulf coastal regions of the United States. Individual estimates are shown in Table 2.

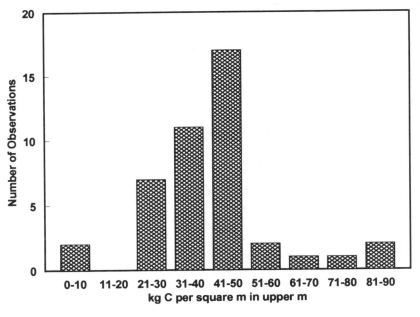

Figure 4. Frequency distribution of carbon stored in *the upper meter* of selected marsh soils from the Atlantic and Gulf coastal regions of the United States. Individual estimates are shown in Table 2.

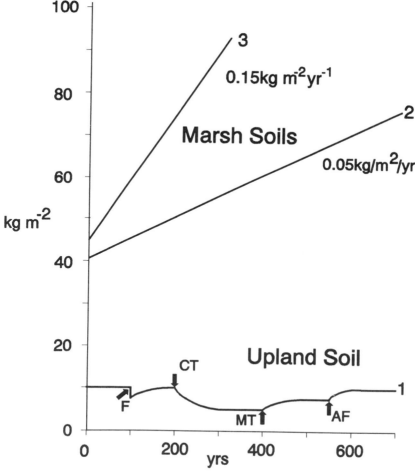

Figure 5. Carbon storage and sequestration in an upland soil (1) under various natural and managed conditions, and in marsh soils (2 and 3) under two rates of carbon sequestration, 0.05 and 0.15 kg C m^{-2} yr^{-1}, respectively. F indicates fire; CT indicates conventional tillage agriculture; MT indicates minimum tillage agriculture; AF indicates agro-forestry.

affect either primary production or rates of decomposition, would be expected to induce some change in the steady state carbon density of a soil.

Work of numerous researchers has demonstrated the significant impact of agricultural management on levels of soil C. Research such as that of Beare and Hendrix (1991) illustrated that land clearing and the cultivation of agricultural crops can cause a decrease in soil carbon by as much as 30 to 70 percent. Figure 5 (point CT) is intended to illustrate the rapid decrease in soil carbon levels which may occur following a shift from a natural system (forest or prairie) to an agricultural system managed under a *conventional tillage* system. Following an initial rapid decline in soil C, the figure shows the system reaching a new steady state level at approximately 50 percent of the original carbon density. If at some point in time, there was a change in agricultural management from conventional tillage to a minimum tillage system (point MT) a shift in the steady state level of soil carbon toward a somewhat higher level would be expected. Additional changes in management or land use from cultivated crops to Agro-Forestry (point AF) might eventually bring the levels of soil carbon near to those sustained under natural conditions (Lal, 1991; Lugo and Brown, 1991). Because of the magnitude of alteration in soil carbon levels which can be induced by management, these concepts have received much recent attention.

This previous discussion concerning management effects on changes in soil carbon is intended to contrast the dynamics of carbon storage in coastal marsh soils. Because coastal marsh accretion generally keeps pace with apparent sea level rise, most coastal marsh soils are accreting at rates of between 2 and 8 mm yr^{-1}. Based on measurements and estimates of soil C, bulk density, and rates of sea level rise, annual rates of carbon sequestration in coastal marsh soils are estimated to be between 0.05 and 0.5 kg C m^{-2} yr^{-1}. Curves representing two rates of carbon sequestration (0.05 and 0.15 kg C m^{-2} yr^{-1}) are illustrated in Fig 5 (curves 2 and 3). The contrast between these two curves and curve 1 (illustrating management induced changes in other soils) is immediately evident. The curves representing the marsh soils are steeply sloping indicating a much more rapid rate of carbon sequestration. Because marsh soils are continually accreting in association with rising sea level, they are therefore continually sequestering carbon at a very high rate. Therefore, on a *per hectare basis*, the long term potential for carbon storage in coastal marsh soils dwarfs that of non-wetland soils.

C. Accretion and Expansion vs Erosion of Marsh Soils

The previous discussion has focused upon the accretionary aspect of marsh dynamics, but the areal extent of marshes must also be considered. Sea level rise generally would be expected to cause lateral landward expansion of coastal marsh areas, especially in regions with extensive low-lying and gently sloping adjacent uplands, such as is common in the lower Delmarva Peninsula and in the Mississippi Delta region. In areas where coastal marshes are circumscribed by more steeply sloping terrain, they would not expand significantly toward the land as sea level rises.

A second important factor which must be addressed is the degree to which coastal marshes are being eroded. Coastal marshes have always been valued as a protective buffer against coastal and shoreline erosion. But it is precisely because the marshes often take the brunt of storm energy that in some areas the marshes themselves are being eroded. In certain marshes of the Blackwater and Nanticoke drainage areas of the Chesapeake Bay, the area covered by marsh vegetation has decreased during the last 50 years and has been replaced by enlarging inland ponds (Stevenson et al., 1985). Similar situations with even more rapid losses of coastal wetlands have been noted in certain Louisiana marshes (Delaune et al., 1983). To understand the significance of eroding marsh soils, the fate of the C-rich sediment must be known. If it is transported into aerobic coastal or estuarine waters, it can be oxidized and the carbon returned to the atmosphere. If the sediment is simply transported to where it accumulates in anaerobic bottom sediments, then the carbon will remain sequestered.

Consideration must also be given to the health and growth of coastal marshes under various sea level rise scenarios. If coastal marshes can keep pace through vertical accretion with a moderately accelerated sea level rise, then they might help to function as a self-regulating mechanism within the global system by continuing to sequester large amounts of C. Some workers, however, have reported that marshes are not effectively keeping pace with present rates of sea level rise and that the areal extent of marshes is actually diminishing (Stevenson et al., 1985). Higher rates of sea level rise, such as those predicted as a result of global warming, might aggravate the situation and cause a more dramatic decline in the extent of coastal marshes.

V. Conclusions

Tidal marsh soils represent a significant store of terrestrial carbon and the carbon density of these soils can be more than an order of magnitude greater than for proximate non-marsh soils. Of perhaps greater significance is the potential for carbon sequestration in accreting tidal marsh soils, where carbon may be stored at rates ranging between 0.05 and 0.5 kg C m^{-2} yr^{-1}. Wherever marsh soil development keeps pace with sea level rise, these soil systems can be expected to function as major carbon sinks. Sea level rise, however has also been implicated in degradation and erosion of tidal marsh soils. Evaluation of the future impact of sea level rise on tidal marsh sequestration of soil carbon must address marsh accretion, erosion, and the final disposition of eroded marsh sediments.

References

Adam, P. 1990. *Saltmarsh ecology*. Cambridge Univ. Press. Cambridge and NY.

Beare, M.H. and P.F. Hendrix. 1991. Possible mechanisms for the accumulation and loss of soil organic carbon in agroecosystems on the Southern Piedmont. p. 43-51. In: Sequestering carbon in soils: A workshop to explore the potential for mitigating global climate change. US EPA Report EPA/600/3-91/031. Corvalis, OR.

Darmody, R.G. and J.E. Foss. 1979. Soil-landscape relationships of the tidal marshes of Maryland. *Soil Sci. Soc. Am. J.* 43:534-541.

Delaune, R.D., R.H. Baumann, and J.G. Gosselink. 1983. Relationships among vertical accretion, coastal submergence, and erosion in a Louisiana Gulf Coast Marsh. *J. Sed. Petrol.* 53:147-157.

Field, D.W., A.J. Reyer, P.V. Genovese, and B.D. Shearer. 1991. Coastal wetlands of the United States: An accounting of a valuable national resource. US Dept. Commerce, NOAA.

Gornitz, V., S. Lebedeff, and J. Hansen. 1982. Global sea level trend in the past century. *Science* 1611-1614.

Grossman, R.B., E.C. Benham, J.R. Fortner, S.W. Waltman, J.M. Kimble, and C.E. Branham. 1992. A demonstration of the use of soil survey information to obtain areal estimates of organic carbon. Am. Soc. Photogram. Remote Sens./Am. Cong. Surv. Map. 92:457-465.

Hicks, S.D., H.A. Debaugh Jr., and L.E. Hickman. 1983. Sea level variations for the United States 1855-1980. U.S. Dept. of Commerce, National Oceanic and Atmospheric Administration, Rockville, MD.

Lal, R. 1991. Managing soil carbon in tropical agro-ecosystems. p. 56-74. In: Sequestering carbon in soils: A workshop to explore the potential for mitigating global climate change. US EPA Report EPA/600/3-91/031. Corvalis, OR.

Lugo, A.E. and S. Brown. 1991. Management of tropical forest lands for maximum soil carbon storage. p.75-77. In: Sequestering carbon in soils: A workshop to explore the potential for mitigating global climate change. US EPA Report EPA/600/3-91/031. Corvalis, OR.

Mitsch, W.J. and J.B. Gosselink. 1993. *Wetlands*. 2nd Ed. Van Nostrand Reinhold, NY.

Mitsch, W.J. and X. Wu. 1994. Wetlands and global carbon budgets - A two-edged sword in ecosystem function and management. *Advan. Soil Sci.* (This Volume).

Pons, L.J., N. Van Breeman, and P.M. Driessen. 1982. Physiography of coastal sediments and development of potential soil acidity. p. 1-18. In: J.A. Kittrick, D.S. Fanning and L.R. Hossner (ed.). Acid sulfate weathering. SSSA Spec. Publ. 10. ASA, Madison, WI.

Post, W.M., W.R. Emanuel, P.J. Zinke, and A.G. Stangenberger. 1982. Soil carbon pools and world life zones. *Nature* 298:156-159.

Post, W.M., T.H. Peng, W.R. Emanuel, A.W. King, V.H. Dale, and D.L. DeAngelis. 1990. The global carbon cycle. *American Scientist* 78:310-326.

Redfield, A.C. and M. Rubin. 1962. The age of salt marsh peat and its relation to recent changes in sea level at Barnstable, Massachusetts. *Proc. Natl. Acad. Sci. U.S.A.* 48:1728-1735.

Redfield, A.C. 1972. Development of a New England salt marsh. *Ecological Monographs* 42:201-237.

Rickard, D.T. 1973. Sedimentary iron sulphide formation. p. 28-65. In: H. Dost (ed.). Acid Sulfate Soils, Vol.I. ILRI Pub. 18. Wageningen, The Netherlands.

Rabenhorst, M.C. and B.R. James. 1992. Iron sulfidization in tidal marsh soils. p. 203-217. In: R.W. Fitzpatrick and H.C.W. Skinner (eds.). Iron and Manganese Biomineralization Processes in Modern and Ancient Environments. Catena Supplement No. 21. Catena Verlag, Cremlingen-Destedt, Germany.

Reimold, R.J. 1977. Mangals and salt marshes of Eastern United States. p. 157-166. In: V.J. Chapman (ed.). *Ecosystems of the world. 1. Wet coastal ecosystems*. Elsevier, New York.

Soil Survey Staff. 1992. Keys to Soil Taxonomy. SMSS Technical Monograph No. 9. 5th Ed. Pocahontas Press, Inc. Blacksburg, VA.

Soil Survey Staff. 1975. Soil Taxonomy: A basic system of soil classification for making and interpreting soil surveys. USDA Agric. Handbook 436. US Govt. Print. Office, Washington, D.C.

Stevenson, J.C., M.S. Kearney, and E.C. Pendleton. 1985. Sedimentation and erosion in a Chesapeake Bay brackish marsh system. *Mar. Geol.* 67:213-235.

Stuiver, M. and J.J. Daddario. 1963. Submergence of the New Jersey coast. *Science* 142:951.

CHAPTER **8**

Spatial Modeling Using Partially Spatial Data

R.B. Jackson IV, A.L. Rowell, and K.B. Weinrich

I. Introduction

This paper describes the properties of four digital data sets that were used in a computer modeling research project. The properties of the data had unexpected implications in data formatting for model input, creation of a modeling methodology, and analysis and interpretation of results. These issues prompted our modeling team to study the data characteristics more closely and ultimately to define individual data (both input and output) as either fully spatial or partially spatial. Fully spatial data consist of the coordinate information that completely defines unique regions of geographic features and their associated attributes. In contrast, partially spatial data define a region of one or more features and their associated attributes, where the feature locations within the region are not known (Figure 1).

A. Project Context

Conversion of land from forest or grassland to agricultural cropland causes releases of carbon (C) to the atmosphere as carbon dioxide (CO_2) (Bouwman, 1990), a loss of the vegetation sink for CO_2 that was present on the land, and other changes in the natural exchange of CO_2, methane (CH_4), and nitrous oxide (N_2O) between the atmosphere and the soil. Continued agricultural operations (tillage, planting, fertilization, irrigation, etc.) further perturb and subsequently control the direction and or magnitude of the CO_2, CH_4, and N_2O gas exchange. Unfortunately, all three of these gases can adversely affect the earth's radiative cooling ability, which in turn affects global climate. CO_2 is of particular concern. It is believed to be responsible for more than half of the predicted changes in the earth's radiative balance (IPCC, 1990). The current increases in atmospheric CO_2 are attributable to fossil fuel use (IPCC, 1990); however, before the 1950's, C losses to the atmosphere (as CO_2) from land use change and crop production on agricultural lands are believed to have exceeded even emissions from fossil fuels (Houghton et al, 1983). Indeed, in many long-term agricultural research plots and commercial croplands, large soil carbon decreases have been reported. Nevertheless, in other situations, soil carbon increases also have been reported (Jenkinson et al.,1987). Since soil C losses or gains on agricultural lands, after the initial loss attributable to the change in land use, are primarily a function of agricultural management practices (Barnwell et al., 1992), two critical questions arise: first, which crops and lands can be managed (rotations, fertilization, tillage, amendments, etc.) to increase C or to not lose C, and second, how much C could be sequestered in this way. To answer these questions, we used a computer modeling approach that relied heavily on available data, two soils models, CENTURY (Parton et al., 1988) and DNDC (Li et al., 1992), and a Geographical Information System (GIS) (ARC/INFO, 1992). Both models simulate soil C and N levels, and emissions of gases containing C and N, from croplands, as a function of climate, soil properties, crops, and agricultural management practices. A detailed description of the research results and conclusions will not be presented here. However, they can be found in Donigian et. al., 1994.

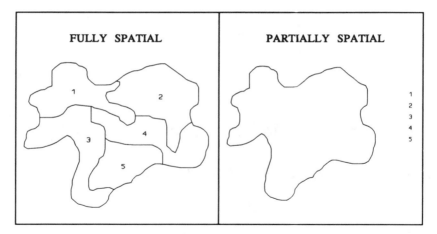

Figure 1. Graphical representation of fully and partially spatial data concepts.

B. Scale of Analysis

Because the soil models we used assume a modeled area of one square meter, fully spatial data would, in theory, be specified on a m^2-by-m^2 basis. In practice however "site-level" analyses have been performed with data at the resolution of a single farm. In keeping with this terminology, this regional study is one whose data are at a coarser level of resolution than the farm. Furthermore, the data are only partially spatial if they represent all farms (or other designated features) within a region and links to a specific farm cannot be derived from the data alone. Since the U.S. Environmental Protection Agency requires answers to global-change-relevant questions (e.g. how is soil C affected by agricultural management ?) for large areas (e.g., entire states or countries) within reasonable time frames and since data at the resolution of the farm are inaccessible at this time, there was really no possibility of modeling all the individual farms (site modeling) within a large region. So we gathered the available databases for the model's input variables and one model output variable (observed crop yields) for the study region. We found that all but one of the databases were partially spatial, that all were at different resolutions, and that the implications of using partially spatial data extended beyond limitations on the GIS to certain limitations on the simulation methods and application of the results. Detailed discussion and presentation of the research results and conclusions from the agroecosystem regional modeling project can be found in Donigian et. al., 1994.

II. Objectives

This paper has two objectives: 1) to describe formatting strategies and techniques employed to obtain a single level of spatial detail in the input and output data, and 2) to introduce the concept of fully versus partially spatial data types, and the implications of their use in spatial modeling.

III. Input Data

As inputs, both models required data on monthly temperatures and precipitation, soil texture and other properties, the specific crop rotations and tillage practices (hereafter referred to as CRT), and other management operations in use in the region (tillage type and frequency, irrigation, fertilization, etc.).

Spatial Modeling Using Partially Spatial Data

A. Climate

The climate information was point data, (weather stations) and therefore was not originally associated with an area, per se. This was the one fully spatial data set. The data were taken from "A Daily Precipitation and Temperature Data Set for the Continental United States" (Wallis et al., 1991). The climate station data were retrieved for the 589 stations that fell within the latitude and longitude limits of our central U.S. study area (lat. 49.00.00 to 32.00.00 and lon. −105.00.00 to −75.00.00). The retrieved data covered 41 years (1948-1988) of monthly total precipitation and monthly minimum and maximum temperature.

B. Soils

All of the soils data were at the resolution of a single county, and all were partially spatial by our previous definition since every county contained more than one soil and the locations of the individual soils within the county could not be discerned from the data alone. Additionally, even if a county had only reported one soil, that county's data would still be partially spatial by our definition unless 1) the soil occupied the entire county area, or 2) the boundaries of that soil within the county could be distinguished. The soil data came from three separate databases: Data Base Analyzer and Parameter Estimator (DBAPE) (Imhoff et al., 1990), The 1982 National Resources Inventory (82-NRI) (Goebel and Dorsch, 1982), and The 1987 National Resources Inventory/Soils-5 database (87-NRI) (Goebel, 1987). Three databases were used because the version of DBAPE we used did not have information for every county in the study region. We next used the 82-NRI, which had data for more of the missing counties, and finally the 87-NRI which had data for the remaining missing counties. The last two databases became available over a time period of 1 year, which meant that, once their data were included, many initial model runs and decisions had already been made based on the DBAPE soil data. For any given county, only DBAPE data were used, or only 82-NRI data were used or only 87-NRI data were used. Thus combining the three soils databases entailed only the sorting of the county's identification code and the federal information processing (fips) codes. Soil data were ultimately available for all 1194 counties in the study region (at county resolutions).

C. Crop Rotations and Tillage

The CRT practices were available to us in much the same format as the soils data, except at a much coarser scale. The scale was an aggregation of counties into 27 agricultural producing regions known as production areas (PA). In each production area, the areas of the various CRTs were known but their respective locations within the PA were not discernable from the data. Thus the CRT data (the distribution of CRTs in the PAs) were also partially spatial. CRT data came from the Center for Agricultural and Rural Development (CARD) in Ames, Iowa. The information was the result of runs of their CEEPES model (CEEPES, 1992) for various agricultural policies that included increases in the use of cover crops, increases in the areas of conservation reserve lands, changes in tillage use, and a status quo scenario.

D. Observed Crop Yields

One of the outputs from CENTURY -- crop yields -- was compared with observed yields for the simulation region for the historical time period of the simulation (≈ 1900 to 1990). Since we used the actual weather for the region for the period 1948-1988, and a representation of that region's climate from 1900 to 1947 (stochastically generated from mean and skewness of the 41-year data record), we expected to be able to simulate the crop yields accurately using the two models. Unfortunately, historical crop yields for the region for the simulation period were not available in electronic form. The data from 1900 to about 1970 were gathered at the state level and were entered manually. The state level data came from the USDA Bureau of Agricultural Economics and the Department of Commerce's Census of Agriculture (USDA/BAE, 1920-1970 ; U.S. DOC/COA, 1920-1970). The yield data from 1971 to 1985 were in electronic form at county resolution. They were obtained from USDA/NASS reel tapes ASB-101 and ASB-102 from (USDA/NASS, 1992).

The status quo information about CRTs would have been more useful if it had been a part of the crop yield information that is collected by the U.S. Department of Agriculture's National Agricultural Statistics Service (USDA/NASS) or the Department of Commerce's Census of Agriculture (U.S. DOC/COA). In that way we would have had a historical CRT database at county resolutions that could have been used to exercise the CEEPES model if we chose to. In any event, USDA/NASS and U.S. DOC/COA receive the necessary information during their data collections to produce fully spatial databases of their agricultural information. However, federal law prevents the transfer of that information as it would disclose the management practices on an individual farm.

IV. Data Congruity

Of the four input data sets just described, only the climate data are fully spatial. The other three are all partially spatial, and are at either county, state, or PA resolutions. Since the models run for an area, we had to determine at which resolution the modeling exercise should be performed, and the best way to transform all of the data layers to that resolution. We knew that, due to the variability in climate, we would have to run the models on areas smaller than a PA but at least as large as our finest resolution database, soils at the county level (Figure 2). Thus all of the data layers would be at a single resolution and be superposable and coincident throughout their feature boundaries. The requirement that the boundaries coincide is a practical restriction of the partially spatial data. For instance, since we do not know where in a region the particular features are, any overlap among regions in different data layers would necessitate the consideration of all possible intersections, which would exponentially increase the required resources and effort with each overlay operation.

A. Area Representation of Point Data (Climate)

The weather station point data had to be transformed to represent areas for the models. Three options for associating an area with a climate station were discussed. Our first option assumed that the model would be run for each of the 1194 counties (the highest resolution of data we had obtained). This prompted the use of a "near" GIS procedure, which selected the climate station nearest to the area centroid of each county for that county's climate station (Figure 3). We then grouped all counties that were assigned to the same climate station, and called this county grouping a climate division (CD). The near procedure resulted in 823 climate divisions (CDs), each of which was still fully spatial since the climate prevailed for the entire area of the county. We were concerned, however, that the number of model runs required would become unmanageable because of the creation of scheduling files. Scheduling files control model flow such as planting and harvesting month, tillage timing, etc. Producing the scheduling files for the model runs was a manual, time-consuming, and error-prone task. The number of model runs per policy scenario then is a product of the number of CDs multiplied by the number of soils present in the CD, multiplied by the number of CRT practices present in the CD. Consequently, the resolution of the analysis was constrained to never reach the resolution of the county level because it was limited by the number of scheduling files that could be created. Three other considerations were the large amount of output that would be generated by runs at this level, the fact that climate would be assigned based on the distance between the county and the weather station, and the fact that all of the CDs were not contiguous along longitude lines or, if contiguous, spanned too wide a latitude range. It was important that the number of scheduling files to be created be kept to a minimum since, if the CRT practices were the same and if the same soils were present in the CD those situations would represent only one model run. However, if the CDs were not contiguous along longitude lines, or if they spanned too wide a latitude range, the potential evapotranspiration calculations (latitude dependent) would require multiple runs.

The second option for transforming the point data to areas was to create climate divisions from the weather data itself. We determined that the models were most sensitive to annual precipitation and maximum temperatures (Burke et al., 1991). Therefore, we computed an average value for precipitation and temperature for each climate station. We then used the GIS to create two contour plots, one plot of the 41-year average monthly maximum temperature and one of the 41-year average annual precipitation. These contour maps were overlaid and areas bounded by either type of contour were the new CDs (areas that we assumed had the same climate) (Figure 4). Various contour intervals were used in the overlay before 2 °C

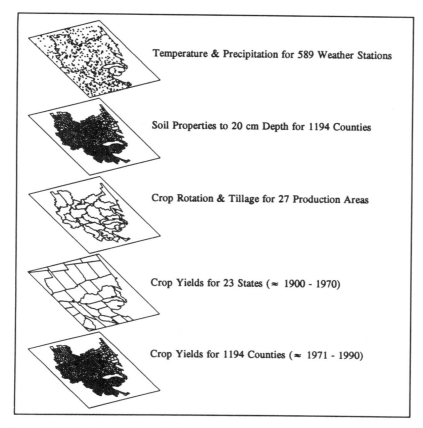

Figure 2. Data layers for analysis.

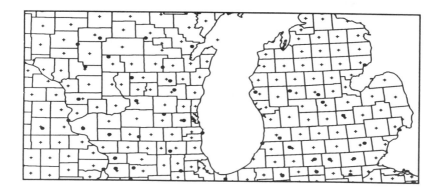

Figure 3. Weather stations (◯) assigned to the nearest county centroids (+); entire study region contains 589 weather stations, 1194 counties.

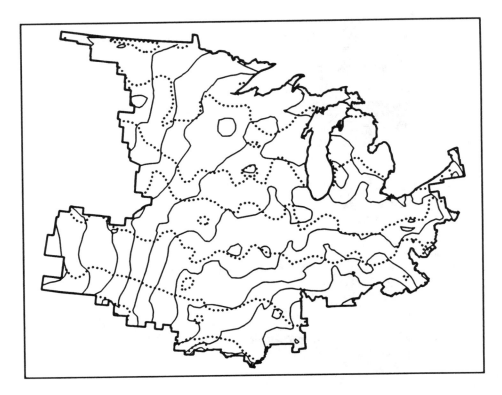

Figure 4. Overlay of average maximum temperature (••••) contour (°C); and average annual precipitation (———) contours (cm).

intervals for temperature and 10-cm intervals for precipitation were chosen (Donigian et al., 1993). This method produced 106 longitudinally noncontiguous CDs of various sizes, and the contours themselves were produced based on distance as had been the case with the "near" procedure. The difference was that the contour distances now were between two points of temperature or precipitation, whereas the near distances were between the weather recording station and the county area centroid. We soon realized that because of the nonspatial nature of our other data layers (i.e., the need to avoid unnecessary multiplication of effort with each overlay), the boundaries of these new CDs had to be forced to follow the boundaries of one of our other data layers. County boundaries were the logical choice, and we had a couple of options to accomplish this. One option was that each county receive the climate files of the CD that occupied a majority of its area, and the other was to manually allocate counties to the CDs based on other criteria (Donigian et al., 1993). Still, however, the regions were longitudinally noncontiguous, and a weather station did not fall within the boundaries of every CD, which meant that every CD did not have a climate file.

Ultimately, our procedure was a variation on the second option. The CDs that were forced to follow county boundaries were drawn manually to ensure contiguous, longitudinally short, roughly equal-sized CDs that did not cross PA boundaries and had at least one set of weather station data that could be assigned to the CD (Donigian et al., 1993) (Figure 5). Thus the CDs were established as the most basic unit to be modeled for this research, which subsequently required that all other input data layers be at CD resolutions.

B. Aggregation (Soils)

The primary soil parameter for input to the models was texture (percent sand, silt, and clay). Both models however, only run for one soil type and CRT scenario at a time, and the soil databases specified not only the various areas of different textures in a county (partially spatial data) but also each texture's different

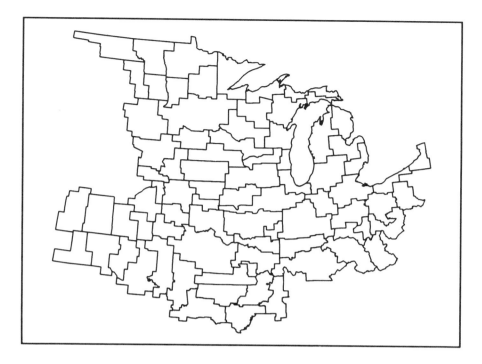

Figure 5. Final 80 climate divisions (CD).

horizon depth and its associated percentage of sand, silt, and clay. The DNDC model was designed to run at a 30-cm depth, while the CENTURY model was designed to represent a 20-cm depth. It was decided that the simulation should model the first 20 cm of soil (Donigian et al., 1993). We accommodated this requirement when more than one layer was needed to reach the 20-cm depth by averaging the high and low ranges on sand, silt, or clay within a layer, and then weighting the respective sand, silt, or clay values in each horizon based on the horizon thickness. We believe the depth-weighted average is a sufficient representation of the soil to 20 cm, particularly if one assumes the soil would be well mixed at that depth as a result of the agricultural operations. The resulting sand, silt, or clay values were then reclassed into one of the twelve texture categories on the Soil Conservation Service texture triangle. Next, the twelve SCS texture categories were reduced to six (Donigian et al., 1993) in order to reduce the total number of runs, and because it was recognized that the models may not significantly distinguish between very similar soil textures (Table 1).

Using the six combined texture categories, and the areas of the individual textures in a county, we aggregated the soil textures up to the CD level simply by adding the areas of like textures for all the counties in the CD. The actual average percentages of sand, silt, and clay that were input to the model for, say, a LOAM soil were determined from a large database of soil samples of LOAM soils that had been collected in the central United States (Flach, 1992).

C. Disaggregation (Production Areas)

The production areas are regions consisting of aggregations of counties. They are the spatial units at which CARD produces CRT distributions according to their economic/policy CEEPES model. The PA resolution data contained the CRT distributions from CARD that resulted under each policy scenario run of their CEEPES model. This information had to be distributed to the CD level for each CRT and policy scenario. The assumption was that the percent distributions of CRTs within the CDs were the same as in the PA in which the CD was a part. This required multiplication by the ratio of a CD area to a PA area. We

Table 1. USDA soil texture categories

12 Texture categories	6 Combined texture categories
Sand	
Loamy sand	Sand/loamy sand
Sandy loam	
Sandy clay loam	Sandy loam/sandy clay loam
Loam	Loam
Silty loam	Silty loam
Silty clay	
Silty clay loam	Silty clay/silty clay loam
Clay	
Clay loam	Clay/clay loam
Silt and sandy clay were not present in the study region	

The study region contains soils of the following suborders: Aquipts, Boralfs, Ochrepts, Orthods, Udalfs, Udults, Udolls, Ustalfs, and Ustolls.

considered ratios of total area, land area, and cropland area, before deciding on CD cropland area divided by PA cropland area.

D. Aggregation and Weighted Disaggregation (Historical Crop Yields)

County-level crop yields (wheat (genus: *Triticum*), grain corn (*Zea mays*), corn silage (*Zea mays*), legume hay, non-legume hay, and soybeans (*Glycine max*)) were averaged for each of the counties in the CD. State-level crop yields were distributed to the CDs based on an area-weighted distribution of the percent of the CD that overlapped the particular states. These data were then combined for each CD (small aggregations of counties) into a continuous record of yield. Initially the disparities in some of the yields between the averaged county data and the distributed states data caused great concern. However, later we realized that our expectations that the data should match were not consistent with the spatial limitations on the data. For instance, why should we have expected crop yields at the county level to be the same as crop yields at the state level? Furthermore, why should county yields match state yields that have been apportioned to counties based on the areas of the counties, as opposed to other criteria such as the counties' weather, or soil type or elevations, etcetera? Eventually, we realized that this discontinuity was simply an artifact of the data that we would have to live with unless we intended to create an entire project out of determining how to disaggregate the information.

V. Summary

We performed an overlay operation to combine the soils and CRT datasets (which were now at CD resolutions also) at the CD level, and to determine the total number of runs. Since both datasets were partially spatial, the overlay became a union (in the mathematical sense, not the GIS command sense) of all potential combinations. Thus the input data required a model run for every possible soil and CRT combination in every CD. Furthermore, once all the data layers had a common boundary, the usefulness of the GIS was greatly diminished, because any subsequent operations could be completed using database software.

The three primary outputs from the models were: soil C storage, crop yield, and GHG emissions. The units on all of the outputs were variable value per area (e.g. g/m^2). Furthermore, there was a value for each of the variables, for every CRT and soil type combination, within each CD, for each year. For display or analysis, the output data could be ranked according to any of the criteria and reported as ranges, or as straight averages for the individual CDs, or as weighted averages. If the output were reported as weighted averages, the weights could be the soil areas or the CRT areas. For example, sequestration potential per area could be reported for a particular soil type weighted across all CRTs, or for a particular CRT across all soils. The simulated crop yields for each CRT and soil in a CD, were in fact weighted across the individual soil areas in the CD for each particular CRT. There was excellent agreement among the simulated and observed yields at the CD level (Donigian et. al., 1993).

The per area results are useful in determining the direction and relative magnitudes of C and N storages and fluxes under various combinations of agricultural management, soil, climate, and CRT. Nevertheless, one important goal of the research project was to estimate the total amounts of C and N stored in the soil, or lost in gaseous emissions. To compute total fluxes, total storages, or total crop weights, the respective areas of CRT and soils would have to be reconciled, or area data would have to be collected at the individual farm level, or one of the partially spatial datasets would have to be eliminated through some type of averaging or weighting (as in the weighting described for the per area values). The reconciliation of areas refers to determining whether, for instance, the total soil area in the soils database for the study region is larger than the CRT areas given by CEEPES in the study region.

We also wanted to extract from the output the impacts of CRT and soil combinations on the model outputs. This was an important consideration since we believe that implementation of new agricultural management, based on this research, would need to be able to locate those farms whose combination of CRTs and soils, are now, or could be, conducive to C storage and GHG emissions abatement. Fortunately, since every available combination of soil and CRT within each CD was modeled, the influences of an individual soil and CRT combination can be determined, but only on a per area basis. The relationship of the totals to their respective CRT and soil combination depends on the method used to compute the totals in the first place. If the totals were computed using new area data that had been collected at the resolution of the farm, the CRT and soil areas would be equal, and therefore identifiable. If the totals had been computed using a weighting across one of the partially spatial datasets, then identification of the influences of an individual CRT and soil combination would not be possible. Finally we are not certain that any type of reconciliation of the soil and CRT areas that we have (without weighting out one of the partially spatial datasets) would allow us to compute totals for which we could identify both the CRT and the soil. Certainly another interpretation for application of the model output is that the grand total C storages and reductions in GHG emissions are valid for the entire CD or study region irrespective of what information can be used at the individual farm level, or delineated for other studies.

VI. Conclusions

For the exercise discussed in this paper, site modeling was the application of the computer models CENTURY and DNDC at the level of an individual farm; whereas, regional modeling was the application of the models for large areas for which data at the level of the individual farm were not available. We have defined the data that were available as either partially or fully spatial, where fully spatial data fill a region of one feature (i.e. region of climate) or a region of more than one feature where each individual feature can be a delineated subregion (not present in our data). In contrast, partially spatial data do not fill a region of one feature or is a region of more than one feature where each feature cannot be delineated into a defined subregion (i.e. different soil types within a county whose locations within the county are not discernable from the data). In general, partially spatial data are often available in tabular form, and may have been produced for purposes that did not require exact locations or a one-to-one correspondence among all of the variables, or laws may restrict the reporting of the parameters necessary -- for instance, a complete location -- to make the data fully spatial. Nevertheless, the information can be valuable and is often all that is available, for regional modeling.

The primary GIS analysis tool that was impacted in this research was the overlay operation. As all of the input data sets were partially spatial except for the climate divisions, any data layer that overlapped any part of another had to be assumed to intersect with all features in both data layers; whether the overlapping layers had one feature or one thousand. Since each unique combination required a model run, a fully spatial soil dataset would have reduced the number of intersections, resulting in fewer runs of the models with the data we had. Since fully spatial data were not available, we required that all of the data layers have common boundaries (county boundaries) in order to minimize the number of unique intersections (model runs).

The inclusion of even one partially spatial dataset can essentially drive the analysis methods, since it has significant impacts on the GIS operations (Table 2). Conversely the inclusion of only one fully spatial data set (as in our analysis), can significantly reduce or increase the necessity of GIS in the overall analysis. The

Table 2. Changes postulated for GIS analysis functions operating on partially spatial data

GIS operation	Postulated changes
Edit	None
Query	None
Overlay	Must consider all possible intersections
Neighborhood	Must execute for an entire region; cannot be executed within regions
Connectivity	Must execute for an entire region; cannot be executed within regions
Output formatting	Varies

motivation for certain methods of analysis then are, in essence, predefined by the type of data available, and the requirements for use of the output.

References

ARC/INFO - *Geographical Information Systems*. 1992. ESRI, Inc. Redlands, Ca. USA.

Barnwell, T.O. Jr., R.B. Jackson, IV, E.T. Elliott, E.A. Paul, K. Paustian, A.S. Donigian, A.S. Patwardhan, A. Rowell, and K. Weinrich. 1992. An Approach to Assessment of Management Impacts on Agricultural Soil Carbon. *Water, Air, and Soil Pollution*. 64:423-435..

Bouwman, A.F. 1990. *Soils and The Greenhouse Effect*. Wiley Publishers. Chichester, New York.

Burke, I.C., T.G. Kittel, W.K. Lauenroth, P. Snook, C.M. Yonker, and W.J. Parton. 1991. Regional Analysis of the Central Great Plains - Sensitivity to Climate Variability. *Bioscience* 41(10):685-692.

CEEPES - *Comprehensive Environmental and Economic Policy Evaluation System*. 1992. Iowa State University Center for Agricultural and Rural Development (CARD). Ames, Iowa 50011.

Donigian, A.S., A.S. Patwardhan, R.B. Jackson, T.O. Barnwell, K.B. Weinrich, and A.L. Rowell. 1994. Modeling the Impacts of Agricultural Management Practices on Soil Carbon in the Central U.S. p. 121-135. In: R. Lal, John Kimble, Elissa Levine, and B.A. Stewart (eds.). *Soil Management and Greenhouse Effect*. Advances in Soil Science, CRC Press, Inc., Boca Raton, FL.

Flach, K. and Aqua Terra Consultants. 1992. Personal communications.

Goebel, J.J. and R.K. Dorsch. 1982. *National Resources Inventory: A Guide for Users of 1982 NRI Data Files*. U.S. Soil Conservation Service, GPO Washington, D.C.

Goebel, J.J. 1987. *National Resources Inventory*. U.S. Soil Conservation Service. Resource Inventory Division. USDA Soil Conservation Service - RIGIS Division. Washington, D.C. 20013.

Houghton, R., J. Hobbie, and J. Melillo. 1983. Changes in the Carbon Content of Terrestrial Biota and Soils Between 1860 and 1980: A Net Release of CO_2 to the Atmosphere. *Ecol. Monogr.* 53:235-262.

Imhoff, J.C., R.F. Carsel, J.L. Kittle Jr., and P.R. Hummel. 1990. *Data Base Analyzer and Parameter Estimator (DBAPE) Interactive Program User's Manual*. EPA/600/3-89/083. U.S. Environmental Protection Agency, Athens, GA.

IPCC - Intergovernmental Panel on Climate Change. 1990. *Climate Change, The IPCC Scientific Assessment*. Cambridge University Press. Cambridge, Great Britain.

Jenkinson, D.S., P.B.S. Hart, J.H. Rayner, and L.C. Parry. 1987. Modelling the Turnover of Organic Matter in Long-Term Experiments at Rothamsted. In: J.H. Cooley (ed.). *Soil Organic Matter Dynamics and Soil Productivity*. INTECOL Bull.

Li, C.S., S.E. Frolking, and T. Frolking. 1992. DNDC, A Model of Nitrous Oxide Evolution from Soil Driven by Rainfall Events: I and II. Model Structure and Sensitivity. *Journal of Geophysical Research* 97(D9):9759-9783.

Parton, W.J., J.B. Stewart, and C.V. Cole. 1988. Dynamics of C, N, P and S in Grassland Soils: A Model. *Biogeochemistry* 5:109-131.

USDA/BAE -U.S. Department of Agriculture/Bureau of Agricultural Economics. 1920-1970. State Crop Yield Estimates. Washington, D.C. 20250.

USDA/NASS - U.S. Department of Agriculture/National Agricultural Statistics Service. 1992. Crop Estimates Reel Tapes ASB-101 & ASB-102. USDA Rm 5809-S Building, Washington, D.C. 20250.

U.S. DOC/COA - U.S. Department of Commerce/Bureau of the Census. 1987. *Census of Agriculture*. U.S. Summary & Data, Vol. 1-2. Geographic Series. U.S. Department of Commerce, Washington, D.C. 1987.

Wallis, J.R., D.P. Lettenmaier, and E.F. Wood. 1991. A Daily Hydroclimatological Data Set for the Continental United States. *Water Resources Research* 27(7):1657-1663.

CHAPTER 9

Effects of Global Change on Carbon Storage in Cold Soils

Walter C. Oechel and George L. Vourlitis

I. Introduction

Northern ecosystems (arctic, boreal forest, and northern bogs) contain an estimated 350 to 455 Pg of C (10^{15} g) in the permafrost and soil active layer (Miller, 1981; Miller et al., 1983; Billings, 1987; Post, 1990; Gorham, 1991; Oechel and Billings, 1992), which is equivalent to 25-33% of the total world soil C pool (Billings, 1987). Arctic tundra ecosystems alone contain about 192 Pg of soil C, or approximately 14% of the global soil C pool (Billings, 1987).

Net carbon balance is the difference between the C accumulated by photosynthesis minus losses due to plant and soil respiration and relatively small losses to herbivory. Net primary productivity often exceeds heterotrophic respiration in northern ecosystems, due to the predominance of cold, wet soils (Gorham, 1991; Oechel and Billings, 1992). As a result, these ecosystems in the historic and recent geologic past have been net sinks for carbon with respect to the atmosphere of about 0.1 to 0.3 Pg C per year (Schell, 1983; Schell and Ziemann, 1983; Marion and Oechel, 1993; Miller et al., 1983; Post, 1990; Gorham, 1991; Oechel and Billings, 1992). Tussock tundra, which is the largest store of carbon in arctic terrestrial ecosystems (29.1 Pg C), was estimated to be accumulating carbon at the rate of 23 g C m^{-2} y^{-1} (0.02 Pg C per year worldwide, Miller et al., 1983). Wet sedge tundra was estimated to contain 14.4 Pg C worldwide, and to be accumulating carbon at the rate of 27 to 120 g m^{-2} y^{-1}, or approximately 0.03-0.13 Pg C worldwide (Miller et al., 1983; Oechel and Billings, 1992; Chapin et al., 1980; Coyne and Kelley, 1975). Although northern ecosystems are thought to have been on balance, net sinks for carbon throughout history, they are generally known to be net sources of CH_4 to the atmosphere due to the predominantly waterlogged soils (Sebacher et al., 1986; Whalen and Reeburgh, 1992; Vourlitis et al., 1993a).

High-latitude ecosystems are anticipated to undergo the most significant increases in surface temperature with a doubling of atmospheric CO_2. Surface temperature increases of 4 °C in summer and as much as 17 °C in winter have been predicted for northern latitudes (Mitchell et al., 1990; Schlesinger and Mitchell, 1987). Although clouds and water vapor may feed back and limit the increase by as much as 50%, there is a general consensus of significantly higher temperatures in the arctic.

Global warming may already be occurring in arctic Alaska. Thermal profiles of permafrost and surface temperature records from remote weather stations indicate a temperature rise across the north slope of Alaska of 2-4°C within the last century (Lachenbruch and Marshall, 1986), and most likely within the last few decades (Oechel et al., 1993; Chapman and Walsh, 1993). Weather records also indicate a distinct warming trend in much of the Canadian arctic and boreal forest (Hengeveld, 1991) and in the former USSR (Kukla and Karl, 1992).

High-latitude ecosystems are particularly vulnerable to climate change due to the large C stocks in northern latitude soils and the predominance of permafrost. Climate change could result in higher temperatures and altered precipitation which could increase the depth of the soil active layer and water table, resulting in warmer soil temperatures, greater soil aeration, and higher rates of soil decomposition (Silvola, 1986; Post, 1990; Gorham, 1991; Billings et al., 1982, 1983, 1984; Billings and Peterson, 1992). If soil decomposition increases more rapidly than net primary production, the system ceases to be a carbon sink and becomes a carbon source. Alternatively, a concomitant increase in nutrient mineralization should be realized from enhanced soil decomposition (Post, 1990), which should lead directly to greater plant growth (Shaver and Chapin, 1980; Chapin and Shaver, 1985) and ecosystem productivity (Oechel and Billings, 1992).

There are many uncertainties in predicting the long-term effect of global change on arctic ecosystems. The new equilibrium level of ecosystem productivity, and the resulting C balance, will depend upon a variety of factors such as plant genetic constraints, plant competition, resource availability, and site soil water content (Billings et al., 1982, 1983, 1984; Oechel and Strain, 1985; Chapin and Shaver, 1985; Oberbauer et al., 1986b; Tissue and Oechel, 1987; Oechel and Billings, 1992; Oechel et al., 1993). Over longer time intervals, invasion of shrub or tree species may act to increase above-ground C storage (Oechel and Billings, 1992, Oechel et al., 1993; Marion and Oechel, 1993).

This paper discusses recent changes in ecosystem CO_2 flux in arctic Alaska, and environmental controls on both CO_2 and CH_4 flux in tussock and wet sedge tundra ecosystems. The short- and long-term effects of elevated CO_2 and temperature are discussed in regard to the possible future C flux dynamics in arctic ecosystems subjected to high-latitude warming and associated climate change.

II. Carbon Balance Measurements

A. Locations of Research Sites

Carbon balance measurements were made between 1990 and 1992 along a 200 km long latitudinal transect on the North Slope of Alaska between Prudhoe Bay and Barrow in the North and Toolik Lake to the South (Table. 1). Sites to the south (Toolik Lake, Happy Valley, and Sagwon Hills) represent *Eriophorum vaginatum*-dominated tussock tundra, while sites to the north represent wet sedge tundra dominated by *Eriophorum angustifolium, Carex aquatilis,* and *Dupontia fischeri* (Walker and Acevedo, 1987). Soils are classified as Ruptic-Histic-Pergelic Cryaquepts (Rieger et al., 1979) with an Oe horizon of partially decomposed organic matter. The maximum depth to permafrost (i.e. active layer depth) varies between 25-40 cm in the wet sedge sites and 40-60 cm in the tussock tundra sites.

B. CO_2 Flux Measurements

Flux measurements were made with a closed gas exchange system utilizing a 257 L cuvette and a LI-COR 6200 photosynthesis system (Vourlitis et al., 1993 b). Each site was affixed with 4-6, 0.5 m^2 polycarbonate chamber bases during mid-May of each year using a chain saw. Plants were still dormant during the base installation procedure, thus minimizing damage to roots and rhizomes. The bases provided the air-tight seal required between the chamber and the plot during net flux measurements.

Plots were sampled every 1-1.5 hours over a 24- or 48-hour period, and diurnal flux measurements were repeated every 5-7 days throughout the growing season at each site. Seasonal flux was calculated by integrating daily totals over the period of snow melt to the first significant snowfall, which is assumed to correlate with soil freezing. Measurements were made between early June and mid-September, corresponding to the beginning and end of season dates.

Recent data indicate that up to 13.8 g C m^{-2} is lost between December and February from arctic ecosystems in NE Russia (Zimov et al., 1993). Because the measurements reported here were made during the snow-free period only, they probably underestimate the annual CO_2 loss to the atmosphere (Oechel et al., 1993; Zimov et al., 1993).

C. CH_4 Flux Measurements

Whole ecosystem methane flux was sampled at each site using the chamber described above (Vourlitis et al., 1993a and b). During each sampling event, the chamber was sealed to the permanent base for up to 20 minutes, and covered with an opaque tarp to inhibit heating of the soil and chamber atmosphere. Gas

Table 1. Characteristic of research sites and locations on the North Slope of Alaska

Site	Latitude	Longitude	Distance (km)[a]	Tundra type	Moisture
Barrow	71°18'	156°40'	2	Wet sedge	Wet-dry
Prudhoe Bay-wet	70°22'	148°45'	2.5	Wet sedge	Wet
Prudhoe Bay-moist	70°22'	148°45'	2.5	Wet sedge	Moist
APL 133.3-wet	69°50'	148°45'	50	Wet sedge	Wet
APL 133.3-moist	69°50'	148°45'	50	Wet sedge	Moist
APL 133.3-drained	69°50'	148°45'	50	Wet sedge	Moist
Sagwon Hills	69°25'	148°45'	100	Tussock	Moist
Happy Valley	69°08'	148°50'	140	Tussock	Moist
Toolik Lake	68°38'	149°35'	200	Tussock	Moist

[a] Distance refers to the distance to the Arctic Ocean.

samples were taken at 0 (immediately after the chamber was affixed to the base), 5, 10, and either 15 or 20 minutes following chamber installation by withdrawing approximately 10 ml of gas using B-D Glaspak™ syringes affixed with B-D 21 gauge needles. Gas samples were taken every 6-8 hours over a 24-hour period during each sampling event, and immediately returned to our lab located in Prudhoe Bay for analysis. The concentration of methane in each gas sample was determined by injecting exactly 1 ml of gas into a Shimadzu GC Mini-2™ gas chromatograph affixed with a flame ionization detector, and a MS 5A 80/100 column. The column temperature was set at 90 °C, with a carrier gas (N_2) flow rate of 0.5 L min^{-1}. Sample methane concentration was calculated from a standard curve using 10.3 and 9.03 mg kg^{-1} standards (Scott Specialty Gases). Methane flux was calculated as the rate of concentration change over the 15 or 20 minute sampling duration (Vourlitis et al., 1993a and b).

Annual methane flux was calculated by integrating the area under the daily methane flux curve over the duration of the growing season. Daily flux values for the first and last days of the growing season were arbitrarily defined to equal zero, as actual methane flux was not sampled on these dates.

D. Response of Tussock Tundra to Elevated CO_2 and Temperature

The direct effects of elevated CO_2 and temperature on tussock tundra physiology and growth were determined at Toolik Lake between 1983 and 1985 using null-balance, controlled-environment greenhouses (Prudhomme et al., 1982; Oechel et al., 1992). Experimental plots were 1.5 m^2 in area, and were subjected to either ambient CO_2 concentrations (340 ppm), 2-times ambient (680 ppm), or 2-times ambient plus a 4 °C increase in chamber temperature with 3-fold replication (Oechel et al., 1992). Plots were left in the ground during the 3-year period, and a given plot was subjected to the same treatment each growing season. Carbon dioxide flux was measured continuously as the amount of CO_2 injected into the chamber to maintain its set-point minus the amount of CO_2 scrubbed (Oechel et al., 1992).

III. Carbon Flux Dynamics

A. Recent Change in CO_2 Flux

Recent measurements of CO_2 flux in arctic Alaska indicate that tussock and wet sedge tundra are now sources of CO_2 to the atmosphere (Table 2, Oechel et al., 1993). Tussock tundra ecosystems, which were previously estimated to be accumulating 23 g C m^{-2} yr^{-1} (Miller et al., 1983), are now losing approximately 112 g C m^{-2} yr^{-1}, corresponding to a change of 587% over the last 1-2 decades (Oechel et al., 1993). Similarly, wet sedge ecosystems are now either slight sources of CO_2 to the atmosphere, or roughly in balance (Table 2). Compared to earlier estimates of C balance using harvest, cuvette, and aerodynamic techniques (Miller et al., 1983; Chapin et al., 1980; and Coyne and Kelley, 1975), the current sink strength of these ecosystems has diminished by approximately 120% over the last 2 decades.

Table 2. Current and previous estimates of carbon balance in the major tundra ecosystems on the North Slope of Alaska (Data are means ±1 SE; positive flux values indicate CO_2 loss to the atmosphere, while negative values indicate ecosystem accumulation of CO_2)

Ecosystem	Previous	Current	% Change
	------g C m^{-2} yr^{-1}------		
Tussock	-23^a	112 ± 43^c	587
Wet sedge	-62 ± 30^b	11 ± 11^c	118

[a] Data from Miller et al., 1983; [b] data averaged from Miller et al., 1983; Chapin et al., 1980; and Coyne and Kelley, 1975 (n = 3); [c] data from Oechel et al., 1993 and Oechel et al., unpublished data (n = 3 sites).

These data indicate a change in carbon balance of tundra ecosystems with respect to the atmosphere, possibly in response to recently reported trends in high latitude warming (Lachenbruch and Marshall, 1986; Hengeveld, 1991; Oechel et al., 1993; Chapman and Walsh, 1993; Kukla and Karl, 1992). Due to the greater sensitivity of permafrost dominated high latitude ecosystems to climate change (Chapin et al., 1980; Oechel and Billings, 1992; Oechel et al., 1993), the change in C balance observed in arctic Alaska may represent the first indication of the effects of climate change on ecosystem function.

B. Environmental Controls of CO_2 and CH_4 Flux

The decline in ecosystem sink strength may be due more to the indirect effects of elevated temperature on site soil water content than by temperature directly (Oechel et al., 1993). For example, surface soils of the IBP-II sites near Barrow, Alaska were described to be saturated during the IBP effort of the 1970s (Gersper et al., 1980; W. C. Oechel, pers. observ.); however, they are now unsaturated for the majority of the growing season (Oechel et al., unpublished data).

The magnitude of CO_2 efflux on the North Slope of Alaska was found to increase from north to south, primarily due to the temperature and soil water content gradient which is superimposed on this latitudinal transect (Figure 1). Sites with lower soil water content and higher temperature, such as the tussock tundra sites to the south were greater sources of CO_2 to the atmosphere, but weaker sources of CH_4 (Figure 1). In contrast, sites with greater soil water content, such as the wet sedge tundra to the north, were weaker sources of CO_2 (or approximately in balance) but were greater sources of CH_4 to the atmosphere (Figure 1).

Species composition and soil properties differ along the latitudinal gradient, thus making it difficult to attribute the trend in C flux to the soil water content and temperature gradient alone. Inter-annual variability in soil water content and temperature, however, can be used to assess to the effects of these variables on C flux dynamics. The 1991 growing season was generally wetter and cooler than 1990 on the North Slope of Alaska (Figure 2). Inter-annual variability in air temperature and soil water content resulted in significant differences in CO_2 and CH_4 flux in wet sedge ecosystems, and only slight differences in tussock tundra fluxes (Figure 3). Wet sedge ecosystems were net sinks for atmospheric CO_2 and were greater sources of CH_4 during the wetter and cooler 1991 growing season (Figure 3). Tussock tundra ecosystems remained significant sources of CO^2 to the atmosphere and were slightly higher sources of CH_4 during the 1991 growing season (Figure 3). These data suggest that C balance in wet sedge ecosystems is more sensitive to changes in temperature and site soil water content than tussock tundra ecosystems.

Variability in soil water content within a given site produces a pattern similar to the CO_2 and CH_4 flux observed along the latitudinal transect and between consecutive years (Figure 4). At the Franklin Bluffs site, for example, net annual CO_2 flux decreased logarithmically with increased soil water content, while net annual CH_4 flux increased logarithmically with increased soil water content (Figure 4).

High-latitude warming and associated climate change can potentially cause changes in ecosystem water balance which could ultimately affect CO_2 (Billings et al., 1982, 1983, 1984; Post, 1990; Oechel and Billings, 1992; Oechel et al., 1993) and CH_4 flux dynamics (Sebacher et al., 1986; Whalen and Reeburgh, 1992; Vourlitis et al., 1993a; Torn and Chapin, 1993). These indicate the close relationship between C

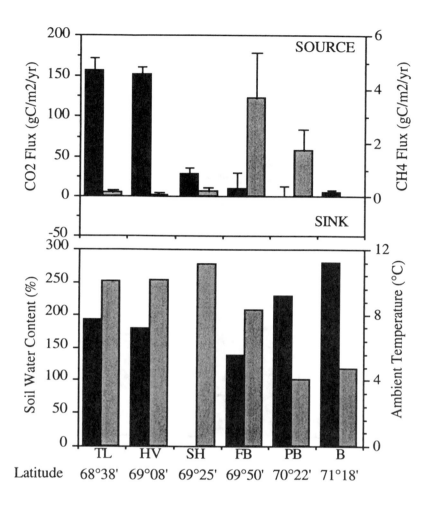

Figure 1. Net ecosystem CO_2 (closed bars) and CH_4 (stippled bars) flux (top panel), and (bottom panel) percent soil moisture (closed bars) and temperature (stippled bars) along a latitudinal gradient on the North Slope of Alaska between 1990-1992. Data for the top panel are means ± 1 SE, n (corresponding to the number of plots measured at each site) = 4-6 for TL (Toolik Lake) and HV (Happy Valley), n = 12-18 for FB (Franklin Bluffs), n = 8-12 for PB (Prudhoe Bay), n = 4 as SH (Sagwon Hills), and n = 30 for B (Barrow). Positive flux values indicate net efflux to the atmosphere. CH_4 flux and soil moisture were not determined at Barrow and Sagwon Hills, respectively.

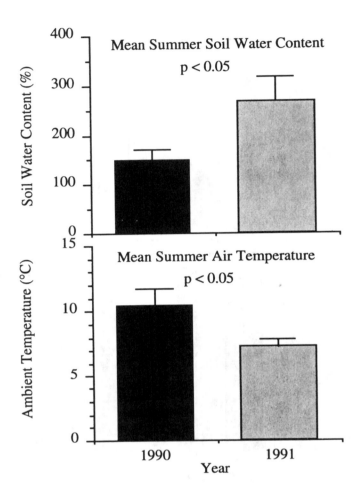

Figure 2. Average seasonal soil moisture (top panel) and air temperature (bottom panel) during the 1990 (closed bars) and 1991 (stippled bars) growing seasons. Data are means ± SE, n (corresponding to the number of sites sampled throughout the growing season) = 7 for soil moisture and 5 for air temperature. Differences between years were calculated using a paired t-test, df = 5 for soil moisture and 3 for air temperature.

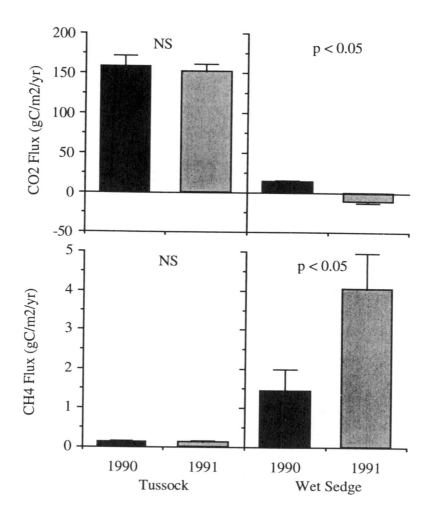

Figure 3. Average seasonal CO_2 (top panel) and CH_4 flux (bottom panel) in tussock tundra and wet sedge ecosystems during the 1990 (closed bars) and 1991 (stippled bars) growing seasons. Data are means ± SE, n (corresponding to the number of sites sampled throughout the growing season) = 2 for tussock and n = 5 for wet sedge tundra. Differences between years were analyzed using a paired t-test, df = 1 for tussock and df = 4 for wet sedge ecosystems. NS = not significant.

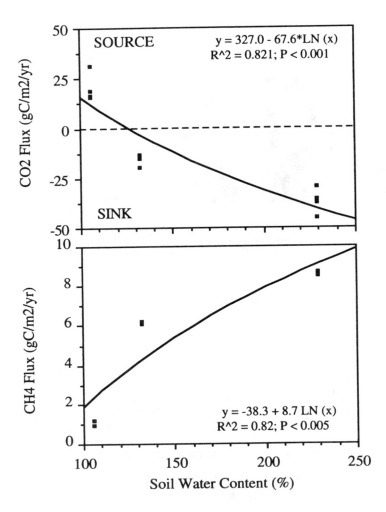

Figure 4. Net annual CO_2 (top panel) and CH_4 flux (bottom panel) as a function of soil moisture at Franklin Bluffs during the 1991 growing season. Positive values indicate net loss of C to the atmosphere. n (corresponding to the number of plots repeatedly measured throughout the growing season) = 12 for CO_2 and 6 for CH_4 flux, df = 10 for CO_2 and 4 for CH_4.

storage and soil water content status in high latitude ecosystems. The relationship between flux and temperature, however, is more difficult to define. Historical data suggest that higher temperatures should result in greater ecosystem productivity (Billings, 1987; Oechel and Billings, 1992; Marion and Oechel, 1993), while the latitudinal and inter-annual data displayed here indicate the opposite trend. Field studies of tussock tundra exposed to elevated temperature with water table held constant indicate that net influx increases under higher temperatures (Oechel and Billings; 1992), however, studies from laboratory microcosms of wet sedge tundra indicate that net CO_2 flux is more a function of water table depth than temperature (Billings et al., 1982, 1983, 1984).

We feel that the pattern of C balance along the latitudinal transect is more a function of site water content and community composition than the direct effects of temperature alone (Oechel et al., 1993). The data describing the effects of inter-annual variability in soil water content and temperature on net C balance indirectly support this hypothesis. If the inter-annual variability of temperature and soil water content are analyzed for tussock and wet sedge tundra separately, we find that only temperature in the tussock tundra is significantly different between seasons (paired t on LN transformed data = 4.515, df = 2, p = 0.05), while only soil water content in wet sedge tundra was significantly different between seasons (paired t on LN transformed data = -3.17, df = 4, p < 0.05). Since C balance in tussock tundra was no different between 1990 and 1991, the reduced temperature observed during 1991 apparently had little effect on net CO_2 flux. In wet sedge ecosystems, however, the significantly greater soil water content observed during 1991 resulted in significantly greater CO_2 influx and CH_4 efflux, which is in support of results from both laboratory microcosms and field observations describing the effects of water balance on both CO_2 (Billings et al., 1982, 1983, 1984) and CH_4 flux (Sebacher et al., 1986; Vourlitis et al., 1993a).

These data have important implications regarding the feedback of arctic ecosystems on the global CO_2 and CH_4 budget. It is widely known that CO_2 and CH_4 are radiatively active gases which directly contribute to global warming by absorbing outgoing tropospheric radiation (Wigley, 1987). While CO_2 is an extremely efficient greenhouse gas, CH_4 is approximately 20 times more reactive than CO_2 (Ramanathan et al., 1985; Lashof and Ahuja, 1990). Changes in annual precipitation, runoff, and soil water content predicted for northern latitudes have the potential to affect the rate of CO_2 and CH_4 efflux, which could ultimately affect the role of arctic ecosystems in the global C budget. Because of the large C stores in arctic soils, if the current efflux of CO_2 from the arctic continues, these ecosystems could represent a strong positive feedback on global atmospheric CO_2 concentration and concomitant climate change.

IV. Ecosystem Response to Elevated CO_2 and Temperature

A. Short-term Response

Tussock tundra exposed to elevated CO_2 and ambient temperatures exhibited a loss of photosynthetic capacity, while plots exposed to both elevated CO_2 and temperature retained the enhanced photosynthetic capacity over the duration of a 3-year experiment in environmentally controlled chambers at Toolik Lake, Alaska (Grulke et al., 1990; Oechel et al., 1992; Oechel et al., unpublished data). Elevated CO_2 significantly increased net ecosystem C accumulation during the first year of exposure, while ambient plots showed net C loss. This stimulation was found to decrease by the second year, and by the third year, plots exposed to elevated CO_2 exhibited rates of net C loss that were similar to ambient plots. Plots that were exposed to both elevated CO_2 and temperature, however, were net sinks for atmospheric CO_2 throughout the 3 years of experimentation.

These results indicate that initial stimulation of tussock tundra ecosystems to elevated CO_2 is transient, with homeostatic adjustment occurring rapidly (within the first 1-3 years) after exposure to elevated CO_2. Prolonged enhancement of photosynthesis by elevated CO_2 generally requires that nutrients are abundant (Oechel and Strain, 1985; Oechel and Billings, 1992). Plant growth and photosynthesis is generally limited by nutrient availability (particularly N and P) in arctic ecosystems (Shaver and Chapin, 1980; Chapin and Shaver, 1985; Oberbauer et al., 1986b). However, fertilization usually acts to stimulate tissue production rather than increasing leaf photosynthesis (Shaver and Chapin, 1980; Chapin and Shaver, 1985; Bigger and Oechel, 1982), and carbohydrate supply may actually exceed plant need (Tissue and Oechel, 1987).

Carbohydrate sufficiency in *Eriophorum vaginatum* and many other arctic plants may explain the rapid ecosystem level homeostatic adjustment to elevated CO_2 (Tissue and Oechel, 1987; Grulke et al., 1990; Oechel and Billings, 1992). Without adequate sinks, accumulated carbohydrates may feed back to reduce the amount of RuBP carboxylase, the enzyme responsible for C fixation, resulting in a concomitant decrease in photosynthesis (Tissue and Oechel, 1987).

Northern latitude warming may lead to greater thaw depth and drainage of upper soil layers (Oechel and Billings, 1992; Oechel et al., 1993), which would act to enhance soil aeration, decomposition, and mineralization (Post, 1990; Marion and Miller, 1982; Marion and Black, 1987). Increased nutrient availability may have stimulated sink activity and strength in tussock plots exposed to elevated CO_2 and temperature (Lawrence and Oechel, 1983a,b; Kummerow and Ellis, 1984), allowing for the utilization of excess carbohydrate reserves, and elevated CO_2 (Tissue and Oechel, 1987; Grulke et al., 1990).

Elevated temperature may also directly increase sink strength and activity, as growth of some arctic plants is stimulated by moderately high temperatures (Limbach et al., 1982; Kummerow and Ellis, 1984; Chapin and Shaver, 1985). Hence, increased temperatures predicted for northern latitude ecosystems may either directly or indirectly stimulate ecosystem net primary production.

B. Long-Term Response

Analysis of carbon content in soil cores indicates that C accumulation of wet sedge tundra ecosystems during the warmer hypsithermal period of the mid-Holocene was significantly greater compared to the cooler late-Holocene (Billings, 1987; Ovendon, 1990; Marion and Oechel, 1993). During the warmer mid-Holocene, C accumulation rates of wet sedge ecosystems near Prudhoe Bay were approximately 6.7 g C m^{-2} yr^{-1} compared to 1.2 g C m^{-2} yr^{-1} during the cooler late-Holocene (Marion and Oechel, 1993). These data suggest that C accumulation should, at least in the longterm, increase with high-latitude warming.

Billings (1987) described vegetation changes in sub-arctic Alaska between the middle and late-Holocene. During the mid-Holocene, the vegetation was composed primarily of a *Picea-Betula-Alnus* forest; however, during the late-Holocene, Picea declined in abundance and the vegetation became more similar to the modern boreal forest flora (Billings, 1987). Colinveaux (1964) documented the presence of *Betula* pollen in cores from the Barrow, Alaska region during the warmer mid-Holocene period, indicating the presence of tussock tundra and/or boreal forest flora in the modern high arctic region.

These data suggest that the observed increase in C accumulation during the warmer mid-Holocene period was accompanied by shifts in species composition, with the migration of some boreal forest flora into modern day arctic regions. Plant species respond differently to elevated CO_2 and temperature, which would result in differential growth (Oberbauer et al., 1986a; Tissue and Oechel, 1987) and significant changes in competitive relationships (Bazzaz, 1990). Over these longer time scales, population changes resulting from these differential responses may alter species composition and whole ecosystem C balance (Oechel and Billings, 1992).

V. Conclusions

There are many uncertainties in predicting the long-term effect of global change on arctic ecosystems. The new equilibrium level of ecosystem productivity, and the resulting C balance, will depend upon a variety of factors such as plant genetic constraints, plant competition, resource availability, and site soil water content. In the medium term, if future conditions are warmer and drier, arctic ecosystems are likely to become even stronger sources of CO_2 and weaker sources of CH_4. In contrast, if future conditions are warmer and wetter, arctic ecosystems could become weaker sources of CO_2 (or even sinks) and greater sources of CH_4 to the atmosphere.

It is clear that high-latitude ecosystems are particularly susceptible to global warming and associated climate change due to the large soil C stocks and the importance of permafrost, temperature, and soil water content in controlling rates of biological activity. High-latitude warming and climate change are likely already occurring, which may explain the recent change in net ecosystem C flux reported in arctic Alaska.

If the present scenario of C loss persists, arctic ecosystems may represent a significant positive feedback on global atmospheric CO_2 and climate change. Alternatively, high latitude warming may increase nutrient mineralization rates and plant sink strength which should result in greater ecosystem productivity. Under the latter conditions, arctic ecosystems could represent a negative feedback on global atmospheric CO_2 and climate change. Over longer time scales, invasion of shrub or tree species may act to increase above-ground C storage. There is currently no evidence that these mechanisms are acting, or will act quickly enough to reduce an initial, large net efflux of CO_2, and it seems likely that significant amounts of C will be lost before other factors reincorporate this lost C back into arctic ecosystems as plant tissue and/or soil organic matter.

VI. Summary

The carbon balance of arctic ecosystems has changed over the last two decades possibly due to the recently reported high latitude warming. Tussock tundra ecosystems are now strong sources of CO_2 to the atmosphere while wet sedge tundra is either a weak source or approximately in balance. In contrast, tussock tundra is a weak source of CH_4 to the atmosphere, while wet sedge tundra is a much stronger source. The carbon balance of these ecosystems is strongly controlled by soil water content and weakly controlled by soil or ambient temperature. Hence, the change in carbon balance is probably due more to the indirect effect of high-latitude warming on soil water content rather than the direct effect of elevated temperature alone.

The future carbon balance of arctic systems will depend on how high latitude warming effects regional climate patterns, and in turn, how the rates of soil and whole ecosystem processes are affected by this climate change. Evidence suggests that ecosystem level productivity should eventually increase due to high-latitude warming; however, increased productivity may have to be accompanied by changes in species composition which occur at much longer time scales than rates of soil microbial processes. As a result, net carbon loss is expected to occur over the short- to mid-term until rates of ecosystem productivity increase enough to counteract the increase in soil decomposition rates and carbon loss.

Acknowledgments

This research was funded by the US Department of Energy (grant number DE-FG03-86ER60479), with additional support provided by the North Slope Borough Department of Wildlife Management, the North Slope Borough Service Area 10, and ARCO Department of Environmental Compliance.

References

Bazzaz, F.A. 1990. The response of ecosystems to global change: research agenda. 117-124. Global Climate Feedbacks. DOE CONF - 90006134.

Bigger, C.M. and W.C. Oechel. 1982. Nutrient effect on maximum photosynthesis in arctic plants. *Holarct. Ecol.* 5:158-163.

Billings, W.D. 1987. Carbon balance of Alaskan tundra and taiga ecosystems: past, present and future. *Quart. Sci. Rev.* 6:165-177.

Billings, W.D. and K.M. Peterson. 1992. Some possible effects of climatic warming on arctic tundra ecosystems of the Alaskan North Slope. In: R.L. Peters and T. Lovejoy (eds.). Consequences of the Greenhouse Effect for Biological Diversity. New Haven, CT, Yale University Press.

Billings, W.D., J.O. Luken, D.A. Mortenson, and K.M. Peterson. 1982. Arctic tundra: A source or sink for atmospheric carbon dioxide in a changing environment? *Oecologia (Berlin).* 53:7-11.

Billings, W.D., J.O. Luken, D.A. Mortenson, and K.M. Peterson. 1983. Increasing atmospheric carbon dioxide: possible effects on arctic tundra. *Oecologia (Berlin).* 58:286-289.

Billings, W.D., K.M. Peterson, J.D. Luken, and D.A. Mortenson. 1984. Interaction of increasing atmospheric carbon-dioxide and soil nitrogen in the carbon balance of tundra microcosms. *Oecologia (Berlin).* 65:26-29.

Chapin, F.S. III. and G. Shaver. 1985. Individualistic growth response of tundra plant species to environmental manipulations in the field. *Ecology.* 66:564-576.

Chapin, F.S. III., P.C. Miller, W.D. Billings, and P.I. Coyne. 1980. Carbon and nutrient budgets and their control in coastal tundra. p. 458-484. In: J. Brown, P.C. Miller, L.L. Tieszen, and F.L. Bunnell (eds.). An Arctic Ecosystem: The Coastal Tundra at Barrow, Alaska. Stroudsburg, PA, Dowden, Hutchinson and Ross.

Chapman, W.L. and J.E. Walsh. 1993. Recent variations of sea ice and air temperatures in high latitudes. *Bull. Amer. Met. Soc.* 74:33-47.

Colinveaux, P. A. 1964. Origin of ice ages: Pollen evidence from arctic Alaska. *Science.* 145:707-708.

Coyne, P.I. and J.J. Kelley. 1975. CO_2 exchange over the Alaskan arctic tundra: Meteorological assessment by an aerodynamic method. *J. Appl. Ecol.* 12:587-611.

Gersper, P.L., V. Alexander, S.A. Barkley, R.J. Barsdate, and P.S. Flint. 1980. The soils and their nutrients. 219-254. In: J. Brown, P.C. Miller, L.L. Tieszen, and F.L. Bunnell (eds.), An Arctic Ecosystem: The Coastal Tundra at Barrow, Alaska. Stroudsburg, Pennsylvania, Dowden. Hutchinson, and Ross.

Gorham, E. 1991. Northern peatlands: role in the carbon cycle and probable responses to climatic warming. *Ecol. Appl.* 1:182-195.

Grulke, N.E., G.H. Riechers, W.C. Oechel, U. Hjelm, and C. Jaeger. 1990. Carbon balance in tussock tundra under ambient and elevated atmospheric CO_2. *Oecologia (Berlin)* 83:485-494.

Hengeveld, H. 1991. A State of the Environment Report. Atmospheric Environment Service, Environment Canada.

Kukla, G. and T.R. Karl. 1992. Recent rise of the nighttime temperatures in the northern hemisphere. *DOE Res. Sum.* 14:1-4.

Kummerow, J. and B.A. Ellis. 1984. Temperature effect on biomass production and root/shoot biomass ratios in two arctic sedges under controlled environmental conditions. *Can. J. Botany* 62:2150-2153.

Lachenbruch, A.H. and B.V. Marshall. 1986. Changing climate: geothermal evidence from permafrost in the Alaskan arctic. *Science* 234:689-696.

Lashof, D.A. and D.R. Ahuja. 1990. Relative contributions of greenhouse gas emissions to global warming. *Nature* 344:529-531.

Lawrence, W.T. and W.C. Oechel. 1983a. Effects of soil temperature of carbon exchange of taiga seedlings. I. Root Respiration. *Can. J. For. Res.* 13:840-849.

Lawrence, W.T. and W.C. Oechel. 1983b. Effects of soil temperature on carbon exchange of taiga seedlings. II. Photosynthesis. *Can. J. For. Res.* 13:850-859.

Limbach, W.E., W.C. Oechel, and W. Lowell. 1982. Photosynthetic and respiratory responses to temperature and light of three Alaskan tundra growth forms. *Holarct. Ecol.* 5:150-157.

Marion, G.M. and C.H. Black. 1987. The effect of time and temperature on nitrogen mineralization in arctic tundra soils. *Soil Sci. Soc. Am. Jour.* 51:1501-1507.

Marion, G.M. and P.C. Miller. 1982. Nitrogen mineralization in a tussock tundra soil. *Arc. Alp. Res.* 14:287-293.

Marion, G.M. and W.C. Oechel. 1993. Mid- to late-Holocene carbon balance in arctic Alaska and its implications for future global warming. *Holocene* 3.3:193-200.

Miller, P.C. 1981. *Carbon Balance in Northern Ecosystems and the Potential Effect of Carbon Dioxide Induced Climate Change.* U.S. Department of Energy, Washington, D.C. 1-109.

Miller, P.C., R. Kendall, and W.C. Oechel. 1983. Simulating carbon accumulation in northern ecosystems. *Simulation* 40:119-131.

Mitchell, J.F.B., S. Manabe, S. Tokioka, and V. Meleshko. 1990. Equilibrium climate change and its implications for the future. 131-178. In: J.T. Houghton, G.J. Jenkins, and J.J. Ephraums (eds.), Climate Change, The IPCC Scientific Assessment. Cambridge, U.K., Cambridge University Press.

Oberbauer, S.F., N. Sionit, S.J. Hastings, and W.C. Oechel. 1986a. Effects of CO_2 enrichment and nutrition on growth photosynthesis, and nutrient concentration of Alaskan tundra plant species. *Canadian Journal of Botany* 64:2993-2998.

Oberbauer, S.F., W.C. Oechel, and G. Riechers. 1986b. Soil respiration of Alaskan tundra at elevated atmospheric CO_2 concentrations. *Plant and Soil* 46:145-158.

Oechel, W.C. and B.R. Strain. 1985. Native species responses to increased atmospheric carbon dioxide concentration. 119-154. Direct Effects of Increasing Carbon Dioxide on Vegetation. Washington, D.C., U.S. Dept. of Energy, Office of Basic Energy Sciences, Carbon Dioxide Research Division.

Oechel, W.C. and W.D. Billings. 1992. Anticipated effects of global change on carbon balance of arctic plants and ecosystems. 139-168. In: F.S. Chapin III, R.L. Jefferies, G.R. Shaver, J.F. Reynolds, and J. Svobada (eds.). *Arctic Physiological Processes in a Changing Climate.* Academic Press, San Diego, CA.

Oechel, W.C., G. Riechers, W.T. Lawrence, T.I. Prudhomme, N. Grulke, and S.J. Hastings. 1992. "CO_2LT", a closed, null-balance system for long-term *in situ* ecosystem manipulation and measurement of CO_2 level, CO_2 flux, and temperature. *Funct. Ecol.* 6:86-100.

Oechel, W.C., S.J. Hastings, G.L. Vourlitis, M.A. Jenkins, G. Riechers, and N. Grulke. 1993. Recent change of arctic tundra ecosystems from a carbon sink to a source. *Nature* 361:520-523.

Ovendon, L. 1990. Peat accumulation in northern peatlands. *Quat. Res.* 33:377-386.

Post, W.M. 1990. Report of a workshop on climate feedbacks and the role of peatlands, tundra, and boreal ecosystems in the global carbon cycle. Oak Ridge National Laboratory. 32.

Prudhomme, T.I., W.C. Oechel, S.J. Hastings, and W.T. Lawrence. 1982. Net ecosystem gas exchange at ambient and elevated carbon dioxide concentrations in tussock tundra at Toolik Lake, Alaska: An evaluation of methods and initial results. The Potential Effects of Carbon-dioxide Induced Climate Change in Alaska: Proceedings of a Conference, School of Agriculture and Land Resources Management, University of Alaska, Fairbanks., Misc. Publication.

Ramanathan, V., H.B. Cicerone, H.B. Singh, and J.T. Kiehl. 1985. Trace gas trends and their potential role in climate change. *Jour. Geophys. Res.* 90:5547-5566.

Rieger, S., D.S. Schoephorster, and C.E. Furbush. 1979. Exploratory soil survey of Alaska. U.S. Department of Agriculture, Soil Conservation Service, Washington D.C.

Schell, D.M. 1983. Carbon-13 and Carbon-14 abundances in Alaskan aquatic organisms: Delayed production from peat in arctic food webs. *Science* 219:1068.

Schell, D.M. and P.J. Ziemann. 1983. Accumulation of peat carbon in the Alaska arctic coastal plain and its role in biological productivity. p. 1105-1110. Permafrost, Fourth International Conference. Washington, D.C., National Academy Press.

Schlesinger, M.E. and J.F.B. Mitchell. 1987. Climate model simulations of the equilibrium climatic response to increased carbon dioxide. *Rev Geophys* 25:760-798.

Sebacher, D.I., R.C. Harriss, K.B. Bartlett, S.M. Sebacher and S.S. Grice. 1986. Atmospheric methane sources: Alaskan tundra bogs, an alpine fen, and a subarctic marsh. *Tellus* 38B:1-10.

Shaver, G.R. and F.S. Chapin III. 1980. Response to fertilization by various plant growth forms in an Alaskan tundra: nutrient accumulation and growth. *Ecology* 61:662-675.

Silvola, J. 1986. Carbon dioxide dynamics in mires reclaimed for forestry in eastern Finland. *Annales Botanici Fennici* 23:59-67.

Tissue, D.T. and W.C. Oechel. 1987. Response of *Eriophorum vaginatum* to elevated CO_2 and temperature in the Alaskan tussock tundra. *Ecology* 68:401-410.

Torn, M.S. and F.S.I. Chapin. 1993. Environmental and biotic controls over methane flux from arctic tundra. *Chemosphere* 26:357-368.

Vourlitis, G.L., W.C. Oechel, S.J. Hastings, and M.A. Jenkins. 1993a. The effect of soil water content and thaw depth on CH_4 flux from wet coastal tundra ecosystems on the North Slope of Alaska. *Chemosphere* 26:329-337.

Vourlitis, G.L., W.C. Oechel, S.J. Hastings, and M.A. Jenkins. 1993b. A system for measuring in situ CO_2 and CH_4 flux from unmanaged ecosystems: An arctic example. *Funct. Ecol.* 7:369-379.

Walker, D.A. and W. Acevedo. 1987. Vegetation and a landsat-derived land cover map of the Beechey Point quadrangle, arctic coastal plain, Alaska. CRREL Report 87-3.

Whalen, S.C. and W.S. Reeburgh. 1992. Interannual variations in tundra methane emission: a 4-year time series at fixed sites. *Global Biogeochem. Cycles* 6:139-159.

Wigley, T.M.L. 1987. Relative contributions of different trace gases to the greenhouse effect. *Clim. Monitor* 16:14-29.

Zimov, S.A., G.M. Zimova, S.P. Daviodova, Y.V. Voropaev, Z.V. Voropaeve, O.V. Prosiannikova, I.V. Semiletova, and I.P. Prosiannikova. 1993. Winter biotic activity and production of CO_2 in Siberian soils: A factor in the greenhouse effect. *Jour. Geophys. Res.* 98:5017-5023.

CHAPTER **10**

Global Soil Erosion by Water and Carbon Dynamics

R. Lal

I. Introduction

Total precipitation received at the earth's surface is about 577×10^3 km^3 or 1130 mm, annual runoff to the oceans is 47×10^3 km^3 or 92 mm, and the differences of 530×10^3 km^3 or 1038 mm is estimated to be the evapotranspiration (USSR National Committee for the International Hydrological Decade, 1974). Out of the total precipitation received, 119×10^3 km^3 or 233 mm is received on the land surface. Total amount of water currently stored in human-made reservoirs is estimated at 5×10^3 km^3 or about 11% of the total annual runoff from the land surface to the oceans. The total irrigated agricultural land area of 300×10^6 ha uses about 4×10^3 km^3 of water annually. About 85% of the radiation balance at the earth's surface is accounted for by evapotranspiration (UNESCO, 1978).

Total runoff has an important impact on global soil erosion, and transport of dissolved and suspended load to the ocean. Raindrops striking the soil surface and flowing runoff have tremendous kinetic energy responsible for soil detachment and sediment transport. Soil erosion is also affected by wind current, gravity, and ice as other principal agents of soil erosion. There is a vast amount of erosive energy associated with hydrologic cycle and other agents of soil erosion. For 233 mm of precipitation received on the land surface, the average kinetic energy is estimated at 20 J m^{-2} mm^{-1} of rain. With total land area of 14.8×10^9 ha, the total kinetic energy falling on land by precipitation is about 0.7×10^{18} J year^{-1}. The total potential energy involved in runoff from the land surface is about 3×10^{20} J year^{-1} (Walling, 1987).

II. Global Extent of Soil Erosion

Although the available statistics is vague and subjective, the land area affected by accelerated soil erosion is estimated at about 1100×10^6 ha (Table 1), with relative affected area in different continents being in the order of Asia > Africa > South America > Europe > North and Central America > Oceania and the Pacific. Wind erosion affects about 550×10^6 ha. Some areas, especially those in the semi-arid regions, are affected by both water and wind erosion. The annual global loss of agricultural land due to accelerated erosion is estimated at three million ha (Buringh, 1981). The total area of productive land presumably destroyed by erosion since the beginning of settled agriculture is about 430×10^6 ha (Kovda, 1983).

In the context of carbon dynamics, it is the magnitude of soil displaced and its carbon contents that are more relevant than the land area affected. Brown (1984) estimated that the world's cropland area is losing about 23×10^9 Mg of soil in excess of new soil formation each year. Walling (1987) revised these data and estimated that total material transport to the oceans by the world's rivers is about 19×10^9 Mg year^{-1}, comprising 14×10^9 Mg year^{-1} as suspended load, 4×10^9 Mg year^{-1} as dissolved load, and 1×10^9 Mg year^{-1} as bed load. These rates of sediment transport to the oceans translate into global denudation rate of about 64 mm per 1000 years.

Table 1. Global area affected by accelerated soil erosion

Region	Area affected (10^6 ha)	
	Water erosion	Wind erosion
Africa	227.3	187.8
Asia	433.2	224.1
Europe	113.9	41.6
North and Central America	106.6	38.8
Oceania/Pacific Islands	83.4	16.4
South America	124.1	41.4
World	1088.5	550.1

(From WRI, 1992-1993.)

The sediment load carried by the world rivers originates over the land due to several processes of soil erosion, e.g., inter-rill or splash erosion, rill erosion, gully erosion, tunnel erosion, mass movement, coastal and river bank erosion, etc. Out of the total sediments detached, only a fraction is transported into the major river system and finally into the ocean. The sediment delivery ratio is a complex problem due to several sinks or storage systems within the watershed (Walling, 1983; Meade, 1982). It is generally believed that the sediment delivery ratio may be as low as 10%, e.g., only 10% of the sediment originating over the watershed are eventually transported to the ocean.

Estimates of sediment transport from different continents to the oceans are shown in Table 2. The total sediment transport for different continents is in the order of Asia > Oceania > South America > North and Central America > Africa > Europe. In comparison, the dissolved load for continents is in the order of Asia > North and Central America > South America > Europe > Oceania > Africa. For the low mean annual runoff, the dissolved load is rather high for the continent of Europe. The last column in Table 2 depicts computations of the total sediment yield. The highest total sediment yield of 645 Mg km^{-2} yr^{-1} is from Oceania and the Pacific Islands followed by that of 286 Mg km^{-2} yr^{-1} for Asia. The least sediment yield of about 48 Mg km^{-2} yr^{-1} is observed for the continent of Africa.

There is a wide range of total load carried in rivers of the world (Table 3). Three rivers carrying the highest total load are Ganges-Brahmaputra, Hwang Ho, and the Amazon. The highest dissolved load is carried by the Amazon river. The total sediment yield in major rivers is in the order of Hwang Ho > Ganges-Brahmaputra > Magdalena > Irrawdy. Some of the lowest suspended and dissolved loads are carried by the rivers in Central and Western Africa, e.g., Niger and Zaire. In most African rivers, the dissolved load exceeds the suspended load. Some rivers with small drainage area carry even higher load than those listed in Table 3, e.g., Huantuchan (China) with a suspended load of 53,500 Mg km^{-2} yr^{-1}, Dali (China) with 25,600 Mg km^{-2} yr^{-1}, Trengwen (Taiwan) with 28,000 Mg km^{-2} yr^{-1}, Perkerra (Kenya) with a load of 19,520 Mg km^{-2} yr^{-1} (Walling, 1987; 1989). Most of these rivers with high suspended load drain highly erodible loess soils in semi-arid environments. Some rivers draining volcanic steeplands in New Zealand also carry high suspended load.

Magnitude of soil erosion from the terrestrial ecosystems is influenced by several interacting factors including climate, relief, soil, vegetation, human perturbations, tectonic instability, and volcanic activity. Because of the deposition within the watershed, sediment yield depends on the land area considered. Sediment yield per unit area often decreases exponentially with increasing size of catchment area (Maruszczak, 1984). Climatic erosivity is a major determinant of soil erosion and sediment transport. Erosivity is influenced by several climatic parameters including rainfall (amount, intensity, kinetic energy, seasonality), snow melt, temperature, evapotranspiration, wind velocity, etc. Sediment transport in relation to climate have been studied by Fournier (1960), Douglas (1967), Walling and Webb (1987) and Jansson (1988). It is a common belief that erosion risks and sediment yield are extremely high for semi-arid climates. Contrary to the popular belief, Jansson (1988) observed that most climates with a dry season (B_s) do not have as high sediment yield as tropical rainy climates with no dry period (A_f). Furthermore, warm temperate climate with a dry season (C_s and C_{wa}) exhibit high sediment yield (Table 4). According to Jansson's analysis, some very erosive climates are A_f, C_{wa}, C_s, C_{fa}, and B_s. The sediment yield in major rivers is also closely related to the landuse intensity and demographic pressure. In general, the more is the demographic pressure the higher is the sediment yield. Abernathy (1987) observed a close relationship

Table 2. Sediment transport from different continents to the oceans

	Land area (10^6 km^2)	Mean annual runoff (10^3 km^3)	Total annual yield (10^6 t yr^{-1})			Total sediment yield (t km^2 yr^{-1})
			Suspended load	Dissolved load	Total	
Africa	15.3	4.1	530	201	731	47.8
Asia	28.1	14.3	6433	1592	8025	285.6
Europe	4.6	2.1	230	425	655	142.4
North and Central America	17.5	7.8	1462	758	2220	126.8
Oceania/Pacific Islands	5.2	2.4	3062	293	3355	645.2
South America	17.9	11.7	1788	603	2391	133.6
Total	88.6	42.4	13,505	3,872	17,377	(196.1) Mean

(From Meybeck, 1979; Milliman and Meade, 1983; and Walling, 1987.)

Table 3. Suspended sediment and dissolved loads of major rivers of the world

River	Drainage area (10^6 km^2)	Load (10^6 t) Suspended	Load (10^6 t) Dissolved	Load (10^6 t) Total	Total sediment yield (t/km^3/yr)
Ganges-Brahmaputra	1.48	1670	151	1821	1230.4
Hwang Ho (yellow)	0.77	1080	34	1114	1446.8
Amazon	6.15	900	290	1190	193.5
Yangtze	1.94	478	166	644	322.0
Irrawdy	0.43	265	91	356	827.9
Magdalena	0.24	220	28	248	1033.3
Mississippi	3.27	210	131	341	124.3
Orinoco	0.99	210	50	260	262.6
Zaire	3.82	43	47	90	23.6
Ob	2.50	16	90	106	42.4
Lena	2.50	12	85	97	38.8
Yenesei	2.58	13	73	86	33.3

(From Walling, 1987.)

Table 4. Median sediment yield in relation to climate

Koppen's classification	Description	Sediment yield (t km^{-2} yr^{-1})
A_f	Tropical rains, no dry period	936
A_m	Tropical monsoon rains	182
A_w	Tropical rains with dry winter	85
B_{sh}	Dry climate, hot	292
B_s	Dry climate, steppe	105
B_{sk}	Dry climate, steppe, cold mean temperature > 18 °C	155
C_f	Warm temperate, desert	1337
C_{wa}	Warm temperate rains, with warmest month > 22 °C	905
C_w	Warm temperate, desert	599
C_{fs}	Warm temperate rains, with warmest month > 22 °C	167
D_{ws}	Boreal climate, warmest month > 22 °C	997
D_{sh}	Boreal desert, with at least 4 months > 10 °C	477
D_s	Boreal desert	400
E_t	Snow climate - Tundra	426
$E_{t(mt)}$	Snow climate - Tundra mountains	129

A Tropical rain climate
 f no dry period
 w dry winter
 s dry summer
 m monsoon rains, dry period

B Dry climates
 s steepe
 w desert
 h hot, mean temp. > 18 °C
 k cold, mean temp. < 18 °C

C Warm temperate rain climates
 a warmest month > 22 °C
 b at least 4 monts > 10 °C
 c 1-4 months > 10 °C
 d coldest month below -38 °C

D Boreal climates

E Snow climate
 T tundra
 F always frost

(After Jansson, 1988.)

between sediment yield and the population pressure, and calculated that in many developing countries annual sediment yields are increasing at a rate equivalent to 1.5 times the rate of population growth.

III. Flux of Organic and Inorganic Carbon in World Rivers

Carbon transport from land to the ocean by rivers is in three forms: (i) particulate organic carbon (POC) comprising leaf litter, woody debris, and soil organic matter, (ii) dissolved organic carbon (DOC) from decomposition of carbon in soil and leaf litter, and (iii) dissolved inorganic carbon (DIC) mostly in the form of HCO_3^-, CO_3^- ions and dissolved CO_2. Bicarbonates (HCO_3^-) constitute the principal inorganic form of carbon, POC, and DOC together are called total organic carbon (TOC).

The data on transport of DOC, POC, DIC are not available for all principal rivers of the world. Some data are available for rivers of the US (Likens, et al., 1981) and New Zealand (Moore, 1987). Load of HCO_3^--C in rivers of major continents are shown in Table 5. Total HCO_3^--C transported to the oceans is estimated at about 0.5×10^{15} g yr^{-1}. These estimates are similar to those of 0.454×10^{15} g yr^{-1} reported by Garrels, et al. (1975) and Kempe (1979). Some estimates of the transport of DOC and POC are also available. Meybeck (1984) estimated that carbon contents of sediments transported in the river is about 20 mg L^{-1} comprising equal proportion of inorganic (10 mg L^{-1}) and organic carbon (10 mg g^{-1}). Meybeck further estimated that about one-third of the total C transported in the rivers is in suspended load, and two-

Table 5. Estimates of annual river discharge and carbon contents

Continent	Discharge 10^3 km^3	HCO$_3$ (ppm)	C (ppm)	Total C discharge as carbonates (10^{15} g yr^{-1})
Africa	4.1	43	8.5	0.035
Asia	14.3	79	15.5	0.222
Europe	2.1	95	18.7	0.039
North and Central America	7.8	68	13.3	0.104
Oceania/Pacific Islands	2.4	316	6.2	0.015
South America	11.7	31	6.1	0.071
Total	42.4			0.484

(Adapted from Livingston, 1963; Baumgartner and Reichel, 1975; and Kempe, 1979.)

Table 6. Dissolved and suspended organic carbon in Ohio rivers for 1984

Location	Latitude/longitude data	Mean organic carbon (mg L^{-1}) DOC	POC
Bloomingdale	40° 19' N, 80° 48' W	4.2	0.4
Clear Fork	40° 21' N, 81° 59' W	3.3	0.8
Opossum Run	40° 10' N, 82° 3' W	2.7	1.1
Muskingum	39° 38' N, 81° 51' W	3.3	0.4
Hocking	39° 19' N, 82° 0' W	3.5	2.1
Sandy Run	39° 21' N, 82° 18' W	1.6	0.4
Big Four Hollow	39° 22' N, 82° 19' W	1.6	0.3
Hull Hollow	39° 21' N, 82° 19' W	2.3	0.6
Scioto R.	39° 12' N, 82° 51' W	6.6	1.2
Maumee R.	38° 38' N, 83° 12' W	3.4	0.2
Sandusky	39° 10' N, 84° 17' W	5.6	0.7
Cuyahoga	41° 30' N, 83° 42' W	8.7	1.2
Grand R.	41° 18' N, 83° 9' W	6.4	7.0
	41° 23' N, 81° 37' W	7.4	1.1
	41° 44' N, 81° 15' W	9.1	0.9
Mean		4.6	0.8

(Unpublished data for U.S. Department of Interior, Geological Survey, Columbus, Ohio.)

thirds is dissolved load. Garrels, et. al. (1975) estimated global mean concentration of DOC at 3.28 ppm, and that of particulate carbon at 1.76 ppm.

Table 6 shows the data of DOC and POC for several streams in Ohio monitored during 1981. The concentration of DOC ranged from 1.6 to 9.1 mg L^{-1} with a mean of 4.6 mg L^{-1} (USGS, 1989; 1993). In comparison, concentration of POC was low, and ranged from 0.2 to 7.0 mg L^{-1} with a mean of 0.8 mg L^{-1}. These streams drain a range of land uses, and from the data available, it is difficult to relate DOC and POC concentrations to land use and soil management. The data from a field plot experiment conducted on a Miamian soil in Ohio are shown in Table 7 (Thomas, 1991; Thomas, et al., 1992). These measurements were made for three land use systems. The concentration of DOC ranged from 3.5 to 5.9 mg L^{-1} with a mean of 4.6 mg L^{-1}. The mean value of 4.6 mg L^{-1} was the same as observed for several streams in Ohio. Furthermore, DOC concentration was not significantly different among land use and soil management treatments (Table 7).

Table 7. Effects of landuse and soil management systems on dissolved organic carbon per season

Treatment	Concentration (mg L^{-1})	Total loss (kg ha^{-1})
Alfalfa	3.5	2.1
Corn, conventional till	3.9	5.3
Corn, ridge till	4.9	5.4
Forest	5.9	1.2
Mean	4.6	3.6
LSD (0.05)	NS	NS

(Adapted from Thomas, 1991; Thomas et al., 1992.)

Table 8. Estimates of global annual transport of dissolved organic carbon (DOC) and particulate organic carbon (POC)

Carbon form	Estimates by Garrels et al., 1975	Revised estimates[a]
DOC (10^{15} g yr^{-1})	0.123	0.170
POC (10^{15} g yr^{-1})	0.066	0.085
Total organic carbon	0.189	0.255

[a] Recalculated with revised runoff rates shown in Table 2 and 5 using mean concentrations of 4 ppm for DOC, and 2 ppm for POC.

Using the mean concentrations of 4.0 mg L^{-1} for DOC and 2.0 mg L^{-1} for POC the revised calculations of carbon transport in World's river are shown by the data in Table 8. These calculations show that total organic carbon transport to the oceans is about 0.255 × 10^{15} g yr^{-1} or slightly more than half of the HCO$_3^-$ - C transport. Therefore, total carbon (organic and inorganic) transport to the oceans is estimated at about 0.739 × 10^{15} g yr^{-1}. These estimates can be updated with recent data on runoff flow and on water contents for organic and inorganic carbon transport in rivers from different continents.

IV. Soil Erosion and C Dynamics

The data in Table 9 present estimates of the impact of global soil erosion by water on C dynamics. World soils contain about 1500 × 10^{15} g C in the top 1 m depth over the total land area of 14.8 × 10^9 hectares or 29% of the earth's surface (Schlesinger, 1984; Buringh, 1984). The global sediment transport to the ocean of 19 × 10^{15} g yr^{-1} is equivalent to 190 × 10^{15} g yr^{-1} of soil displaced from terrestrial ecosystems assuming the mean delivery ratio of 10%. Based on these assumptions, estimates of C transport to the oceans from different continents are shown in Figure 1. These estimates are based on the data on sediment transport computed by Walling (1987). With a mean C content of 3%, total C displaced in soil from the terrestrial ecosystems is 57 × 10^{15} g yr^{-1}. It is assumed that C in soil displaced is easily decomposed, and as much as 20% is mineralized each year and released into the atmosphere as CO$_2$. Therefore, C flux into the atmosphere from soil physically displaced by erosion processes from terrestrial ecosystems is estimated at 1.14 × 10^{15} g yr^{-1}. Assuming mean organic C content of 3%, organic C transported with sediments to the ocean is about 0.57 × 10^{15} g yr^{-1}. With a total global runoff of 42.4 × 10^3 km^3 containing mean C (DOC plus POC) content of 6 mg L^{-1} amounts to sediment-born organic C transport to the ocean at the rate of 0.254 × 10^{15} g yr^{-1}.

A balance sheet based on these estimates is shown in Figure 2. Annual rate of soil displacement by erosion in terrestrial ecosystems accounts for only 0.38% of the C stored in world soils. Organic carbon transported to the world oceans accounts for one-tenth of the displaced carbon or 0.038% of the C in

Table 9. Soil erosion and C dynamics

References	Statistics	Total C (10^{15} g yr^{-1})
1. Soil erosion	* Delivery ratio of 10% * Total sediment displaced $= 190 \times 10^{15}$ g yr^{-1} * Organic carbon content of 3%	5.7
2. Decomposition	* 20% of the C displaced is biodegraded, mineralized, and released as CO over the watershed	1.14
3. Sediment	* Global sediment transport to the oceans is 19×10^{15} g yr^{-1} * Organic carbon content of sediment is 3%	0.57
4. Runoff	* Total runoff is 42.4×10^3 km^3 * Total carbon is 6 mg L^{-1} (mg L^{-1} DOC, 2 mg L^{-1} POC) * Inorganic carbon in runoff	0.255 0.486

Table 10. Organic carbon loss in soil erosion from tropical lands

Landuse	Area (10^6 ha)	Soil erosion rate (Mg ha^{-1} yr^{-1})	Carbon content or eroded sediment (%)	Transport of carbon (10^{12} g yr^{-1})
Arable	418.4	15	2.5	156.9
Permanent crops	51.4	12	3.0	18.5
Permanent pastures	1226.0	15	2.0	367.8
Forest and woodland	1867.5	10	3.5	653.6
Other lands	1320.5	15	2.0	396.2
Total				1593.0

world's soils. It is difficult to conclude whether C stored in world soils is being depleted because the rate of C sequestration due to soil restoration processes is not known (Figure 2).

V. Carbon Dynamics and Soil Erosion in the Tropics

Mean rates of soil erosion over the tropics are difficult to estimate. On the basis of literature surveys (Lal, 1984; 1990), the data presented in Table 10 show average estimated erosion rates for large watersheds. Using these data, the total transport or movement of C displaced with soil erosion is estimated at 1.59×10^{15} g yr^{-1}. These estimates may range from a low of 0.80×10^{15} g yr^{-1} to a high of 2.40×10^{15} g yr^{-1}. However, only a fraction of soil moved from its original place is transported out of the watershed. The delivery ratio for tropical watersheds may also be as low as 10%. This implies that as much as 0.16×10^{15}

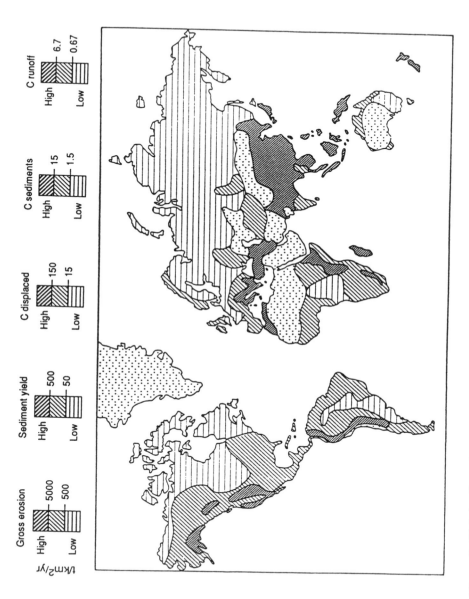

Figure 1. Global soil erosion and its impact on C dynamics. The data on sediment yield is from Walling, 1987.

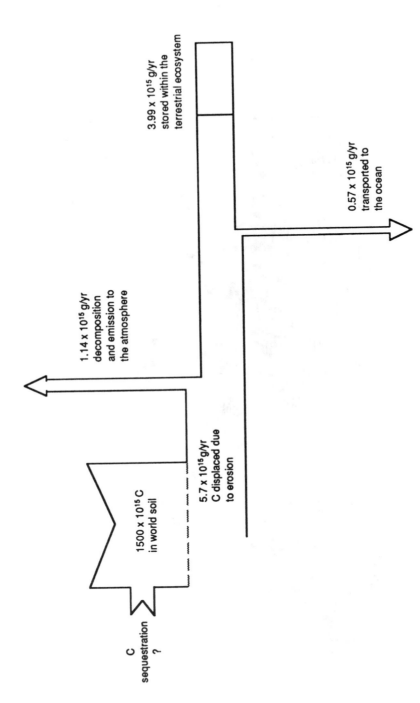

Figure 2. Global soil erosion and dynamics of soil organic carbon.

g C yr^{-1} may be transported out of tropical watersheds with a range of 0.08×10^{15} g C yr^{-1} to 0.24×10^{15} g yr^{-1}. Schlesinger (1981) estimated that world's river transport about 0.37×10^{15} g C yr^{-1}. The estimates for tropical rivers are at 0.16×10^{15} g C yr^{-1} implying thereby that about 40% of C transported in world's rivers is contributed by those draining tropical watersheds.

VI. Conclusion

Soil erosion from terrestrial ecosystems accounts for about 190×10^{15} g soil displaced annually, of which only 19×10^{15} g of sediments are transported to the oceans each year. World soils contains about 1500×10^{15} g C in the top 1 m depth. Of this, 5.7×10^{15} g C is displaced annually due to global soil erosion. Only 0.57×10^{15} g C is transported annually to the oceans, and remaining 3.99×10^{15} g is redeposited within the terrestrial ecosystems. The magnitude of C sequestration within the terrestrial ecosystems due to restorative landuses is not known. However, the magnitude of annual C flux into the atmosphere from carbon displaced by erosion is estimated at 1.14×10^{15} g.

In addition to its on-site and off-site economic impacts, global soil erosion also has a major impact on C dynamics. Ecological and environmental effects of erosion induced changes in soil carbon warrant serious and planned effort to reduce soil erosion risks, and minimize transport of sediments into world's waterways.

References

Abernathy, C.L. 1987. Soil erosion and sediment yield. Unpublished report, cited by Walling, D.C. 1989.
Baumgartner, A. and E. Reichel. 1975. The World Water Balance, Oldenbourge Verlag, Munchen, Wien: 1-179.
Brown, L.R. 1984. Conserving soils. p. 53-75. In: L.R. Brown (ed.). *State of the World*, Norton, NY.
Buringh, P. 1981. An assessment of losses and degradation of productive agricultural land in the world. FAO Working Group on Soils Policy, Rome, Italy.
Buringh, P. 1984. Organic carbon in soils of the world. p. 91-109. In: G.M. Woodwell (ed.). *The Role of Terrestrial Vegetation in the Global Carbon Cycle: Measured by Remote Sensing*. J. Wiley & Sons, NY.
Chin, Y.P. and P.M. Gschwend. 1992. Partitioning of polycyclic aromatic hydrocarbons to marine porewater organic colloids. *Env. Sci. Tech.* 26:1621-1626.
Douglas, I. 1967. Man, vegetation, and the sediment yield of rivers. *Nature* 215:925-928.
Fournier, F. 1960. *Climat et Erosion*. Presses Universitaires de France, Paris.
Garrels, R.M., F.T. Machenzie, and C. Hunt. 1975. p. 1-206. In: *Chemical cycles and the global environment, assessing human influences*. W. Kaufman, Inc., Los Altos, CA. 3rd Edition.
Hobbie, J.E. and G.E. Likens. 1973. Output of phosphorus, dissolved organic carbon, and fine particulate carbon from Hubbard Brook Watersheds. *Limnol. Oeanogr* 18:734-742.
Jansson, M.B. 1988. A global survey of sediment yield. *Geog. Ann.* 70 A (1-2):81-98.
Kempe, S. 1979. Carbon in the fresh water cycle. p. 317-342. In: B. Bolin, E.T. Degens, S. Kempe, and P. Ketner (eds.), *The Global Carbon Cycle*. SCOPE 13. J. Wiley & Sons. U.K.:317-342.
Koppen, W. 1936. Das Geographische System der Klimat. p. 1-44. In: W. Koppen and R. Geiger (eds.), *Handbuch der Klimatologie*. Band 1, Teil C.
Kovda, V.A. 1983. Loss of productive land due to salinization. *Ambio* 12:91-93.
Lal, R. 1984. Soil erosion in tropical arable lands and its control. *Adv. Agron.* 28:183-248.
Lal, R. 1990. *Soil erosion in the Tropics: Principles and Management*. McGraw Hill, NY. 580 pp.
Likens, G.E., F.T. Machenzie, J.E. Richey, J.R. Sedell, and K.K. Turekian. 1981. *Flux of organic carbon by rivers to the oceans*. U.S. Dept. of Energy, Washington, D.C. Conf. 80091490, UC-11. 359 pp.
Livingstone, D.A. 1963. *Chemical composition of rivers and lakes*. U.S. Geological Survey. Prof. Pap. 440 G. 64 pp.
Maruszczak, H. 1984. Spatial and tamporal differentiation of fluvial sediment yield in the Vistula river basin. *Geogr. Pol.* 50:253-269.
Meade, R.H. 1982. Sources, sink, and storage of river sediment in the Atlantic drainage of the United States. *J. Geol.* 90:235-252.
Meybeck, M. 1979. Concentration des eaux fluviales en elements majeurs et apports en solution aux oceans. *Revue de Geologie Dynamique et de Geographie Physicque* 21:215-246.

Meybeck, M. 1984. Les fluxes et le cycle geochimique des elements. These d'Etate, Universite Pierre et Marie Curie, Paris.

Millman, J.D. and R.H. Meade. 1983. World wide delivery of river sediment to the oceans. *J. Geology* 91:1-21.

Moore, T.R. 1987. Dissolved organic carbon in forested and cutover drainage basin, Westland, New Zealand. p. 481-487. In: R.H. Swanson, P.Y. Bernier and P.D. Woodard (eds.). "Forest Hydrology and Watershed Management", Wallingford, Oxfordshire, IAHS Proc.

Schlensinger, W.H. 1981. Transport of organic carbon in world's rivers. *Tellus* 33:172-187.

Schlensinger, W.H. 1984. Soil organic matter: a source of atmospheric CO_2. In: G.M. Woodwell (ed.). *The Role of Terrestrial Vegetation in the Global Carbon Cycle: Measured by Remote Sensing*. J. Wiley & Sons, NY.

Thomas, M.L. 1991. Landuse and management effects on soil properties, runoff, erosion, and water quality. M.Sc. thesis. The Ohio State University, Columbus, OH. 228 pp.

Thomas, M.L., R. Lal, T.J. Logan, and N.R. Fausey. 1992. Landuse and management effects on non-point loading from Miamian soil. *Soil Sci. Soc. Am. J.* 56:1871-1875.

UNESCO. 1978. *World Water Balance and Water Resources of the Earth*. UNESCO Studies and Reports in Hydrology. No. 25, 662 pp.

USGS. 1989. *Ground water level, water quality, and potential effects of toxic-substance spills or cessation of quarry dewatering near a municipal groundwater supply, southern Franklin County, Ohio*. US Geological Survey, Water Resources Inv. Report 88-4138, Columbus, OH 111 pp.

USGS. 1993. Dept. of Interior, US Geological Survey, unpublished data, Columbus, Ohio.

USSR National Committee for the International Hydrological Decade. 1974. Mirovoi Vodnyi Balans i Vodnye Resursy Zemli, Gidrometeoizdat, Leningrad. 638 pp.

Walling, D.E. 1983. The sediment delivery problem. *J. Hydrol.* 65:209-237.

Walling, D.E. 1987. Rainfall, runoff, and erosion of the land: a global view. p. 89-117. In: K.J. Gregory (ed.). *Energetics of Physical Environment*. J. Wiley & Sons, U.K.

Walling, D.E. 1989. The struggle against water erosion and a perspective on recent research. p. 39-60. In: K. Ivanov and D. Pechinov (eds) "Water Erosion". IHP-111 Project 2.6, UNESCO, Paris.

Walling, D.E. and B.W. Webb. 1987. Material transport by the world's rivers: evolving perspectives. *IAHS Publ.* 164:313-329.

World Resources Institute 1992-1993. *Towards sustainable development -- A guide to the global environment*. World Resources Institute, Washington, D.C. 385 pp.

CHAPTER 11

Gaseous Emissions from Agro-Ecosystems in India

I.P. Abrol

I. Introduction

In recent years, agricultural activities have been perceived as emitting a significant fraction of the greenhouse gases, in particular, carbon dioxide, methane and nitrous oxides into the atmosphere. These gases are considered as amongst the principal gases responsible for global warming. Agricultural activities mainly responsible for gaseous emissions include rice cultivation, metabolic activity in the guts of livestock, burning of vegetation and intensive cultivation involving increasing use of fertilizers. The latter also leads to loss of soil organic matter through mineralization and emission of CO_2. A reliable assessment of the amount and kind of gaseous emissions is important and basic for developing and promoting strategies that will reduce gaseous emissions and/or promote their absorption by natural sinks. This paper gives an overview of the available information on the subject in the Indian context.

II. Land Use

Predominant land uses for 1990s are shown in Table 1, and are briefly described below.

A. Forests

The area under forests constitutes only 22 percent of the land area in contrast to the recommended 33 percent considered as minimum for ecological security. India's forest area has been dwindling. According to the National Remote Sensing Agency (FSI 1989) only 36 million ha of the forest area has a crown density of more than 40 percent. The remaining area is in a state of degradation with much of the vegetal cover already lost.

B. Net Cropped Area

This refers to the land actually cultivated, which has remained nearly constant at about 140 million ha over the two decades ending in 1990. In future, there is little or no possibility of expanding cultivated area. In fact there appears to be an urgent need to take out of production a significant fraction of degraded lands of low productivity. Annual cropping is currently practiced on these lands. But these lands should be used to promote conservation and restoration.

About 32 percent of the net cropped area (45.2 million ha) have supplemental irrigation facilities, while in the remaining 68 percent of the area crop production is dependent on rainfall. In about 38.4 million ha more than one crop are grown in a year (Ministry of Agriculture, 1992).

Table 1. Land use in India in 1989-90

Land use category	Area (million ha)
Forests	67.08
Net cropped area	141.73
Area under non-agricultural uses and barren and uncultural land	41.24
Area under permanent pastures and other grazing lands, land under miscellaneous tree crops and groves and culturable wastelands	30.48
Fallow lands	24.30
Total	304.83

Geographical area of India is 328.73 million ha while the reporting area is 304.83 million ha. (Adapted from Agricultural Statistics at a Glance, 1992, Directorate of Economics and Cooperation, Ministry of Agriculture, New Delhi.)

Table 2. Fertilizer use in India for three decades ending in 1990

	Nutrient consumption (as fertilizer)	
	N	N + P + K
Year	(thousand tonnes)	
1970-71	1,487	2,177
1980-81	3,678	5,516
1990-91	7,997	12,546

(Adapted from Fertilizer statistics, 1990-1991, The Fertilizer Association of India, New Delhi.)

Use of chemical fertilizers, an important input for agricultural production has increased nearly 6 times in the two decades ending in 1990. The consumption of major nutrients (NPK) averaged 72.2 kg ha^{-1} of cropped area in the year 1991-92. This level of fertilizer use is relatively low considering the consumption of nutrients from fertilizers in several developed countries e.g. 365 in Japan; 300 in Netherlands etc. Fertilizer use in the Nile valley of Egypt is also high at 400 kg ha^{-1} nutrients (The Fertilizer Association of India, 1992). In some of the intensively cropped area at India, the use of fertilizers is also quite high e.g. 160 kg nutrients per ha cropped area (for double cropped area this value will be 320 kg. nutrients ha^{-1} yr^{-1}) in the state of Punjab; 142 kg in the state of Tamil Nadu, etc. Nitrogenous fertilizers constitute a major fraction (about 65 percent) of the total fertilizers used in agriculture (Table 2).

Amongst principal crops, rice accounts for the largest area under a single crop and occupied 42.6 million ha in the year 1990-91. It is a staple food of a majority of the Indians. Fortyfive percent of the area cropped to rice is irrigated while the crop in remaining areas is rainfed. As discussed in a subsequent section, rice is grown under a wide range of soil and climatic conditions.

About 4.91 million ha of sloping lands is used for shifting cultivation, of which nearly 1.08 million ha are cultivated annually. Shifting cultivation is practiced chiefly in Orissa and the North-eastern states of Nagaland, Manipur, Arunachal Pradesh, and Mizoram. Shifting cultivation is considered a significant contributor to emissions of radiatively-active gases.

India has a large livestock population (Table 3). However, little of the cropped area is used for growing fodder for livestock, and reliable data on the area under fodder crops is not available. For this reason the

Table 3. Livestock population in India

Animal	Numbers (thousands)
Cattle	199,528
Buffloes	79,190
Sheep	45,750
Goats	110,334
Pigs	10,448
Others	3,476
Total	448,726

(Adapted from 14th livestock census conducted in 1987 and reported in Agricultural Statistics at a Glance, 1992, Directorate of Economics and Statistics, Ministry of Agriculture, New Delhi.)

Table 4. Changes in forest cover between 1983 and 1987

Class	Area (thousand hectares)		
	1983	1987	Change
Closed forest	36,141	37,847	+1,706
Open forest	27,658	25,741	-1,917
Managed forest	409.6	422,5	+12.9

(Adapted from the State of Forest Report 1989, Forest Survey of India, Ministry of Environment and Forests, New Delhi.)

vast majority of the livestock are raised on common and grazing lands. Since most of these lands are heavily over-grazed and are in a state of advanced degradation, most livestock are underfed and are in a poor state of health.

III. Gasous Emissions from Agroecosystems

A brief account of the available information on gaseous emissions from principal agroecosystems is presented below:

A. Forest Ecosystems

The loss of carbon due to deforestation is difficult to estimate because of the uncertainties in the extent of area deforested and degraded, and lack of knowledge about the above ground and carbon contents of soils in principal ecosystems. Studies undertaken by the Forest Survey of India (1989) (Table 4) have shown that during the period 1983 to 1987 deforestation rate was 47,500 ha per year. An attempt has recently been made by the Forest Research Institute of India to estimate the CO_2 emissions from forest conversion including deforestation, shifting cultivation, accidental fires, extraction of fuel wood, controlled burning etc. (Table 5) In these computations the relationship suggested by Seiler and Crutzen (1980) was adopted for calculating the amount of biomass burnt annually as a result of clearing the vegetation (Equation 1).

$$M = A \cdot B \cdot a \cdot b \tag{1}$$

Where M is the biomass burnt per year "A" is the area of land cleared per year (ha), "B" is the biomass density (Mg ha^{-1}), "a" is fraction of the biomass above ground and "b" is the fraction of the above ground biomass that is burnt. Values of "a" and "b" for these calculations were adopted from the published literature.

Table 5. Estimate of carbon emission from forest ecosystems

Source	CO_2 released (million tonnes)
Deforestation	0.60
Shifting cultivation	5.35
Accidental fires	27.23
Controlled burning	0.69
Firewood burning	0.90
Total	34.77

(Adapted from the State of Forest Report 1989, Forest Survey of India, Ministry of Environment and Forests, New Delhi.)

1. Deforestation

The emission of carbon as CO_2 from forest conversion is calculated in two stages; (a) above ground, and (b) the sub-soil emissions.

Above ground emission (C1) constitutes two components; M, due to forest conversion and M' due to above ground biomass decay.

$M = A\ (47,500) \cdot B\ (50\ \text{Mg ha}^{-1}) \cdot a\ (0.71,\ \text{taking the forest cover as open}) \cdot b\ (0.35) = \underline{590{,}187.5}$ Mg
M' (Annual above ground biomass decayed)
 $= A(47,500) \cdot B\ (50) \cdot a\ (0.71) \cdot b\ (0.65)/25$
 $= 43{,}842.5$ Mg;
Total emission from above ground = M + M' (C1)
 $= 634{,}030$ Mg.

Sub-soil emissions(C2) due to soil disturbance $(47{,}500 \times 0.20 \times 80 \times 0.30)/25 = 9120$ Mg; Assuming that 20 percent of the area of forest converted is utilized for agriculture and that contributes to carbon sequestration by the biomass that regrows in these converted land and that the remaining 80 percent of such converted land is left as grazing/fallow over the time, negative contribution (C 3) will be = 475,000.

The net carbon released = C(1) + C(2) - C(3)
 $= 634{,}030 + 9120 - 47{,}500$
 $= 595{,}650$ Mg.

2. Shifting Cultivation

Above ground emission (M) $= 995{,}600\ (A) \cdot 25.5(B) \cdot 0.71(a) \cdot 0.50(b)$
 $= 9{,}012{,}669$ Mg

Annual above ground biomass
 decay (M') $= 995{,}600(A) \cdot 25.5\ (B)\ 0.71(a) \cdot 0.50(b)/25$
 $= 360{,}507$ Mg.
Total (M+M') carbon released by above ground biomass
 $= 9{,}373{,}176$ Mg   (C1)
Subsoil emission $= 955{,}776$ Mg (C2)
Carbon sequestration by biomass
 that regrows $= 4{,}978{,}000$ Mg
Net carbon released due to shifting cultivation practice, therefore
 $= C(1) + C(2) - C(3) = 5{,}350{,}952$ Mg

3. Accidental Fires

According to the Forest Survey of India (1989) nearly one-fifth of the closed forests, (37.5 million ha) and half of the open forests (26 million ha) get accidentaly burnt every year. In accidental fires, the carbon is released due to burning of litter, twigs, and branches. Assuming average litter biomass accumulation on the forest floor as 4.5 and 1.5 Mg ha^{-1} yr^{-1} respectively for closed and open forests and that the forest is accidentally burnt at an interval of 5 and 2 years, it can be assumed that on the average 22.5 Mg ha of litter (4.5 Mg × 5 yrs) in closed and 3 Mg ha^{-1} (1.5 Mg × 2 yrs) accumulates in the open forest and would be burnt in case of accidental fire. Gaseous emissions due to accidental fire therefore amount to:

$$M = A\ (7{,}560{,}000) \cdot B\ (22.5) \cdot a(1) \cdot b\ (0.20)$$
$$= 34{,}020{,}000\ \text{Mg}$$
$$M' = A\ (12{,}850{,}000) \cdot B\ (3.0) \cdot a(1) \cdot b(0.26)$$
$$= 10{,}023{,}000\ \text{Mg}$$
$$0.45\ (M + M') = 19{,}820{,}000\ \text{Mg}$$

Carbon released due to annual biomass decay (of 16,460,000) = 7,407,000 Mg
Total carbon released = 27,230,000 Mg

4. Controlled Burning

Total carbon released = 698,365 Mg

5. Extraction of Firewood

Total carbon released = 900,000 Mg

B. Rice Ecosystem

Methane is an important greenhouse gas contributing to global warming. It accounted for nearly 15 percent of the greenhouse effect for the decade of 1980s. For this reason it is important to understand various sources and sinks of methane, and possible causes of its increase in the atmosphere. Rice paddies are considered an important contributor to atmospheric methane. IPCC (1990) observed that almost 90 percent of world's rice paddies is in Asia, of which about 60 percent is in China and India, Ahuja (1989) estimated that rice-ecosystem in India contributes 37.8 Tg of methane annually, as compared to total world emission of methane of 110 Tg. These estimates were based on extrapolation of results of a few measurements in California (25 g CH$_4$/m^2/harvest) and in Italy (54 g CH$_4$/m^2/harvest). Mitra (1991) measured methane efflux in selected rice growing regions during the period 1985-1990. He showed that the annual emission from all rice areas in India may be around 3 Tg, a value much lower than the one estimated by Ahuja (1989). However, since the database for these estimates was extremely limited, a methane measurement campaign was carried out during the rice growing period of 1991. In this campaign methane emission was measured in all the major rice growing areas of the country. Measurements during this campaign were carried out for the entire growing season covering 18 locations in the states of Assam, Bihar, Delhi, Kerala, Orissa, Tamil Nadu, Uttar Pradesh, and West Bengal. Sites were so chosen as to represent the important rice-agroecosystems. Measurement of methane efflux were made using the static chamber technique, and air samples were analysed by a gas chromatograph. Methodology was standardized and calibrated through the National Physical Laboratory. The absolute calibration compatiability at international level was established by exchanging samples with the Division of Atmospheric Physics, CSIRO, Australia and the National Institute of Agro-environment Sciences (NIAES), Japan. The campaign was organized by Dr. A.P. Mitra of the National Physical Laboratory. Based on the compaign, Table 6 gives a summary of the estimate of methane emission from the principal rice ecosystems. Results showed a wide variation in emission rates for rice fields under varying water regimes. Emissions were much higher for rainfed lowland and deep water areas than rice grown under irrigated conditions with a better control of water regime. As expected, emissions were almost negligible for upland rice. Data obtained in these studies, in some instances, showed a net

Table 6. Methane emission estimates from rice fields

Rice ecozone	Area (ha x 10³)	Integrated seasonal methane flux (g m⁻²)			Total methane emission (Tg yr⁻¹)		
		Maximum	Minimum	Mean	Maximum	Minimum	Mean
Rainfed lowland	17,238	60 ± 2	5 ± 1	9 to 46	4.17 ± 0.82	2.74 ± 0.67	3.40
Deep water	2,434	24 ± 2	14 ± 1	19	0.58 ± 0.046	0.342 ± 0.025	0.46
Irrigated	16,500	2 ± 0.6	0.06 ± 0.01	0.74	0.335 ± 0.078	0.061 ± 0.007	0.164
Upland	5,973			Negligible			Negligible
Total	42,235				5.1 ± 0.9	3.1 ± 0.7	4.0

(From Mitra (ed.), 1992.)

Table 7. Estimate of methane emission from livestock in India

Category	Population (million)	Methane released (Tg yr^{-1})
Cattle		3.929
Indigenous	190.0	3.667
Cowbreed	9.0	0.262
Buffalo	79.2	1.734
Sheep and goat	16	0.813
Total		6.476

influx indicating that rice fields may also be a sink for methane, particularly under irrigated conditions. The results show CH$_4$ emission of about 4 Tg yr^{-1}.

IV. N$_2$O Emissions from use of Nitrogenous Fertilizers

Fertilizer use in India is rapidly increasing (Table 2). Urea is the main nitrogen fertilizer. The evolution of nitrus oxide from fertilized soils is dependent upon the type and amount of fertilizer and method of application, soil and ambient conditions, crop grown, and water management practices. Ahuja (1989) estimated N$_2$O emisssion from India at 0.035 Tg yr^{-1}. However, similar to methane compaign there is need for experimental evaluation of the quantitative aspects of N$_2$0 emissions under different soil, crop and management practices.

V. Livestock and Gaseous Emissions

All animals lose some fraction of their feed energy in the form of methane emissions resulting from fermentation of carbohydrates in the digestive system. Ruminants lose more than most others; methane production rate being a function of the quantity and quality of the feed, body weight, energy expenditure and enteric ecology. Emissions increase with body weight, higher feeding level, quality of the feed and work output. India's large livestock population is considered to be an important factor in CH$_4$ emissions. However, experimental data on the quantities of methane emitted by different species under different agroecological conditions are not available.

Based on the emission factor by Crutzen et al. (1986), Ahuja (1989) estimated the annual methane emission of 10.4 Tg from animal population of 1985 in India in comparison to a global emission of 78.2 Tg for the same year.

Mitra (1992) estimated methane emission from cattle and buffalo by adopting the relationship suggested by Blaxter and Clapperton (1965):

$$Y = 1.30 + (0.112 \times D) + L (2.37 - 0.05 \times D)$$

where Y is methane yield in megajoules/100 megajoules of gross energy feed intake, D is the digestibility of the feedstock used and L is level of feeding expressed as multiples of the energy levels needed for maintenance.

Using this relation, annual methane emission for mature animals is estimated at 22.6 and 32.0 kg for the indigenous and cross-bred cows, and 25.7 and 40 kg respectively, for indigenous and cross-bred buffaloes. These values are considerably lower than the values of 58 and 95 kg/animal/yr predicted earlier for cows in USA and Germany, respectively. The low values arrived at for the Indian livestock are due to the relatively lower body weights and low feeding rates. Table 7 presents the estimates of methane for Indian livestock.

VI. Soils and Carbon Sequestration

Soils and vegetation represent an enormous potential for sequestering carbon. The accumulation of carbon in the soils is a function of the balance between carbon deposition and release processes. The equilibrium depends upon several factors including the nature of vegetation, biotic species, precipitation, and temperature etc. When the equilibrium is disturbed, as for example by deforestation, intensive cultivation etc., soil carbon rapidly declines to levels much lower than the soils' carbon carrying capacity. It is estimated that in Indian soils more than 50 percent of the carbon in surface and 10 percent in rest of the profile has been lost. The loss in many Indian soils could be as high as 60 to 70 percent (Jenny and Raychaudhry 1960). Carbon loss in many salt-affected soils may be as high as 80% (personal communication Gupta and Rao, CSSRI, Karnal). Gupta and Rao calculated carbon-pool in major soils in different agroclimatic zones using data base of the benchmark soils (Murthy et al., 1982), and other sites.

According to these calculations the current carbon pool in Indian soils is 24.3 Pg which is nearly 1.7 percent of the total world pool contained in 2.5 percent of the world's land surface. Since a majority of Indian soils are severely depleted of carbon, there is a good potential for restoring the soils to their original carbon carrying capacity. Accordingly it is estimated that nearly 34.9 Pg C could be potentially stored in Indian soils, The gap of 10.6 Pg represents the amount of additional soil carbon that can be sequestered. Soils with the greatest potential for carbon sequestration are likely to be young soils which are severely carbon depleted. However, these soils still retain the requisites for high primary production and can be suitably managed to increase productivity and enhance carbon storage in the profile.

VII. Summary

Agriculture related activities are perceived as contributing a significant fraction of the greenhouse gases, in particular, carbon dioxide, methane, and nitrous oxide. Rice fields, the guts of livestock, burning of vegetation, intensive cultivation involving increasing use of fertilizers and loss of soil organic matter are the principal activities that are attributed as contributing to increased gaseous emissions.

India's net cultivated area, 140 million ha, has remained almost constant over the past two decades and there is little likelihood of extending the cultivated area in future. The required increases in agricultural production to meet the needs of increasing population are likely to be achieved through increased intensity of use of existing cultivated land and through productivity gains. Rice, the staple food for a majority of Indians and grown over some 42 million ha, accounts for 33.4 per cent of the area under foodgrains crops and 42.55 per cent of the food grains. Nearly 45 per cent of the rice is irrigated, the remaining being grown under rainfed conditions. According to preliminary studies by Prasher et al. (1992) there are wide variations in measured methane emission under different agroecological conditions. Based on the limited observations the authors suggested a value of 2.5 to 6.0 Tg yr^{-1} from the rice areas.

Nearly all cultivated soils are low in organic matter and nitrogen. Use of nitrogenous fertilizers has increased rapidly over the past three decades; from 0.21 in 1960-61 to 8.0 million tonnes in 1990-91 and is likely to double by the close of this decade. Urea is the principal fertilizer. Although good estimates of emissions of nitrous oxide are not available research efforts hold promise for evolving management strategies for reducing the emissions by improving the use efficiencies of applied fertilizers.

India has a large livestock population, 448.73 million, including 199.5 million cattle, 79.19 million buffaloes, 110.3 million goats and 45.75 million sheep. Preliminary calculations estimate the methane emissions at 7 to 8 Tg yr^{-1} (Mitra, 1991). Notwithstanding extremely limited data base to permit reliable country-wide extrapolation as of now, there appear good research opportunities to develop management strategies to reduce the emission levels and to enhance the capacity of agroecosystems as sinks for the gaseous emissions.

References

Ahuja, Dilip R. 1989 Regional anthroprogenic emissions of greenhouse gases. Office of Policy Analysis, Environmental Protection Agency, Washington, D.C.

Blaxter, K.L. and Clapperton, J.L. 1965. Prediction of the amount of methane produced by ruminants. *British Journal Nutrition* 19:511-522.

Crutzen P.J., I Aselmann and W Seiler 1986. Methane production by domestic animals, wild ruminants, other herbororeus fauna and humans. *Tellus* 38B:271-284.

The Fertilizer Association of India. 1992. Fertilizer Statistics 1991-1992. The Fertilizer Association of India. New Delhi, India.

Forest Survey of India. 1989. The State of Forest Report 1989. Government of India, Forest Survey of India, Ministry of Environment and Forests, New Delhi.

Jenny, H and S.P. Raychaudhry. 1960. Organic matter in Indian Soils. Indian Council of Agricultural Research, New Delhi.

Ministry of Agriculture. 1992. Agricultural Statistics at a Glance. Ministry of Agriculture. New Delhi, India.

Mitra, A.P. (ed.). 1991. Greenhouse gas emissions in India - A preliminary report, CSIR, New Delhi. 22 pp.

Mitra, A.P. (ed.). 1992. Greenhouse gas emissions in India - 1991 Methane Campaign, CSIR, New Delhi. 91 pp.

Murthy, R.S. et al. 1982. Soils of Benchmark sites. National Bureau of Soil Survey and Land Use Planning. Nagpur, India.

Parashar, D.C., J. Rai, R.K. Gupta, and M. Singh. 1991. Parameters affecting methane emission from paddy fields. *Indian Journal of Radio and Space Physics* 20:12-17.

Seiler, W. and P.J. Crutzen. 1980. Estimates of gross and net fluxes of carbon between the sivsphere and the atmosphere farm biomass burning. *Climate Change* 2:207-247.

CHAPTER 12

Methane Emissions from Canadian Peatlands

T.R. Moore and N.T. Roulet

I. Introduction

It is well established that atmospheric concentrations of methane have been increasing at a rate of about 0.9% yr^{-1}, though there is recent evidence that this rate is decreasing (Steele et al., 1992). There are many sources of atmospheric methane, but wetlands are considered to be the largest natural source, contributing approximately 110 Tg yr^{-1}, of a total of 500-540 Tg yr^{-1} (Fung et al., 1991). As a large proportion of northern (poleward of 40-50 °N) landscapes are covered with wetlands, especially peatlands, and as the largest atmospheric concentrations of methane occur in the northern high latitudes, attention has focussed on northern wetlands as a source of atmospheric methane. Based on limited field measurements, Aselmann and Crutzen (1989), Cicerone and Oremland (1988) and Matthews and Fung (1987) estimated emissions from northern peatlands to be 30 - 80 Tg CH$_4$ yr^{-1}, though a more recent three-dimensional model synthesis has placed this figure at about 35 Tg yr^{-1} (Fung et al., 1991).

About 14% of the Canadian land surface is covered by wetlands, primarily peatlands (Figure 1). Various peatland types can be recognized, such as bogs, fens, and swamps in the subarctic, boreal, and cool temperate regions, as well as areas subject to shallow inundation, such as beaver ponds (National Wetlands Working Group, 1988). These types are differentiated on the basis of plant cover, minerotrophic status and hydrology and are probably similar to those found elsewhere in Europe and Asia, such as the western Siberian Lowland (Botch and Masing, 1983). At the southern margin, drainage for peatland exploitation has occurred, mainly for peat fuel, horticultural peat, and the growth of crops and trees: in southern Ontario and Québec, 20-50 % of the peatlands have been drained (National Wetlands Working Group, 1988). Armentano and Menges (1986) and Gorham (1991) estimated that drained peatlands in the temperate to subarctic zones covered 12-20 × 10^{10} m^2.

In this chapter, we examine rates of methane emission from Canadian wetlands, identify and quantify the important controlling factors, and illustrate how changes in wetlands, such as drainage and climatic change, can affect these rates.

II. Field Methane Emission Rates

In Canada during the past 5 years, methane emission rates have been determined at several locations (Figure 1). These locations include over 120 sites which represent swamps, bogs, and fens (the major peatland types in Canada) in regions ranging from the cool temperate to high subarctic. A simple static chamber technique has usually been employed (Moore and Roulet, 1991) with a sampling frequency of about once per week during the snow-free season. The static chambers allow measurements to incorporate small-scale variability and at remote sites, but give fluxes about 20% lower than those obtained from dynamic chambers in which the air is circulated (Moore and Roulet).

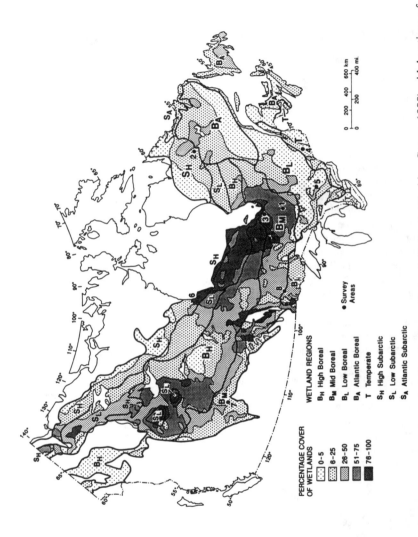

Figure 1. Wetland coverage and regions of Canada (from National Wetlands Working Group, 1988) with locations of methane flux measurements. Locations are: 1 - Cochrane, Ontario (Bubier et al., in press a); 2 - Schefferville, Québec (Moore and Knowles, 1990; Moore et al., 1990); 3 - Moosonee, Ontario (Hamilton et al., in press; Moore et al., in press); 4 - Montreal, Québec (Glenn et al., 1993; Moore and Knowles, 1990; Windsor, 1993); 5 - Dorset, Ontario (Roulet et al., 1992a); 6 - Churchill, Manitoba (Holland 1992); 7 - Alberta (Vitt et al., 1990); 8 - Kenora, Ontario (Kelly and Rudd, pers. comm.).

A major problem with the interpretation of the flux measurements is the high spatial variability in fluxes recorded from chambers placed in apparently ecologically uniform sites. Coefficients of variation on the same sampling date from replicate chambers placed in sites covering about 100 m² commonly range between 30 and 100% (e.g. Bartlett et al., 1989; Moore and Roulet, 1991). Where there is more pronounced variation in topography or vegetation, such in the microtopography created by hummocks and hollows in bogs, within-site variability of flux can be higher, with hollows producing 20 times as much methane as hummocks (e.g. Bubier et al.,1993). Further limitations to these data are the fact that winter emissions are rarely measured, but may be an important component of the annual budget in boreal regions (e.g. Dise, 1993), and that episodic fluxes of methane may be missed, associated with either decreases in atmospheric pressure (Mattson and Likens, 1990) or a falling water table (Windsor et al., 1992). Thus, there can be large uncertainties associated with the estimate of an average annual flux.

Estimates of annual methane flux from over 120 sites range from 0 to 130 g CH_4 m⁻² yr⁻¹, with the majority < 10 g CH_4 m⁻² yr⁻¹ (Figure 2). Whilst there are some overall patterns between the ecological classification of the wetlands and methane emission rates, there is considerable variability within each class. The highest rates are generally found at aggrading peatland sites where the water table remains at or above the surface of the peat and vegetation, such as in *Sphagnum*-sedge lawns, or where the peat is degrading, such as in small pools (e.g. Hamilton et al., 1992). Rates are also large where there is shallow inundation of organic soils, such as that created by beavers: beaver ponds can emit between 5 and 130 g CH_4 m⁻² yr⁻¹ (Table 1). Although this range is at the upper end of the methane flux we have measured for wetlands, because of the small coverage of beaver ponds in many areas (e.g. 0.2% near Cochrane and 1% near Dorset [Bubier et al., in press; Roulet et al., 1992a]), beaver ponds do not appear to be a large source of atmospheric methane, contrary to previous suggestions (e.g. Naiman et al., 1991; Nisbet, 1989).

Using methane emission rates for different types of wetlands, and an areal estimate of their cover derived from aerial photography, satellite images or ground surveys, estimates of regional emission rates can be made. For example, wetlands in the Dorset, Cochrane, and Schefferville areas (from the low boreal, high boreal and subarctic zones, respectively) emit an average of 1.6, 3.4, and 2.2 g CH_4 m⁻² yr⁻¹ (Bubier et al., 1993; Moore et al., 1990; Roulet et al., 1992a), respectively. At a different scale, the Northern Wetlands Study calculated an emission rate of 0.5×10^{12} g CH_4 yr⁻¹ for the entire Hudson Bay Lowland (0.32×10^6 km²), an average of 1.5 g CH_4 m⁻² yr⁻¹ (Roulet et al., in press). Assuming an average emission rate of 2 g CH_4 m⁻² yr⁻¹ from Canadian wetlands, and a coverage of 1.27×10^{12} m² (National Wetlands Working Group, 1988), we can estimate that they emit 2 to 3×10^{12} g CH_4 yr⁻¹. Using this figure, the atmospheric contribution of methane from wetlands between 40 and 70 °N is probably between 20 and 40×10^{12} g yr⁻¹. The value recently calculated recently by Bartlett and Harriss (in press) lies within this range.

III. Controls on Methane Emission Rates

Methane emissions from wetland soils to the atmosphere are dependent on the rates of methane production and consumption and the ability of the soil and plants to transport the gas to the surface. Three major environmental controls on emission rates can be recognized: temperature (which influences microbial activities of methane production and consumption), water table position (a surrogate for aerobism/anaerobism and thus methane production and consumption), and substrate characteristics (the ability of the soil to promote methane production and consumption). In addition, there is evidence that plant type and cover can strongly influence methane emission rates, through both the ability of some plants to transport methane from the soil to the atmosphere and the creation of micro-environments for methane production and consumption around plant roots. In a subarctic fen, Whiting and Chanton (1992) have been able to correlate methane emission rates with plant cover, and in a tundra wetland Bartlett et al. (1992) observed larger methane fluxes from stands of *Carex rostrata* and *Arctophila fulva* than from open water.

A. Temperature

Several laboratory studies have shown the importance of temperature in controlling rates of methane production and consumption under anaerobic and aerobic conditions, respectively (e.g. Svensson, 1984). Using samples collected from a range of Canadian peatlands incubated as slurries, Dunfield et al. (1993) have shown that methane consumption rates had optima in the range 20-25 °C, with generally low

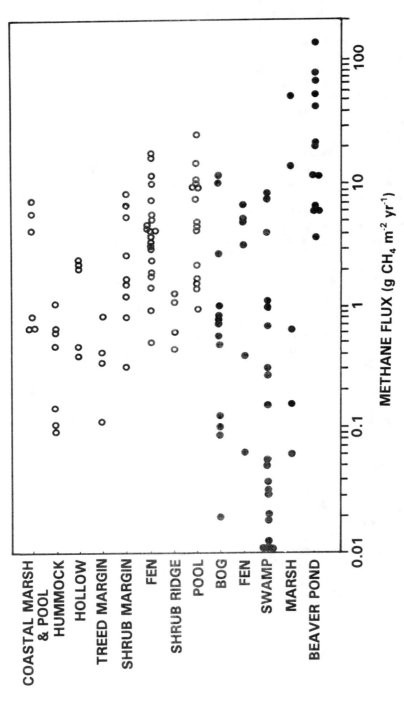

Figure 2. Estimated annual methane emission rates from wetlands in Canada, based on broad ecological grouping for subarctic (open circles) and boreal-temperate (shaded circle) regions. Data for subarctic sites derived from Holland (1992), Moore et al. (1990), and Moore et al. (in press); data for boreal-temperate sites derived from Bubier et al. (1993a), Moore and Knowles (1990), Roulet et al. (1992a), and Windsor (1993).

Table 1. Methane emissions from beaver ponds and margins

Source	Location	Ecological site	Annual flux g CH_4 m^{-2} yr^{-1}
Bubier et al. (1993a)	Ontario	Marsh (-18 cm water)	0.7[a]
		Marsh (-3 cm water)	13.4[a]
		Submerged (5 cm water)	52.5[a]
		Deep water (50-150 cm water)	43.7[a]
Ford and Naiman (1988)	Québec	Undisturbed stream	0.1-0.4
		Beaver pond	5.9
Naiman et al. (1991)	Minnesota	Upland forest	0.3
		Sedge meadow	0.5
		Submergent (45 cm water)	14.3
		Deep pond (125 cm water)	11.6
Roulet et al. (1992a)	Ontario	Beaver ponds	7.6
Vitt et al. (1990)	Alberta	Beaver pond	76.2
Weyhenmeyer (1992)	Ontario	Beaver pond	6.2
Windsor (1993)	Québec	Submerged margin	78
		Shallow water (15-50 cm)	130
Yavitt et al. (1990)	W. Virginia, Maryland	Beaver pond	300, 250[b]
Yavitt et al. (1992)	New York	Beaver ponds	34-40

[a] Assuming a season of 150 d, for which the rates from early-June to mid-August were representative, and no significant emission of methane during the winter; [b] daily flux (g CH_4 m^{-2} yr^{-1}) reported at two dates, expressed as mean and median.

activation energies (20-80 kJ mol^{-1}) and Q_{10} values. The response of methane production to temperature was stronger, with optima in the range 25-30 °C and high activation energies (123-271 kJ mol^{-1}) and Q_{10} values (5.3 - 16), especially at the lower end of the temperature range (i.e. 5-15 °C).

Although the temperature response of methane production and consumption rates is clearly defined in the laboratory, evidence for major temperature controls on field fluxes of methane is less obvious. In some cases, thermal regime evolution explains much of the seasonal pattern of flux (e.g. Crill et al., 1988). In many of our own studies, simple correlations of the seasonal pattern with soil temperatures are weak (e.g. Moore et al., 1990).

B. Water Table Position

Through its control on the creation of aerobic and anaerobic zones in the soil profile, and thus the potential for methane consumption and production, as well as limiting exchange between the atmosphere and soil gases, water table exerts a major control on methane flux from wetlands. Using laboratory columns of peat soils, Moore and Knowles (1989) established a strong linear relationship between the logarithm of methane flux and the water table depth. Using similar columns of bog, fen, and swamp soils, Moore and Dalva (1993) were able to separate the influence of temperature and water table position on methane flux, though there was a significant statistical interaction between temperature and water table position. Columns in which the water table was at the peat surface emitted an average of 5 times the amount of methane of columns where the water table was at a depth of 40 cm. Raising the temperature from 10 to 23 °C increased the average methane flux by 6.6 times, though most treatments revealed a smaller increase.

The vertical movement of the water table is also important: lowering and raising the water table in these laboratory peat columns revealed a strong hysteresis between the falling and rising limbs (methane flux in falling > rising) and that as the water table fell from the surface to a depth of 20 cm, there was an increase in methane flux, counter to the anticipated decrease with a lowered water table. It appears that methane trapped in pore-water is released as the water table is dropped, and this effect is most pronounced in the upper layers, where the gas diffusivity also increases after drainage. As pore-water methane concentrations in peats can be large (<10 mg CH_4 L^{-1}, equivalent to < 10 g CH_4 m^{-2}, Moore et al., 1990; Moore and Dalva, 1993), and a lowering of the water table can cause the rapid release of a portion of this stored methane.

As with temperature, clear statistical identification of the control of water table position on the seasonal pattern of methane flux at wetland sites is often confounded by coincident changes in temperature and plant activities. Thus, we have been generally unsuccessful in being able to clearly explain times series of methane flux from field sites in terms of either water table position, temperature, or a combination of both. Also affecting this poor correlation are the high spatial variability of flux measurements at one date (and resulting imprecision of average flux estimate) and the occurrence of episodic fluxes of methane associated with small drops in water table position (Windsor et al., 1992). Other researchers have been more successful (e.g. Crill et al., 1988; Dise et al., 1993).

The importance of water table in controlling methane flux becomes more important than temperature when comparing average summer flux among sites (Roulet et al., 1992a). This can be illustrated by using the means of flux, water table depth and temperature at 10-20 cm for sites in 5 locations in the boreal and subarctic regions (Figure 3). Although the overall relationship log mean CH_4 flux:mean water table depth was statistically significant but not very powerful for predictive purposes ($r^2 = 0.33$, $p < 0.001$), analysis of data from individual regions produced regressions with almost identical slopes (0.29-0.37), yet significantly different constants (0.47-1.89). Temperature played a secondary role, and its inclusion in the regressions did little to improve the predictive capacity. Water table position and temperature are often closely related in peat profiles. A key control on methane flux is the ability of the peat profile to consume the methane produced, as has been found by Swedish workers (Sundh et al., 1992). The difference in the regression constants of the regional analysis may be related to the relative permanance of wetness in the regions: the smallest constant was observed in the Dorset region, which is the southernmost and most susceptible to water table drawdown in the summer. The occurrence of oxygenated water at rich fen sites with rapid groundwater movement may explain the small methane flux from sites which have a shallow water table.

The flux:water table relationships provide the basis for a simple predictive model of methane flux, based on average water table position. As there is a strong correlation between some plant species and water table position in many peatlands, an alternative technique is to use plant community and individual species to predict methane flux. Bryophytes show the strongest correlation and could provide the same predictive power as water table, with the advantage that plant communities and species are more easily mapped than water table depth (Bubier, pers. comm.). Whalen and Reeburgh (1992), based on a four year study of methane flux from tundra sites, suggest that the system is too complex to allow the successful use of a single variable as a predictor of methane flux, and that predictors of methane flux should focus on parameters that integrate factors important in methane production and consumption.

C. Substrate and Microbial Controls

Methane flux rates from wetland soils depend on the ability of the microbial populations to both produce and consume methane in the soil profile, and the controls on these rates. Although the processes of methane production and consumption are complex, progress in understanding has been made (see reviews by Boone, 1991; King, 1992; Knowles, 1993; and Topp and Hanson, 1991).

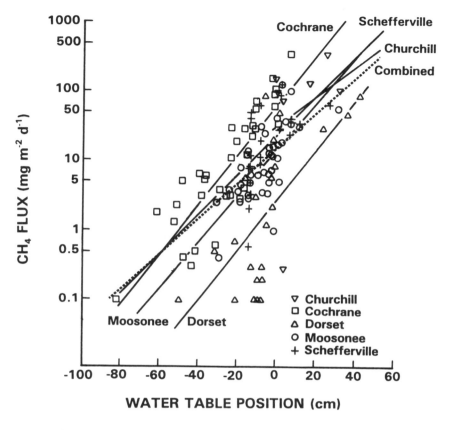

Figure 3. Relationship between the mean daily methane flux (as logarithm) and mean water table position for wetland sites in 5 locations from boreal and subarctic Canada. Each point represents between 50 and 360 methane flux measurements at each wetland site during the snow-free season, generally May to September. Data were derived from the references indicated in Figure 1. (From Moore and Roulet, 1993.)

The nutritional requirements for methanogens and methanotrophs are complex. Studies of potential substrates used by the methanogenic microflora in Canadian peat samples showed that there was no response to acetate amendment at the higher pH values and in most cases acetate inhibited methane production at low pH values (Knowles, 1993). On the other hand, most peats showed a positive response to H_2 supplement, and the evidence suggests that the dominant methanogens are $H_2 + CO_2$ users. In at least two peat types, incubation in 2-10% H_2 caused the optimum pH for methanogenesis to decrease by 1-3 units. This might suggest that the limitation of methane production at low pH values is caused by effects of low pH, not on methanogens *per se*, but on other organisms which can control H_2 availability.

Most of the detailed laboratory work on controls of methanogens and methanotrophs is not readily transferrable to field conditions, and relationships between methane flux and more broadly based characteristics of soils are required. Using samples from a Minnesota bog, Williams and Crawford (1985) were able to illustrate the importance of pH on methanogenesis in laboratory incubations, with methanogenic activity down to pH 3.1. Using laboratory incubations of peat slurries in which the pH was changed, Dunfield et al. (1993) showed that methane producing and consuming organisms were not well adapted to *in situ* pH values (Figure 4). The optimum pH for methane production and consumption was about 2 pH

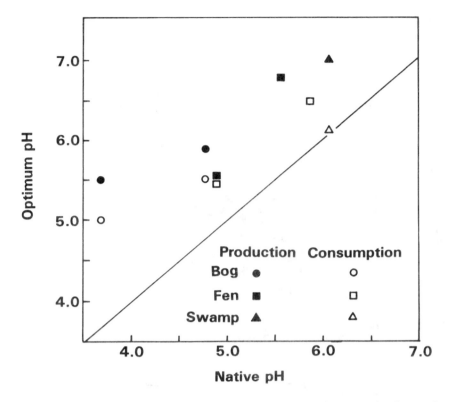

Figure 4. Relation between the optimum pH for methane production and consumption under laboratory conditions and the native pH of the peat samples. (From Dunfield et al., 1993.)

units higher than *in situ* pH at low pH values, but only 0-1 units higher in the less acid samples. There was no indication of the presence of acidophilic organisms in the samples. The significance of this relationship to the understanding of field methane fluxes is not clear, as, in some areas, both large and small methane fluxes have been reported from very acid peats (e.g. Moore and Knowles, 1990). Although soil properties, such as pH, affect the rates of methane production and consumption, their control on field fluxes may be overriden by variations in temperature, water table, and plant type.

The importance of the microbial populations of methane producers and consumers in controlling flux can be illustrated by measurements of methane production and consumption potentials derived from incubations of peat slurries under anaerobic and methane-enriched aerobic conditions (Yavitt et al., 1988). In the Moosonee area of the Hudson Bay Lowland, Moore et al. (in press) observed variations in methane production and consumption potentials in the surface horizons of peat profiles covering 3 orders of magnitude. Anaerobic methane production potentials ranged from 0.02 to 10 μg CH_4 g^{-1} d^{-1}, whereas aerobic methane consumption potentials ranged from 0.1 to 50 μg CH_4 m^{-1} d^{-1}. Hummock samples showed very small potential rates of consumption and production, samples collected from above the water table exhibited large consumption but small production rates and those collected from beneath the water table had large rates of both consumption and production. Variations in methane consumption and production potentials help explain differences in field methane fluxes, in addition to the role of water table position and temperature. Similarly, changes in methane production and consumption potential from drained peatland soils also explain the change in these soils from methane sources to sinks upon drainage (e.g. Glenn et al., 1993; Roulet et al., 1993).

IV. The Effect of Drainage and Climatic Change on Methane Emissions

A. Drainage

Lowering of the water table in peatlands has occurred over $12\text{-}0 \times 10^{12}$ m^2 in temperate, boreal, and subarctic regions (Armentano and Menges, 1986; Gorham, 1991). It would be expected that this would increase the thickness of the zone of methane consumption and decrease the thickness of the zone of methane production in the peat profile, resulting in a reduction in methane emission rates.

At the Wally Creek Experimental Forest, near Cochrane, forested peatlands have been subjected to a water table lowering of 5 to 30 cm, depending on distance from the drainage ditch (Roulet et al., 1993). Peatlands were converted from a source (5-15 mg m^{-2} d^{-1}) when undisturbed to a very small sink of methane (-0.1 to -0.3 mg m^{-2} d^{-1}) when drained, with the conversion from source to sink occurring when the water table was at a depth of 25-30 cm beneath the peat surface. This change was induced by the increased methane consumptive capacity and the elimination of methane production in the oxidized unsaturated zone. However, there was substantial methane emitted from the drains, showing that overall methane emission rates from drained peatlands depend on drain spacing (Roulet and Moore, pers. comm., 1993). Similar conclusions have been reached in Scandinavian studies of peatland drainage for forestry (Martikainen et al., 1991).

The effect of a second, more drastic, perturbation was examined on swamp soils drained for horticultural crop production near Napierville, Québec (Glenn et al., 1993). Compared to undrained swamps, which emit 1-4 g CH_4 m^{-2} yr^{-1} (Moore and Knowles, 1990), the lowering of the water table to a depth of between 75 and 150 cm resulted in the conversion of the soils to small methane sinks (-0.1 to -0.4 g CH_4 m^{-2} yr^{-1}). There was no evidence of methane consumption by the peat at sites where the chambers were located on soil between the crops, and thus did not contain roots, suggesting that the rhizosphere plays an important role in soil methane consumption.

Assuming that, prior to drainage, peatlands have an average methane flux of 2-5 g m^{-2}, then drainage of $12\text{-}20 \times 10^{10}$ m^2 will have resulted in a reduction of methane emission to the atmosphere of between 0.2 and 1.0×10^{12} g yr^{-1}, a small proportion of the overall emission rates presently estimated to be about 500×10^{12} g (Fung et al., 1991). Drained peatlands have very small methane consumption rates (generally <2 mg CH_4 m^{-2} d^{-1}), similar to the rates of methane consumption observed in forest, grassland and upland tundra soils (e.g. Adamsen and King, 1993; Steudler et al., 1990; Striegl et al., 1992; Whalen and Reeburgh, 1990; Whalen et al., 1991).

B. Climate Change

Although there are some disagreements on specific patterns, all the General Circulation Models predict substantial increases in temperature (particularly during the winter) in the areas of major Canadian wetland coverage, associated with a doubling of atmospheric carbon dioxide concentrations. Changes in precipitation are less substantial, but suggest a slight increase. How are these changes in climate likely to affect emission of trace gases from peatlands? The three controls of temperature, water table position, and plant production and tissue quality are likely to change in response to changing climate and the resulting change in flux will be the balance between the competing influences of these three factors.

One way to answer this question is to measure methane flux over a number of years at the same site, and use a climate analogue approach to estimate fluxes under the GCM scenario conditions. Research projects are rarely designed to measure fluxes over many years, but there are studies which show that substantial year-to-year variations in flux can occur, for example in temperate swamps and subarctic fens (Moore and Knowles, 1990) and in tundra (Whalen and Reeburgh, 1992).

A second approach is to use the observed dependence of methane fluxes on temperature and water table position derived from field measurements and to apply changes in thermal and hydrologic regimes from GCM scenarios to predictions of changes in methane flux. We have used these GCM summer air temperature and precipitation changes (+ 3 °C and + 1 mm d^{-1}) for several sites across Canada to estimate changes in peatland thermal regime and position of the water table, utilizing a one-dimensional water and heat flux model (Roulet et al., 1992b). The increase in evapotranspiration, derived from higher summer temperatures, was greater than the increase in precipitation, resulting in a lowering of the water table of between 8 and 15 cm. Peat temperatures at 10 cm increased by 0.8-2.0 °C at Schefferville. We then applied these changes to our regression model of temperature and water table control on methane flux, derived from Moore et al. (1990). The temperature increase resulted in <15% increase in methane flux, but the lowered water table resulted in a decrease of 50 - 80 %, suggesting that the overall response of northern fens to climate change will be a reduction in methane emissions (Table 2). A greater sensitivity of emissions to water table than temperature is also indicated by the laboratory column experiment (Moore and Dalva, 1993). Harriss and Frolking (in press) used mean summer temperature anomalies for the 20th century of +- 2 °C and the methane flux:temperature relationship of Crill et al. (1988) to estimate what the inter-annual variability in methane emissions for northern wetlands might be. They estimate a maximum variation of 5 Tg CH_4 yr^{-1}; with northern wetlands contributing approximately 30 Tg CH_4 yr^{-1}, this amounts to < 20%. However, none of the above approaches adequately takes into account the changes in vegetation that are likely to result from climatic change. These changes may have a more pronounced long-term influence on methane fluxes, compared to the simple mechanistic model described above.

V. Conclusions

Methane fluxes from wetlands are characterized by large temporal and spatial variability, which inhibit the interpretation and extrapolation of the results. Measurements of methane emissions from over 120 wetland sites in Canada during the snow-free season suggest that annual fluxes of methane range from 0 to 130 g CH_4 m^{-2} yr^{-1}, most being <10 g m^{-2} yr^{-1}. This translates in to an estimate of 2-3 Tg CH_4 yr^{-1} from Canadian wetlands and 20-40 Tg CH_4 yr^{-1} from wetlands between 40 and 70° N. Emission rates are controlled by rates of methane production and consumption in the soil and the transport of the methane to the surface, through soil or plants. Emission rates are correlated with soil temperature and water table position, or a combination of both, though the relationship may be complex. Water table depth and vegetation type offer opportunities for predicting methane emission rates within wetland regions. Although the substrate controls on methane production and consumption have been identified in the laboratory, their influence in field soils is probably overriden by variations in temperature, water table position and plant cover. Drainage of peatlands converts them from a source into a small sink of methane, with uptake rates < 2 mg CH_4 m^{-2} d^{-1}. Combination of temperature and water table changes associated with 2 x CO_2 climate change scenarios suggests that methane flux will be lowered from some northern Canadian peatlands.

References

Adamsen, A.P.S. and G.M. King, 1993. Methane consumption in temperate and subarctic forest soils: rates, vertical zonation, and responses to water and nitrogen. *Appl. Environ. Microbiol.* 59:485-490.

Armentano, T.V. and E.S. Menges. 1986. Patterns of change in the carbon balance of organic-soil wetlands of the temperate zone. *J. Ecol.* 74:755-774.

Aselmann, I. and P.J. Crutzen. 1989. Global distribution of natural freshwater wetlands and rice paddies, their net primary productivity, seasonality and possible methane emissions. *J. Atmos. Chem.* 8:307-358.

Bartlett, D.S., K.B. Bartlett, J.M. Hartman, R.C. Harriss, D.I. Sebacher, R. Pelletier-Travis, D.D. Dow, and D.P. Brannon. 1989. Methane emissions from the Florida Everglades: patterns of variability in a regional wetland ecosystem. *Global Biogeochem. Cycles* 3:363-374.

Bartlett, K.B., P.M. Crill, R.L. Sass, R.C. Harriss, and N.B. Dise. 1992. Methane emissions from tundra environments in the Yukon-Kuskokwim Delta, Alaska. *J. Geophys. Res.* 97:16645-16660.

Bartlett, K.B. and R.C. Harriss. Review and assessment of methane emissions from wetlands. *Chemosphere* 26:261-320.

Botch, M.S. and V.V. Masing. 1983. Mire ecosystems of the U.S.S.R. p. 95-152. In: A.J.P. Gore (ed.). *Ecosystems of the World 48 - Mires: Swamp, Bog, Fen and Moor.* Elsevier, Amsterdam.

Boone, D.R. 1991. Ecology of methanogenesis. p. 57-70. In: J.E. Rogers and W.B. Whitman (eds.). *Microbial Production and Consumption of Greenhouse Gases: Methane, Nitrogen Oxides, and Halomethanes.* American Society for Microbiology, Washington.

Bubier, J., T.R. Moore and N.T. Roulet. 1993a. Methane emissions from wetlands in the boreal region of northern Ontario, Canada. *Ecology* 74:2240-2254..

Bubier, J., A. Costello, T.R. Moore, N.T. Roulet and K. Savage. 1993b. Microtopography and methane flux in boreal peatlands, northern Ontario, Canada. *Can. J. Bot.* 71:1056-1063.

Cicerone, R.J. and R.S. Oremland. 1988. Biogeochemical aspects of atmospheric methane. *Global Biogeochem. Cycles* 2:299-328.

Crill, P.M., K.B. Bartlett, R.C. Harriss, E. Gorham, E.S. Verry, D.I. Sebacher, L. Madzar, and W. Sanner. 1988. Methane flux from Minnesota peatlands. *Global Biogeochem. Cycles* 2:371-384.

Dise, N.B. 1993. Methane emission from Minnesota peatlands: spatial and seasonal variability. *Global Biogeochem. Cycles* 7:123-142..

Dise, N.B., E. Gorham, and E.S. Verry. 1993. Environmental factors controlling methane emissions from peatlands in northern Minnesota. *J. Geophys. Res.* 98:10583.

Dunfield, P., Knowles, R., R. Dumont, and T.R. Moore. 1993. Methane production and consumption in temperate and subarctic peat soils: response to temperature and pH. *Soil Biol. Biochem.* 25:321-326.

Ford, T.E. and R.J. Naiman. 1988. Alteration of carbon cycling by beaver: methane evasion rates from boreal forest streams and rivers. *Can. J. Zoo.* 66:529-533.

Fung, I., J. John, J. Lerner, E. Matthews, M. Prather, L. Steele, and P. Fraser. 1991. Global budgets of atmospheric methane: results from a three-dimensional global model synthesis. *J. Geophys. Res.* 96:13033-65.

Glenn, S., A. Heyes, and T.R. Moore. 1993. Carbon dioxide and methane fluxes from drained peat soils, southern Quebec. *Global Biogeochem. Cycles* 7:247-258..

Gorham, E. 1991. Northern peatlands: role in the carbon cycle and probable responses to climatic warming. *Ecological Applications* 1:182-195.

Hamilton, J.D., C.A Kelly, J.W.M. Rudd, and R.H. Hesslein. Flux to the atmosphere of CH_4 and CO_2 from wetland ponds in the Hudson Bay Lowland. *J. Geophys. Res.* (In press.)

Harriss, R.C. and S. Frolking. IThe sensitivity of methane emissions from northern freshwater wetlands to global change. In: P. Firth and S. Fisher (eds.). *Global Warming and Freshwater Ecosystems.* (In press.)

Holland, S. 1992. *Methane Emission in a Subarctic Wetland Environment.* M.Sc. thesis, McMaster University, Hamilton, Ontario.

King, G.M. 1992. Ecological aspects of methane oxidation, a key determinant of global methane dynamics. *Adv. Microb. Ecology* 12:431-468.

Knowles, R. 1993. Methane: processes of production and consumption. p. 145-156. In: *Agricultural Ecosystem Effects on Trace Gases and Global Climate Change.* ASA Special Publication no. 55, Madison, Wisconsin.

Martikanen, P.J., H. Nykanen, J. Silvola, and P. Crill. 1991. Methane and carbon fluxes, and nitrous oxide production at some virgin and drained peat sites in Finland. *Proceedings 10th. Int. Symp. Environmental Biogeochem.*, 25-26.

Matthews, E. and I. Fung. 1987. Methane emission from natural wetlands: global distribution, area, and environmental characteristics of sources. *Global Biogeochem. Cycles* 1:61-86.

Mattson, M.D. and G.E. Likens. 1990. Air pressure and methane fluxes. *Nature* 347:718-719.

Moore, T.R. and M. Dalva. 1993. The influence of temperature and water table position on CO_2 and CH_4 emissions from laboratory columns of peatland soils. *J. Soil Sci.* 44:651-664.

Moore, T.R., A. Heyes, and N.T. Roulet. Methane emissions from wetlands, southern Hudson Bay Lowland. *J. Geophys. Res.* (In press.)

Moore, T.R. and R. Knowles. 1989. The influence of water table levels on methane and carbon dioxide emissions from peatland soils. *Can. J. Soil Sci.* 69:33-38.

Moore, T.R. and R. Knowles. 1990. Methane emissions from bog, fen and swamp peatlands in Quebec. *Biogeochem.* 11:45-61.

Moore, T.R. and N.T. Roulet. 1991. A comparison of dynamic and static chambers for methane emission measurements from subarctic fens. *Atmosphere-Ocean* 29:102-109.

Moore, T.R. and N.T. Roulet. 1993. Methane flux:water table relations in northern peatlands. *Geophys. Res. Lett.* 20:587-590.

Moore, T.R., N.T. Roulet, and R. Knowles. 1990. Spatial and temporal variations of methane flux from subarctic/northern boreal fens. *Global Biogeochem. Cycles* 4:29-46.

Naiman, R.J., T. Manning, and C.A. Johnston. 1991. Beaver population fluctuations and tropospheric methane emissions from boreal wetlands. *Biogeochem.* 12:1-15.

National Wetlands Working Group. 1988. *Wetlands of Canada.* Polyscience Publications, Montreal, Quebec.

Nisbet, E.G. 1989. Some northern sources of atmospheric methane: production, history, and future implications. *Can. J. Earth Sci.* 26:1603-1611.

Roulet, N.T., R. Ash, and T.R. Moore. 1992a. Low boreal wetlands as a source of atmospheric methane. *J. Geophys. Res.* 97:3739-3749.

Roulet, N.T., R. Ash, W. Quinton, and T.R. Moore. 1993. Methane flux from drained peatlands: the effect of a persistent water table lowering on flux. *Global Biogeochem. Cycles.* 7:749-769.

Roulet, N.T., A. Jano, C.A. Kelly, L. Klinger, T.R. Moore, R. Protz, and W.R. Rouse. The role of the Hudson Bay Lowland as a source of atmospheric methane. *J. Geophys. Res.* (In press.)

Roulet, N.T., T.R. Moore, J. Bubier, and P. Lafleur. 1992b. Northern fens: methane flux and climatic change. *Tellus* 44B:100-105.

Steele, L.P., E.J. Dlugokencky, P.M. Lang, P.P. Tans, R.C. Martin, and K.A. Masarie. 1992. Slowing down of the global accumulation of atmospheric methane during the 1980s. *Nature* 358:313-316.

Steudler, P.A., R.D. Bowden, J.M. Melillo, and J.D. Aber. 1990. Influence of nitrogen fertilization on methane uptake in temperate forest soils. *Nature* 341:314-316.

Striegl, R.G., T.A. McConnaughey, D.C. Thorstenson, and J.C. Woodward. 1992. Consumption of atmospheric methane by desert soils. *Nature* 357:145-147.

Sundh, I., C. Mikkela, M. Nilsson, and B. Svensson. 1992. Potential methane oxidation in a Sphagnum peat bog: relation to water table level and vegetation type. *Int. Peat Congr.* 142-51.

Svensson, B.H. 1984. Different temperature optima for methane formation when enrichments from acid peat are supplemented with acetate or hydrogen. *Appl. Env. Microbiol.* 48:389-394.

Topp, E. and R.S. Hanson. 1991. Metabolism of radiatively important trace gases by methane-oxidizing bacteria. p. 71-90. In: J.E. Rogers and W.B. Whitman (eds.). *Microbial Production and Consumption of Greenhouse Gases: Methane, Nitrogen Oxides, and Halomethanes.* American Society for Microbiology, Washington.

Vitt, D.H., S. Bayley, T. Jin, L. Halsey, B. Barker, and R. Craik. 1990. *Methane and Carbon Dioxide Production from Wetlands in Boreal Alberta.* Report of Contract No. 90-0270, Alberta Environment Ministry, Edmonton, Alberta.

Weyhenmeyer, C.E. 1992. *Methane Emissions from a Boreal Beaver Pond.* M.Sc. thesis, Trent University, Peterborough, Ontario.

Whalen, S.C., W.S. Reeburgh, and K.S. Kizer. 1991. Methane consumption and emission by taiga. *Global Biogeochem. Cycles* 5:261-273.

Whalen, S.C. and W.S. Reeburgh. 1990. Consumption of atmospheric methane by tundra soils. *Nature* 346:160-162.

Whalen, S. and W.S. Reeburgh. 1992. Inter annual variation in tundra methane emissions: A 4-year time series at fixed sites. *Global Biogeochem. Cycles.* 6:139-159.

Whiting, G.J. and J.P. Chanton. 1992. Plant-dependent CH_4 emission in a subarctic Canadian fen. *Global Biogeochem. Cycles* 6:225-231.

Williams, R.T. and R.L. Crawford. 1985. Methanogenic bacteria, including an acid-tolerant strain, from peatlands. *Appl. Env. Microbiol.* 50:1542-1544.

Windsor, J. 1993. Methane emissions from the eastern temperate wetland region and spectral characteristics at subarctic fens. M.Sc. thesis, McGill University, Montreal, Ontario.

Windsor, J., T.R. Moore and N.T. Roulet. 1992. Episodic fluxes of methane from subarctic fens. *Can. J. Soil. Sci.* 72:441-452.

Yavitt, J.B., G.E. Lang, and D.M. Downey. 1988. Potential methane production and oxidation rates in peatland ecosystems of the Appalachian Mountains, United States. *Global Biogeochem. Cycles* 2:253-268.

Yavitt, J.B., G.E. Lang, and A.J. Sexstone. 1990. Methane fluxes in wetland and forest soils, beaver ponds, and low-order streams of a temperate forest ecosystem. *J. Geophys. Res.* 95:22463-22474.

Yavitt, J.B. L.L. Angell, T.J. Fahey, C.P. Cismo, and C.T. Driscoll. 1992. Methane fluxes, concentrations, and production in two Adirondack beaver impoundments. *Limnol. Oceanogr.* 37:1057-1066.

CHAPTER 13

Decomposition of Organic Matter and Carbon Emissions from Soils

Darwin W. Anderson

I. Introduction

A recent paper has said that "Soil organic matter is characterized by a three-dimensional, heteropolymeric, complex structure and polydisperse distribution in the mineral matrix of soil" (Hempfling and Schulten, 1991 as quoted in Schulten et al., 1992, p. 237). This description of organic matter or at least the characteristics of organic matter certainly points to the complexity of its structure, its association with mineral soil and, perhaps intuitively, its varied role within soils. Understandably, and supported by the above description, it has been difficult for those of us who have been trained to think linearly and deductively to appreciate the function and nature of organic matter in all their complexities. Generally, I surmise that there have been two schools of thought, those who have studied what organic matter is, and those who are concerned with what organic matter does. Today, with improved technologies for examining the chemical structures, biological processes and functions of organic matter, coupled with the urgency brought on by an increased interest in global carbon cycling, we appear at a juncture where we may be able to work collectively in order to understand both what soil organic matter is, and what it does. The subject of the paper is to look at processes of organic matter decomposition and carbon dioxide evolution from soils, using both classical and recent work on the structure and function of organic matter.

II. Carbon Stores in Soils

The soils of the world contain more organic carbon (C) than either the biota or that exists as carbon dioxide (CO_2) in the atmosphere. A recent estimate of the global soil C pool is 1200 to 1500 Pg, with an additional 150 Pg in peatlands (Bolin and Fung, 1992). The latter estimate is almost certainly an underestimate. For example, Canada has nearly 1 million Km^2 of peatlands, with a conservative estimate of depth being one meter. Considering an average bulk density of 0.15 Mg m^{-3} and a C content of 45%, Canada peatlands represent a store of about 75 Pg of organic C.

Largest amounts of soil C, up to 2000 Mg C ha^{-1}, occur in the peatlands, where C decomposition is slowed by anaerobic and often cold soil environments, and in cryic or pergelic soils where C is mechanically moved below the permafrost table. Another group of soils with large C stores are the organic soils that form under productive forests in cool, high rainfall areas. They are classified as Folisols in Canada or Folists in Soil Taxonomy. Fox et al. (1987) describe 11 Folisols with depths from 0.27 to 1.03 m overlying rock, fragmented rock or mineral soil, and supporting productive forest in coastal British Columbia. Bulk densities are not given, but assuming values of about 0.4 Mg m^{-3} (Witty and Arnold, 1970) indicates a C store of about 1,000 Mg ha^{-1}. If the storage of large C reserves in wet or permafrost soils can be ascribed mainly to limiting physical conditions for decomposition, the Folisols which occur in moist (but not anaerobic) environments with high productivity must be considered to represent a "biochemical" accumulation of C.

The peat of Folisols usually is well-decomposed or humified material with strongly acidic pH values. Radiocarbon dates indicate the accumulation of about 0.5 m depth of peat in 500 to 1000 yr under productive forest (Fox et al., 1987). The well-drained peat soils under productive forest appear to represent a build-up of highly recalcitrant humic materials which are biochemically resistant to further decomposition.

Most mineral soils contain about 50 to 200 Mg C ha^{-1}, with slightly more in grasslands than in forests. A comparison of related soils in Alberta, Canada indicates a C store of about 80 Mg ha^{-1} in semi-arid grassland soils, to about 150 Mg ha^{-1} in soils of the humid grasslands, with somewhat less C in soils under the Boreal Forest (Figure 1). There are several points not evident in Figure 1. One concerns the magnitude of storage of carbon in B horizons and, to a degree, C horizons. Carbon in subsoils cycles much more slowly than the organic matter of surface horizons and, therefore, can be considered a more secure store of soil C (Table 1). A second point deals with the storage of C in the forest floor, which usually accounts for up to one-half of the soil C stored in forest soils, and must be considered a less secure C pool. Finally, there is the effect of clay on the amount of and nature of C stored in soils, which must be considered.

III. Theoretical Considerations, Biological Recalcitrance, and Structure of Humic Substances

A. Theoretical Considerations

Theoretical considerations based on the maximum power principle, as formulated by Lotka (1922) and discussed further by Odum (1983) postulate that systems will prevail that maximize the flow of useful energy. Veizer (1988) suggests that natural systems have a tendency to self-organize in order to maximize useful power -- that is store more energy that can be fed back or depreciated to catalyze the inflow of additional energy. Applying the maximum power principle to soil ecosystems does not imply, at least in my opinion, that soils will accumulate ever-increasing amounts of organic C. Rather, the C balance in soils will reflect the storage of energy in organic matter countered by the requirements for energy and nutrients by the biota of succeeding generations. Soil C stores, therefore, will reflect a balance between organic inputs mainly from plant residues, and decomposition driven by the energy and nutrient demands of the microbial population, and plant requirements for nutrients. A corollary of this idea will be that organic matter in a particular soil will be tough (recalcitrant) enough to exist or endure in that particular edaphic environment, but not too tough to severely limit decomposition and the release of nutrients.

A detailed comparison of the amount and structural characteristics of the organic matter in a sequence of soils from the semi-arid Brown (Aridic Boroll) to humid Black (Udic Boroll) soil zones across Saskatchewan supports the contention that organic matter will be as tough as required (Anderson et al., 1974b). Organic matter in Brown soils has relatively more of the partially decomposed residue (mainly roots), simple, weakly humified humic materials with a high proportion of hydrolyzable or labile components. Black soils of more humid environments have higher humus contents, more strongly decomposed or humified materials, a strong presence of condensed alkyl aromatic structures, and a larger proportion of non-hydrolyzable or chemically resistant components. These differences are consistent with the idea that humus will be as resistant to decomposition as required to exist at a particular equilibrium level in a soil, but not so resistant to decomposition that nutrients will be unavailable to the biota.

B. Radiocarbon Age of Humus and Humus Fractions

The degree of biological recalcitrance or resistance to decomposition of soil C is supported by the long mean residence times for humus and humic fractions. Anderson and Paul (1984) used radiocarbon dating, and report a mean age of 1910 ± 45 years for the organic C in the A horizon of a clayey Black soil in Saskatchewan. Even older ages were characteristic of the strongly condensed humic acid components, and materials resistant to acid hydrolysis. Most surface soils, however, have mean ages of organic C in the range of a few hundred to perhaps one thousand years (Campbell et al., 1967).

Radiocarbon ages of paired, native and cultivated, silty Black soils in the parkland region of Saskatchewan indicate several interesting points (Table 1). The methods of ultrasonic dispersion and size fractionation, and organic C contents of the soils and size fractions are described in Tiessen and Stewart (1983). One is that the mean age of C in B horizons is much greater than surface horizons, reflecting the

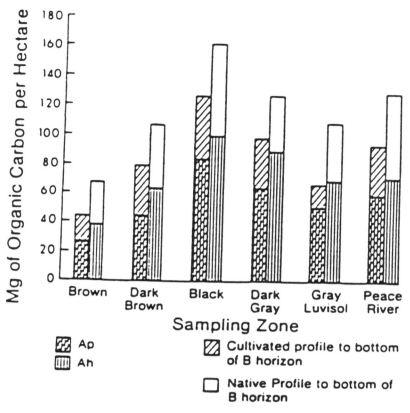

Figure 1. Organic carbon in native and cultivated soils of the different soil zones of Alberta. (After McGill et al., 1988.)

Table 1. Radiocarbon age and $\delta^{13}C$ of organic C of native and cultivated silty Black (Boroll) soils and component size fractions

Soil	Fraction	$\delta^{13}C$ ‰	Equivalent age, yr
Native, Ah horizon	Whole soil	-25.1	385 ± 110
	Coarse silt	-26.3	345 ± 115
	Fine silt	-25.2	440 ± 115
	Coarse clay	-24.8	530 ± 115
	Fine clay	-24.3	400 ± 115
Native, B horizon	Whole soil	-24.2	2420 ± 140
Cult-60 yrs, Ap horizon	Whole soil	-25.0	1100 ± 145
	Coarse silt	-25.6	385 ± 100
	Fine silt	-25.5	900 ± 120
	Coarse clay	-25.0	905 ± 120
	Fine clay	-23.3	965 ± 145
Cult-60 yrs, B horizon	Whole soil	-24.4	2560 ± 155
Cult-90 yrs, Ap horizon	Whole soil	-25.5	735 ± 125

generally slower processes to accumulate the C at depth, and reduced decomposition pressure resulting in longer turnover times. Secondly, the reduction in organic C content that results from cultivation generally results in an increased mean age of the C of Ap horizons. The more biologically active and labile materials decompose preferentially, and the resistant and, therefore, older components become more concentrated. Finally, there are the differences in mean age among humic materials associated with different size fractions, with age generally increasing from coarser to finer fractions. This is consistent with the ideas of Baldock et al. (1992) and others, that material associated with clay is more strongly humified or older than the mainly particulate, organic materials of the coarse silt or fine silt fractions. The increase in $\delta^{13}C$ values also is consistent with increased humification or microbial processing of humus in finer fractions. The older ages for fine clay humus are, however, a departure from earlier work which showed that humic substances associated with the fine clay are younger than the mean age for the soil, or the age of the coarse clay-associated humus (Anderson and Paul, 1984).

C. Studies of Humus Structure Based on Whole Soils

Early work on soil humic substances was limited by the absence of technologies for studying organic matter *in situ* and the difficulty of extracting or isolating a large and representative fraction from the mineral part of the soil. In addition, there is the concern that the process of chemical extraction itself changes the structural characteristics of the humic substances extracted. For example, it was established several decades ago that the humic substances extracted by dilute (0.1 or 0.5 M) NaOH were more aliphatic and of higher molecular weight than the more aromatic materials extracted with sodium pyrophosphate solutions (Theng et al., 1967). The latter extractant isolated strongly condensed, more humified materials, presumably because of the complexing by the pyrophosphate of cations such as Ca or Fe that formed cation bridges between humus and negatively charged sites on clay minerals.

Work at the University of Saskatchewan progressed from using mainly alkali extractants, to the use of alkali extractants in association with pyrophosphate, and then to a method that isolated selectively the materials associated with fine mineral colloids (Anderson et al., 1974a). The clay-associated humic materials were of high molecular weight, largely aliphatic with proportionately more N in hydrolyzable, mainly amino forms rather than being an integral part of the humic acid structure. The material associated with fine clay also was enriched in the microbial metabolites resulting from the decomposition of added, labelled substrates (McGill and Paul, 1976).

Subsequent work in our laboratories, and consistent with developments elsewhere, involved size fractionation following dispersion of the soil with ultrasonic energy in water suspension (Christensen, 1992). About one-half to two-thirds of the humic material is associated with the clay and fine silt fractions, and can be characterized by radiocarbon dating, chemical analyses, and other methods (Anderson and Paul, 1984; Anderson et al., 1981, for example). Although chemical extractions, and even size fractionations have some deficiencies related to possible changes or redistributions of humic materials during the processes in the laboratory, the work did contribute substantively to our ideas about soil humus (Anderson, 1979). These were ideas that have been modified and refined based on the new study methods of the 1980s and 1990s, but remain as a sound basis for subsequent developments.

The new developments of the past decade that I want to mention are those that allow the study of the structure and nature of humus while it remains in the soil. They include, among others, the use of solid state cross-polarization/magic angle spinning ^{13}C nuclear magnetic resonance spectroscopy (CP/MAS ^{13}C NMR) as typified by a major study by Baldock et al., 1992, and pyrolysis-field ionization mass spectrometry (Py-FIMS) as illustrated by several studies of Schulten and coworkers (Schulten et al., 1992; Schulten and Schnitzer, 1993). The studies, which are described in some detail, are but a selection of the considerable work going on in this area of humus research today. They are mentioned not just because of the elegance of the research methodologies, but the significance of findings in term of understanding organic matter processes.

The initial applications of solid state ^{13}C NMR spectroscopy to soils produced featureless spectra that were difficult to interpret, due to strong ^2H-^{13}C dipolar interactions, low sensitivity and chemical shift anisotropy (Baldock et al., 1992). The recent application of high-power dipolar-decoupling, cross polarization (CP) and magic angle spinning (MAS) techniques has resulted in high resolution spectra which provide detailed information on the chemical structure of organic matter. CP/MAS ^{13}C NMR is limited by low C contents (and even lower ^{13}C contents), and by the presence of paramagnetic species such as Cu^{2+}

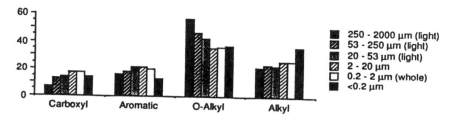

Figure 2. Contribution of the total acquired GP/MAS ^{13}C NMR signal intensity from each type of carbon for a Mollisol from Australia. (After Baldock et al., 1992.)

and Fe^{3+} in many soils. The former limitation can be overcome in soils with <5% organic C by size fractionations that segregate the C-rich, fine fractions.

Studies based on CP/MAS ^{13}C NMR have indicated generally that humic materials have a higher proportion of aliphatic or straight-chain structures and less aromatic or condensed aromatic materials than was indicated by earlier studies using other methods. For example, a summation of the findings for a Mollisol from Australia (Figure 2), based on signal intensities, indicate fewer aromatic, and much stronger aliphatic signals. The signal for O-alkyl is considered to result mainly from carbohydrate and protein materials in plant residues, and decreases in intensity from coarse to fine size fractions. Strong signals indicative of alkyl C, mainly in long chain aliphatics dominated by polymethylene structures, increase with decrease in particle size. The alkyl C is considered to be the most recalcitrant material found in soils, due both to its chemical structure and strong interaction with clays. Aromatic structures make up a surprisingly low proportion of the humic materials, as indicated by the CP/MAS ^{13}C NMR spectroscopy.

Based on their findings using CP/MAS ^{13}C NMR, elemental analysis of the size fractions and previous work, Baldock et al. (1992) have proposed a simple model for the oxidative decomposition of plant materials in mineral soils (Figure 3). The model proposes that the extent of decomposition follows a continuum from fresh plant residues that occur as fragments in the larger size fractions (>20 μm), to partially degraded residues in the intermediate (20-2 μm) fractions, with the most degraded or humified residues, and microbially synthesized components, associated with the clay fractions.

The initial stages of the decomposition include the microbial breakdown of plant materials, carbohydrates and protein, with a proportion of the C evolved as CO_2, and the remainder of the C assimilated and converted into microbial tissue and metabolites. The initial decomposition of O-alkyl C from plant sources will not result in a complete disappearance of its signal, in that microbially produced O-alkyl C will be formed.

After an initial concentration of more recalcitrant aliphatic and aromatic structures, the second stage results in the degradation of the lignin molecules previously surrounded by O-alkyl C in plant fragments, and a reduction in aromatic C. With advancing decomposition (now perhaps more aptly termed the microbial synthesis of humic materials) alkyl C dominates the fine fractions. The alkyl structures are thought to be mainly polymethylene structures. The accumulation of alkyl C is thought to result from both selective preservation by association with clay, and *in situ* synthesis by the microbial population.

Pyrolysis-field ionization mass spectrometry (Py-FIMS), like NMR spectroscopy, can be used on whole soils that contain at least 1.5 to 2% organic matter (Schulten et al., 1992). The method requires the even heating of the sample in a pyrolysis chamber, coupled with a FI mass spectrometer to measure the mass of the compounds released as humic substances are degraded thermally. The assignment of mass signals to most probable compounds relies on complicated computer-based calculations. Py-FIMS appears superior to the CP/MAS ^{13}C NMR in that exact identification of compounds is possible, but the structure of the more complex humic constituents from which the smaller compounds have been released by pyrolysis can not be determined directly. Schulten and Schnitzer (1993) evaluated the data from Py-FIMS, as well as from NMR spectroscopy, oxidative-reduction degradation, colloid chemical studies and electron microscopy, and have proposed a state-of-the-art structural complex for humic substances. The humic acid structure is basically linked alkylaromatic structures, incorporates N in heterocyclic and nitrile forms, as well as the functional groups where most of the oxygen occurs. The elemental composition is $C_{308}H_{328}O_{90}N_5$, with a molecular weight of 5540 Daltons. Linking one molecular weight of humic acid with 10% carbohydrate and 10%

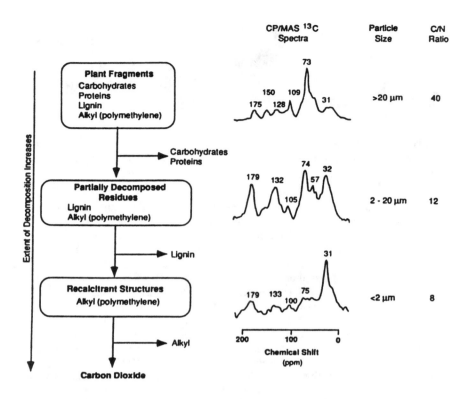

Figure 3. A simple model describing the oxidative decomposition of plant materials in mineral soils, based on the data for an Australian Mollisol. (After Baldock et al., 1992.)

protein (associated but not part of the basic structure) by weight results in an elemental composition of $C_{342}H_{388}O_{124}N_{12}$, and an analysis of 61.8% C, 5.9% H, 29.8% O, and 2.5% N. The theoretical analysis is quite similar to humic acids extracted from many soils.

The Py-FIMS method has contributed substantively to concepts of the chemical structure of humic materials, and to a better understanding of process at the level of measuring sequential changes in the molecular composition of humus. Schulten et al. (1992) used the method to evaluate humus formation in a pot experiment that had been sampled over 34 years. The study included the growth of two grass species in loamy calcareous soil, and the annual incorporation of all plant residues into the soil. The FI mass spectra of stems and leaves, and roots were measured, followed by spectra on the soil in years 2, 7, 13, 19, 25, 29 and 34 of the experiment. There were differences between aboveground plant parts and roots. Lignin dimers were abundant in stems, and sugars and suberin-derived phytosterols dominated the spectra of roots. Sterols appear to be important to the formation and stability of macroaggregates, those larger than 0.25 mm (Dr. C. Monreal, personal communication).

Organic matter contents in the soil increased to about 1.5% in year 7, then levelled off at about 2% in years 13 and 34. The rapid increases in the first decade were attributed mainly to an enrichment of lignin dimers (Table 2). Phenols and lignin monomers as well as the lignin dimers account for about one-quarter of the signal intensities. That, plus the 10% of the signal intensity from alkylaromatic structures indicates a major role for aromatics in the formation of recalcitrant humus. Alkanes and alkenes appear to be minor constituents, a result not consistent with the previously described NMR work (Baldock et al., 1992). The gradual change in the character of the N compounds in humus is of particular interest. Initially most of the N is in amino and amide forms probably associated with protein residues. Heterocyclic forms of N increase with time until they make up about two-thirds of the N in humic materials. This finding is in agreement with data on the proportion of non-hydrolyzable N in humic materials, which increases with degree of

Table 2. Summed ion intensities of mass signals characteristic for classes of molecular subunits in samples from 2nd through 34th year (% of total ion intensity) and distribution of N compounds (%, sum of N compounds = 100%), of a pot experiment documenting humification

Compound class	Loamy marl[a]	\multicolumn{7}{c}{Sampling years}						
		2nd	7th	13th	19th	25th	29th	34th
Phenols/lignin monomers	<1	9	2	10	20	12	13	20
Lignin dimers	<1	1	4	6	4	11	9	6
Alkylaromatics	<1	9	4	8	10	9	9	10
Mono-/poly-saccharides	1	8	9	12	15	14	15	16
Alkanes/alkenes	<1	1	1	1	1	1	2	2
Fatty acids	<1	1	1	2	1	3	3	2
N compounds	<1	4	2	4	11	6	6	11
Amino-N/amides	39	36	84	10	7	8	13	7
Nitriles	55	44	12	28	23	32	36	20
Heterocycles	6	20	4	62	70	60	51	72

[a] Calcareous silt loam used as parent material.
(From Schulten et al., 1992.)

humification of organic matter (Anderson et al., 1974b). Heterocyclic N is thought to be associated with chemically and/or mineralogically stabilized humic materials.

The increase in alkylaromatics, particularly between years 13 and 19, and the maintenance of a relatively high concentration until the end of the experiment is consistent with the proposal that alkylaromatics are responsible for the basic skeletal structure of humic substances (Schulten and Schnitzer, 1993).

A comparison using Py-FIMS, differential thermal analysis and thermo gravimetry of soils cultivated intensively with no fertilization with otherwise similar soils receiving annual additions of farmyard manures (FYM) has provided information relating to humus structure and process (Leinweber et al., 1992). The organic C and C:N ratios were 1.24% and 15.1 for the unamended soil, 2.09% and 13.8 for the FYM soil. The three analytical methods all indicate similar changes in the nature of humus due to management. Py-FIMS yields additional information on the molecular chemical composition of the humic materials. The FYM soil contained more thermo-labile organic substances, largely attributed to mono- and polysaccharides. As well, the FYM-amended soil contained more organic matter liberated at higher temperatures, postulated to be decomposition products of lignin, nitrogen compounds and probable long-chained alkanes and alkenes. These components, cross-linked with each other and strongly associated with the mineral matrix represent a very stable humic fraction. The presence of similar but less intense signals in the spectra of the intensively cultivated, unfertilized soil implies a considerable resistance to biodegradation of this group of humic materials.

IV. Influence of Clay on Soil Organic Matter

Organic C contents generally increase with the clay content of the soil, other factors being equal. Nichols (1984), for example, found a strong correlation between organic C content and clay content for Mollisol soils in warm (thermic) soils of the southern Great Plains. In contrast, in the Northern Plains, correlations between organic C and clay contents were evident for mesic (warm by comparison), but not for frigid or cold soils (McDaniel and Munn, 1985; Sims and Nielsen, 1986). It appears that the role of clay in stabilizing organic matter in soils is more important in the warmer soils where decomposition pressure can be expected to be stronger. In cold soils, cool temperatures may be the main factor slowing decomposition, and clay content is less important.

It is probable that in certain environments, especially those where moisture limits productivity, that clay plays an indirect role by influencing moisture storage and fertility and, thereby, the production of organic residues, the initial stocks for humus formation. Clay has a direct influence, as well, by adsorbing humic substances and stabilizing them against further decomposition or humification as mentioned earlier. The large proportion (50 to 60%) of organic C (and nutrients such as N and S) associated with clay and fine silt fractions indicates the importance of the association (Anderson et al., 1981). Radiocarbon ages older than the mean for the whole soil, particularly of material in the C-rich, coarse clay and fine silt fractions, are evidence for the role of clay in stabilizing C in soils (Anderson and Paul, 1984). However, evidence from studies where labelled residues have been added to soil indicates that nutrient-rich humic materials such as microbial metabolites, polysaccharides or amino acids quickly become associated with clay, especially fine clay (McGill and Paul, 1976; Christensen, 1992). These and other reports indicate that clay may play a dual role in influencing organic matter processes and turnover. Much of the clay-associated humic materials, particularly in the coarse clay and fine silt, appears to be biologically recalcitrant with long turnover times. On the other hand, material linked to fine clay is often enriched in recent, nutrient-rich substrates of high biologically activity, stabilized by adsorption to clay (Anderson and Paul, 1984).

V. Cultivation-induced Carbon Emissions from Soils

Conversion to agriculture has nearly always resulted in a reduction in the C stored in an ecosystem. In forested regions where organic matter contents of mineral soils are low (and may increase in agriculture) the C stored in the forest stand and in the forest floor results in a major efflux to the atmosphere, particularly if the material is burned. In grasslands, where there is much less C in the aboveground compartments, C reserves are reduced more gradually. There is, however, fairly rapid initial declines in organic matter as roots, organic residues and labile humus fraction are decomposed. Following the initial decline there is a long period of slower decline, and then a new equilibrium level provided that loss of organic matter-rich topsoil by erosion is not a factor.

It is likely that in most agricultural areas that maximum CO_2 emission from organic matter decomposition occurs at, and shortly after the start of agriculture. A preliminary analyses of agriculture in Western Canada (Manitoba, Alberta, and Saskatchewan) gives an apt description of the most probable changes in CO_2 emission with time and increase in cultivated land area (Figure 4). Emissions in the first three decades of this century were large, declining with time to much lower levels. Cumulatively, an estimated 5×10^8 Mg or 0.5 Pg has been evolved to the atmosphere. Although this amount is not large in comparison to total stores, it does represent a very substantial addition, particularly in early decades when Canada's industrial emissions were much smaller than today.

Many people have reported a 50% decrease in the organic matter content of cultivated soils in comparison to their virgin equivalents. Analyses of data from a variety of sources indicates that the <u>concentration</u> of C in surface horizons has declined markedly with cultivation, often by more than 50%. The decrease, however, represents mineralization losses to the atmosphere, plus decreases in concentration due to the erosion of organic matter-rich topsoil, and the incorporation of subsoil of lower C content into the plow layer. More detailed, pedon-based calculations that include the A and B horizons and changes in bulk density of surface horizons indicate that losses are considerably less than 50%. Tabulation of data (Table 3) from twelve paired, native and cultivated hillslopes in Saskatchewan (66 paired comparisons of pedons) indicates a probable loss of about 20 Mg ha^{-1} of C due to cultivation. (Additional details of sampling design, analytical methods and complete data are available from the author in an unpublished report, S.I.P #M76). This study considered the entire hillslope, both strongly eroded and depositional areas. Except for the sites in the Dark Brown soil zone where net decreases in cesium (^{137}Cs) contents indicate net erosion losses and therefore much greater losses of C, most of the decreases can be attributed to mineralization. Interestingly, a comprehensive analysis of previously published data from 625 paired soil samples of many different environments determined a C loss with cultivation of 1.5 Kg m^{-2} (15 Mg ha^{-1}) (Mann, 1986). For most soils, the decrease was 30% of the C in the 0 to 30 cm depth.

The above analysis indicates a probable loss of about 20 Mg ha^{-1} for the cultivated land area of Western Canada, or a total loss of 300 to 400×10^6 Mg. Although the data are sparse and further refinement is critical to better estimates, several estimates by a variety of methods consistently suggest a net C efflux from Western Canadian agriculture of 0.5 to perhaps one Pg over this century.

Of greater interest is the question of whether today's agriculture in the American and Canadian plains region is a net source or sink for C. Several analyses have shown that early losses were large (Mann, 1986), and that modern agricultural methods may be adding C back to once-depleted soils. For example, an analysis

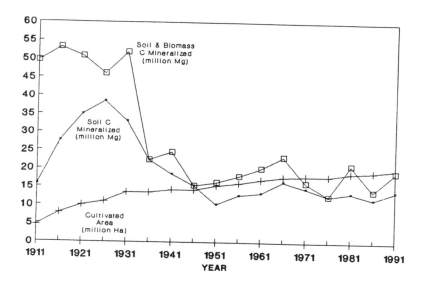

Figure 4. Estimates of the amounts of C mineralized from Western Canadian lands converted to agriculture from 1911 to 1991. (M.M. Boehm, unpublished data analyses.)

Table 3. Amounts of organic C and cesium (^{137}Cs) in paired native and cultivated soil profile within toposequences the Brown, Dark Brown and Black soil zones of Saskatchewan

	Native		Cultivated	
	Organic C (Mg/ha)	Cesium (Beq/m^2)	Organic C (Mg/ha)	Cesium (Beq/m^2)
Brown (21)[a]	105.4 ± 12.9	2570 ± 181	108.5 ± 17.8	2501 ± 490
Dark Brown (23)	101.0 ± 11.7	2555 ± 81	57.8 ± 9.6	1784 ± 168
Black (22)	115.0 ± 11.2	2472 ± 131	101.4 ± 15.9	2738 ± 473
All (66)	107.0 ± 6.8	2533 ± 78	88.7 ± 8.8	2332 ± 232

[a] The number of paired profile comparisons.

by Cole et al. (1990) indicates small net losses and even net gains of C in recent decades as compared to earlier periods. The estimated losses ranged from 12 to 28 Mg ha^{-1} in the first forty years of cultivation. The slight gains are estimated at a few Mg ha^{-1} and are attributed to higher production because of improved crop varieties, fertilizers and weed control, reduced tillage, and a higher proportion of organic residues returned to the soil. Similarly, a comparison of conventional and zero-tillage fields indicates greater storage of C in the latter (Arshad et al., 1990). Bauer and Black (1981) report that organic C contents of soils under stubble-mulch tillage appear to be equivalent to organic C contents on similar soils sampled decades ago, indicating a stabilization of organic matter levels with good agricultural management.

There are, however, mild concerns of the security of the additional C stores in soil due to zero tillage. The additional C is stored on or near the soil surface and may be more easily lost by decomposition in comparison to organic material deeper in the soil, particularly the material complexed with clay. In addition, highly mineralizable forms of C and N, and a high potential for nitrate mineralization suggest that denitrification potential may be higher in zero tillage, raising a concern for nitrous oxide emission.

In summary, agriculture has been historically a net contributor to CO_2 levels in the atmosphere. Today, agricultural lands in areas such as the plains region of North America are a reduced source, and possible sink for CO_2. Research is required to relate basic ideas on the structure and nature of soil humic materials to

VI. Summary and Conclusions

The considerable storage of organic matter in well drained or oxidized soils may be explained by theories such as the maximum power principle (Odum, 1983). The principle postulates that natural systems will self-organize in order to maximize useful power -- that is to store more energy (carbon) that can be fed back or depreciated to catalyze the inflow of additional energy. Equilibrium levels of soil C reflect a balance between decomposition, driven by nutrient demands of the biota, and inputs of organic residues. Both are influenced by the edaphic environment, resulting in soil organic matter that is just resistant enough to decomposition to exist in that particular soil and environment.

The factors contributing to biological recalcitrance include the chemical structures of the humic materials themselves, the role of clay in adsorbing or protecting potentially labile substrates, and factors of inaccessibility. Some organic C is inaccessible to microbes because of its protection within microaggregates, and other organic C stores are in B horizons and subsoils where turnover rates are much reduced.

Despite good progress with analytical methods to evaluate chemical composition of humic materials *in situ* (without extraction) there still remains different ideas about the chemical factors contributing to recalcitrance. Recent work with CP/MAS ^{13}C NMR spectroscopy indicates that strongly humified humus is dominated by straight-chain or polymethylene type structures, probably closely linked to mineral colloids (Baldock et al., 1992).

Other studies, based mainly on Py-FIMS spectroscopy indicate a larger role for aromatic structures, and postulate a complex structure of linked alkylaromatic units that includes increasingly more heterocyclic N with increasing humification (Schulten et al., 1992).

It appears probable that both aliphatic and aromatic components are integral part of the biologically resistant humus, with adsorption to clay an important factor in most soils. Clay protection of humus may act to both enhance storage of C, by slowing the decomposition of labile, energy- and nutrient-rich substrates; and to slow the progression to highly humified humic materials of very limited biological importance. Continuing research in this area is important to an understanding of C cycling in soils, and the impact of relatively new agronomic practices such as zero-tillage on long-term C storage.

Acknowledgement

The author gratefull acknowledges the discussion, review, and material provided by Mane M. Boehn to this paper.

References

Anderson, D.W. 1979. Processes of humus formation and transformation in soils of the Canadian Great Plains. *J. Soil Sci.* 30:77-84.

Anderson, D.W. and E.A. Paul. 1984. Organo-mineral complexes and their study by radiocarbon dating. *Soil Sci. Soc. Am. J.* 48:298-301.

Anderson, D.W., E.A. Paul, and R.J. St. Arnaud. 1974a. Extraction and characterization of humus with reference to clay associated humus. *Can. J. Soil Sci.* 54:317-323.

Anderson, D.W., D.B. Russell, R.J. St. Arnaud, and E.A. Paul. 1974b. A comparison of humic fractions of Chernozemic and Luvisolic soils by elemental analyses, UV and ESR spectroscopy. *Can. J. Soil Sci.* 54:447-456.

Anderson, D.W., S. Saggar, J.R. Bettany, and J.W.B. Stewart. 1981. Particle size fractions and their use in studies of soil organic matter. I. The nature and distribution of forms of carbon, nitrogen, and sulfur. *Soil Sci. Soc. Am. J.* 45:767-772.

Arshad, M.A., M. Schnitzer, D.A. Angers, and J.A. Ripmeester. 1990. Effects of till vs no-till on the quality of soil organic matter. *Soil Biol. Biochem.* 22:595-599.

Baldock, J.A., J.M. Oades, A.G. Waters, X. Peg, A.M. Vasallo, and M.A. Wilson. 1992. Aspects of the chemical structure of soil organic materials as revealed by solid-state ^{13}C NMR spectroscopy. *Biogeochemistry* 16:1-42.

Bauer, A.M. and A.L. Black. 1981. Soil carbon, nitrogen and bulk density comparisons in two cropland tillage systems after 25 years and in virgin grassland. *Soil Sci. Soc. Am. J.* 45:1166-1170.

Bolin, Bert and Inez Fung. 1992. The carbon cycle revisited. p. 151-164. In: D. Ojima (ed.). *Modeling the earth system*. UCAR, Boulder, Co.

Campbell, C.A., E.A. Paul, D.A. Rennie, and K.J. McCallum. 1967. Applicability of the carbon-dating method of analysis to soil humus studies. *Soil Sci.* 104:217-224.

Christensen, B.T. 1992. Physical fractionation of soil and organic matter in primary particle size and density fractions. *Adv. Soil Sci.* 20:1-90.

Cole, C.V., I.C. Burke, W.J. Parton, D.S. Schimel, D.S. Ojima, and J.W.B. Stewart. 1989. Analysis of historical changes in soil fertility and organic matter level of the North American Great Plains. p. 1-10. In: *Proc. Soils and Crops Workshop*. Univ. Saskatchewan, Saskatoon, SK.

Fox, C.A., R. Trowbridge, and C. Tarnocai. 1987. Classification, macro-morphology and chemical characteristics of Folisols from British Columbia. *Can. J. Soil Sci.* 67:765-778.

Leinweber, P., H.-R. Schulten, and C. Horte. 1992. Differential thermal analysis, thermogravimetry and pyrolysis field ionisation mass spectrometry of soil organic matter in particle size fractions and bulk soil samples. *Thermochimica Acta* 194:175-187.

Lotka, A.J. 1922. Contributions to the energetics of evolution. *Proc. Natl. Acad. Sci.* 8:151-154.

Mann, L.K. 1986. Changes in soil carbon storage after cultivation. *Soil Sci.* 142:279-288.

McDaniel, P.A. and L.C. Munn. 1985. Effect of temperature on organic carbon-texture relationships in Mollisols and Aridisols. *Soil Sci. Soc. Am. J.* 49:1486-1489.

McGill, W.B., and E.A. Paul. 1976. Fractionation of soil and ^{15}N nitrogen turnover to separate the organic and clay interactions of immobilized N. *Can. J. Soil Sci.* 56:203-212.

McGill, W.B., J.F. Dormaar, and E. Reinl-Dwyer. 1988. New perspectives on soil organic matter quality, quantity and dynamics on the Canadian prairies. p. 30-48. In: *Land degradation and conservation tillage*. Proc. 34th Meeting, Can. Soc. Soil Sci., Calgary, AB.

Nichols, J.D. 1984. Relation of organic carbon to soil properties and climate in the southern Great Plains. *Soil Sci. Soc. Am. J.* 48:1382-1384.

Odum, H.T. 1983. *Systems ecology*. John Wiley & Sons, New York. 644 pp.

Schulten, H.-R. and M. Schnitzer. 1993. A state of the art structural concept for humic substances. *Naturwischenschaften* 80:29-30.

Schulten, H.-R., P. Leinweber, and G. Reuter. 1992. Initial formation of soil organic matter from grass residues in a long-term experiment. *Biol. Fert. Soils* 14:237-245.

Sims, Z.R. and G.A. Nielsen. 1986. Organic carbon in Montana soils as related to clay content and climate. *Soil Sci. Soc. Am. J.* 50:1269-1271.

Theng, B.K.G., J.R.H. Wake, and A.M. Posner. 1967. The humic acids extracted by various reagents from a soil. II. Infrared, visible and ultra-violet absorption spectra. *J. Soil Sci.* 18:349-363.

Tiessen, H. and J.W.B. Stewart. 1983. Particle size fractions and their use in studies of soil organic matter: II. Cultivation effects on organic matter composition in size fractions. *Soil Sci. Soc. Am. J.* 47:509-514.

Veizer, Jan. 1988. The evolving exogenic cycle. p. 175-220. In: C.B. Gregor, R.M. Garrels, F.T. Mackenzie, and J.B. Maynard (eds.). *Chemical cycles in the evolution of the Earth*. John Wiley & Sons, New York.

Witty, J.E. and R.W. Arnold. 1970. Some Folists on Whiteface Mountain, New York. *Soil Sci. Soc. Am. Proc.* 34:653-657.

CHAPTER 14

Impact of Fall Tillage on Short-Term Carbon Dioxide Flux

D.C. Reicosky and M.J. Lindstrom

I. Introduction

The possibility of global greenhouse warming due to rapid increase of carbon dioxide, is receiving increased attention (Wood, 1990; Post et al., 1990). This concern is warranted because potential climatic changes could result in increased temperature and drought over present agricultural production areas (Wood, 1990). Direct measurements to quantify CO_2 flux as impacted by agricultural management practices (Houghton et al., 1983; Post et al., 1990) are needed. Agriculture's role in sequestering carbon is not clearly understood. The crop root system can be used to sequester and redistribute carbon deeper in the soil profile where it tends to be protected and less susceptible to decomposition. Management practices need to be developed to optimize CO_2 utilization from soil and plants in photosynthesis to increase crop yields. There is a definite need for information on the impact of tillage on the CO_2 fluxes from soil and how these can be managed to minimize impact on global climate change.

Elliot and Cole (1989) presented a conceptual framework for assessing agriculture's impact on global climate change. Detailed information is needed on the carbon balance within the soil and how it may be affected by agricultural practices. There is limited information on the fluxes of carbon dioxide between agricultural ecosystems and the atmosphere. Information is needed on both the short term effects of agricultural management decisions and the long term effects as it may affect global climate change. Direct evidence on the effect of tillage method on carbon dioxide flux rates is limited.

Gliński and Stępniewski (1985) presented data of long-term CO_2 flux from soil, based on soil oxygen measurements and demand throughout the year. Seasonal CO_2 evolved ranged from 1800 to 47,200 kg CO_2 ha^{-1} yr^{-1} across various soils and climates. Daily respiration under field conditions ranged from 0 to 56 g CO_2 m^{-2} d^{-1} depending on soil type and time of year. Relatively small differences between bare and cropped soils due to root respiration have been observed (Pritchard and Brown, 1979). These results do not consider the short-term effects of tillage method on the CO_2 flux from soil.

Over the past two decades, conservation tillage has evolved primarily for erosion control. However recent concern for global climate change puts a new light on the importance of conservation tillage and how it can be implemented on many soils to help reduce soil carbon losses. Conservation tillage has the potential for converting many soils from carbon sources to sinks. Wide-spread implementation of conservation tillage practices from the current 27% to 76% of the total cropland could enlarge the soil carbon pool (Kern and Johnson, 1991). Implementing conservation tillage practices implies a build up of carbon and organic matter in agricultural soils.

While tillage and cultivation results in loss of soil carbon and nitrogen (Campbell and Souster, 1982; Campbell et al., 1976; Mann, 1986), the direct influence of tillage on CO_2 flux is varied and highly interactive. Soil loosening should increase the CO_2 flux due to better accessibility of oxygen necessary for the organic matter decomposition and respiration resulting in CO_2 release. However, other consequences of tillage that may affect soil water and temperature are not easy to predict. Grabert (1968) observed an increase in total soil respiration rate with increased plowing depth. Conversely, Richter (1974) found higher

CO_2 evolution in three soils for zero-tillage in comparison with rototillage. De Jong (1981) compared CO_2 flux on a loamy soil under cereals, grassland, and fallow and found the highest CO_2 evolution rate under grassland and lowest under fallow conditions. He also studied the effect of relief and observed a two-fold increase in soil respiration at the foot of the hill as compared to the top of the hill. Even on uniform, level soils, spatial variability can be large. Rochette et al. (1991) found spatial variability, as described by the coefficient of variation for each series of measurements, was highest at 69% in May and decreased to 25% near season's end. The number of measurements required to estimate soil respiration within 10% of the mean at a 0.05 probability level was 190 before crop emergence and 30 measurements 70 days after emergence.

Roberts and Chan (1990) used simulated tillage techniques to examine the importance of tillage-induced increases in soil respiration as a mechanism for organic matter loss. They measured the CO_2 evolution from soil cores after applying a simulated tillage and found the total carbon losses that could be directly attributed to tillage ranged from .0005% to .0037%. They concluded that the increase in microbial respiration due to tillage was probably not a major factor leading to losses of soil organic matter in the soils under intensive cultivation.

There have been a few reported attempts to measure carbon dioxide evolution immediately after tillage in the field. Hendrix et al. (1988) were unable to detect any stimulation of carbon dioxide release immediately after plowing using 0.1 m aluminum cylinders and the alkali-absorption method. Rovira and Greacen (1957) studied the effect of aggregate disruption on the activity of micro organisms in the soil and found that by breaking apart aggregates they could release 21 kg C ha^{-1}, which was close to the maximum loss reported in the work by Roberts and Chan (1990) using simulated tillage methods. Rovira and Greacen (1957) concluded that an increase in the decomposition of organic matter was induced by tillage and was a factor in decline of organic matter in tilled soils. However, this mechanism is reportedly small compared to other mechanisms of C loss (Roberts and Chan, 1990).

To better understand the carbon flow dynamics within an agricultural production system, information is needed on the short-term impacts of various tillage methods. Canopy gas exchange techniques allow short-term measurements of fluxes that can contribute to a better understanding of the underlying processes that cause decreases in soil carbon (Kanemasu et al., 1974). Our hypothesis is that different fall tillage methods will affect the short-term CO_2 flux. The objective of this work was to measure the effect of several fall tillage methods on the short-term CO_2 flux from soil relative to no-till. A large portable field chamber was used to measure gas exchange and to minimize problems due to spatial variability. Various combinations of soil disturbance and residue incorporation were established using conventional tillage equipment commonly used in the northern Corn Belt.

II. Materials and Methods

This experiment was conducted in the fall of 1991 on a Hamerly clay loam (fine, loamy, frigid Aeric Calciaquoll) on the West Central Experiment Station of the University of Minnesota, Morris, Minnesota (45° 35' N. lat. 95° 55' W. long.). The surface soil is characterized by 330 mm of clay loam or silty clay loam and is nearly level (slope < 1%) moderately well-drained formed on a calcarious loam and clay loam ground moraine. The Ap horizon has a bulk density of approximately 1.30 Mg m^{-3}, and entire soil profile has available water-holding capacity of 105 mm m^{-1} soil.

The study area was planted to spring wheat (*Triticum aestivum* L. cv. Marshall) on April 22 (DY 112), 1991, and harvested on August 22 (DY 224), 1991. During this period seasonal rainfall was slightly above normal with 300 mm accumulative rain from planting to harvest. The average wheat yield was 2354 kg ha^{-1} with normal amount of above-ground residue ($\approx$3531 kg ha^{-1}) and assuming a shoot:root ratio of 10:1 at anthesis yields 589 kg ha^{-1} root residue (Klepper, 1991). Following grain harvest, the stubble was left at an average height of about 0.18 m and the remaining residue was spread uniformly over the entire area. Following harvest and prior to tillage (Sept. 4, 1991, DY 247) the rainfall was normal for the area and totalled 50 mm. In order to minimize the effect of weeds and volunteer wheat plants on CO_2 exchange rate, the area was sprayed with a herbicide, Ranger[1], (glyphosate) at 0.8 kg a.i. ha^{-1} on August 30, 1991. After

---

[1] Names of products are included for the benefit of the reader and do not imply endorsement or preferential treatment by USDA.

Table 1. Soil disturbance and residue incorporation resulting from tillage methods

Tillage Method	Average depth of tillage (m)	Degree of soil disturbance (%)	Degree of residue incorporation (%)
1. Moldboard plow	0.250	100	100
2. Moldboard plow + disk twice	0.250	100	100
3. Disk harrow once	0.075	100	30
4. Chisel plow once	0.150	50	30
5. No-tillage	0	0	0

the tillage and subsequent rainfalls, additional herbicide was again applied to control weeds and volunteer wheat as they emerged. Herbicide applications were made on Sept. 13, Sept. 17, and Sept. 20 at the rate of 0.8 kg ha^{-1} a.i. of glyphosate. The area was kept weed free so there was no uptake of CO_2 by any live plant material.

Soil gravimetric water content profiles were obtained in close proximity to the measurement areas on the day of tillage. Soil samples for organic carbon were taken on Sept. 24, 1991 (DY 267) in proximity to the measurement areas. The surface 76 mm was sampled using a 20 mm i.d. soil probe. Ten soil samples around the measurement area were composited, air dried, mixed, sieved through 500 μm screen, and subsampled for organic carbon determinations using standard laboratory techniques (Yeomans and Bremner, 1988).

Commercially available tillage implements were used to evaluate the effect of various methods of tillage on CO_2 flux from the surface of different combinations of depth of soil disturbance and residue incorporation. Tillage treatments are summarized in Table 1 showing depth of tillage, estimated degree of disturbance, and estimated degree of residue incorporation with no-tillage (NT) as a check treatment. The four tillage treatments included moldboard plow (MP), using a 3 bottom plow, 0.46 m wide bottoms, to a depth of 0.25 m that resulted in complete inversion of the surface layer and nearly 100% incorporation of the residue. The second treatment was moldboard plow to the 0.25 m depth followed by a disk harrow twice (MP + D). This resulted in the same depth and degree of soil disturbance with smaller aggregates, a less porous surface and recompaction of soil around the buried residue. The third treatment was disk harrow (DH) that resulted in a shallow soil disruption (0.075 m) and partial incorporation of the residue. The chisel plow (CP) treatment used a standard chisel plow with 11 shanks on 0.30 m centers, 0.076 m twisted shovels staggered on three bars for complete soil disruption. The primary difference between DH and CP was the depth of soil disturbance, 0.15 m with CP compared to 0.075 m with DH. The disk and chisel tillage methods are commonly used for overwinter wind and water erosion control. The check treatment was NT with soil and residue as left by harvest equipment from the preceding wheat crop.

The tillage equipment was pulled by a medium-size farm tractor ($\approx$70 kw). Where the tillage implement was narrow, several adjacent passes were made to achieve the necessary plot width. The tillage plots were 6 m wide x 110 m long strips to allow the tillage implement to operate at the appropriate depth. Chamber measurements were made about 20 m from plot ends to minimize border effects. A 6-m alley allowed tractor movement between each plot and access from two directions.

Initial tillage was completed on Sept. 4, 1991 (DY 247) in the order MP, MP + D, DH, and CP. The portable chamber measurements for CO_2 exchange rates were initiated within 5 min of the last tillage pass. The tillage was done when the surface soil water potential was approximately -45 kPa, a suitable water content for tillage of this soil.

The CO_2 flux from the tilled soil surfaces was measured using a large portable chamber described previously (Reicosky, 1990; Reicosky, et al. 1990). Briefly, the chamber (volume of 3.25 m^3 covering a horizontal land area of 2.67 m^2) with mixing fans running was moved over the tilled surface until the chamber reference points align with plot reference stakes, lowered, and data rapidly collected at 2-s intervals for a period of 80 s to determine the rate of CO_2 and water vapor increase. The chamber was then raised, calculations completed and the results stored to computer disc. Data included time, plot identification, solar radiation, photosynthetically active radiation, air temperature, wet bulb temperature, and output of the infrared gas analyzer measuring CO_2 and water vapor concentration. After the appropriate lag times, data

Table 2. Summary of organic carbon content and initial soil water content

DY	Tillage method	Post-tillage[a] organic C (%)	Gravimetric water content		
			0-76 (mm)	76-152 (mm)	152-305 (mm)
			----------kg kg^{-1}----------		
247	Moldboard plow	3.19	.263	.280	.268
247	Moldboard plow + disk	3.42	.242	.273	.281
247	Disk	3.65	.272	.304	.302
247	Chisel	3.44	.247	.292	.289
247	No tillage	3.14	.245	.286	.266
254	No tillage	----	.333	.339	.295
267	No tillage	----	.260	.302	.291

[a] Samples on Sept. 24, 1991 (DY 267) to 76 mm-depth.

for a 30-s calculation window was selected to convert the volume concentration of water vapor and CO_2 to a mass basis which was linearly regressed as a function of time (Reicosky et al., 1990). The slopes of these regression lines reflect the rate of CO_2 and water vapor increase within the chamber and are then expressed on a unit horizontal land area basis. These measurements are expressed on "land area" basis and differentiated from "exposed soil surface area" basis that results from differences in surface roughness.

Total time for a single measurement required for data collection and computation was approximately 3 minutes. Triplicate measurements were made on each tillage treatment before moving to the next plot. Within any one day, as many as five measurement cycles were made on each tillage treatment to provide limited data on the diurnal dynamics of CO_2 flux. Later in the study, after the fluxes on the tilled plots had decreased substantially, only two measurement cycles per day were made.

III. Results and Discussion

The soil water content at the time of tillage and organic C shortly after tillage were similar for all the five tillage areas (Table 2). Thus, there should have been little effect of initial soil water content and organic C on the initial CO_2 flux rate. Any differences measured should be a result of interaction between soil disturbance and residue incorporation resulting from each tillage method. Because of the importance of temperature and water on soil respiration and CO_2 flux from the soil surface, the data will be discussed in two groups based on time after tillage and a significant rainfall four days after tillage. Daily maximum air temperatures during the study period ranged from 26.5°C early to as low as 6.3°C later, while daily minimum air temperature ranged from 20.0°C to 0.8°C. The rainfall event provided a 3-day total of 49 mm with a maximum intensity of 20 mm h^{-1} on Day 250. This rainfall intensity was sufficient to cause settling of the soil and some re-consolidation of the loosened soil. Thus the results will be discussed as short term for the first three days (55 hours) after tillage and before the rainfall event and intermediate term for the 15 days following the rainfall event.

The short-term (55 hours after tillage) effect of tillage method on CO_2 flux is summarized in Figure 1. Relatively large initial fluxes (29 g CO_2 m^{-2} h^{-1}) from MP were observed. Each data point is the mean of 3 replicates. Error bars represent ± one standard deviation. Those data points without visible error bars had low standard deviation such that the error bars were contained within the symbol. The sign convention is positive indicating CO_2 flux from the soil surface to the atmosphere. The magnitudes of the CO_2 flux from MP and MP + D were largest immediately after tillage reflecting a "flush" of microbial CO_2 and CO_2 released from large voids caused by tillage (Blevins et al., 1984; Buyanovsky and Wagner, 1983; Buyanovsky, et al., 1986; Hendrix et al., 1988). Noteworthy was the rapid decrease in the CO_2 flux from MP to about 2 g m^{-2} h^{-1} 55 hours after tillage. Immediately after tillage, the MP + D treatment had a flux as large as 7 g m^{-2} h^{-1} that decreased to 2 g m^{-2} h^{-1} within 3 h. The CP treatment showed a similar trend as MP and MP + D only smaller in magnitude. These tillage methods are compared with NT where there was little change in the CO_2 flux from 0.7 to 0.2 g m^{-2} h^{-1} during the same 55 hour period.

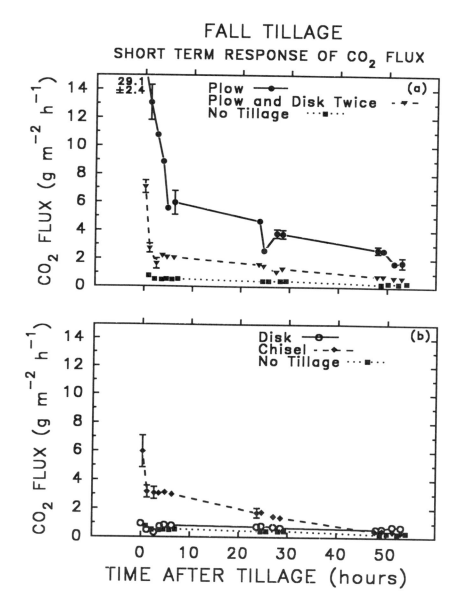

Figure 1. Short-term effect of fall tillage method on carbon dioxide flux versus time: a) Moldboard plow, moldboard plow plus disk twice, and no tillage; b) Disk harrow, chisel plow, and no tillage. Tillage was done on September 4, 1991 (DY 247).

The high CO_2 flux from MP measured 5 min after tillage requires further discussion on potential sources of CO_2. The potential contribution of free carbonates in the subsoil (>0.33m) was not evaluated but was assumed to be below the depth of tillage. It is possible a slight negative pressure inside the chamber drew CO_2 from the large voids, however independent pressure measurements showed no change when the chamber was lowered. The air-filled pore space before tillage may have contained a high concentration of CO_2. An average bulk density of 1.3 Mg m^{-3} with a volumetric water content at the time of tillage of 0.35 m^3 m^{-3} yields an air-filled pore space of 0.16 m^3 m^{-3} in the top 0.25 m of soil. If the CO_2 concentration in the soil air was 3% CO_2 (Buyanovsky and Wagner, 1983), then 2.36 g CO_2 m^{-2} was potentially available for release to the atmosphere ($\approx$ 8% of the initial flux). If the CO_2 in the soil air was 6%, then the CO_2 released could account for about 16% of the initial flux. Contributions of CO_2 through macropores below the tillage depth could also add to the high initial flux as long as the gas permeability of the disturbed soil was sufficiently high.

If we assume the total CO_2 contribution from soil pores can be estimated from the difference in the first and second measurements (about 1 h later) from MP, then 8 g CO_2 m^{-2} came from the soil pores. This estimate assumes a linear decrease in flux and uses the trapezoidal rule to approximate the total CO_2 lost during that 1 h interval. Back calculating, this requires the CO_2 concentration of the air-filled pore space to be about 10%. This is a slightly higher value compared to field measurement of 6 to 8% (Buyanovsky and Wagner, 1983), but not totally unrealistic when compared with 17% CO_2 measured by Yamaguchi et al. (1967) where soil temperature was controlled at 30°C.

The release of dissolved CO_2 from soil water as a result of the change in partial pressure of CO_2 when the soil was inverted and exposed to atmospheric levels may partially explain the high initial flux. This phenomena becomes more important when we consider the air-water interfacial area as a factor controlling gas transport (Skopp, 1985; Skopp et al., 1990). The 1.5-fold increase in the air-water interfacial area associated with the tillage-induced change in bulk density (and air-filled porosity) would contribute to the release of dissolved CO_2 if all air-water interfaces are exposed to atmospheric pressures and concentrations. This effect would increase as the soil dried until a critical low water content is reached. If we assume a CO_2 concentration of 3% or a partial pressure of 3 kPa, Henry's Law suggests nearly a 100-fold decrease in the partial pressure of CO_2 in water to 0.03 kPa. At 20°C and standard pressure (100 kPa), the solubility of CO_2 is 0.1688 g CO_2 (100 g H_2O)$^{-1}$ (Chemical Rubber Company, 1959). Using a volumetric water content of 0.35 m^3 m^{-3} in the top 0.25 m of soil yields 87.5 l H_2O m^{-2} soil that contained 4.43 g CO_2 m^{-2} soil ($\approx$ 15% of the initial flux) released on exposure to the atmosphere. Assuming 6% CO_2 in the soil air, 8.86 g CO_2 m^{-2} soil ($\approx$ 30% of the initial flux) could have been released at tillage. If the initial "flush" of CO_2 can be estimated from the CO_2 lost between the first and second measurements (a difference of 8 g CO_2 m^{-2} in $\approx$ 1 h) and the primary sources of CO_2 were from soil pores and degassing of soil water, then the CO_2 concentration in the pores was about 3.5%. Both contributions to the initial flux would be mediated by diffusion of CO_2 in water films and through air-filled pores to an exposed surface and cause the CO_2 flux to lag.

Dissolved CO_2 released during water evaporation could contribute to the initial flux. The measured evaporation 5 min after tillage was 0.54 mm h^{-1} or 0.54 kg H_2O m^{-2} h^{-1}. Using the CO_2 solubility at 20°C and 3 kPa partial pressure would yield less than 0.03 g CO_2 m^{-2} h^{-1}, an insignificant amount. Similarly the change in solubility due to estimated soil water temperature changes resulting from tillage were negligible.

Virtually little is known how the abrupt change in air-filled porosity can affect the CO_2 release after moldboard plowing. Prior to tillage, the air-filled pore space was 0.16 m^3 m^{-3} in the surface layer. If we assume an "average bulk density" after tillage of 0.9 Mg m^{-3} (Burwell et al., 1963), the air-filled pore space increased to 0.31 m^3 m^{-3} assuming no change in gravimetric water content. The pre-tillage air-filled porosity was near the critical level for adequate aeration observed by Grable and Siemer (1968). The abrupt change in air-filled porosity replenished or doubled the O_2 volume in the voids. An even larger unknown is the freshly exposed "soil surface area" available for reaction with the increased O_2.

Another possible explanation for the high flux immediately after plowing (MP) involves the microsite concept around pieces of organic matter and in the inter-aggregate gas concentration. The spatial variability of anaerobic microsites in "hot-spot" denitrification has been discussed by Parkin (1987) who showed rates of CO_2 production of particulate carbon materials were several orders of magnitude greater than rates of CO_2 production from the soil. Tillage would disrupt the microsites and could release a lot of CO_2. A similar "burst" of CO_2 can be visualized as large soil aggregates (radius > 9mm) are broken apart during tillage. The low inter-aggregate oxygen concentrations measured by Sexstone et al. (1985) suggest high inter-aggregate CO_2 concentrations would be released at tillage. Both potential sources of CO_2 would be spatially averaged in the initial chamber measurements and then decrease with time.

Probably more important in the high initial flux is a readily available substrate for rapid oxidation. Some approximations of the relative contribution of plants to soil respiration are provided by Sauerbeck and Johnen (1976). Using $^{14}CO_2$ techniques, they found the amount of soil carbon mineralized during vegetative growth of wheat and mustard was three times greater than that remaining at harvest time. About 20% of the CO_2 evolved from the soil was attributed to living root respiration, about 20% was attributed to native organic matter in the soil and the remaining 60% was a result of decomposition of root exudates and dead roots. The warm, wet field conditions preceding tillage in this work should have primed the soil for rapid release of CO_2 upon plowing. Although soil temperature was not measured in the tillage plots, approximations were obtained for the 10-cm depth of bare soil at a weather station 1.5 km away. On the day of tillage, the maximum and minimum soil temperatures at 10 cm were 30.0 and 18.9°C, respectively. The average daily maximum and minimum for the 10 days preceding tillage was 32.8 and 23.3°C, respectively. These temperatures were well within the range for ample microbial activity under field conditions (Yamaguchi et al., 1967). Roots and root exudates from the previous wheat crop and fresh roots from the volunteer wheat killed with herbicide 6 days before tillage were probably readily available for both biological and chemical oxidation. The relatively long-time response for microbial decomposition and release of CO_2 in soil suggests some other form of rapid chemical oxidation yet to be determined.

In the short term, MP had the largest initial CO_2 flux followed by the CP, apparently reflecting the depth of soil disturbance and increased void fraction in both tillage treatments. The third highest initial flux was MP + D which also decreased rapidly during the first 55 h that had the same depth of tillage but surface porosity reduced by the two diskings. The DH had a relatively small initial CO_2 flux and throughout the entire period and was only slightly larger than the NT plot. The cumulative CO_2 flux for the 55 h period in Figure 1 was estimated by calculating the area under the curves and resulted in 247, 78, 37, 88, and 22 g CO_2 m^{-2} for MP, MP + D, DH, CP and NT, respectively.

Measurement of the CO_2 flux from the tillage surfaces during the intermediate term was continued after a major rainfall event (3 day total of 49 mm) for up to 19 days after the initial tillage (Figure 2). During this period the rates continued the earlier trends for each tillage method but at lower values. Daily variation reflected the effect of air temperature on surface soil temperature prior to freezing the soil. Throughout the remaining period, MP consistently had the highest CO_2 flux followed by MP + D. The DH and the CP had relatively low fluxes that were not different from NT during the remainder of the study. The cumulative CO_2 flux from each tillage surface for the 19 days after tillage was 913, 475, 391, 366, and 183 g m^{-2} for MP, MP + D, DH, CP, and NT, respectively. Although not a part of this study, the seasonal variation related to temperature would suggest higher cumulative CO_2 flux if the tillage had been done in mid-summer (Currie, 1975; Buyanovsky et al., 1986; Rochette et al., 1991).

Differences in the cumulative CO_2 flux between the short and intermediate terms (Figure 3) reflect the difference in soil disturbance where the four tillage methods were referenced to NT. The MP was the most disruptive tillage treatment with 100% residue incorporation, greatest surface roughness, largest soil surface area, and probably largest void fraction. This combination of factors all promoted CO_2 flux from the tilled surface in both the short and intermediate terms. The MP + D, while tilled just as deep as MP, had reduced surface roughness and a reduced void fraction that resulted in lower CO_2 flux.

Part of the differences in cumulative CO_2 flux due to tillage are probably related to air permeability, spatial distribution and mixing of the residue, and continuity and duration of macropores. Mielke et al. (1986) showed the air permeability in the surface layer of no-till soil was less than that of plowed soils and was related to higher organic carbon contents. Staricka et al. (1991a, b) showed the depth distribution of crop residue was consistently related to tillage tool and was not uniformly distributed with tillage depth. They found chisel plow and disc harrow had most of the residue mixed in the top 10 cm while the depth of incorporation with moldboard plow was 22 to 28 cm with 75% of the residue buried below 10 cm. More important was the tendency of the crop surface residue to become localized between furrow slices which also happen to be areas of high porosity that may also result in preferential water flow (Staricka et al., 1991b; Staricka, 1993 personal communication). The presence of residue in close proximity to the continuous macropores between the furrow slices provide an open pathway for CO_2 to escape and atmospheric O_2 to aerate the soil. Secondary tillage on MP+D virtually eliminated the large pore continuity and caused a smaller CO_2 flux. The next question to be addressed is whether the secondary tillage restricted the CO_2 flux and caused a build-up of CO_2 in the air-filled pore space that may decrease the rate of organic matter decomposition.

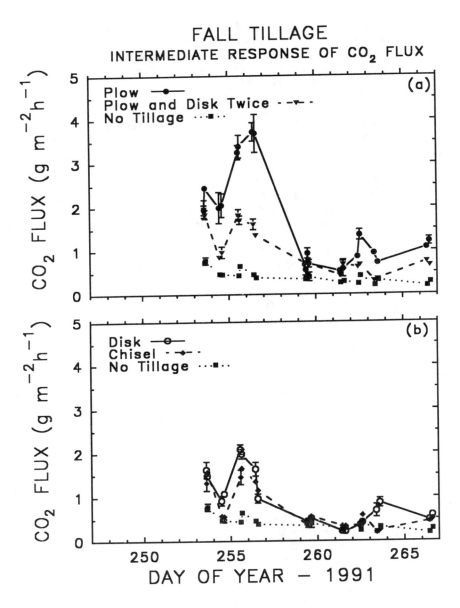

Figure 2. Intermittent-term effect of fall tillage method on carbon dioxide flux versus time following a major rainfall event: a) Moldboard plow, moldboard plow plus disk, and no tillage; b) Disk harrow, chisel plow, and no tillage. Note the change in scales from Figure 1.

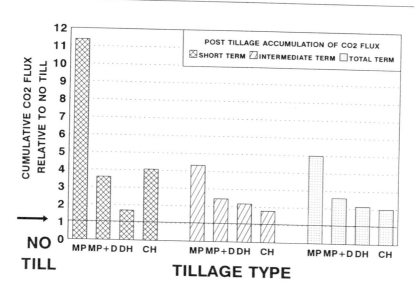

Figure 3. Cumulative CO_2 flux from tillage treatments relative to no-till for three different periods.

The total CO_2 released during the 19 days of this study can be related to the soil organic matter and top residue and roots of the previous wheat crop mixed into the soil during tillage. If we assume the 4120 kg ha^{-1} of wheat residue was 45% carbon, then the carbon release as CO_2 could account for 134, 70, 58, 54, and 27% of the C in the residue for MP, MP+D, DH, CP, and NT, respectively. The excessively large percentage on MP suggests other readily available sources of CO_2 from soil organic matter and/or forms of chemical oxidation in the field following moldboard plowing that need further research. The possible contributions of organic matter decomposition and CO_2 from below the tillage layer was not evaluated.

The degree to which residue disappeared from the soil surface on DH and the CP was essentially the same. Differences in CO_2 flux between these two treatments in the short term probably reflect the void fraction and surface roughness. The results suggest that the depth of soil disturbance and the resultant macropore continuity was more important than residue incorporation in determining the magnitude of short term CO_2 flux. After the heavy rainfall, differences between DH and CP were not significant, however both had higher cumulative CO_2 flux than NT. These results are in general agreement with the work of Grabert (1968), who observed an increase in total respiration rate with increased plowing depth. The results are contrary to those of Richter (1974) who found higher CO_2 evolution in three soils at zero tillage in comparison to rototillage and Hendrix et al. (1988), who observed greater CO_2 output from no-tillage than conventional tilled soils. An uncertainty in these studies is the impact of crust formation on CO_2 flux. The timing in the short-term dynamics measured with field gas exchange techniques may not reflect the long-term carbon balance on soils because of soil surface changes and re-consolidation that occur over time.

IV. Summary and Conclusions

In summary, fall tillage methods affected short and intermediate-term CO_2 flux from a wheat field. Immediately after tillage, moldboard plow had the largest rate > moldboard plow plus disk twice > chisel plow > disk > no-tillage. The high initial fluxes showed no lag often associated with microbial activity and were apparently related to gaseous diffusion and/or rapid direct oxidation of carbon substrates. The cumulative CO_2 evolved for the 19 days of this study was the largest on moldboard plow only > moldboard plow plus disk twice > disk > chisel plow > no-tillage. Results suggest that high initial flush of CO_2 was more related to surface soil roughness and depth of soil disturbance that released CO_2 in soil pores and from solution than to residue incorporation. The CO_2 flux from all tillage plots showed small but consistent differences 19 days after tillage and 64 mm of rain with rates substantially lower than immediately after tillage. The data suggest lower short-term soil CO_2 fluxes are associated with tillage methods that limit soil disturbance. Differences in CO_2 flux between tillage methods suggest that reduced tillage may minimize agriculture's contribution to global CO_2 increase. Tillage methods that minimize depth and extent of soil disturbances will have the least impact on global climate.

References

Blevins, R.L., M.S. Smith, and G.W. Thomas. 1984. Changes in soil properties under no-tillage. p. 190-230. In: R.E. Phillips and S.H. Phillips (eds.). No-tillage Agriculture: Principles and Practices. Van Norstrand Reinhold, New York.

Burwell, R.E., R.R. Allmaras, and M. Amemiya. 1963. A field measurement of total porosity and surface microrelief of soils. *Soil Sci. Soc. Am. Proc.* 27:697-700.

Buyanovsky, G.A. and G.H. Wagner. 1983. Annual cycles of carbon dioxide level in soil air. *Soil Sci. Soc. Am. J.* 47:1139-1145.

Buyanovsky, G.A., G.H. Wagner, and C.J. Gantzer. 1986. Soil respiration in a winter wheat ecosystem. *Soil Sci. Soc. Am. J.* 50:338-344.

Campbell, C.A., E.A. Paul, and W.B. McGill. 1976. The effect of cultivation and cropping on the amounts and forms of soil nitrogen. p. 7-101. In: Proc. Alberta Soil Science Workshop, Western Canadian Nitrogen Symposium, Calgary, Alberta, Canada.

Campbell, C.A. and W. Souster. 1982. Loss of organic matter and potentially mineralizable nitrogen from Saskatchewan Soils due to cropping. *Can. J. Soil Sci.* 62:651-656.

Chemical Rubber Company. 1959, *Handbook of Chemistry and Physics.* 41st ed. p. 1706.

Currie, J.A. 1975. Soil respiration, Tech. Bull., Min. Agric., Fish. Food 29:461.

De Jong, E. 1981. Soil aeration as affected by slope position and vegetative cover. *Soil Sci.* 131:34-43.

Elliot, E.T. and C.V. Cole. 1989. A perspective on agroecosystem science. *Ecology* 70:1597-1602.

Gliński, J. and W. Stępniewski. 1985. Soil aeration and its role for plants. CRC Press, Inc. Boca Raton, FL, USA. 229 pp.

Grabert, D. 1968. Measurements of soil respiration in a model experiment on the deepening of the arable layer. (German, English Summary). *Albrecht-Thaer-Arch.* 12:681-689.

Grable, A.R. and E.G. Siemer. 1968. Effects of bulk density, aggregate size, and water suction on oxygen diffusion, redox potentials, and elongation of corn roots. *Soil Sci. Soc. Am. Proc.* 32:180-186.

Hendrix, P.F., H. Chun-Ru, and P.M. Groffman. 1988. Soil respiration in conventional and no-tillage agroecosystems under different winter cover crop rotations. *Soil Tillage Res.* 12:135-148.

Houghton, R.A., J.E. Hobbie, J.M. Melillo, B. More, B.J. Peterson, G.R. Shaver, and G.M. Woodwell. 1983. Changes in the carbon content of terrestrial biota and soils between 1860-1980: A net release of CO_2 to the atmosphere. *Ecological Mono.* 53:235-262.

Kanemasu, E.T., W.L. Powers, and J.W. Sij. 1974. Field chamber measurements of CO_2 flux from soil surface. *Soil Sci.* 118:233-237.

Kern, J.S. and M.G. Johnson. 1991. The impact of conservation tillage use on soil and atmospheric carbon in the contiguous United States. US EPA Report EPA/600/3-91/056. 28 pp.

Klepper, B. 1991. Root-shoot relationships. p. 265-286. In: Y. Waisel, U. Kafkafi, and A. Eshel (eds.). *Plant Roots: The Hidden Half.* Marcel Dekker, New York.

Mann, L.K. 1986. Changes in soil carbon storage after cultivation. *Soil Sci.* 142:279-288.

Mielke, L.N., J.W. Doran, and K.A. Richards. 1986. Physical environment near the surface of plowed and no-tilled soils. *Soil Tillage Res.* 7:355-366.

Parkin, T.B. 1987. Soil microsites as a source of denitrification variability. *Soil Sci. Soc. Am. J.* 51:1194-1199.

Post, W.M., T.H. Peng, W.R. Enamuel, A.W. King, V.H. Dale, and D.L. DeAngelis. 1990. The global carbon cycle. *Am. Scientist* 78:310-326.

Pritchard, D.T. and N.J. Brown. 1979. Respiration in cropped and fallow soil. *J. Agric. Sci.*, Cambridge, 92:45-51.

Reicosky, D.C. 1990. Canopy gas exchange in the field: closed chambers. *Remote Sensing Reviews* 5(1):163-177.

Reicosky, D.C., S.W. Wagner, and O.J. Devine. 1990. Methods of calculating carbon dioxide exchange rate for maize and soybean using a portable field chamber. *Photosynthetica* 24(1):23-38.

Richter, J. 1974. A comparative study of soil gas regime in a soil tillage experiment with different soils. I. Field measurements. *Z. Pflanzenernaehr. Bodenkd.* 137:135-147.

Roberts, W.P. and K.Y. Chan. 1990. Tillage-induced increases in carbon dioxide loss from soil. *Soil Tillage Res.* 17:143-151.

Rochette, P., R.L. Desjardins, and E. Pattey. 1991. Spatial and temporal variability of soil respiration in agricultural fields. *Can. J. Soil. Sci.* 71:189-196.

Rovira, A.D. and E. L. Greacen. 1957. The effect of aggregate disruption on the activity of microorganisms in the soil. *Aust. J. Agric. Res.* 8:659-673.

Sauerbeck, D. and B. Johnen. 1976. The turnover of plant roots during the growth period, its influence on "soil respiration." (German, English summary). *Z. Pflanzenernaehr. Dueng. Bodenkd.* 3:315-328.

Sexstone, A.J., N.P. Revsbech, T.B. Parkin, and J.M. Tiedje. 1985. Direct measurement of oxygen profiles and denitrification rates in soil aggregates. *Soil Sci. Soc. Am. J.* 49:645-651.

Skopp, J. 1985. Oxygen uptake and transport in soils: analysis of the air-water interfacial area. *Soil Sci. Am. J.* 49:1327-1331.

Skopp, J., M.D. Jawson, and J.W. Doran. 1990. Steady-state aerobic microbial activity as a function of soil water content. *Soil Sci. Soc. Am. J.* 54:1619-1625.

Staricka, J.A., R.R. Allmaras, D.R. Linden, and W.W. Nelson. 1991a. A spatial pattern of tillage influences on agrichemical movement in macroporous pathways. p. 338-346. In: T.J. Gish and A. Shirmohammadi (eds.). Preferential Flow. Proc. Workshop Amer. Soc. Agric. Eng., Chicago, IL. 16-17 Dec. 1991. Amer. Soc. Agric. Eng., St. Joseph, MI.

Staricka, J.A., R.R. Allmaras, and W.W. Nelson. 1991b. Spatial variation of crop residue incorporated by tillage. *Soil Sci. Soc. Am. J.* 55(6):1668-1674.

Wood, F.P. 1990. Monitoring global climate change: The case of greenhouse warming. *Am. Meter. Soc.* 71(1):42-52.

Yamaguchi, M., W.J. Flocker, and F.D. Howard. 1967. Soil atmosphere as influenced by temperature and moisture. *Soil Sci. Soc. Am. Proc.* 31:164-167.

Yeomans, J.C. and J.M. Bremner. 1988. A rapid precise method for routine determination of organic carbon in soil. *Commun. Soil Sci. Plant Anal.* 19(13):1467-1476.

CHAPTER 15

Organic Matter Inputs and Methane Emissions from Soils in Major Rice Growing Regions of China

J.S. Kern, D. Bachelet, and M. Tölg

1. Introduction

Irrigated rice (*Oryza sativa*) paddies are a major source of methane (CH_4) (Khalil and Rasmussen, 1990; Schütz et al., 1990a). The total area of the world's harvested rice fields in 1989 was 143 x 10^6 ha, of which 52% were irrigated, 27% rainfed, 8% deepwater, and 13% upland. Deep water and upland rice systems do not have significant CH_4 emissions (Neue et al., 1990). Rice has been cultivated in China for over 7000 years and comprises half of the grain output of China (Gong and Xu, 1990). Data from Huke (1982) shows that in China the area of harvested rice fields (double crop land counts twice) was 31 x 10^6 ha of which 93% was irrigated. Khalil et al. (1991) estimated CH_4 emissions from rice fields in China at around 30 Tg CH_4 yr^{-1} (Tg = 10^{12} g), Bachelet and Neue (1993) estimated from 9 to 22 Tg CH_4 yr^{-1}, and Lin (1993) estimated 11 Tg CH_4 yr^{-1}.

The microbially-mediated reduction of organic matter to CH_4 is a complex interaction of soil redox potential, pH, temperature, microbial competition, and the geochemistry of N, S, Mn, and Fe (Zehnder and Stumm, 1988; Neue et al., 1990; Bouwman, 1991). The effect of temperature on CH_4 emissions is complex. Methane emissions increased diurnally with increased soil temperatures in the surface 10-cm in Italy (Seiler et al., 1984; Holzapfel and Seiler, 1986). In the Sichuan province of China, Khalil et al. (1991) reported that the flux of methane increased with higher soil temperatures which followed changes in seasonal microbial ecology and the development stage of rice plants. Schütz et al. (1990b) concluded that CH_4 emissions followed diurnal changes in soil temperature and that seasonal variations were not related to soil temperature. They suggested that other soil processes, such as root exudation and CH_4 oxidation in the rhizosphere, also affect CH_4 emissions. Fertilization also affects emissions. Rice fields fertilized with rice straw had higher CH_4 emissions than untreated fields near Vercelli, Italy (Schütz et al., 1989). In greenhouse experiments, Mariko et al. (1991) found that additions of composted straw resulted in a six fold reduction in CH_4 emission than additions of uncomposted straw. In Texas, organic fertilizer additions (Sass et al., 1990) and high solar radiation (Sass et al., 1991b) increased CH_4 emissions.

The objective of this study was to update the estimate of CH_4 emission from irrigated rice in China taking into account organic matter inputs, climate, and type of soil. Field measurements of CH_4 emissions from irrigated rice fields worldwide, with information about organic matter inputs, type of soil, and climate, were compiled. The effect of C and N inputs from fertilizers and the effect of temperature on CH_4 emission rates were evaluated using linear regression techniques. The results of the regression analyses were used in a geographic information system with geographic databases of the location of rice paddies and estimated fertilizer inputs to calculate total annual CH_4 emissions from major rice growing areas of China.

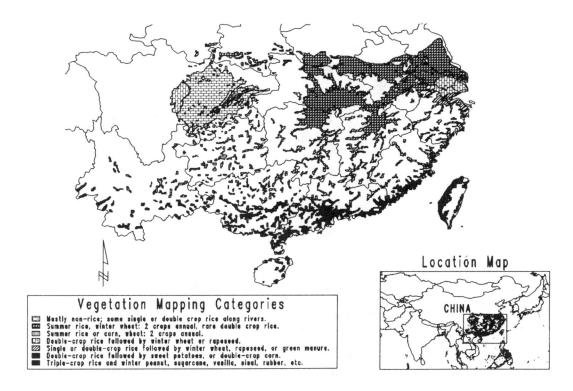

Figure 1. Rice categories from vegetation map by province for China.

II. Materials and Methods

A. Rice Location in China

Rice location was estimated from a 1:4M scale vegetation map produced by the Chinese Academy of Sciences (Hou, 1979). A provisional digital file of the vegetation map was provided by D. Hastings of the National Geophysical Data Center, Boulder, Colorado, and G. Wen of the Chinese Institute of Atmospheric Physics, Beijing, China. The vegetation map (Figure 1) does not show any rice north of about 35 °N latitude, where rice production was identified by statistics compiled by the International Rice Research Institute (IRRI, 1991) and Huke (1982). The soil map of China (Li and Sun, 1990), at 1:10M scale, differed from the vegetation map in that it indicated minor areas of rice cultivation in northeast China north of 35 °N latitude. The vegetation map indicated that there was a large rice growing area east of Chengdu in east-central Sichuan province, but the soil map of China indicated that the area consists of "purplish soil" with smaller areas of soil cultivated to rice along some of the rivers (Li and Sun, 1990).

B. CH_4 Emissions and Fertilizer Inputs

We reviewed the literature to obtain data on organic matter inputs, type of soil, climate, rice cultivar, rice yield, and CH_4 emissions. The study by Bouwman (1991) that summarized published measurements of CH_4 emissions with different fertilizer treatments provided much of the information we used. Data from field studies were selected exclusively for this study because greenhouse experiments commonly use soils that have been dried and sieved that have different redox properties than soils *in situ* (Bartlett, 1986). Only studies conducted over an entire growing season were used because emissions may vary widely throughout

Table 1. Location of CH_4 emission measurements and type of soil

Reference	Location	Year(s)	Type of soil(s)
Butterbach-Bahl, 1992	Vercelli, Italy	1990, 1991	Udifluvent[1]
Chen et al., 1993	Beijing, China	1990	Udifluvent
Cicerone et al., 1983	Davis, California	1982	Typic Chromoxert
Dai et al., 1990	Hangzhou, China	1987	Udifluvent[1]
Khalil et al., 1991	TuZu, China	1988, 1989	Udifluvent[1]
H. Neue, pers. comm.	Los Baños, Philippines	1991	Aquandic Eqiaqualf
Sass et al., 1990	Beaumont, Texas	1989	Typic Pelludert
Sass et al., 1990	Beaumont, Texas	1989	Entic Pelludert
Sass et al., 1991a	Beaumont, Texas	1990	Typic Pelludert
Sass et al., 1991b	Beaumont, Texas	1990	Vertic Ochraqualf
Schütz et al., 1989	Vercelli, Italy	1984-1986	Udifluvent[1]
Schütz et al, 1990	Hangzhou, China	1987	Udifluvent[1]
Seiler et al., 1984	Andalusia, Spain	1982	Udifluvent[1]
M. Tölg, pers. comm.	TaoYuan, China	1992	Endoaquept[1]
R. Wassmann, pers. comm.	TaoYuan, China	1991	Endoaquept[1]
Yagi and Minami, 1990	Ibaraki, Japan	1988	Udifluvent[1]
Yagi and Minami, 1990	Ibaraki, Japan	1988	Udand[1]

[1] Classification inferred from available data from the soil map of the world (FAO, 1974-1978).

Table 2. C, N, and dry matter content of organic fertilizers

Material	% C	% N	% Dry matter	Reference
Rice straw	36	0.7	100	R. Lantin, pers. comm.
Green manure	40	4.0	23	Wen, 1984
Rapeseed cake	40	4.6	100	Wen, 1984
Manure	40	0.4	30	Wen, 1984
Aerobic compost	26	2.2	25	Inoko, 1984
Anaerobic compost	40	3.4	25	Inoko, 1984; Wen 1984

studies conducted over an entire growing season were used because emissions may vary widely throughout a season. Field measurements for 4 sites in China, 4 sites in Japan, 2 sites in the USA, and 1 site in Italy, 1 site in the Philippines, and 1 site in Spain were compiled (Table 1). There were a total of 85 measurements of annual CH_4 emissions including 26 from China, 30 from Italy, 12 from Japan, 1 from Spain, 6 from the Philippines, and 10 from the USA. All soil classifications were converted to the US Soil Taxonomy system (Soil Survey Staff, 1992; 1975). If soil classification was not given in a study, the classification was obtained from the United Nations Food and Agriculture Organization (FAO) soil map of the world (FAO, 1974-1978).

Fertilizer treatments reported for each study were converted to C and N inputs using the relationships shown in Table 2. Waterlogged composted material was assumed to retain 33.8% of the initial C and 82.9% of the initial N (Wen, 1984). More aerobic compost was assumed to retain 19.4% of the initial C and 53% of the initial N (Wen, 1984). Thus, if aerobic compost is 26% C and 2.2% N (Inoko, 1984), then waterlogged compost would be 45% C and 3.8% N. Biogas fermentation residues were assumed to have the same C and N as waterlogged compost while, in fact, they may have slightly more C and N (Wen, 1984).

Data from rice paddies near Hangzhou, in the Zhejiang province of China, were reported by Schütz et al. (1990a), and Dai (1988) for different mineral and organic fertilizer treatments. The soil in Hangzhou was reported as being formed in fluvial deposits and, based on soil climate from the FAO soil map of the world (FAO, 1974-1978), the soil is a Udifluvent. Khalil et al. (1991) presented data from rice paddies near TuZu, in the Sichuan province of China where biogas residues were used. The soils are Udifluvents, based on the FAO soil map of the world. Chen et al. (1993) reported field measurements from rice paddies in Beijing and Nanjing where mineral fertilizers, manure, and rapestraw were added. Based on the FAO soil map of the world, soils used for rice cultivation in Beijing and Nanjing are Udifluvents. The soil at the site in TaoYuan in the Hunan province of China apparently lacks an argillic horizon (M. Tölg, pers. comm.) and is probably an Endoaquept.

The study from Japan for the Ibaraki prefecture (Yagi and Minami, 1990) had measurements from 4 different sites which were 2 Udifluvents and 2 Udands. Methane emissions data for rice fields of Aquandic Epiaqualfs in the Philippines that were treated with mineral fertilizer, manure, and rapestraw additions were provided by H. Neue (pers. comm.). In the USA, Cicerone et al. (1983) reported CH_4 measurements for rice grown in Davis, California on fine, thermic Typic Chromoxerts using mineral N fertilizer. In Texas, Sass et al. (1990; 1991a; 1991b) conducted field studies on Typic Pelluderts, Vertic Pelluderts, and Vertic Ochraqualfs using mineral fertilizer and straw. Low CH_4 emissions for an Entic Pelludert may have been due to damage to the levee that changed the redox conditions (Sass et al., 1990). Seiler et al. (1984) measured emission rates near Sevilla, Andalusia province in Spain, from saline Udifluvents (based on the FAO soil map of the world) and found that mineral fertilizers inhibited CH_4 emissions. Much research has been conducted for Udifluvent rice fields in Vercelli, Italy (Schütz et al. 1989) with different rates of rice straw and/or mineral fertilizer additions. Yield data were not available for many of the measurement sites so none of these data were included in the regression analyses.

Data on organic amendments to rice paddies are scarce, but the apparent global trend is for a decrease (Neue et al., 1990). Winchester et al. (1988) studied agricultural practices in the Anhui and Fujian provinces of China and reported that organic matter inputs were used extensively as fertilizer which is contrary to the global trend. A commonly used organic fertilizer in China is composed of half human and animal excreta with the other half consisting of plant clippings that are composted in family-sized fermentation pits or biogas generators (Winchester et al., 1988). The use of high levels of organic fertilizers in China appears to be currently economically feasible and rates are commonly 60×10^3 kg ha^{-1} of compost (Wiens, 1984). Rice yields of 7.5×10^3 kg ha^{-1} require the equivalent of 0.8 to 1.1×10^3 kg C ha^{-1} if rice straw is applied as the sole source of N (Chen, 1990).

C. Growing Degree Days and Crop Calendar

Temperature data were compiled from a global dataset of mean monthly temperatures (Leemans and Cramer, 1992) with a 30-minute latitude-longitude spatial resolution. The location of the study sites (Table 1) and their crop calendars were used to calculate the daily sum of degrees above 10°C. The season of single crop rice was assumed to be the same as the "intermediate and single late rice crop" of Matthews et al. (1991) which was early May through mid-September. The "early rice crop" season of Matthews et al. (1991) was assumed to correspond to the first season of double crop rice which is mid-March through early-July. The second crop of the double crop rice was assumed to be the same as Matthews et al. (1991) "double late rice crop" which is early-June through the end of October. The triple crop rice season was assumed to be 14 February to 20 May, 21 May to 31 August, and 1 September to 30 November based on data from De Datta (1981) for Nan Hai county, Guangdong province, China. Consequently, the single crop season length was assumed to be 144 days, the double crop season lengths 114 and 138 days, and the 3 triple crop rice season lengths 95, 103, and 91 days respectively.

D. Total Annual CH_4 Flux

Assuming to occur 24 hours a day, CH_4 emissions for each rice crop were calculated using regression equations for C and N inputs. Total CH_4 flux was calculated by summing emissions for each of the three crops (single, double, and triple). The rate of mineral versus organic fertilizer use was based on the total N fertilizer production in China divided by area of major crops harvested in China. In 1986, the total N fertilizer production in China was 12.1×10^6 t (Wang, 1991). The amount of N fertilizers imported by China annually in the late 1980s was 10×10^6 t. The total area of the major crops harvested in China was 128×10^6 ha (Xu and Peel, 1991) which gives an average of 96 kg N ha^{-1} for domestically produced fertilizers. A rice yield of 7500 kg ha^{-1} requires about 125 kg N ha^{-1} (Chen, 1990) and an average of 125 kg N ha^{-1} is used for rice in Jiangsu province of China. Thus, 77% of the N fertilizer demand for rice in

China may be met by mineral fertilizers produced in China. This is in fairly close agreement with Wen (1984) who states that in 1979 organic matter inputs provided 33% of the N requirements in the main rice growing regions. When imported N fertilizers are included, the average use in China is 173 kg N ha^{-1} that agrees almost exactly with Geng (1992) and could possibly meet the N requirements of rice agriculture. Thus, the organic matter inputs may be below the 33% suggested by Wen (1984), 23% if domestic mineral fertilizer production is considered, and somewhat less if imported fertilizers are included. Total annual CH_4 flux was calculated assuming that 75% of the harvested rice received 125 kg N ha^{-1} from mineral fertilizers only. The remaining 25% were assumed to receive 125 kg N ha^{-1} and 1000 kg C ha^{-1} that corresponds to the 825 or 1125 kg C ha^{-1} required to obtain a rice yield of 7500 kg ha^{-1} when only rice straw is used as a fertilizer (Chen, 1990).

III. Results

A. Regression Analyses of Factors of CH_4 Emissions

Data from Table 1 were compiled and used for regression analyses. Multiple linear regression techniques (Statistical Analysis System Institute, 1989) were used to study the relationships among the measured variables and CH_4 emissions. Results of the regression analyses are presented in Table 3. Regression analysis could not be performed for the data from Spain alone because there was only one sample. Carbon inputs as a predictor of CH_4 emissions was most effective in Italy ($R^2 = 0.38$) and least effective in the USA ($R^2 = 0.03$). Nitrogen by itself was not an effective predictor of CH_4 emissions except for the USA data where an R^2 value of 0.47 was obtained. Better results were obtained when N was combined with C for the data from China, the Philippines, and the USA. The inclusion of the sum of degree days improved the relationships for every country except China and Japan. The regression equation for the data from China using C and N was selected for this study. The equation predicts that land that received only mineral N fertilizers at 125 kg N ha^{-1} would have an emission rate of 8.67 mg CH_4 hr^{-1}, while soils that received 1000 kg C ha^{-1} inputs, as well as 125 kg N ha^{-1}, emit 15.67 mg CH_4 hr^{-1}.

B. Rice Location

When all of the areas identified by the vegetation map as being rice growing areas were summed, the result was nearly twice the total harvested area reported by IRRI (1991). Different values for percent areal extent of single, double, and triple crop rice were tried for each of the six classes of rice cultivation shown by the vegetation map until the total was close to the total area harvested according to IRRI (1991). "Single or double crop rice followed by winter wheat and green manure" was found to be much too extensive in Sichuan province compared to IRRI (1991) statistics and was relabeled as "mostly non-rice; some single or double crop rice along rivers" in Figure 1, and the area was assigned 5% single and 2% double crop rice. The "summer rice, winter wheat" category was assigned 30% single and 2% double crop rice. "Summer rice or corn" was assigned 40% single crop rice. "Double-crop followed by winter wheat" was assigned 30% double crop rice. "Single or double crop rice followed by winter wheat" was assigned 25% single and 40% double crop rice. "Double crop rice followed by sweet potatoes" was assigned 45% double crop rice. "Triple crop rice" was estimated to have 45% triple crop rice.

The results of the reclassification of the vegetation map to area of rice cultivation are shown in Table 4. Rice is also grown in the provinces of Beijing, Hebei, Heilongjiang, Jilin, Liaoning, Nei Mongol, Ningxia, Qinghai, Shandong, Shanxi, Tianjin, Xinjang, and Xizang (IRRI, 1991) but these provinces are not shown as rice-growing areas on the vegetation map (Hou, 1979). The rice production of these provinces could correspond to an additional 2010 (IRRI, 1991), 1562 (Huke, 1982) or 1557 (Matthews et al., 1991) x 10^3 ha. Table 4 shows that the total area for the remainder of the provinces obtained using the vegetation map agreed well with other sources of data.

Table 3. Results of the linear regression analyses of C and N versus CH_4 emissions

Country	DF^a	C			C, N			C, N, degree days		
		Coeff.[b]	b^c	R^2	Coeff.[b]	b^c	R^2	Coeff.[b]	b^c	R^2
All	84	0.003	9.618	0.20	0.003 0.007	9.618	0.20	0.002 0.007 -0.065	18.422	0.25
China	25	0.003	12.543	0.14	0.007 -0.082	18.822	0.22	0.007 -0.083 0.010	17.691	0.32
Italy	29	0.002	12.214	0.38	0.002 0.005	11.683	0.39	0.002 0.011 -0.543	83.356	0.46
Japan	11	0.002	1.943	0.22	0.002 -0.012	2.079	0.22	0.002 -0.012 0	2.079	0.22
Philippines	5	0.002	4.260	0.36	0.002 0.048	0.537	0.85	0.002 -0.040 -0.063	10.877	1.0
USA	9	0.001	12.864	0.03	-0.002 0.096	-3.880	0.47	0 0.139 -0.048	-2.37	0.57

[a] DF = degrees of freedom; [b] Coeff. = regression coefficient in order of heading; [c] b = intercept.

Table 4. Rice area harvested from vegetation mapping compared with reported rice production by province

Province	Vegetation mapping (10³ ha)				IRRI (1991) (10³ ha)	Huke (1982) (10³ ha)	Matthews et al. (1991) (10³ ha)
	Single	Double	Triple	Total			
Anhui	1,621	236	0	1,857	2,273	1,847	2,152
Fujian	261	1,649	0	1,910	1,509	1,807	1,670
Gansu	45	6	0	51	0	11	4
Guangdong	174	4,291	0	4,465	3,583	4,187	4,107
Guangxi	292	3,238	0	3,530	2,494	2,762	2,825
Guizhou	588	291	0	879	732	733	779
Hainan[a]	0	0	532	532	0	0	0
Henan	1,024	137	0	1,161	427	443	400
Hubei	1,895	338	0	2,233	2,606	2,083	2,620
Hunan	1,070	1,329	0	2,399	4,354	4,258	4,468
Jiangsu	1,452	677	0	2,129	2,420	3,026	2,628
Jiangxi	1,091	1,598	0	2,689	3,298	2,653	3,402
Shaanxi	292	39	0	331	158	179	163
Shanghai	19	266	0	285	258	427	276
Sichuan	669	1,430	0	2,099	3,113	3,344	3,163
Yunnan	341	1,096	322	1,759	1,008	875	1,091
Zhejiang	504	1,864	0	2,368	2,380	2,693	2,529
Totals:	11,338	18,485	1,476	30,677	30,613[b]	31,328	32,277

[a] Hainan was grouped with Guangdong for reported rice statistics; [b] totals are for provinces where the vegetation map category was rice.

Table 5. Estimated annual CH_4 emissions by province for China

Province	CH_4 emissions (10^9 g CH_4 yr^{-1})	Province	CH_4 emissions (10^9 g CH_4 yr^{-1})
Anhui	652	Hunan	796
Fujian	616	Jiangsu	728
Gansu	18	Jiangxi	886
Guangdong	1,404	Shaanxi	116
Guangxi	1,114	Shanghai	90
Guizhou	300	Sichuan	684
Hainan	100	Yunnan	522
Henan	408	Zhejiang	780
Hubei	780	Total	9,974

C. CH_4 Fluxes from Soils Used for Irrigated Rice in China

The total annual CH_4 emissions for major rice growing areas in China are estimated to be 10 Tg CH_4 yr^{-1}. The CH_4 emissions for each province that had rice cultivation identified by the vegetation map are shown in Table 5. The provinces in the south (Guangdong and Guangxi) with large amounts of double crop rice had the highest emissions. Provinces with large contiguous areas of single rice (Anhui, Hubei, Hunan, and Jiangxi) had somewhat lower emissions.

IV. Summary and Conclusions

The approach used in this study provided algorithms to estimate annual CH_4 emissions from C and N inputs. These algorithms will be particularly useful when more detailed data about mineral and organic fertilizer use are collected from across Asia. The uncertainty of the current estimates is high because the regression analysis of C and N effects on CH_4 emissions had relatively low R^2 values (0.32 for China and 0.47 for the USA). Bouwman (1991) used a subset of the measurements included in this paper to conclude that additions of organic matter increases CH_4 emissions. The trend for N fertilizers to decrease CH_4 emissions agrees with the observations of Lin (1993). Low R^2 values reflect the large variability of emissions rates among and within studies for different years. Production and emission of CH_4 are very complex and many other factors than C and N inputs are important. The inclusion of sum of degree days data from long term averages in the regression analyses seemed to give some promising results. Future work will include compiling as much actual measured temperature data as possible to try improve the regression estimates. More spatially and temporally detailed temperature data are required to reduce uncertainties.

The estimate of 10 Tg CH_4 made here for China is at the low range of other published studies. Bachelet and Neue (1993) estimated CH_4 emissions using several different approaches at between 9 Tg CH_4 yr^{-1} (as a proportion of net primary productivity) and 22 Tg CH_4 yr^{-1} (assuming a constant emission rate). Lin (1993) estimated 11 Tg CH_4 yr^{-1} using 5 different emission rates of 7.8 to 60 mg CH_4 hr^{-1} m^2. Clearly more CH_4 emission measurements and more detailed data about fertilizer inputs, soil organic C, air temperature, solar radiation, and land use are needed to reduce the uncertainties in these estimates.

Acknowledgements

David Hastings of the National Geophysical Data Center, Boulder, Colorado, in cooperation with Wen Gang from the Chinese Institute of Atmospheric Physics in Beijing, China was very helpful in providing use with a digital vegetation map. Ronald Sass, Rice University, Texas provided a helpful technical review.

The information in this document has been funded wholly by the U.S. Environmental Protection Agency (EPA) under contract 68-C8-0006 to ManTech Environmental Technology, Inc. It has been subjected to the Agency's peer and administrative review, and it has been approved for publication as an EPA document.

V. References

Bachelet, D. and H.U. Neue. 1993. Methane emissions from wetland rice areas of Asia. *Chemosphere* 26:219-237.

Bartlett, R.J. 1986. Soil redox behavior. Chapter 5, p. 79-177. In: D.L. Sparks (ed.). *Soil physical chemistry*. CRC Press, Boca Raton, FL.

Bouwman, A.F. 1991. Agronomic aspects of wetland rice cultivation and associated methane emissions. *Biogeochem.* 15:65-88.

Butterbach-Bahl, K. 1992. *Mechanisms of the production and emission of methane from ricefields* (in german). Ph.D dissertation at the Institute of Botany and Microbiology of the Technical University of Munich, Munich, Germany.

Chen, J. 1990. Properties, cultivation and management of fertile paddy soils. Chapter 35, p. 673-704. In: C. Li and O. Sun. *Soils of China*. Science Press. Beijing, China.

Chen, Z., D. Li, S. Kesheng, and B. Wang. 1993. Features of CH_4 emission from rice paddy fields in Beijing and Nanjing. *Chemosphere* 26:239-245.

Cicerone, R.J., J.D. Shetter, and C.C. Delwiche. 1983. Seasonal variation of methane flux from a California rice paddy. *J. Geophys. Res.* 88:11022-11024.

Dai, A. 1988. *Emission rates from rice paddies in HangZhou China during fall of 1987*. M.S. thesis, Institute of Atmospheric Physics, Academia Sinica, Beijing, China.

De Datta, S.K. 1981. *Principles and practices of rice production*. J. Wiley & Sons, NY.

Food and Agricultural Organization, UNESCO 1974-1978: Soil map of the world, Vol. 1-10, Paris, France.

Geng, S. 1992. Sustainability of agriculture in China. *J. Prod. Agric.* 5:176-180.

Gong, Z. and Q. Xu. 1990. Paddy soils. Chapter 14, p. 237-260. In: C. Li and O. Sun. *Soils of China*. Science Press. Beijing, China.

Holzapfel-Pschorn, A., and Seiler, W. 1986. Methane emission during a cultivation period from an Italian rice paddy. *J. Geophys. Res.* 91:11803-11814.

Hou, H.Y. 1979. *Vegetation map of China, scale 1:4M*. Lab. of Plant Ecol. and Geobot., Inst. of Bot. Chinese Acad. of Sci., Beijing, China.

Huke, R.E. 1982. Agroclimatic and Dry-Season Maps of South, Southeast, and East Asia. International Rice Research Institute, Los Baños, Philippines.

Inoko, A. 1984. Compost as a source of plant nutrients. p. 137-145. In: International Rice Research Institute (ed.). *Organic matter and rice*. International Rice Research Institute. Los Baños, Philippines.

International Rice Research Institute (IRRI). 1991. World Rice Statistics, 1990. IRRI, Manila, Philippines.

Khalil, M.A.K. and R.A. Rasmussen. 1990. Constraints on the global sources of methane and an analysis of recent budgets. *Tellus* 42B:229-236.

Khalil, M.A.K., R.A. Rasmussen, M.-X. Wang, and L. Ren. 1991. Methane emissions from rice fields in China. *Environ. Sci. Tech.* 25:979-981.

Leemans, R. and W.P. Cramer. 1992. IIASA database for mean monthly values of temperature, precipitation, and cloudiness on a global terrestrial grid. In: J.J. Kineman and M.A. Ochrenschall. 1992. *Global ecosystems database version 1.0:Disc A, Documentation manual*. US Dept. of Comm./National Oceanic and Atmospheric Administration, National Geophysical Data Center, Boulder, CO.

Li, C. and O. Sun. 1990. *Soils of China, 2nd Ed*. Chinese Academy of Science. Beijing Science Press.

Lin, E. 1993. Agricultural techniques: factors controlling methane emissions. p. 120-126. In: L. Gao, L. Wu, D. Zheng, and X. Han (eds.). *Proceedings of the international symposium on climate change, natural disasters, and agricultural strategies*. China Meteorological Press, Beijing, China.

Mariko, S., Y. Harazono, N. Owa, and L. Nouch. 1991. Methane in flooded soil water and the emission through rice plants to the atmosphere. *Environ. and Exper. Botany*, 31:343-350.

Matthews, E., I. Fung, and J. Lerner. 1991. Methane emission from rice cultivation: geographic and seasonal distribution of cultivated areas and emissions. *Global Biogeochem. Cycles* 5:3-24.

Neue, H.U., P. Becker-Heidmann, and H.W. Sharpenseel. 1990. Organic matter dynamics, soil properties, and cultural practices in rice lands and their relationship to methane production. In: A.F. Bouwman (ed.). *Soils and the Greenhouse Effect*. J. Wiley & Sons, NY.

Sass, R.L., F.M. Fisher, P.A. Harcombe, and F.T. Turner. 1990. Methane production and emission in a Texas rice field. *Global Biogeochem. Cycles* 4:47-68.

Sass, R.L., F.M. Fisher, P.A. Harcombe, and F.T Turner. 1991a. Mitigation of methane emissions from rice fields: possible adverse effects of incorporated rice straw. *Global Biogeochem. Cycles* 5:275-287.

Sass, R.L., F.M. Fisher, F.T. Turner, and M.F. Jund. 1991b. Methane emission from rice fields as influenced by solar radiation, temperature, and straw incorporation. *Global Biogeochem. Cycles* 5:335-350.

Schütz, H., W. Seiler, and H. Rennenberg. 1990a. Soil and land use related sources and sinks of methane (CH_4) in the context of the global methane budget. Chapter 12, p. 269-285. In: A.F. Bouwman (ed.). *Soils and the greenhouse effect*. J. Wiley & Sons, NY.

Schütz, H., W. Seiler, and R. Conrad. 1990b. Influence of soil temperature on methane emission from rice paddy fields. *Biogeochem.* 11:77-95.

Schütz, H., A. Holzapfel-Pschorn, R. Conrad, H. Rennenberg, and W. Seiler. 1989. A 3-year continuous record on the influence of daytime, season, and fertilizer treatment on methane emission rates from an Italian rice paddy. *J. Geophys. Res.* 94:16405-16416.

Seiler, W., A. Holzapfel-Pschorn, R. Conrad, and D. Scharffe. 1984. Methane emissions from rice paddies. *J. Atmos. Chem.* 1:241-268.

Soil Survey Staff. 1975. *Soil Taxonomy: a basic system of soil classification for making and interpreting soil surveys*. Agric. Handb. 436. U.S. Gov. Print. Office, Washington, D.C. Soil Survey Staff. 1992. *Keys to soil taxonomy, 5th Ed*. Soil Management Support Services technical monograph. no. 19, Pocahontas Press, Blacksburg, VA.

Statistical Analysis System (SAS) Institute, 1989. *SAS/STAT user's guide, v. 6, 4th Ed., Vol. 1*, SAS Institute, Cary, NC.

Wang, W. 1991. Infrastructure and agricultural inputs. p. 144-178. In: G. Xu and L.J. Peel (eds.). *The agriculture of China*. Oxford University Press, NY.

Wen, Q. 1984. Utilization of organic materials in rice production in China. p. 46-56. In: International Rice Research Institute (ed.). *Organic matter and rice*. International Rice Research Institute. Los Baños, Philippines.

Wiens, T.B. 1984. Microeconomic study of organic fertilizer use in intensive farming in Jiangsu province, China. p. 533-556. In: International Rice Research Institute (ed.). *Organic matter and rice*. International Rice Research Institute. Los Baños, Philippines.

Winchester, J.W., F. Song-miao, and L. Shao-meng. 1988. Methane and nitrogen gases from rice fields of China - possible effect of microbiology, benthic fauna, fertilizer, and agricultural practice. *Water, Air, and Soil Pollut.* 37:149-155.

Xu, G. and L.J. Peel (eds.). 1991. *The agriculture of China*. Oxford University Press, NY.

Yagi, K. and Minami, K. 1990. Effects of organic matter applications on methane emissions from Japanese paddy fields. p. 467-473 In: A.F. Bouwman, (ed.). *Soils and the Greenhouse Effect*. J. Wiley & Sons, NY.

Zehnder, A.J.B. and W. Stumm. 1988. Geochemistry and biogeochemistry of anaerobic habitats. Chapter 1, p. 1-38. In: A.J.B. Zehnder (ed.). *Biology of anaerobic microorganisms*. J. Wiley & Sons, NY.

CHAPTER 16

Soil CO_2 Flux in Response to Elevated Atmospheric CO_2 and Nitrogen Fertilization: Patterns and Methods

J.M. Vose, K.J. Elliott, and D.W. Johnson

I. Introduction

The evolution of carbon dioxide (CO_2) from soils is due to the metabolic activity of roots, mycorrhizae, and soil micro- and macro-organisms. Although precise estimates of carbon (C) recycled to the atmosphere from belowground sources are unavailable, Musselman and Fox (1991) propose that the belowground contribution exceeds 100 Pg y^{-1} globally. This represents a major component of C flux in the global C cycle. Belowground C cycling processes and subsequent soil CO_2 fluxes are equally important at ecosystem scales; however, we have limited knowledge of the magnitude of fluxes within and across ecosystems. Increased knowledge of the magnitude of C fluxes, as well as the factors which regulate these fluxes is critical for understanding ecosystem C cycling and potential responses to factors such as climatic change. In this study, we quantified soil CO_2 flux from soils growing ponderosa pine (*Pinus ponderosa* L.) under conditions of elevated atmospheric CO_2 and soil nitrogen (N).

Separating the contributing sources (i.e., roots vs. microbes) of soil surface CO_2 flux has proven difficult. The relative contribution of roots versus other soil components has been estimated to vary between 35 to 65% of the total CO_2 evolved (Edwards and Harris, 1977; Ewel et al. 1987). Factors influencing the rate of CO_2 evolution include soil temperature and moisture (through their influence on metabolic activity of both roots and microbes) (Schlentner and Van Cleve, 1985; Wiant, 1967), available soil carbon, and root biomass (Behera et al., 1990). Hence, changes in root biomass and/or activity related to elevated CO_2 should directly influence the total CO_2 evolution from forest soils. Indirect effects related to carbon source and amount (e.g., fine root turnover, exudates) should also influence CO_2 evolution by altering microbial activity. Finally, increased soil N availability could alter soil CO_2 flux by changing root (Ryan, 1991) and microbial activity and/or biomass, particularly if soil or litter C:N ratios are substantially altered.

Several techniques are available for measuring CO_2 evolution from soils. Static chamber methods include soda lime or bases (KOH or NaOH) which measure CO_2 "trapped" over the measurement interval (see Cropper et al., 1985). Static measures of CO_2 evolution may also be made by gas chromatograph analysis of air samples collected from sealed chambers on the soil surface (Raich et al. 1990). de Jong and Schappert (1972) describe a variation of the static method by using a chamberless technique based on CO_2 profiles (pCO_2) in the soil. Dynamic chamber methods quantify CO_2 evolution by continuously monitoring CO_2 levels in chambers with either a closed or flow-through system and an infrared gas analyzer (IRGA). Studies comparing measurement techniques have found wide disparity between static chamber, static chamberless, and dynamic chamber methods (Edwards and Sollins, 1973; Cropper et al., 1985; Raich et al., 1990; Rochette et al., 1992, Norman et al., 1992). In general, static chamber techniques provide lower estimates of CO_2 evolution than dynamic chamber techniques, while pCO_2 techniques provide higher CO_2 evolution estimates than dynamic chamber techniques (de Jong et al., 1979). Although more difficult and expensive to conduct, dynamic, IRGA-based techniques are considered more reliable (Ewel et al., 1987) and they can be configured to quantify diurnal patterns.

The objectives of this study were: (1) to quantify diurnal patterns in soil CO_2 evolution using a dynamic, IRGA based measurement system, (2) to examine the impacts of elevated atmospheric CO_2 and nitrogen fertilization on soil CO_2 evolution, and (3) to compare estimates using the IRGA based system with pCO_2 measurements.

II. Methods

A. Site Description

The study was conducted at the USDA Forest Service Institute of Forest Genetics in Placerville, CA (longitude = 121°W, latitude = 39°N). The elevation of the site is 843 m, receives an average of 1000 mm of annual precipitation, and has a mean annual temperature of 18 °C. The soil is Aiken clay loam (Xeric Haplohumult) derived from andesite. Extensive sampling prior to study establishment indicated uniform soil chemical and textural characteristics across the study area. Bulk density of the soil averaged 1.14 g cm^{-3}, porosity was 54%, reaction was moderately acidic (pH_{CaCl} = 5.1 in upper 18 cm), and base saturation (1 $\underline{M}$ NH_4Cl extraction) was 50 to 60%.

B. Experimental Design and Treatments

The experiment utilized open top chambers (3 m diameter; hexagonal shape) as a means of elevating CO_2 concentration (Ball et al. 1991). Air was delivered to the chambers at three air changes per minute and was distributed using 45 cm diameter plastic tubing perforated with 2.54 cm holes on 15 cm centers. The experimental design consisted of three levels of N (ambient, 10, and 20 g m^{-2} yr^{-1} of N as ammonium nitrate, applied in early spring), and four continuous CO_2 treatments (ambient, no chamber; ambient chamber; +175 ppm; and +350 ppm). Each of the chambered treatments was replicated three times, and the unchambered treatment was replicated twice. Due to cost limitations, the 10 g m^{-2} yr^{-1} N, +175 CO_2 treatment was excluded. Hence, there were a total of 11 treatments. Each chamber contained 21 ponderosa pine seedlings (grown from seed) equally spaced at about 0.3 m in all directions. At the time of sampling, seedlings had been grown in the chambers under treated conditions for two years. Soils were irrigated weekly with sufficient water to maintain soil water potential at $\geq$ -0.07 MPa.

C. CO_2 Sampling

1. IRGA Measurements

We measured diurnal patterns of soil CO_2 flux using an automated, flow-through, IRGA based measurement system (Figure 1). The system measured flux sequentially from ten soil chambers (10 cm diameter, 10 cm height, 785 cm^3 volume) constructed of PVC pipe. Soil chamber edges were sharpened on the open end and driven approximately 2 cm into the soil surface with a rubber mallet. All tubing was 5 mm (i.d.) flexible PVC. Air was passed through the chambers via inlet and outlet fittings attached to the upper sides of the chamber. Air flow through the chambers was regulated with a dual-sided air pump (Spec-Trex Corp.) which balanced flow into and out of the chambers. Actual flow rate (ml min^{-1}) was controlled by varying voltage (0-12 VDC) supplied to the pump and was measured and logged electronically with a flow meter and data logger (Campbell 21X). An air flow rate of 1000 to 1500 ml min^{-1} provided stable readings within 7 to 8 minutes. Chamber sampling was controlled with a multiplexer, data logger, and solenoids which opened sequentially (chambers 1-10) at ten minute intervals. Carbon dioxide concentrations of air entering and exiting the chambers was measured and logged electronically with an IRGA (ADC LCA3) operating in differential mode and a data logger (Campbell 21X), respectively. Soil CO_2 flux (mg CO_2 m^{-2} min^{-1}) was calculated based on the difference in CO_2 entering and exiting the chamber, the soil area sampled beneath the chamber, and the flow rate. Only data from the last minute of sampling were used in flux calculations.

Sampling was conducted over a six day period in mid-October, 1992. On each day, two soil respiration chambers were randomly placed in each of five treatment-replication combinations, with the restriction that chambers could be no closer than 2.5 cm from a seedling. This restriction was imposed to ensure that

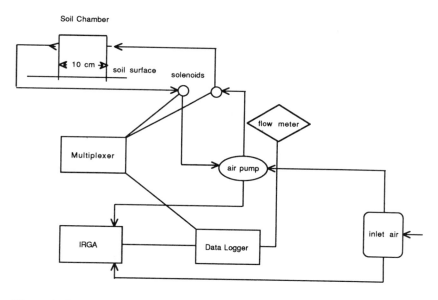

Figure 1. Schematic representation of automated sampling system. Only one of ten soil sampling chambers is shown in the diagram.

seedlings were not damaged in the course of installing the chambers. Soil CO_2 flux was measured for 20-22 hrs (i.e., a diurnal cycle) on each day. On each successive day, the chambers were moved to a new set of treatment-replication combinations and the diurnal measurements repeated until all treatments and replications were sampled. Using this sampling approach, we assumed that there would be minimal day to day variation in diurnal soil CO_2 flux. Because climatic conditions were generally constant (i.e., no rain, cloudless skies, and stable temperatures) throughout the six day measurement period, we are confident that this assumption is valid.

2. pCO_2 Measurements

Gas wells were established at 15 and 30 cm depths within each open top chamber. During the same week as the IRGA based sampling, gas wells were sampled during daytime hours with a syringe, and CO_2 concentration was determined with a LICOR 6262 analyzer. Soil CO_2 flux was estimated using a procedure based on the concentration gradient (pCO_2) between the two depths and a CO_2 diffusion coefficient. This is described in the following formula:

$$q = -D(\partial c/\partial z),$$

where q = flux (mg CO_2 m^{-2} min^{-1}), D = diffusion coefficient (m^2 min^{-1}), c = concentration (mg CO_2 m^{-3}), and z = depth (m) (from deJong and Schappert 1972). The diffusion coefficient, D, was calculated using the equation (Rolston 1986):

$$D = (P_{eff}^{10/3}/E^2),$$

where D = diffusion coefficient of CO_2 in soil (m^2 min^{-1}), P_{eff} = effective soil porosity (air-filled soil pores, a function of soil porosity and water content), and E = total porosity. Soil moisture release curves were constructed to determine the relationship between soil water tension and water content. The relationship between soil water tension and P_{eff} was determined by subtracting soil water content from total porosity across a range of soil water tensions. Hence, for each sample period soil water content was determined gravimetrically and the corresponding soil water tension was determined from soil moisture release curves. Effective soil porosity at a given soil water content was then determined from the relationship between soil water tension and P_{eff}.

D. Soil Moisture and Temperature

Soil moisture (averaged over 15 cm depth) was measured within 10 cm of each soil CO_2 flux chamber location using a "TRASE" time domain reflectometry measurement system (Soil Moisture Instruments, Inc.) with sample rods installed vertically in the soil. Soil moisture measurements were taken before and after the flux measurements were conducted. To characterize diurnal soil temperature variation, soil temperature at 10 cm depth was measured for one complete diurnal cycle at five locations within and outside open-top chambers. Measurements were made with Type-T thermocouples connected to a data logger (Campbell 21X) and multiplexer.

E. Statistical Analysis

We used integrated values of diurnal measurements to estimate soil CO_2 flux on a daily basis for each open-top chamber-treatment combination. These integrated values were used in analysis of variance (ANOVA) to test for treatment effects (General Linear Models Procedure, SAS Institute, 1987). A reduced error term {chamber(treatment)}, which accounted for the subsample of two soil CO_2 flux chambers per open-top chamber, was used in the ANOVA to test for treatment effects. Because we had an unbalanced experimental design, contrast statements (Snedecor and Cochran, 1980) were constructed to determine the effects of CO_2, nitrogen, and CO_2 x nitrogen interaction on soil CO_2 flux.

III. Results and Discussion

A. Diurnal Patterns

Across all treatments, soil CO_2 flux varied by as much as seven-fold over a diurnal measurement period (Table 1). Maximum rates ranged from 7.78 to 17.00 mg CO_2 m^{-2} min^{-1} and minimum rates ranged from 1.11 to 3.74 mg CO_2 m^{-2} min^{-1}. These rates are comparable to those determined with other continuous techniques; e.g., in slash pine (*Pinus elliotii* Engelm.), Ewel et al. (1987) found average annual rates of 9 mg CO_2 m^{-2} min^{-1}. Kucera and Kirkman (1971) found October rates in tall-grass prairie on the order of 7.5 mg CO_2 m^{-2} min^{-1}. In a yellow poplar (*Liriodendron tulipifera* L.) forest, Edwards and Sollins (1973) found values ranging from 5 to 14 mg CO_2 m^{-2} min^{-1} for spring and summer, respectively.

The diurnal patterns observed in our study were correlated with diurnal variation in soil temperature (Figure 2). Maximum rates of soil CO_2 flux occurred in the late afternoon ($\approx$ 1500 hr) when soils were warmest, and minimum rates occurred in the morning ($\approx$ 0900 hr) when soils were coolest. The apparent temperature influence is not surprising. Temperature has a major impact on respiratory processes (Ryan 1991) and relationships between temperature and soil CO_2 flux are well established (e.g., Schlentner and Van Cleve, 1985; Naganawa et al., 1989). However, there was a noticeable surge in soil CO_2 flux between 2100 hr and 0100 hr and other periods when patterns in soil temperature and soil CO_2 flux did not agree (e.g., 0900 hr). This indicates that other factors are also contributing to the diurnal patterns we observed. Soil moisture stayed consistent throughout the measurement cycle (Table 2) so the diurnal variation can not be attributed to moisture.

In contrast to our study, Edwards and Sollins (1973) found that forest floor CO_2 flux rates were greatest at night and most of the nighttime increase came from the litter fraction. Two factors may explain the differences between our study and Edwards and Sollins (1973). First, they performed their experiments in a closed canopied forest which substantially dampened diurnal temperature variation. In fact, litter temperature in their study remained essentially constant. In contrast, conditions at our site (e.g., clear skies, no forest canopy) promoted large diurnal variations in soil temperature. Second, in our study there was no litter layer. Hence, any phenomena related to elevated night respiration from litter (e.g., increased moisture content due to dew formation) would not have been observed in our study.

Table 1. Average, maximum, and minimum soil CO_2 evolution (mg m^{-2} min^{-1}) measured over a 24 hour period; values in parentheses are standard errors; n=2 for chamberless (OPEN) treatments and n=3 for all others

Treatment	24 hour average	Daily maximum	Daily minimum
350 CO_2 + 0N	7.15 (2.54)	12.72 (5.90)	3.20 (0.77)
350 CO_2 + 10N	6.26 (2.82)	12.86 (7.63)	2.79 (0.94)
350 CO_2 + 20N	4.98 (1.01)	7.78 (1.73)	2.34 (0.35)
525 CO_2 + 0N	10.18 (2.73)	17.00 (4.65)	2.52 (0.98)
525 CO_2 + 20N	8.20 (3.22)	13.87 (3.70)	2.06 (0.95)
700 CO_2 + 0N	5.50 (0.85)	9.93 (1.57)	1.99 (0.91)
700 CO_2 + 10N	8.16 (1.73)	10.86 (2.45)	1.11 (0.58)
700 CO_2 + 20N	6.05 (1.71)	14.26 (3.57)	3.74 (0.74)
Open + 0N	5.72 (0.03)	9.01 (3.08)	3.27 (0.44)
Open + 10N	6.88 (3.19)	11.04 (5.30)	2.89 (0.89)
Open + 20N	5.80 (0.47)	9.46 (0.81)	2.57 (0.22)

Similar to our study, Edwards and Sollins (1973) observed a noticeable surge in soil CO_2 flux between 2100 hr and 0100 hr and this phenomena has also been observed in other studies (e.g., Witkamp, 1969; Witkamp and Frank, 1969). Witkamp (1969) observed surges in soil CO_2 flux when the soil was warmer than the air, which he attributed to convectional forces flushing soil CO_2 into the atmosphere. Convectional processes may partially contribute to the diurnal patterns observed in our study because there was a rapid decline in air temperature in the early morning (0100 hr). The elevated soil CO_2 flux between 2100 hr and 0100 hr observed in our study and by Edwards and Sollins (1973) is surprising and difficult to explain. We are confident that these patterns are not artifacts of our sampling system because in recent studies we have included closed blank chambers which show no diurnal patterns (Vose and Elliott, unpublished data), while data from chambers on the soil surface show a distinct diurnal pattern. Edwards and Sollins (1973) hypothesized that physical/climatic factors, such as changes in vapor pressure, were responsible for the elevated late evening rates. Biological factors may also be important. For example, the contribution of root respiration could increase due to: (1) increased root activity following cessation of aboveground light driven processes, and/or (2) an increase in "apparent" root respiration due to decreased dissolution of respired CO_2 in the transpiration stream. We have no data on diurnal patterns in root respiration and know of no data from the literature. Hence, a complete understanding of factors driving the diurnal patterns we observed will require further study. It is clear from our results that soil CO_2 flux sampling techniques in ecosystems with highly variable diurnal soil temperature need to sample over the entire diurnal cycle. For example, sampling only during warm, afternoon hours will bias any extrapolations upward and sampling during cool, morning hours will bias extrapolations downward. Although less well understood, other physical and biological factors may also contribute to diurnal variability and increase the importance of diurnal sampling for accurate quantification of fluxes. Our automated system enabled us to sample over the entire diurnal cycle and integrate fluxes. While static techniques provide cumulative flux measurements over a 24-hr period, there can be substantial differences in flux estimates between static and dynamic methods (see Edwards and Sollins, 1973; Kucera and Kirkman, 1971; and Cropper et al., 1985; Rochette et al., 1992).

B. Treatment Effects

We tested the impacts of elevated CO_2 and N fertilization on daily flux rates (Table 3). The daily value integrated the shorter term data collected over the diurnal cycle. Differences in rates varied as much as two-fold (i.e., 7.1 vs. 15.5 g CO_2 m^{-2} day^{-1}); however, there was considerable variability which precluded detection of significant differences ($p < 0.10$) among treatments (Table 4). While diurnal soil temperature variation was substantial (e.g., Figure 2), there was little variation in soil temperature among treatments (Vose and Elliott, unpublished data) and no correlation between mean soil CO_2 flux rates and soil moisture (Table 2). Although our system included only two-year-old seedlings, the range of soil CO_2 flux values observed in our study is at the upper end of values found in the literature for a variety of forests. The high soil CO_2 fluxes observed in our study may be due to favorable environmental conditions (i.e., warm and wet soils); however, some of these differences may be related to the method of measurement. For example,

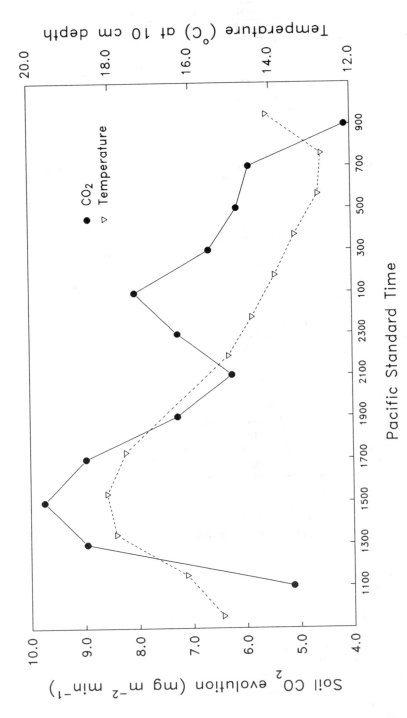

Figure 2. Example of temporal variation in soil CO_2 flux and soil temperature at 10 cm depth. Soil CO_2 flux values are averaged across all treatments.

Table 2. Average percent soil moisture measured at 15 cm depth at the beginning and end of each 24-hour period for each treatment; values in parentheses are standard error

Treatment	Initial soil moisture (m³ m⁻³)	Final soil moisture (m³ m⁻³)
350 CO_2 + 0N	26.25 (0.45)	25.08 (0.64)
350 CO_2 + 10N	27.58 (1.55)	26.43 (1.33)
350 CO_2 + 20N	28.30 (1.64)	28.12 (1.92)
525 CO_2 + 0N	28.55 (1.63)	28.82 (1.56)
525 CO_2 + 20N	29.65 (1.23)	28.17 (0.92)
700 CO_2 + 0N	24.48 (3.97)	23.52 (3.68)
700 CO_2 + 10N	23.47 (3.06)	21.73 (2.41)
700 CO_2 + 20N	23.63 (3.81)	23.00 (3.83)
Open + 0N	26.88 (0.32)	26.45 (0.30)
Open + 10N	30.02 (0.52)	29.30 (1.00)
Open + 20N	23.58 (8.28)	23.28 (7.78)

OPEN = Chamberless

Table 3. Average integrated soil CO_2 evolution (g m⁻² day⁻¹) per treatment measured in October 1992. Average values are least square means with standard errors in parentheses (n=3)

Treatment	Means
350CO_2 + 10N	9.84 (1.65)
350CO_2 + 20N	8.66 (1.90)
525CO_2 + 0N	7.06 (2.13)
525CO_2 + 20N	15.52 (1.65)
700CO_2 + 0N	15.52 (1.65)
700CO_2 + 10N	8.02 (1.90)
700CO_2 + 20N	8.69 (1.90)
OPEN + 0N	11.73 (1.65)
OPEN + 10N	8.05 (2.02)
OPEN + 20N	10.52 (2.47)
	8.32 (2.02)

OPEN = Chamberless.

in aspen (*Populus tremuloides* Michx.) stands, soil respiration rates in October ranged from about 4 to 6 g CO_2 m⁻² day⁻¹ using soda lime (Jurik et al., 1991; Weber, 1990). In jack pine (*Pinus banksiana* Lamb.) stands, Weber (1985) found November values of 4 g CO_2 m⁻² day⁻¹ using soda lime. Static techniques, such as soda lime or base traps, have been shown to underestimate soil CO_2 flux relative to IRGA systems (Ewel et al., 1987; Edwards and Sollins, 1973). In contrast, Edwards and Sollins (1973) found flux rates ranging from 3 to 26 g m⁻² day⁻¹ (mean = 16.7) in a yellow poplar (*Liriodendron tulipifera* L.) stand using a continuous IRGA system. Because there is no litter layer, the CO_2 evolved from the soil is exclusively from root and soil microbial respiration. Hence, some of the factors contributing to the flux variability observed among treatments may be related to variation in root biomass/activity and soil microbial populations among treatments. Several studies have shown increased root biomass in response to elevated CO_2 (e.g., Rogers et al., 1992; Norby et al., 1987) and plant respiration is positively correlated with tissue nitrogen (Ryan, 1991). In our study, there was a trend toward greater soil CO_2 evolution from the elevated CO_2 treatment (particularly the 525 CO_2 treatment) which coincides with greater root biomass on these treatments (determined from destructive sampling; R. Walker, unpublished data). As seedlings grow and continue to increase their belowground biomass, differences in soil CO_2 evolution among treatments may become more apparent.

Table 4. Analysis of variance table for test of treatments with contrasts for CO_2 and N comparisons (October 1992)

Source	df	SS	MS	F	p>F
Treatment	10	309.053	30.905	0.63	0.767
Contrast "Chamber"	1	1.135	1.135	0.02	0.880
Contrast "CO w/CC"	2	200.867	100.434	2.06	0.155
Contrast "N"	2	33.144	16.572	0.34	0.716
Contrast "N control vs N high"	1	18.301	18.301	0.38	0.547
Error Cham(Trt)	19	925.022	48.6854		

Note: **Contrast "Chamber"** tests for a difference between 350CO$_2$ and chamberless treatments across levels of N. Since there is no significant difference between 350CO$_2$ and chamberless treatments, then **Contrast "CO$_2$ w/CC"** tests for a difference between the average of 350CO$_2$ and chamberless treatments vs. the average of the 525CO$_2$ & 700CO$_2$ treatments across nitrogen level; +10N nitrogen level is ignored. **Contrast "N"** tests for a difference between +0N vs. average of the +10N and +20N treatments across CO$_2$ level; 525CO$_2$ level is ignored. **Contrast "N control vs N high"** test for a difference between +0N vs. +20N treatments across CO$_2$; +10N is ignored because it doesn't occur in every possible level of CO$_2$, i.e., no 525CO$_2$ + 10N treatment.

C. pCO$_2$ vs. IRGA Measurements

Comparisons between soil CO$_2$ flux using the pCO$_2$ method versus the dynamic IRGA method are shown in Figure 3. In this comparison, only IRGA based fluxes from time periods and soil chambers corresponding with the time and location of pCO$_2$ measurements were used. Results of the two methods were significantly correlated ($r^2 = 0.50$; $p < 0.05$); however, there were substantial differences in the magnitude of the flux estimates. In addition, the relationship was strongly influenced by the two most extreme values. The pCO$_2$ estimates were two- to three-fold greater than the IRGA based measurements and substantially greater than values reported in the literature. deJong et al. (1979) also found consistently higher values in soil CO$_2$ flux using the pCO$_2$ method compared with a continuous IRGA system. They attributed the majority of those differences to pressurizing the soil chambers and clipping aboveground plant material in the IRGA measurements. Pressurized chambers and clipping of plant material were not relevant factors in our study. Instead, differences between the methods in our study may be related to errors in determining the CO$_2$ diffusion coefficient for the pCO$_2$ technique. Because the pCO$_2$ technique is substantially easier and less expensive to conduct than the IRGA based technique, we will continue to explore its utility by improving upon the diffusion estimates using laboratory studies.

IV. Summary and Conclusions

We quantified significant diurnal patterns in soil CO$_2$ flux using an automated IRGA based system. Most of this variation corresponded to diurnal variation in soil temperature. From these patterns, it is clear that methods which sample only a portion of this diurnal period may produce misleading results, particularly if

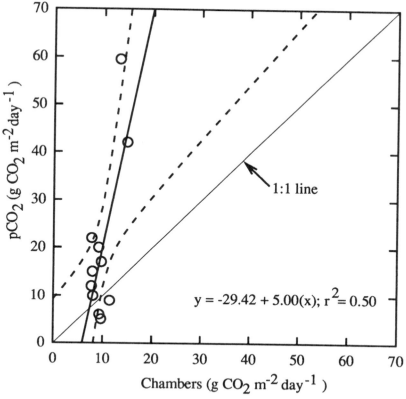

Figure 3. Comparison of soil CO_2 flux predicted with the pCO_2 method versus the IRGA based method. Dashed lines represent 95% confidence intervals of the regression.

data are extrapolated temporally. Although there was substantial variation in average daily fluxes (e.g., flux rates varied from 7 to 15 g CO_2 m^{-2} day^{-1}), integrated values of soil CO_2 flux were not significantly different among treatments. Fluxes were greatest for the midlevel CO_2 treatment, which was the treatment with the greatest root biomass. Differences between treatments may be detectable in subsequent years assuming that differential root biomass patterns continue. Our fluxes were comparable to studies where continuous IRGA systems were used, but substantially greater than studies using static techniques. A major limitation in comparing flux rates among studies is the uncertainty of methods. For example, we found that pCO_2 flux estimates were two to three times greater than IRGA based estimates.

Acknowledgements

This research was funded by Southern California Edison and is a component of the Electric Power Research Institute Forest Response to CO_2 Study. Helpful reviews were provided by W. Swank, J. Knoepp, and two anonymous reviewers. R. Souter provided valuable assistance in the statistical analysis and P. Clinton provided valuable assistance in data collection.

References

Ball, J.T., D.W. Johnson, B.R. Strain, R. Thomas, and R.F. Walker. 1991. *Effects of CO_2 on forests*. First Annual Report to EPRI. Desert Research Institute, Reno, NV.

Behera, N., S.K. Joshi, and D. Pati. 1990. Root contribution to total soil metabolism in a tropical soil from Orissa, India. *For. Ecol. Manage.* 36:125-134.

Cropper, W.P. Jr., K.C. Ewel, and J.W. Raich. 1985. The measurement of soil CO_2 evolution in situ. *Pedobio.* 28:35-40.

de Jong, E. and H.J.V. Schappert. 1972. Calculation of soil respiration and activity from CO_2 profiles in the soil. *Soil Sci.* 113:328-333.

de Jong, E., R.E. Redmann, and E.A. Ripley. 1979. A comparison of methods to measure soil respiration. *Soil Sci.* 127(5):300-306.

Edwards, N.T. and W.F. Harris. 1977. Carbon cycling in a mixed deciduous forest floor. *Ecology* 58:431-437.

Edwards, N.T. and P. Sollins. 1973. Continous measurement of carbon dioxide from partitioned forest floor components. *Ecology* 54:406-412.

Ewel, K.C., W.P. Cropper, Jr., and H.L. Gholz. 1987. Soil CO_2 evolution in Florida slash pine plantations. I. Importance of root respiration. *Can. J. For. Res.* 17:325-329.

Jurik, T.W., G.M. Briggs, and D.M. Gates. 1991. Soil respiration from five aspen stands in Northern lower Michigan. *Am. Midl. Nat.* 126:68-75.

Kucera, C.L. and D.R. Kirkman. 1971. Soil respiration studies in tall-grass prairie in Missouri. *Ecology* 52:912-915.

Musselman, R.C. and D.G. Fox. 1991. A review of the role of temperature forests in the global CO_2 balance. *J. Air Waste Manage. Assoc.* 41(6):798-807.

Naganawa, T., K. Kyuma, H. Yamamoto, Y, Yamamoto, H. Yokoi, and K. Tatsuyama. 1989. Measurement of soil respiration in the field: Influence of temperature, moisture level, and application of sewage sludge compost and agrichemical. *Soil Sci. Plant Nutr.* 35(4):509-516.

Norby, R.J., E.G. O'Neill, W.G. Hood, and R.J. Luxmoore. 1987. Carbon allocation, root exudation, and mycorrhizal colonization of *Pinus echinata* seedlings grown under CO_2 enrichment. *Tree Phys.* 3:203-210.

Norman, J.M., R. Garcia, and S.B. Verma. 1992. Soil surface CO_2 fluxes and the carbon budget of a grassland. *J. Geophysical Res.* 97:18845-18853.

Raich, J.W., R.D. Bowden, and P.A. Steudler. 1990. Comparison of two static chamber techniques for determining carbon dioxide efflux from forest soils. *Soil Sci. Soc. Am. J.* 54:17541757.

Rochette, P., E.G. Gregorich, and R.L. Des Jardins. 1992. Comparison of static and dynamic closed chambers for measurement of soil respiration under field conditions. *Can. J. Soil Sci.* 72:605-609.

Rogers, H.H., C.M. Peterson, J.N. McCrimmon, and J.D. Cure. 1992. Response of plant roots to elevated carbon dioxide. *Plant Cell Environ.* 15:749-752.

Rolston, D.E. 1986. Gas diffusivity. Chap. 46. In: *Methods of soil analysis, Part I. Physical and mineralogical methods*. American Society of Agronomy-Soil Science Society of America.

Ryan, M.G. 1991. The effect of climate change on plant respiration. *Ecol. App.* 1:157-167.

SAS Institute. 1987. *SAS/STAT guide for personal computers, version 6*. SAS Institute Inc., Cary, NC.

Snedecor, G.W. and W.G. Cochran. 1980. *Statistical methods, 7th Ed.*. Iowa State University Press, Ames, IA.

Schlentner, R.E., and K. Van Cleve. 1985. Relationships between CO_2 evolution from soil, substrate temperature, and substrate moisture in four mature forest types in interior Alaska. *Can. J. For. Res.* 15:97-106.

Weber, M.G. 1985. Forest soil respiration in eastern Ontario jack pine ecosystems. *Can. J. For. Res.* 15:1069-1073.

Weber, M.G. 1990. Forest soil respiration after cutting and burning in immature aspen ecosystems. *For. Ecol. Manage.* 31:1-14.

Wiant, H.V., Jr. 1967. Influence of moisture content on "soil respiration". *J. For.* 65: 902-903.

Witkamp, M. 1969. Cycles of temperature and carbon dioxide evolution from litter and soil. *Ecology* 50: 922-924.

Witkamp, M. and M.L. Frank. 1969. Evolution of CO_2 from litter, humus, and subsoil of a pine stand. *Pedobio.* 9: 358-365.

CHAPTER **17**

Soil Respiration and Carbon Dynamics in Parallel Native and Cultivated Ecosystems

G.A. Buyanovsky and G.H. Wagner

I. Introduction

Respiration, in general, is an energy-yielding oxidative reaction in living matter. Oxygen is used during this process and carbon dioxide produced. Soil respiration (SR) is usually understood as evolution of biologically generated CO_2 from the soil surface. It is expressed as a flux (e.g. gCO_2[or C] m^{-2} d^{-1}). This emission is one of the major contributors to greenhouse gas accumulation. To assess the importance of SR to greenhouse effects, one needs to know not only the rate of evolution of carbon dioxide from the soil surface, but also the role of the main contributors to its flux. In this paper we have tried to distinguish among several sources of CO_2 in SR such as live root respiration, litter decomposition and humus breakdown.

The rate of SR depends upon availability to soil organisms of organic carbon and oxygen and on the capacity of the environment to absorb (or disperse) CO_2. In non-calcareous soils SR is determined by the activity of soil biota and plant roots. In soils with a high content of $CaCO_3$ this process is obscured by the dynamic equilibrium between carbonate and bicarbonate of calcium (Buyanovsky, 1980).

Soil respiration has been studied for many years, long before carbon dioxide was included in the list of greenhouse gases. A significant amount of data was accumulated and the role of SR in global carbon balance was assessed (see, for instance, Houghton and Woodwell, 1989; Raich and Schlesinger, 1992). Most of the data on SR concerns natural ecosystems, while respiration of cultivated soils has been the objective of few studies (Singh and Gupta, 1977; Raich and Schlesinger, 1992).

It is well known that the transition of virgin soils to arable status causes up to 50% reduction in organic matter content during the first years of cultivation (Lee and Bray, 1949; Puhr and Worzella, 1952; Haas et al. 1957; Coleman et al., 1984; Wagner, 1990). Following initial sharp decreases, the reduced level of humus is finally stabilized after several decades of agricultural use (Bauer and Black, 1981), and the ecosystem usually regains steady-state status.

Although transitional carbon losses are well documented, the differences in the carbon cycle between native ecosystems and its cultivated counterparts, after they reached an equilibrium, are unclear. Even when such ecosystems are located in close proximity to each other, dynamics of all biological processes differ dramatically due to tillage and cultivation, application of fertilizers and pesticides, removal of a significant part of net primary production for human consumption, and others.

II. Components of Soil Respiration

In our studies in central Missouri we have accumulated data on carbon flow and storage pools in agroecosystems (wheat, *Triticum aestivum*, soybeans, *Glycine max* [L.] Merril, corn, *Zea mays*) and their native counterpart, tallgrass prairie. The native plant cover is characterized by several dominant warm season grasses including big bluestem (*Andropogon gerardi* Vitman), little bluestem (*Schizachyrium scoparium* Nash), prairie dropseed (*Sporobolus heterolepis* [A. Gray] A. Gray), and Indian grass (*Sorghastrum nutans* [L.] Nash) (Buyanovsky et al., 1987).

We considered SR to be the total CO_2 production from the soil surface, which includes live root respiration and mineralization of plant-accumulated carbon within the soil profile and on the soil surface. Because plant carbon is the primary source of energy for soil-inhabiting organisms, we assumed that in a steady-state (balanced) ecosystem total SR roughly equals net primary production plus root respiration. This assumption, as well as the use of ^{14}C-labeling in the field, allowed us to distinguish among the principal sources of SR: surface and root litter mineralization, stable organic matter degradation and live root respiration.

A. Live Root Respiration

Assessment of root respiration of plants growing in the field is a major challenge to those studying carbon balance in ecosystem. Earlier (Buyanovsky et al., 1987) we estimated root respiration using a simplified model for carbon balance in a steady-state ecosystem. In the model, the following pathways of plant C accumulated in above- and belowground biomass during a given year were considered:

1. Soil respiration
   a. mineralization of litter and humus
   b. live root respiration
2. Transfer to above- and belowground litter
3. Transfer to humus pool.

According to this model, root respiration is represented as the difference between the total soil respiration and total net production on an annual basis. Aboveground productivity for the prairie was assessed at approximately monthly intervals through the growing season. Net annual production was estimated as the positive sum of incremental changes in green biomass recorded between successive samplings. For winter wheat, all remnants excluding grain were weighed and returned to the soil at harvest. Root biomass was estimated by taking soil cores at certain periods during growing season.

Under tallgrass prairie, from 25 to 30% of total SR was attributed to root systems, and the balance to heterotrophic activity. Root respiration of wheat contributed 12-15% of the total carbon evolved as CO_2, which was approximately one-half of the share of root respiration for prairie (Table 1).

The respective roles of autotrophs and heterotrophs as CO_2 sources for prairie are engaged concurrently for most of the year. Root respiration of the multispecies plant community depends upon the same variables as total biological activity. By contrast, the contribution of root respiration in agroecosystem is limited in time by the growing period of a single crop. The live root component of total SR was approximated on the basis of $^{14}CO_2$ evolution from labeled material during different stages of degradation. After accounting for decomposition of plant materials and estimating the degree of mineralization of native soil organic matter, the remainder of the SR was assigned to live root respiration. For winter wheat the contribution of this source of CO_2 varied dramatically during the course of a year (Figure 1, A). Starting late June and through October, SR is generated primarily by decomposition of residues with a small contribution from soil organic matter. Estimates of live root respiration during late October-November showed contributions between 20% and 35% of the total soil respiration. After winter dormancy the role of this source gradually increased and in April-May respiration of living roots supplied about 40% of the total CO_2 evolved (Buyanovsky and Wagner, 1987).

Root respiration of soybeans, a warm season crop, is limited by a short period from June to the end of September (Figure 1, B). In early September soybean roots are mostly "cannibalized", and much of energy-valuable content is transferred to the seeds with the result that only empty root shells remain in soil. During the period of active vegetative growth, however, root respiration represents a very significant portion of CO_2 evolution from the soil. Direct measurements in the field with the use of ^{14}C showed that approximately 14% of the carbon fixed photosynthetically each day was respired by the root system (Wagner and Buyanovsky, 1989). The ratio of specific activity of carbon evolved from the soil surface and that of carbon coming from root respiration over the same period of time indicates that in August-September, 62% of the total respiration was from the root system with the remainder from heterotrophic activity. Applying this partitioning to the measured total SR suggest a maximum root respiration rate of 3.2 g m^{-2} day^{-1} of carbon, or about 4 mg h^{-1} of CO_2 1 g of roots.

Table 1. Role of major sources of CO_2 in total soil respiration

	Tallgrass prairie	Winter wheat	Soybeans
	----------Percent of total----------		
Litter			
Aboveground	15-20	35-40	30-35
Belowground	20-25	40-45	12-15
Root respiration	25- 0	12-15	35-40
Soil organic matter	30-35	10-12	12-15

Estimated on the basis of difference between carbon input by plant residues and total SR, root respiration of soybeans represented about 35-40% of total carbon lost from the soil surface as CO_2 (Table 1).

B. Decomposition of Dead Roots

Roots may represent up to 50% of total net annual production of crops, a significant source of available energy for soil biota (Coleman et al., 1983). During the growing season they provide available carbon in the form of exudates, sloughed-off material, and dead roots, whereas after harvest the whole root biomass becomes available for heterotrophs.

In the native ecosystem with prevailing perennial plants only part of the root system (about 25%) can be consumed by heterotrophs annually (Dahlman and Kucera, 1965; 1969). Mineralization in prairie takes place mostly during the period July to January. Annual input of C from underground prairie plant residues to SR was approximated at 120 g m^{-2} (Buyanovsky et al., 1987).

The total annual input of C from the roots of wheat, soybean, and corn, averaged for 3 yr, was 154, 126, and 340 g m^{-2}, respectively (Buyanovsky and Wagner, 1986). This structural root C does not include input of C of root exudates nor sloughed-off material.

Although soybean and wheat root biomass at maximum is almost similar (126 and 154 g m^{-2} of C respectively) their input to the soil respiration differs both in total amount of carbon evolved and in the distribution of input through the year. In winter wheat, total root biomass became available to soil biota in the middle of summer. Dead roots were consumed actively from July through September (Figure 1A), the most favorable period for biological activity within the soil profile, with soil temperatures up to 24 °C at the depth of 25 cm and a sufficient water supply. During these 3 months, about 50% of ^{14}C from belowground material had been lost.

Soybean roots became available to the soil heterotrophs only in late September, when the temperature had dropped and the soil was desiccated by actively transpiring plants. During September, soybean roots lost approximately 19 g m^{-2} of C, or less than 25% of total C evolved from soil (Figure 1B). With further cooling of soil, mineralization processes decreased, and during winter no more than 1.5-2 g m^{-2} of C were released from roots. During the following spring and summer, the soybean root C was mineralized at a stable rate of about 10-12 g m^{-2} of C per month.

C. Aboveground Residue Decomposition

Input of aboveground residues to the total release of CO_2 is very important. Based on data from long-term field experiments, Barber (1979) assumed that 18% of root carbon and only 10% of stem carbon enters the humus pool annually. Thus, carbon from the residue left on the soil surface is less susceptible to sequestration than that which accumulated within the soil. Corn can leave more than 15 t ha^{-1} of stalks and cobs on the surface, soybeans, up to 7 t ha^{-1}; and wheat, 6 t ha^{-1} annually (Buyanovsky and Wagner, 1986). Native prairie accumulates 4-6 t ha^{-1} yr^{-1} of standing litter by the end of each growing period (Kucera et al., 1967). This litter becomes available for mineralization only in spring when it comes in contact with the soil.

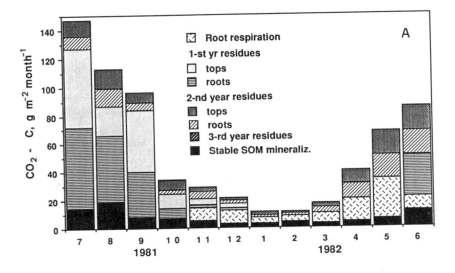

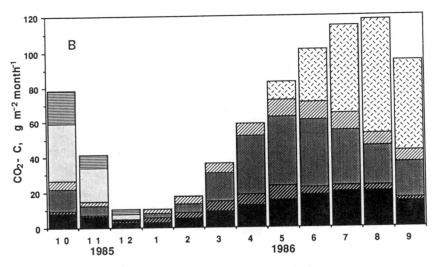

Figure 1. Relative contributions of various sources of CO_2 (g C m^{-2} $month^{-1}$) to the total soil respiration. A: under winter wheat (labeled in spring of 1981). B: under soybeans (labeled in summer of 1985). (from Wagner and Buyanovsky, 1987.)

In agroecosystems the aboveground residues are usually chopped during harvest and left on the surface. In July, following the harvest of winter wheat, straw on the surface lost 55 g m^{-2} of C, which represents 37% of C respired from soil during the month (Figure 1A). Later, in August, mineralization of straw decreased, presumably due to low water content of the surface soil (Buyanovsky et al., 1986). In September, straw was incorporated into the soil by turning over the upper 15 cm layer, and the decomposition rate increased again (to 44 g m^{-2} of C). Beginning at that time the decomposition rate of straw was higher than that of roots. At the end of the first year, 49% of straw carbon had been mineralized. With the resumption of intense mineralization in March, the activity of respired CO_2 indicated that a significant decomposition of residues continued with the specific activity for straw being higher than that for roots. High activity was observed through July, during which time the balance of the previous year residues (second-year residues) steadily supplied significant amounts of carbon (15-25 g m^{-2} $month^{-1}$). Their relative role decreased in

August, however, when current year residues became available. During the first 12 months, more than 70% of straw carbon was mineralized, while during the initial 18 months, it was more than 80%.

Under soybeans, only 50 g m^{-2} of carbon was released from aboveground residues and 27 g m^{-2} from roots during the period from harvest through December, when biological activity ceased (Figure 1B). It was about 20% of total residues, and approximately half of what was released by this time from wheat residues. Evolution of CO_2 from soil during winter months (December, January, February) was very low (less than 10 g m^{-2} of C per month). Degradation increased significantly in March, mainly due to aboveground residues which supplied 30-40 g m^{-2} of C per month during April, May and June.

Prairie litter which remains relatively intact through winter undergoes rapid depletion during the spring season (March-May), when more than 60% of the dead aboveground biomass was mineralized. The principal increase in surface litter occurs in late summer as growth rates in green biomass decline and amounts of dead residues increase. A peak in litter production is reached in late fall, when decay potentials have diminished as well, thus delaying the breakdown process until the following year (Buyanovsky et al., 1987).

D. Humus Breakdown

A mean mineralization rate of 2% per year has been accepted for cultivated silt loam soils in our climatic region (Graham, 1959). Chahal and Wagner (1965) have determined annual loss of soil organic carbon in the laboratory using soils from long-term plots at Sanborn Field. The rate is slower for sod crops and more rapid for crops such as corn where multiple tillage operations aerate the soil several times a year. Rates of mineralization were approximated for long-term experimental plots and are shown in Table 2. Because straw, stalks, and forage were removed from the plots, C input to unmanured plots was limited to the roots and stubble. The latter was approximated on the basis of our data on post-harvest residues (Buyanovsky and Wagner, 1986). Addition of easily mineralizable manure significantly increased root biomass production and the release of carbon from soil organic matter.

In our detailed analysis of carbon dynamics in tallgrass prairie and winter wheat (Buyanovsky et al., 1987) we found that for wheat about 20% of plant accumulated carbon was transferred into soil organic matter, whereas in prairie it was 44%. Assuming equilibrium or near-equilibrium conditions for carbon accumulation in the soil profile, a similar amount of carbon was mineralized and lost from SOM (110 and 196 g m^{-2} for winter and native prairie, respectively).

Our data on $^{14}CO_2$ evolution from experimental plots with ^{14}C labeled plants during the 4th and 5th years after labelling, provided data on annual dynamics of CO_2 which arises from mineralization of SOM. In general, the largest amounts of CO_2 from organic matter oxidation were observed during periods of highest temperatures (June, July, August). During these three months, the soil loses about 45 g of C m^{-2} from SOM (Wagner and Buyanovsky, 1987).

III. Dynamics of Total Soil Respiration

The activity of soil organisms and root metabolism are sources of carbon emission from the soil. The relative role of each of the major sources of CO_2 in SR on an annual basis is illustrated by Table 1. These processes and their relative rates strongly influence the seasonal dynamics of carbon dioxide evolution from the soil and the distribution of residual carbon in an ecosystem. Generally the seasonal trends in CO_2 evolution were similar in all ecosystems considered, with the maximum occurring in July and August when air and soil temperatures were highest. During these 2 months, approximately 35-40% of the annual carbon in the respective systems was released. However, seasonal patterns show different rates for the systems at any given time. An earlier and steeper peak characterizes the cultivated systems compared to the more gradual rise for prairie vegetation dominated by warm season grasses (Buyanovsky et al., 1987). For winter wheat, following winter dormancy, live root respiration is an important source of seasonal increase in CO_2 evolution occurring from March through May. This is succeeded by a brief period when plants are maturing and the curve remains mostly static. In the post-harvest period there is a sharp increase in CO_2 evolution, reaching maximum values as the root system dies and becomes immediately available to soil decomposers. Heterotrophic respiration thus is the single most important source of CO_2 until a new crop is established.

Table 2. Annual carbon flow through soil organic matter in historical plots at Sanborn field (with all aboveground biomass removed)

	Total soil C after 1940 kg/ha	Annual mineralization rate %	C in roots and stubble (C in manure) kg/ha	Humification coefficient	Annual replacement of carbon in SOM kg/ha
Wheat - no treatment	15,800	1.0	600		144
Wheat - manure	34,800	2.0	1,500 (3,300)	0.24 0.10	360 330
Corn - no treatment	14,600	1.5	900	0.24	216
Corn - manure	28,000	2.5	1,760 (3,300)	0.24 0.10	380 330
Timothy	33,750	1.2	975	0.22	390
Timothy - manure	49,750	2.0	1,800 (3,300)	0.40 0.37 0.10	666 330

(From Buyanovsky et al., 1993a.)

A secondary rise in the respiration curve for wheat is attributed to improving soil moisture conditions after fall rains which stimulate decomposer activity.

Soybeans planted in May have root respiration as a major component of total carbon efflux throughout summer. Up to 50-60% of C lost during July-September came from this source. In contrast to wheat, mineralization of aboveground residues occurs mainly after winter. Due to decreasing soil temperature, only about 20% of the soybean residue carbon is mineralized immediately after harvest (in October-November), with the remainder left on the soil surface where it is the subject to severe winter climatic conditions.

The multispecies prairie community which is dominated by slower growing perennials and extended phenological activity, in the aggregate exhibits a less punctuated CO_2 curve. Here, the respective roles of autotrophs and heterotrophs run concurrently for most of the year and are distinguished with more difficulty.

IV. Principal Pathway of Carbon Transfer within Soil Compartment

Carbon flow charts (Buyanovsky et al., 1987) demonstrated significant differences between native and cultivated ecosystems in the accumulation in various compartments and in flow rate. The relatively larger share of total heterotrophic decay of wheat litter components compared to prairie litter serves to explain significant differences in soil organic matter accumulation. In the cultivated system approximately 80% of the initial carbon (433/543 g m^{-2} yr^{-1}) is terminated at the litter level and the balance is transformed subsequently into soil organic matter. For the prairie the system the apportionment is more nearly equal, with only 56% of total carbon fixed in photosynthesis (254/450 g m^{-2} yr^{-1}) being respired from the litter stages. Calculated on the basis of carbon flow, mineralization rates of the total carbon pool for both agricultural and prairie ecosystems are very similar, 0.015 and 0.018 yr^{-1}, which represent a half-life of 46 and 38 yr, respectively.

These numbers characterize average behavior of the soil organic matter pool, within which we presumably have several subordinate pools each with a different rate of carbon flow. There are many presumptions about the number of pools and their ages. Two models most often cited in the literature developed by Jenkinson and Rayner (1977) and Parton et al. (1988), are in general agreement about the pools. Little, however, is known about physical embodiment of those SOM fractions.

We differentiated some of the SOM pools based on our data on ^{14}C residence time in different aggregate fractions of cultivated soil (Wagner et al., 1993). These fractions were separated by wet sieving from the soil which was periodically sampled during 4 years after field labeled soybeans were harvested. Changes in ^{14}C content with time show that these fractions can be grouped into the pools used in organic matter models (Buyanovsky et al., 1994).

The least stable pool (readily decomposable, or "metabolic" plant material) is concentrated in the short lived fraction of vegetative fragments 2-0.2 mm. A smaller fraction, 0.2-0.053 mm, consists of more resistant, or structural plant material. A significant amount of this material was separated from the soil even after 3 years from the time the labeled material was incorporated. It also should be noted that large aggregates (2-1 mm) are formed around vegetative fragments and have the same short life-span (1-5 years) (Wagner et al., 1993). Soil biomass (Jenkinson, 1990) or active soil C (Parton et al., 1988) concentrated in smaller aggregates (0.1-1 mm) or in non-aggregated soil.

As for the very stable "slow" and "passive" carbon pools, which comprise up to 60-70% of total soil carbon (Balesdent et al., 1988), they are associated with silt and clay fractions contained within aggregates.

V. Conclusions

1. Seasonal trends of soil respiration in different ecosystems in Missouri were found to be almost similar, with maximum CO_2 evolution in July-August, when air and soil temperatures are highest.
2. Despite the likeness of general character of soil respiration in different ecosystems, sources of CO_2 emitted, such as live root respiration, plant residue mineralization and humus breakdown play different role in respiration dynamics of each particular ecosystem. Plant life cycles are the most important factor defining CO_2 supply from these sources.

3. Live root respiration has large input to annual CO_2 emission (up to 30-35%) in ecosystems dominated by warm-season plants. In ecosystems with cool-season plants, like winter wheat, root respiration plays relatively lesser role.
4. Plant residue decay in cool-season ecosystems takes place during period most favorable for biological processes (late summer - early fall). In warm-season ecosystem only small part of annual debris can be mineralized during fall, the remainder undergoes harsh winter conditions before it can be used by soil organisms. The relative role of this source of mineralized C is much more significant in cool-season ecosystem.
5. In agroecosystems, as compared with native ecosystems, increased aeration due to tillage and cultivation causes higher mineralization at the litter level and lower inclusion of plant accumulated C into stable SOM (no more than 20% of annual production). In contrast, more than 40% of plant C passes through SOM in native prairie.
6. Comparison of ecosystems with minimal disturbance (native prairie) and cultivated lands shows that significant change of the main reserve of SOM cannot be achieved in a short period of time. Our efforts to increase the role of soils as a sink for excess of carbon probably will result first in a buildup of the more labile fractions.

References

Balesdent, J., G.H. Wagner, and A. Mariotti. 1988. Soil organic matter turnover in long-term field experiments as revealed by carbon-13 natural abundance. *Soil Sci. Soc. Amer. J.* 52:118-124.

Barber, S.A. 1979. Corn residue management and soil organic matter. *Agron. J.* 71:625-627.

Bauer, A. and A.L. Black. 1981. Soil carbon, nitrogen and bulk density comparisons in two cropland tillage systems after 25 years and in virgin grassland. *Soil Sci. Soc. Am. J.* 45:1166-1170.

Buyanovsky, G.A. 1980. Distribution of alkaline-earth carbonates in soils of an arid constructional plain, Transcaucasus. *Geoderma* 24:177-190.

Buyanovsky, G.A., C.L. Kucera, and G.H. Wagner. 1987. Comparative analysis of carbon dynamics in native and cultivated ecosystems. *Ecology* 68(6):2023-2031.

Buyanovsky, G.A., and G.H. Wagner. 1986. Post-harvest residue input into cropland. *Plant and Soil* 93:57-65.

Buyanovsky, G.A. and G.H. Wagner. 1987. Carbon transfer in a winter wheat (*Triticum aestivum*) ecosystem. *Biol. Fertil. Soils* 5:76-82.

Buyanovsky, G.A., G.H. Wagner, and M. Aslam. 1994. Carbon turnover in soil aggregates. *Soil Sci. Soc. Am. J.* (in press).

Buyanovsky, G.A., G.H. Wagner, and C.J. Gantzer. 1986. Soil respiration in a winter wheat ecosystem. *Soil Sci. Soc. Am J.* 50:338-344.

Chahal, K.S. and G.H. Wagner. 1965. Decomposition of organic matter in Sanborn Field soils amended with ^{14}C glucose. *Soil Sci.*, 100:96-107.

Coleman, D.C., C.V. Cole, and T.T. Elliot. 1984. Decomposition, organic matter turnover, and nutrient dynamics in agroecosystems, p. 83-104. In: R. Lowrance et al. (eds.). *Agricultural ecosystems: unifying concepts*. John Wiley and Sons.

Coleman, D.C., C.P.P. Reid, and C.V. Cole. 1983. Biological aspects of nutrient cycling in soil systems. *Adv. Ecol. Res.* 13:1-55.

Dahlman, R.C. and C.L. Kucera. 1965. Root productivity and turnover in native prairie. *Ecology* 46:84-89.

Dahlman, R.C. and C.L. Kucera. 1969. Carbon -14 cycling in the root and soil components of a prairie ecosystem. p. 652-660 In: *Proc. of the 2nd Symp. on Radioecology*. Oak Ridge National Laboratory, Oak Ridge, Tennessee.

Graham, E.R. 1959. An explanation of theory and methods of soil testing. University of Missouri, Agric. Expt. Sta. Bulletin 834.

Haas, J.H., C.F. Evans, and E.F. Miles. 1957. Nitrogen and carbon changes in Great Plains soils as influenced by cropping and soil treatments. United States Technical Bulletin Number 1164, United States Department of Agriculture, Washington, D.C., USA.

Houghton, R.A. and G.M. Woodwell. 1989. Global climatic change. *Sci. Am.* 260:36-44.

Jenkinson, D.S. 1990. The turnover of organic carbon and nitrogen in soil. *Phil. Trans. R. Soc. Lond.* B:329-368.

Jenkinson, D.S. and J.H. Rayner. 1977. The turnover of soil organic matter in some of the Rothamsted classical experiments. *Soil Sci.* 123:298-305.

Kucera, C.L., R.C. Dahlman, and M.R. Koelling. 1967. Total net productivity and turnover on an energy basis for tallgrass prairie. *Ecology* 49:536-541.

Lee, C.K. and R.H. Bray. 1949. Organic carbon and nitrogen contents of soils as influenced by management. *Soil Sci.* 68:203-212.

Parton, W.J., S.W.B. Stewart, and C.V. Cole. 1988. Dynamics of carbon, nitrogen, phosphorus and sulfur in cultivated soils. A model. *Biogeochemistry* 5:109-131.

Puhr, L.J. and W.W. Worzella. 1952. Fertility maintenance and management of South Dakota soils. South Dakota Agricultural Experiment Station Circular 92.

Raich, J.W. and W.H. Schlesinger. 1992. The global carbon dioxide flux in soil respiration and its relationship to vegetation and climate. *Tellus* 44B:81-99.

Singh, J.S., and S. R. Gupta. 1977. Plant decomposition and soil respiration in terrestial ecosystems. *Bot. Rev.* 43:449-528.

Wagner, G.H. 1990. Lessons in soil organic matter from Sanborn Field. p. 64-70. In: *Proceedings of Sanborn Field Centennial*, University of Missouri, Columbia.

Wagner, G.H., and G.A. Buyanovsky. 1987. Sources of CO_2 in soil respiration during winter wheat grows and subsequent residue decay. In: *Soil Biology and Conservation of the Biosphere*, Sopron, Hungary.

Wagner, G.H. and G.A. Buyanovsky. 1989. Soybean root respiration assessed from short-term ^{14}C-activity changes. *Plant and Soil* 117:301-303.

Wagner, G.H., G.A. Buyanovsky, and M. Aslam. 1993. Carbon transfer among soil fractions during soybean residue degradation. In: *N. Senesi (ed.). Proceedings of 6th IHSS meeting, Monopoli, Italy* Elsevier, Amsterdam (in press).

CHAPTER **18**

Biosphere-Atmosphere Exchange of Gaseous N Oxides

G.L. Hutchinson

I. Introduction

Despite the reservations of a few remaining skeptics, there is increasingly conclusive evidence that the "global change" processes induced by anthropogenically perturbed atmospheric trace gas concentrations are very real and potentially important issues. On a time scale of years to possibly a few decades, stratospheric O_3 depletion is probably the biggest human threat among the changes believed to be occurring. However, intense study of the so-called "ozone hole" over Antarctica has identified the origin of this problem (Rowland, 1989), and it appears that with relatively minor socioeconomic sacrifices, it can be overcome. Over the longer term, climate warming is a much bigger threat, because it is more difficult to control and because it has so many ramifications, e.g., changes in the geographic range to which many plant and animal species are adapted, movement across political boundaries of the land areas most suitable for food and fiber production, and even changes in mean sea level that threaten to inundate low-lying coastal areas, eventually including portions of several large cities around the globe.

But has the atmosphere yet warmed significantly? In a recent letter to the editor of *Physics Today*, Kepros (1992) proposed that the simplest approach to addressing this question may be to consider the entire atmosphere as a single system, apply the Ideal Gas law ($PV = nRT$), and then examine each term for evidence of change. He presented logical arguments that P, n, and of course, R are not changing and concluded that to maintain equality, the volume of the atmosphere must expand if it is warming, as the only constraint is gravity. He then argued that "One place to look for such an increase is in space satellite orbit data." When he did, he found that initial announcements of the Hubble Space Telescope launch, 24 April 1990, included a statement that the satellite's "designed lifetime of 15 years would be reduced to 5 years because the atmosphere had expanded." Because of the controversy generated by that statement, it was soon deleted from the announcements, and its basis apparently remains classified (Kepros, 1992).

Regardless whether this approach is eventually proven to have utility, it illustrates the important point that we need to depend less on point measurements and pay more attention to finding macroscopic measures of physical, chemical, and biological phenomena at landscape, regional, and global scales. In this example a single estimate of the change in total atmospheric volume has the potential to provide a better measure of climate warming than millions of surface temperature measurements that vary in time as a function of cloud cover, particulate loading, diel cycles, seasonal cycles, etc., and vary from one point to the next depending on surface color, cover, slope, and a host of other factors. In other words, it makes sense that a global effect should be measured on a global scale, and the "altitude" of the atmosphere is such a global measure. Similarly, a field or landscape effect should be measured on a field or landscape scale, a regional effect on a regional scale, etc. I'll revisit this thought in section VI, but first focus more narrowly on gaseous N oxides by summarizing their relation to global change, then commenting on the global atmospheric budgets of NO_x ($NO + NO_2$) and N_2O, and finally describing processes for production and consumption of the two gases in soil, environmental controllers of those processes, and recent attempts to model their magnitudes and distributions.

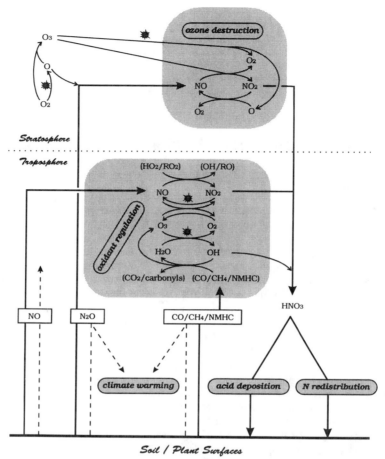

Figure 1. Relation of gaseous N oxides to the five shaded global change processes. Rectangles indicate emissions from soil/plant surfaces; arrows with light, heavy, and dashed lines indicate reaction, transport, and surface reradiation, respectively. Briefly, N_2O is radiatively active and thus contributes to *climate warming*; following transport to the stratosphere, it is oxidized to NO which catalyzes the *destruction of O_3* produced there via O_2 photolysis. NO is not radiatively active, but plays a key role in tropospheric *oxidant regulation*. NO, NO_2, and O_3 form a photostationary state in sunlight, while the alternative oxidation of NO by peroxy radicals results in net O_3 production. Photolysis of tropospheric O_3 produces the OH responsible for oxidizing the greenhouse gases CO, CH_4, and NMHC via reactions that regenerate O_3 if they occur in the presence of adequate NO. OH also oxidizes NO_2 to HNO_3, the wet and dry removal of which contributes to *acid deposition* and *N redistribution*.

II. Relation of Gaseous N Oxides to Global Change

Gaseous N oxides have multiple critical influences on the chemistry and physics of the atmosphere, and therefore, global change processes (Figure 1). Nitrous oxide, for example, is a radiatively active gas that contributes 5-6% of the anthropogenic forcing of the global energy balance (Rodhe, 1990), and is also a player in the issue of stratospheric O_3 depletion. Its persistence or long lifetime in the atmosphere permits transport to the stratosphere where it is photochemically oxidized to NO, which catalyzes a complex set of gas-phase and heterogeneous reactions that results in O_3 destruction. Although NO is not itself radiatively active, it influences the concentrations of important greenhouse gases by regulating the level of atmospheric oxidants involved in removal of CO, CH_4, nonmethane hydrocarbons (NMHC), and other gases. NO_x is itself removed from the atmosphere by oxidation to HNO_3, which is the fastest growing component of acidic deposition, another global change issue not presently in the limelight, but one that will undoubtedly surface again. In addition, the emission, transport, and subsequent redeposition of NO_x accomplishes substantial N redistribution both within and among natural and disturbed ecosystems, which can, in turn, initiate quite complex interactions among various ecosystem processes. For example, because ecosystem primary productivity is often governed by N supply, redistribution of this element may importantly influence the potential for C sequestration in both living biomass and soil organic matter. The atmospheric chemistry of NO_x and N_2O summarized in Figure 1 was recently described in much greater detail by Williams et al. (1992b).

III. Global Atmospheric NO_x and N_2O Budgets

Because of the critical relation of NO_x and N_2O to contemporary environmental issues and other biospheric and atmospheric processes, it is essential to identify and characterize important source and sink terms in the global budgets of these important trace atmospheric constituents. For NO_x, five natural sources have been identified -- microbial processes in soil, lightning, oxidation of atmospheric NH_3, photolytic and biological processes in the oceans, and stratospheric injection (Table 1). The characteristics and estimated magnitudes of the last four of these have changed little since they were summarized in several review articles that appeared in the mid-1980's (Crutzen, 1983; Ehhalt and Drummond, 1982; Homolya and Robinson, 1984; Logan, 1983; Placet and Streets, 1987; Stedman and Shetter, 1983). At that time magnitude of the biogenic soil source had to be estimated from the single data set available, which described NO_x emission from a pasture in Australia (Galbally and Roy, 1978). Differences among review authors in the interpretation and extrapolation of these data led to considerable disagreement regarding the magnitude of this process, but all agreed that soil represents a significant source of global atmospheric NO_x. For example, Logan (1983) estimated that it contributes 4 to 16 Tg N (most likely about 8 Tg N) to the global NO_x budget, equal to and perhaps greater than the lightning source, and much larger than the other natural sources (Table 1). Using additional more recent data, Davidson (1991) estimated the global soil NO source to be about 20 Tg N yr^{-1}, which he computed by summing the products of the mean measured NO emission rate, area, and length of the growing season for each of the Earth's major biomes. His estimate was dominated by emissions from three savanna sites all in the same Latin American country (7.7 Tg N yr^{-1}), so it is subject to the possibility of downward revision once more savanna sites are studied, but also subject to certain upward revision because nearly half the Earth's land surface is covered by biomes for which no NO exchange estimates were available, including deserts, semi-deserts, polar deserts, peatland, mixed forest, tiaga, and tundra. Total soil NO emissions computed by Davidson (1991) is comparable to the Table 1 estimate of 21 Tg N yr^{-1} for global energy-related combustion emissions and exceeds the 12 Tg N yr^{-1} estimate of emissions from biomass burning. Thus, the soil source of NO may represent as much as 40% of the global NO_x budget.

The relatively long 150-year estimated atmospheric lifetime of N_2O (Hao et al., 1987) and the smaller number of important natural sources of this gas make its global budget somewhat simpler, but no more certain (Table 2). In the apparently balanced budget presented by McElroy and Wofsy (1986) and frequently cited by others (e.g., Davidson, 1991), the planetary sum of N_2O sources was estimated to be about 15 Tg N yr^{-1}, of which more than half resulted from microbial activity in soil. The 4 Tg N_2O-N that they estimated was produced each year by fossil fuel combustion nearly matched the amount required to support the observed 0.2-0.3% annual growth rate of atmospheric N_2O over the past few decades (Prinn et al., 1990; Rasmussen and Khalil, 1986), so a cause-and-effect relationship was generally assumed. However, as pointed out by Davidson (1991), this hypothesis was recently discounted by discovery of an artifact in flask sampling procedures that reduced the combustion source estimate more than tenfold (Muzio and Kramlich, 1988). At about the same time, Matson and Vitousek (1990) reported that the tropical terrestrial source must also be reduced 3-4 Tg from the 7.4 Tg N yr^{-1} reported by McElroy and Wofsy (1986), thus further upsetting the apparent balance between known sources and sinks in their budget. Finally, adopting a more recent 110-year estimate of N_2O's atmospheric lifetime (Ko et al., 1991) implies an even larger global sink strength, so the unknown source needed to restore balance to the budget may be quite large indeed.

The relative contributions of the Earth's major biomes to the total soil source of N_2O are not known with any certainty, although it is generally believed that humid tropical forests dominate. Studies of the impact

Table 1. Global atmospheric NO_x budget showing the range and (most probable value) of individual source and sink terms

Source or sink		NO_x (Tg N yr^{-1})	
Sinks			
Precipitation		12-42	
Dry deposition		12-22	
	Total	24 - 64	
Sources			
Fossil fuel combustion		14-28	(21)
Biomass burning		4-24	(12)
Microbial activity in soils		4- 6	(8)
Lightning		2- 20	(8)
Oxidation of ammonia		1- 0	
Oceans (photolytic/biological)		<1	
Input from the stratosphere		0.5	
	Total	25 - 99	

(From Logan, 1983.)

Table 2. Global atmospheric N_2O budget of McElroy and Wofsy (1986) with updates by Davidson (1991)

		N_2O (Tg N yr^{-1})	
Source or sink		McElroy & Wofsy (1986)	Davidson (1991)
Sinks and accumulation			
Stratospheric photolysis		10.6	
Atmospheric accumulation		3.5	
	Total	14.1 ± 3.5	
Sources			
Ocean		2	2
Combustion			
Coal and oil		4	<0.1
Biomass		0.7	0.1-0.3
Fertilized agricultural land		0.8	0.2-2.1
Temperate grasslands		0.1	0.1
Boreal and temperate forests		0.1-0.5	0.3-1.5
Tropical/subtropical forests		7.4	4.1
	Total	15.3 ± 6.7	7.9 ± ?

of deforestation on these systems have yielded ambiguous results, with some indicating up to threefold enhancement (Luizao et al., 1989), and others, a net reduction in N_2O emissions integrated over time (Robertson and Tiedje, 1988). Immense variability in the measurements of N_2O emission from temperate and boreal forests, grasslands, and agricultural lands hampers quantifying or even ranking their contributions to the global budget of the gas. Watson et al. (1990) summarized the present state of knowledge, or lack thereof, regarding the relative importances of all terrestrial, aquatic, and atmospheric N_2O sources and sinks.

IV. NO_x and N_2O Production and Consumption in Soil

Both biotic and abiotic processes are involved in the production of NO_x and N_2O in soils (Hutchinson and Davidson, 1993). For both process types, it is generally agreed that most of the NO_x emitted by soil is NO, with direct soil emission of NO_2 accounting for substantially less than 10% of the total. Most microbial processes that involve oxidation or reduction of N through the +1 or +2 oxidation state probably yield at least trace amounts of NO and N_2O (Firestone and Davidson, 1989), but the bacterial processes of nitrification and denitrification are generally accepted to be the principal biotic sources in soil. The relative unimportance of nonnitrifying/nondenitrifying processes to total biogenic NO and N_2O production was recently reviewed by Tiedje (1988). Abiotic production of N_2O, and particularly NO, occurs primarily through a set of reactions collectively termed chemodenitrification. The most important of these reactions is the disproportionation of nitrous acid (HNO_2) known to occur in acid soils (Nelson, 1982), especially those high in organic matter content (Blackmer and Cerrato, 1986). Although this reaction has not been demonstrated in neutral or alkaline soils in the laboratory, the required accumulation of NO_2^- and low pH may occur at microsites in undisturbed soils as a result of solute concentration in thin water films during freezing or drying, or because of proximity to a colony of NH_4^+ oxidizers, etc. (Davidson, 1992).

A. Nitrification

It is widely accepted that the preponderance of nitrification in soil is accomplished by a few genera of aerobic chemoautotrophic bacteria, e.g., *Nitrosomonas* and *Nitrosospira*, which oxidize NH_4^+ to NO_2^-, and *Nitrobacter*, which converts NO_2^- to NO_3^-, but the biochemical pathway of these transformations is not clearly established. There is good evidence that NH_2OH (N oxidation state -1) is the first intermediate product of NH_4^+ oxidation, but subsequent intermediates with N oxidation states +1 and +2 are not known with any certainty (Hooper, 1984). All intermediates formed during the conversion of NH_2OH to NO_2^- are believed to remain bound to the complex enzyme hydroxylamine oxidoreductase. The oxidation of NO_2^- to NO_3^- by *Nitrobacter* is a simple two-electron shift in N oxidation state from +3 to +5 and involves no intermediates (Schmidt, 1982).

There is abundant evidence that both NO and N_2O are usually included among the products of chemoautotrophic nitrification. Studies using specific inhibitors have demonstrated that both the N_2O (Aulakh et al., 1984; Blackmer et al., 1980) and NO (Davidson, 1992; Tortoso and Hutchinson, 1990) produced during chemoautotrophic nitrification are a direct result of the activity of those organisms responsible for the first step of this process, i.e., oxidation of NH_4^+ to NO_2^-. Recent evidence suggests that production of N_2O by NH_4^+ oxidizers results from a reductive process in which the bacteria use NO_2^- as an electron acceptor, especially when O_2 is limiting (Poth and Focht, 1985). This mechanism not only allows the organisms to conserve limited O_2 for the oxidation of NH_4^+ (from which they gain energy for growth and regeneration), but also avoids the potential for accumulation of toxic levels of NO_2^-. Our knowledge of nitrifier physiology is not sufficient to predict whether their *in situ* production of NO also results from NO_2^- reduction, or if it represents decomposition of an intermediate in the oxidation pathway from NH_4^+ to NO_2^-. Some evidence indicates that NO production during nitrification also increases as O_2 availability decreases (Lipschultz et al., 1981), but Anderson and Levine (1986) and Remde and Conrad (1990) observed no dependence on the partial pressure of O_2, and it is generally accepted that N_2O production by nitrifiers is more sensitive to this condition. As a result, the $NO:N_2O$ ratio of nitrification products, normally of the order of 10 to 20 in fully aerobic environments (Tortoso and Hutchinson, 1990), decreases along with O_2 partial pressure.

The total amounts of NO and N_2O generated by autotrophic nitrifiers are regulated by two separate, but interdependent, sets of controllers-- those that establish the overall rate of NH_4^+ oxidation and those that determine the $NO:NO_3^-$ and $N_2O:NO_3^-$ ratios of nitrification products (Firestone and Davidson, 1989). For example, the importance of N_2O as a product of nitrification increases as O_2 availability decreases, but whether that increased importance translates into higher total N_2O production depends on how much the overall process rate is reduced by the limited availability of O_2. The NO and N_2O yields of nitrification are normally relatively small. For N_2O, it is typically less than 1 percent, particularly in well-aerated soil, but more NO than N_2O is produced under these conditions (Tortoso and Hutchinson, 1990). Hutchinson et al. (1993b) found NO yields in the range 1 to 4 percent of the NH_4^+ oxidized for coarse-textured soils subjected

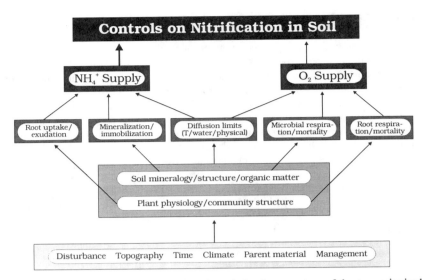

Figure 2. Schematic diagram showing the relative importances of the two principal proximal environmental controls on nitrification in soil (darkest shaded rectangles) and their regulation by increasingly distal factors (increasingly lighter shaded rectangles).

to several N and water treatments, but values as high as 10 percent (Hutchinson and Follett, 1986; Shepherd et al., 1991) and as low as 0.1 percent (Davidson et al., 1993) have also been reported.

Chemoautotrophic NH_4^+-oxidizing bacteria are widely distributed in soil and require only CO_2, O_2, and NH_4^+ to proliferate. Carbon dioxide is essentially never absent, and O_2 is usually adequate (except for brief intermittent periods) in all but a few very anaerobic environments (e.g., sediments, bogs, sludge), so NH_4^+ availability is the factor that most frequently limits the overall rate of nitrification. Regulation of these two proximal controllers by increasingly distal factors is summarized in Figure 2.

B. Denitrification

Denitrification is defined here as respiratory reduction of NO_3^- or NO_2^- to gaseous NO, N_2O, or N_2 that is coupled to electron transport phosphorylation. Unlike the narrow species diversity of organisms responsible for nitrification in soil, denitrification capacity is common to several taxonomically and physiologically different bacterial groups. Denitrifiers, which are basically aerobic bacteria with the alternative capacity to reduce N oxides when O_2 becomes limiting, are so widely distributed in nature that the restriction of the denitrifying activity in a given habitat can usually be assumed to result not from lack of enzyme, but rather from limited substrate availability or the environmental conditions that regulate the process (Firestone and Davidson, 1989). Denitrifying enzyme activity persists for months, even in very dry soil (Smith and Parsons, 1985), and the activation of these enzymes, as well as their *de novo* synthesis, begins almost immediately following soil wetting by precipitation or irrigation (Rudaz et al., 1991).

Although the identity of N compounds involved in the biochemical pathway of denitrification is well established, the nature of NO's relation to the process, as well as the mechanism for formation of the N-N bond during reduction of NO_2^- to N_2O, are the subjects of considerable current controversy (e.g., Goretski and Hollocher, 1990; Heiss et al., 1989). It is generally agreed that N_2O is an obligatory intermediate, and that NO behaves as if it were an intermediate, or at least in rapid equilibrium with an intermediate in the reductive sequence (Averill and Tiedje, 1982). Requirements for the process to occur include availability of suitable reductant (usually organic C), restricted O_2 availability, and presence of N oxides (NO_3^-, NO_2^-, NO, or N_2O). The relative importances of these three denitrification controllers varies among habitats, but for soil and other habitats exposed to the atmosphere, O_2 availability is nearly always the most critical;

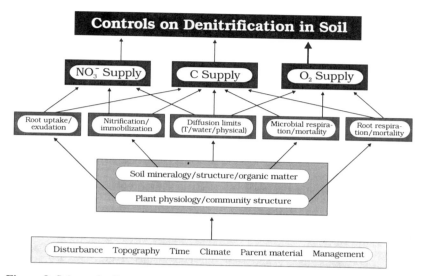

Figure 3. Schematic diagram showing the relative importances of the three principal proximal environmental controls on denitrification in soil (darkest shaded rectangles) and their regulation by increasingly distal factors (increasingly lighter shaded rectangles).

Tiedje (1988) discussed this topic in detail. Regulation of the three proximal process controllers by increasingly distal factors is summarized in Figure 3.

As for nitrification, the total amounts of NO and N_2O produced by denitrification in soil depend not only on factors that determine the overall rate of the process, but also on parameters that control the ratios of its potential products (Firestone and Davidson, 1989). Dinitrogen and N_2O are considered to be the usual end products of denitrification, and that fraction of the process interrupted at N_2O ranges from almost none to the preponderance of N reduced, depending principally on the relative availability of oxidant vs. reductant. When the availability of oxidant overshadows the supply of reductant, then substrate N oxide may be incompletely reduced resulting in a larger $N_2O:N_2$ ratio of end products. Conversely, when the overall rate of denitrification is limited by the supply of oxidant, most of the N oxide is converted to N_2. Although NO is not usually considered to be a major end product of denitrification in soil or water, it is often the principal product in laboratory studies where the process is initiated without restricting rapid equilibration of gaseous products with the ambient atmosphere. In the natural soil environment, however, denitrification generally occurs only when the soil's water content is high enough to restrict O_2 availability, which also restricts the diffusion rates of other gases in soil. The resulting increase in time required for NO diffusion to the soil surface, combined with its instability toward further reduction, allows very little of this gas to escape.

C. Relative Contributions

Abiotic chemodenitrification is generally believed to be responsible for a much smaller fraction of soil NO and N_2O exchange than either nitrification or denitrification, but the relative contributions of these two microbial processes differ widely. Based on the correlations of their measured emission rates of the two gases with soil temperature, soil NO_3^- concentration, and soil water content, Anderson and Levine (1987) concluded that most of the NO was produced by nitrifiers, and most of the N_2O by denitrifiers. Laboratory soil incubation studies employing acetylene (Davidson et al., 1993) or nitrapyrin (Tortoso and Hutchinson, 1990) to inhibit chemoautotrophic nitrifiers have provided additional strong evidence that this microbial group is primarily responsible for NO production, an interpretation that is also consistent with most results from pure culture studies of *Nitrosomonas europae* (Anderson and Levine, 1986; Lipschultz et al., 1981). Other studies are contradictory to the attractively simple hypothesis that NO and N_2O arise from separate microbial sources. For example, Davidson et al. (1993) concluded from acetylene inhibition experiments in a seasonally dry Mexican tropical forest that nitrification was the dominant source of both gases during the dry season and during initial soil wetting that marked the beginning of the wet season, but that denitrification was probably the most important N_2O source during the remainder of the wet season. Similarly, Hutchinson and Brams (1992) found that nitrifiers were responsible for production of both NO and N_2O in a grass pasture on sandy loam in humid, subtropical southern Texas. Laboratory studies (e.g., Johansson and Galbally, 1984; Remde et al., 1989) often indicate much larger NO emission rates under anaerobic than aerobic conditions, but Galbally and Johansson (1989) showed that elevated rates observed in the laboratory match those measured from the same soils in the field only in the aerobic case, probably because of the transport-dependence of the NO and N_2O yields of denitrification described earlier.

V. Environmental Controls on Soil NO_x and N_2O Exchange

Despite the intimidating variability of soil NO_x and N_2O exchange rates (Williams et al., 1992b, Tables 1 and 2) and the diversity of biotic and abiotic processes involved in the production, consumption, and transport of these N gases, various patterns emerge immediately from the inspection of field-measured exchange rates. Warm dry soils produce more NO than wet cool soils, so grassland and savannah soils tend to be stronger sources than forest soils at the same latitude. For example, Johansson et al. (1988) observed larger average NO emission from savannah areas than forested areas in Venezuela. They also noted that the NO fluxes measured from both areas were 3 to 30 times larger than from analogous ecosystems in temperate regions. The greater importance of tropical vs. temperate and boreal biogenic sources also holds for N_2O (Watson et al., 1990), but because denitrification has more significance as a source of this gas, the C-rich and usually wetter forest soils at each latitude typically support larger N_2O fluxes than their grassland or savannah counterparts. These and other patterns in the available data indicate that soil NO and N_2O exchange are strong functions of vegetative cover, site history (fertilization, burning, domestic animal grazing, etc.), and especially soil temperature, soil N availability, and soil water content, which are discussed in greater detail below.

A. Soil Temperature

Figure 4 illustrates the strong dependence of soil NO emissions on temperature. Although emission rates included in the graph vary substantially among measurement locations, their dependence on soil temperature over the range 15 to 35 °C is similar at all locations. The approximate doubling of the NO emission rate for each 10 °C rise in temperature over this range matches the temperature dependence of soil N_2O emissions (Blackmer et al., 1982), as well as most microbial processes including nitrification (Focht and Verstraete, 1977) and denitrification (Galbally, 1989). Declining emissions at temperatures greater than about 35 °C may result partially from soil desiccation and its consequent reduction in the mobility of available nutrients, but is probably due primarily to the failure of autotrophic nitrifiers to grow above 40 °C (Focht and Verstraete, 1977). The heterotrophs responsible for denitrification have higher tolerance to elevated temperature, so the greater significance of this process as a source of N_2O causes soil emission of this gas to exhibit less sensitivity to high temperature. In fact, the Arrhenius equation that describes this exponential temperature dependence generally applies only over the range 15 to 35 °C for nitrification, but over the much broader range 15 to 75 °C for denitrification (Focht and Verstraete, 1977). Changes in temperature below about 15 °C typically have much greater effect on the rates of all biological processes than changes above this threshold and are not well characterized by the Arrhenius equation (Ingraham, 1962). However, the potential for important soil emission of NO and N_2O at low temperature cannot be discounted because the organisms responsible for both nitrification and denitrification are known to possess a significant capacity for adaptation to extreme climates (Focht and Verstraete, 1977; Haynes and Sherlock, 1986).

In addition to its effect on the microbial production of NO and N_2O in soil, temperature also has a strong influence on the physical and chemical parameters that regulate transport of the gases through soil and their subsequent exchange with the atmosphere. Unfortunately, these parameters (e.g., diffusion coefficients, solubilities) also depend on soil texture, soil water content, the composition of aqueous and nonaqueous soil phases, etc., which significantly complicates achieving a predictive understanding of the net effect of temperature on soil NO and N_2O exchange.

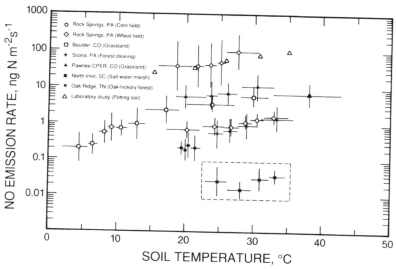

Figure 4. Relation of soil NO emission rate to soil temperature measured in diverse ecosystem types across the U.S.A.; vertical lines represent the SD of fluxes measured over the soil temperature range spanned by horizontal lines. (From Williams and Fehsenfeld, 1991.)

B. Soil N Availability

There is abundant evidence that the availability of organic and inorganic N in soils also strongly influences their rates of NO and N_2O emission. For example, Robertson and Tiedje (1984) found that a net nitrification assay correlated well with N_2O production in intact cores of forest soils from Michigan, while Matson and Vitousek (1987) reported a good correlation between a net mineralization assay and N_2O flux from tropical forest soils (Figure 5), but not from nearby pastures. Because both nitrification and denitrification are often substrate-limited (Figures 2 and 3), soil NH_4^+ or soil NO_3^- pool sizes might be expected to serve as useful indicators of the rate of N transformation, and thus to forecast NO and N_2O exchange rates. Examples supporting this concept include the correlations of NO fluxes with soil NH_4^+ levels reported by Slemr and Seiler (1984), Anderson et al. (1988), Levine et al. (1988) and Hutchinson and Brams (1992), as well as similar correlations of N_2O fluxes with soil NH_4^+ concentrations reported by Mosier et al. (1982, 1983) and Hutchinson et al. (1993b). However, because the turnover rates of soil inorganic N pools are sometimes very rapid, pool sizes may not accurately reflect the prevailing rate of N cycling or, more specifically, the rate of substrate supply to nitrifying or denitrifying bacteria. For example, Davidson et al. (1990) estimated that turnover time of the soil NH_4^+ pool in a dry California grassland was about one day, so low measured NH_4^+ concentrations were a poor indicator of the rate of NH_4^+ supply.

Unfortunately, all reported correlations tend to be site-specific or study-specific, and no single predictive parameter or suite of parameters has emerged that accurately reflects the effect of N availability on soil NO and N_2O emissions across all sites and studies. Failure to find common predictors probably reflects that (1) two very different processes are involved, i.e., nitrification and denitrification, (2) other process-limiting factors that interact with N availability may be more important at some sites than others, and (3) the scale chosen for investigation influences the nature of the predictors likely to be found useful. For example, soil NO_3^- concentration was a good predictor of NO emissions when hardwood forests were compared to fertilized corn fields, but it did not account for substantial variation within each location (Williams and Fehsenfeld, 1991); at the latter scale, modest topographic gradients or local scale effects of crop residues may be important contributors to the observed variability. It is not clear whether the relation of NO emissions to soil NO_3^- concentration at the larger scale reflects the importance of this ion as a substrate for denitrification, as a product of nitrification, or simply that NO_3^- tends to accumulate in soils where N is abundant compared to the availability of readily-oxidizable C, which generally results in a leaky N cycle.

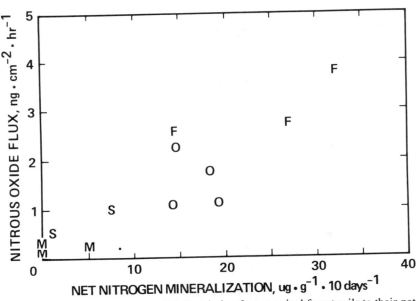

Figure 5. Relation of the rates of N_2O emission from tropical forest soils to their net N mineralization rates; F = fertile forest soils, O = oxisols and ultisols, S = sandy soils, and M = montane forest soils. (From Matson and Vitousek, 1990.)

C. Soil Water Content

The complex dependence of soil NO_x and N_2O exchange on temperature and N availability is further confounded by the effects of soil water on production, consumption, and transport of the two gases. Except for its universal requirement by all life processes, soil water's most important effect results from its strong influence on both gas-phase and solution-phase diffusive transport rates. Higher water contents increase the ratio of water-filled to air-filled soil pore space and result in thicker water films lining the remaining air-filled pores, thus enhancing the transport of species in solution, but retarding that of gases in the soil atmosphere. The consequences of these two opposing effects are illustrated conceptually in Figure 6, which suggests that for aerobic microbial processes like nitrification, the overall process rate is probably limited by solution-phase substrate diffusion through thin water films in dry soil, and by gas-phase O_2 diffusion in wet soil. The optimum soil water content for aerobic processes is about 60% water-filled pore space (WFPS) (Linn and Doran, 1984), but it is important to realize that this value applies to overall process rates, rather than the production of specific end products (Davidson, 1993). For example, the optimal WFPS for N_2O production by nitrifiers may be somewhat higher than for NH_4^+ oxidation, because N_2O is produced by these bacteria only when NO_2^- is used as an electron acceptor, apparently in response to incipient O_2 deficiency. Because it is an anaerobic process, the optimum WFPS for denitrification exceeds 60%, but maximum production of N_2O by the responsible bacteria may occur at a WFPS that is somewhat lower than the optimal value for the denitrification process in general, because the $N_2O:N_2$ ratio of denitrification products is determined by the relative availabilities of oxidant vs. reductant.

In addition to these transport-related effects of a soil's water content on its NO and N_2O evolution rates, several authors have reported a large burst of emissions concurrent with the flush of CO_2 that typically follows wetting of very dry soil (e.g., Anderson and Levine, 1987; Davidson, 1992; Hao et al., 1988; Slemr and Seiler, 1984; Williams et al., 1987). Emission rates during such an event may be up to three orders of magnitude higher than preceding or following rates, so the quantity of soil N lost during its short duration may exceed the total amount emitted during the relatively long periods between times that the soil dries enough to support another emissions burst following the next addition of water. Interestingly, subsequent additions of water by irrigation or rainfall may produce further significant increases in emissions above the levels measured from dry soil, but the amount of increase is small compared to that observed for a single watering of a very dry soil (Johansson et al., 1988; Williams et al., 1987). Results from a laboratory soil incubation study shown in Figure 7 reveal that a second burst of NO similar to the one that

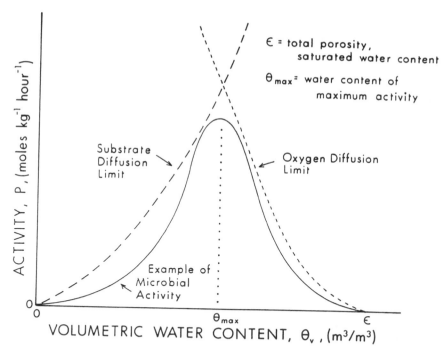

Figure 6. Conceptual model of aerobic microbial activity as a function of volumetric soil water content; dashed lines represent the limits imposed by solution-phase substrate diffusion and gas-phase O_2 diffusion. (From Skopp et al., 1990.)

followed initial wetting occurred only in the treatment where desiccation had reduced the evolution rate to near zero prior to rewetting. Reasons for the unusually large response of N oxide emissions to wetting of dry soil remain unclear.

VI. Modeling Soil NO_x and N_2O Exchange

Despite our rapidly increasing cellular-level understanding of the biotic and abiotic processes responsible for gaseous N oxide production, consumption, and transport in soil, and of the mechanisms for environmental regulation of those processes, applying that knowledge at regional to global space scales and seasonal to annual time scales remains troublesome. Much of the difficulty reflects scaling problems associated with the need to draw inferences about large scale exchanges from the highly variable fluxes typically measured over small areas and short times. Budget calculations like those in Tables 1 and 2 compare best estimates of the magnitudes and distributions of known sources with known sinks, and then attempt to reconcile the net source of each gas with the magnitude, distribution, and trend in its atmospheric concentration. This approach has already resulted in greater understanding of the relative importances of known sources and sinks, and established the potential for existence of additional unknown sources and sinks. Biogenic soil source terms in these budgets were derived by dividing the Earth's surface into a few major biomes and then characterizing each by the product of its area and the mean exchange rate from randomly-selected sites assumed to be representative of that biome. However, the criteria for establishing that a particular site (or measurement time) is representative are not straightforward (Matson et al., 1989), which renders the entire approach hopelessly inadequate for achieving a predictive understanding of the immense variability in soil NO and N_2O exchange rates across time and space domains larger than the scale of available measurements.

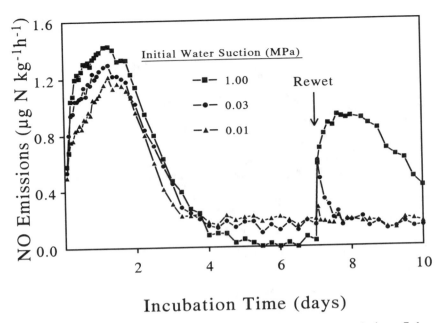

Figure 7. NO emission rate from soil under three water treatments during a 7-day drying cycle, following which each treatment was rewet to its initial water content. (From Hutchinson et al., 1993a.)

Accordingly, the long-range goal of soil NO and N_2O exchange measurements should be to capture the exchange rates in terms of their basic physical, chemical, and biological controllers, so that dependence of the flux on its controllers can be described by simulation models parameterized by variables observable at the scales of interest. Of the three principal environmental controllers of gaseous N oxide exchange identified in section V, soil temperature and soil water content have been remotely sensed and have been successfully modeled at a variety of scales using only readily available climate data, but relating soil water content to its most important effect, i.e., control of O_2 availability, depends on knowledge of soil parameters that are less readily available, e.g., bulk density, drainage, organic matter additions, etc. In addition, identification of an appropriate measure of the third principal environmental controller, soil N availability, has been even more elusive. It has long been recognized that at the cellular level the rate of NH_4^+ supply is an important controller of nitrification (Figure 2), as is the rate of NO_3^- supply for denitrification (Figure 3), but there has been little progress toward using that knowledge to develop a universally applicable measure of N availability at larger time and space scales.

Despite these difficulties, recent literature contains several examples where the adoption of surrogate variables, empirical relations observed in large data sets, or other model parameterization approaches has led to progress toward the goal of process-level simulation of soil NO and N_2O exchange. For example, Li et al. (1992a) recently published a rainfall-driven model for simulating the evolution of N_2O, CO_2, and N_2 from agricultural soils. The model includes thermal-hydraulic, decomposition, and denitrification submodels that generate soil temperature and water profiles and shifts in soil aeration. Inputs include basic climate data, soil physical properties, initial soil chemical properties, and agronomic management practices. Between rain events oxidative organic matter decomposition and N transformation processes dominate, but during rain events denitrification produces N_2O and N_2. Responses of the model to changes in external parameters correspond with conventional wisdom, and the trends and magnitudes of simulated N_2O and CO_2 emission rates were consistent with the results of field experiments conducted at three U.S.A. sites and two in Europe (Li et al., 1992b).

Another recent example is the NO emission inventory for the U.S.A. published by Williams et al. (1992a), who equated soil NO emission to the product of an empirically-derived constant and an exponential soil temperature function based on the data in Figure 4. They suggested that the empirical constant was related in a broad way to the remaining controllers of soil NO exchange, including soil N availability and water content, which differ for each ecosystem type, so they generated a different constant for each. The inventory predicted that in agricultural areas where large amounts of N fertilizers are used, soil is the dominant NO_x source during spring and summer when photochemical activity in the atmosphere is also greatest, so regional O_3 production and acid deposition are likely influenced substantially.

In my ongoing small plot research at Akron, Colorado (Hutchinson et al., 1992) we observed that within seasons and in the absence of perturbations, each plot was characterized by different, but relatively constant, NO and N_2O emission rates that varied among plots in concert with various measures of soil N availability. Superimposed on these background rates were emission peaks that occurred in response to precipitation or fertilization, and especially a combination of these two. Maximal emission rates from each plot following perturbation were proportional to its background emission rates and were unusually large when the precipitation fell on very dry soil. From these and other observations, we hypothesized that the NO and N_2O exchanges might be best simulated by the product of a constant that somehow reflects soil N availability, an exponential function of soil temperature, and a dual-slope linear function of soil WFPS that captures the concepts diagrammed in Figure 6, thereby accounting for all three of the principal environmental controls on soil gaseous N oxide exchange identified in section V. Evidence supporting the potential of this approach includes that approximate order-of-magnitude differences between two sets of fluxes measured on sampling dates with widely divergent conditions were reduced to less than a factor of 2 after normalization to the same temperature and percentage WFPS, and the remaining differences between the two sets were correlated with measured changes in various soil N availability indices. Understanding dependence of the model constant on the sizes and/or transformation rates of identifiable soil N pools and developing an effective method of adjusting the effect of precipitation for soil water content prior to wetting have been the most difficult; much additional work is needed before this approach can be extended over the much larger areal and temporal domains to which we think it may apply.

All three of these modeling approaches employ parameterization schemes that satisfy to some extent the need for using macroscopic measures of large-scale effects that I identified in the Introduction. Compared to simple linear extrapolation of point measurements from randomly-selected sites assumed to be representative, such models (1) provide the basis for more precise inferences about regional/global and seasonal/annual scale phenomena using either ground-based or remotely-sensed data, (2) provide information concerning the timing and distribution of emission events, which in turn helps to define the nature and dynamics of source ecosystems, and (3) permit predicting how future natural and anthropogenic disturbances might influence the soil source of atmospheric NO_x and N_2O (Matson et al., 1989).

VII. Summary and Perspective

Gaseous N oxides are radiatively, chemically, and ecologically important trace atmospheric constituents. Their exchange across the soil-atmosphere boundary is directly or indirectly related to acid deposition, global warming, stratospheric O_3 depletion, groundwater contamination, deforestation, and biomass burning, which include most of the major environmental issues facing society today. Because of the high reactivity of NO_x, its exchange strongly influences local and regional atmospheric photochemistry, while the important global atmospheric consequences of N_2O exchange result from its long lifetime and spectral properties. Despite the vastly different atmospheric chemistries of NO_x and N_2O, biochemical processes involved in the production, consumption, and transport of the two gases in soil are similar. Both biotic and abiotic processes are involved, but the bacterial processes of nitrification and denitrification are of paramount importance and represent a major source of atmospheric N oxides. For both these reaction sequences, the yields of NO and N_2O depend not only on factors that determine the overall process rates, but also on parameters that control their ratios of potential products. The available data suggest that net production of both gases, as well as the NO:N_2O emissions ratio, are strongly dependent on soil temperature, soil N availability, and soil water content.

Although I have emphasized the emission of gaseous N oxides from soil, their wet and dry deposition has equal environmental importance. Nitrous oxide is relatively inert in the troposphere, so it is not readily deposited to the surface, but the opposite is true of NO_x. Its short atmospheric lifetime and the absence of evidence that it accumulates in the atmosphere dictate that surface NO_x sources and sinks must balance. While soil-emitted NO is not itself readily deposited, the first product of its photochemical oxidation in the troposphere (NO_2) is strongly absorbed by both soils and plants, and following further oxidation, the resulting HNO_3 is extremely efficiently deposited to most all surfaces. Transport distances vary from local to regional scales. An example of the former is plant canopy uptake and metabolism of NO_2 formed via oxidation by O_3 of NO emitted from the underlying soil (Rogers et al., 1979), while regional scale transport is exemplified by the deposition to Northeastern U.S.A. forests of HNO_3 formed by photochemical oxidation of NO_x emitted from industrial and agricultural sources in the Midwestern states.

Quite complex interactions may result from this redistribution of biospheric N. For example, in its initial stages chronic low-level deposition of gaseous and particulate N oxides to temperate forest ecosystems have been shown to increase their primary productivity without perturbing their relatively conservative N cycles, but as N availability eventually begins to overwhelm C availability to soil microorganisms and the availability of water and other nutrients to higher plants, the importance of nitrification in the N cycle probably increases (Melillo et al., 1989). This process contributes directly to net production of NO and N_2O and results in accumulation of NO_3^-, which enhances the potential for additional N_2O emissions via denitrification and contributes to leaching losses that enrich groundwater with biologically-available N. Other examples of ecosystem functions influenced by N deposition include (1) the capacity of aerobic soils for microbial oxidation of atmospheric CH_4 was reduced by addition of N to temperate zone forests (Steudler et al., 1989) and grasslands (Mosier et al. 1991), (2) chemical oxidation of CH_4 in the troposphere requires OH, the concentration of which is critically dependent on NO_x levels (Williams et al., 1992b), and (3) COS and CS_2, which exert important influences on the Earth's heat budget as well as the chemistry of the stratosphere, are produced mainly via soil biological processes that are sensitive to N availability (Melillo et al., 1989; Melillo and Steudler, 1989).

Because of these and other known and unknown linkages to atmospheric processes and properties that control the habitability of planet Earth, it is imperative that the exchange of NO_x and N_2O between the Earth's surface and its atmosphere be understood at all scales from cellular to global. Our present capability to predict biosphere-atmosphere interactions at global/annual scales is severely limited by immense variability in the spatial and temporal distributions of the exchanges and remaining uncertainties surrounding the mechanisms that control them. Because direct monitoring of the exchange rates at such large scales is improbable with existing or foreseeable technologies, development of applicable trace gas exchange models driven by variables observable at these scales is imperative. This goal requires not only better understanding of the interacting physical, chemical, and biological processes that regulate gaseous N oxide exchange across the surface-atmosphere interface, but also acquisition of comprehensive data sets against which proposed models can be tested. Thus, there is also urgent need for additional measurement programs that must include intercalibrations and intercomparisons of both sampling techniques and analytical procedures (Fehsenfeld, 1993), and reports of which must describe explicitly the sample allocations and data analyses employed to characterize the exchange rates (Livingston and Hutchinson, 1993). In addition, integrating and extrapolating these data over regional-to-global space scales and seasonal-to-interannual time scales requires improved and updated databases of climatological, ecological, and land use information (Stewart et al., 1989), although it is generally true that even existing databases cannot be adequately utilized without additional flux measurements and better simulation models.

Mechanistic models that reproduce small-scale basic soil N cycling processes have been proposed, but they require input data with resolution never likely to become available in global ecological data bases. The larger-scale models based on ecosystem-level variables that I described earlier have had some success in predicting gaseous N oxide exchange rates, but their parameterization schemes are often based at least partially on empirical relationships that may not be valid outside the climate, soil type, or plant community in which they were determined. Nevertheless, recent recognition of relationships between the most important cellular controllers and their field, landscape, and global scale manifestations reflects the progress made in understanding the regulation of soil NO and N_2O exchange. The challenge facing future research is to integrate increasing knowledge of process controls at all scales with the rapidly-expanding data base of exchange measurements to explain their existing temporal and spatial distributions, and then to predict how future natural and anthropogenic disturbances might influence those distributions. For this, linear extrapolation of a few point measurements across continents or throughout the globe is clearly inadequate.

References

Anderson, I.C. and J.S. Levine. 1986. Relative rates of nitric oxide and nitrous oxide production by nitrifiers, denitrifiers, and nitrate respirers. *Appl. Environ. Microbiol.* 51:938-945.

Anderson, I.C. and J.S. Levine. 1987. Simultaneous field measurements of biogenic emissions of nitric oxide and nitrous oxide. *J. Geophys. Res.* 92:965-976.

Anderson, I.C., J.S. Levine, M.A. Poth, and P.J. Riggan. 1988. Enhanced biogenic emissions of nitric oxide and nitrous oxide following surface biomass burning. *J. Geophys. Res.* 93:3893-3898.

Aulakh, M.S., D.A. Rennie, and E.A. Paul. 1984. Acetylene and N-serve effects upon N_2O emissions from NH_4^+ and NO_3^- treated soils under aerobic and anaerobic conditions. *Soil Biol. Biochem.* 16:351-356.

Averill, B.A. and J.M. Tiedje. 1982. The chemical mechanism of microbial denitrification. *FEBS Lett.* 138:8-12.

Blackmer, A.M., J.M. Bremner, and E.L. Schmidt. 1980. Production of nitrous oxide by ammonia-oxidizing chemoautotrophic microorganisms in soil. *Appl. Environ. Microbiol.* 40:1060-1066.

Blackmer, A.M. and M.E. Cerrato. 1986. Soil properties affecting formation of nitric oxide by chemical reactions of nitrite. *Soil Sci. Soc. Am. J.* 50:1215-1218.

Blackmer, A.M., S.G. Robbins, and J.M. Bremner. 1982. Diurnal variability in rate of emission of nitrous oxide from soils. *Soil Sci. Soc. Am. J.* 46:937-942.

Crutzen, P.J. 1983. Atmospheric interactions--homogeneous gas reactions of C, N, and S containing compounds. p. 67-112. In: B. Bolin and R.B. Cook (eds.). *The major biogeochemical cycles and their interactions.* J. Wiley & Sons, NY.

Davidson, E.A. 1991. Fluxes of nitrous oxide and nitric oxide from terrestrial ecosystems. p. 219-235. In: J.E. Rogers and W.B. Whitman (eds.). *Microbial production and consumption of greenhouse gases: Methane, nitrogen oxides, and halomethanes.* Am. Soc. Microbiol., Washington, D.C.

Davidson, E.A. 1992. Sources of nitric oxide and nitrous oxide following wetting of dry soil. *Soil Sci. Soc. Am. J.* 56:95-102.

Davidson, E.A. 1993. Soil water content and the ratio of nitrous oxide to nitric oxide emitted from soil. p. 369-386. In: R.S. Oremland (ed.). *Biogeochemistry of global change: radiatively active trace gases.* Chapman and Hall, NY.

Davidson, E.A., P.A. Matson, P.M. Vitousek, R. Riley, K. Dunkin, G. GarcìaMèndez, and J.M. Maass. 1993. Processes regulating soil emissions of NO and N_2O in a seasonally dry tropical forest. *Ecology* 74:130-139.

Davidson, E.A., J.M. Stark, and M.K. Firestone. 1990. Microbial production and consumption of nitrate in an annual grassland. *Ecology* 71:1968-1975.

Ehhalt, D.H. and J.W. Drummond. 1982. The tropospheric cycle of NO_x. p. 219-251. In: H.W. Georgii and W. Jaeschke (eds.). *Chemistry of the unpolluted and polluted troposphere.* D. Reidel Publ. Co., Dordrecht, Holland.

Fehsenfeld, F.C. 1994. Measurements of chemically reactive trace gases at ambient concentrations. Chapter 7. In: P.A. Matson and R.C. Harriss (eds.). *Methods in ecology: biogenic trace gas emissions from soil and water.* Blackwell Scientific Publ., London. In press.

Firestone, M.K. and E.A. Davidson. 1989. Microbiological basis of NO and N_2O production and consumption in soil. p. 7-21. In: M.O. Andreae and D.S. Schimel (eds.). *Exchange of trace gases between terrestrial ecosystems and the atmosphere.* J. Wiley & Sons, Chichester.

Focht, D.D. and W. Verstraete. 1977. Biochemical ecology of nitrification and denitrification. *Adv. Microb. Ecol.* 1:135-214.

Galbally, I.E. 1989. Factors controlling NO_x emissions from soils. p. 23-37. In: M.O. Andreae and D.S. Schimel (eds.). *Exchange of trace gases between terrestrial ecosystems and the atmosphere.* J. Wiley & Sons, Chichester.

Galbally, I.E. and C. Johansson. 1989. A model relating laboratory measurements of rates of nitric oxide production and field measurements of nitric oxide emissions from soils. *J. Geophys. Res.* 94:6473-6480.

Galbally, I.E. and C.R. Roy. 1978. Loss of fixed nitrogen from soils by nitric oxide exhalation. *Nature* 275:734-735.

Goretski, J. and T.C. Hollocher. 1990. The kinetic and isotopic competence of nitric oxide as an intermediate in denitrification. *J. Biol. Chem.* 265:889-895.

Hao, W.M., D. Scharffe, P.J. Crutzen, and E. Sanhueza. 1988. Production of N_2O, CH_4, and CO_2 from soils in the tropical savanna during the dry season. *J. Atmos. Chem.* 7:93-105.

Hao, W.M., S.C. Wofsy, M.B. McElroy, J.M. Beer, and M.A. Togan. 1987. Sources of atmospheric nitrous oxide from combustion. *J. Geophys. Res.* 92:3098-3104.

Haynes, R.J. and R.R. Sherlock. 1986. Gaseous losses of nitrogen. p. 242-302. In: R.J. Haynes (ed.). *Mineral nitrogen in the plant-soil system.* Academic Press, Sydney.

Heiss, B., K. Frunzke, and W.G. Zumft. 1989. Formation of the N-N bond from nitric oxide by a membrane-bound cytochrome bc complex of nitrate-respiring (denitrifying) *Psuedomonas stutzeri*. *J. Bacteriol.* 171:3288-3297.

Homolya, J.B. and E. Robinson. 1984. Natural and Anthropogenic Emission Sources. In: A.P. Altshuller and R.A. Linthurst (eds.). *The acidic deposition phenomenon and its effects: critical assessment review papers*. Rep. EPA-600/8-83-016AF, U.S. Environ. Prot. Agency, Washington, D.C.

Hooper, A.B. 1984. Ammonia oxidation and energy transduction in the nitrifying bacteria. p. 133-167. In: W.R. Strohl and O.H. Tuovinen (eds.). *Microbial chemoautotrophy*. Ohio State Univ. Press, Columbus.

Hutchinson, G.L., W.E. Beard, M.F. Vigil, and A.D. Halvorson. 1992. NO and N_2O emissions from perennial grass and winter wheat in the semiarid Great Plains. p. 260. In: *Agron. Abstr.* Am. Soc. Agron., Madison, WI.

Hutchinson, G.L. and E.A. Brams. 1992. NO vs. N_2O emissions from an NH_4^+-amended Bermuda grass pasture. *J. Geophys. Res.* 97:9889-9896.

Hutchinson, G.L. and E.A. Davidson. 1993. Processes for production and consumption of gaseous nitrogen oxides in soil. p. 79-93. In: L.A. Harper, A.R. Mosier, J.M. Duxbury, and D.E. Rolston (eds.). *Agricultural ecosystem effects on trace gases and global climate change*. ASA Spec. Publ. no. 55. Am. Soc. Agron., Madison, WI.

Hutchinson, G.L. and R.F. Follett. 1986. Nitric oxide emissions from fallow soil as influenced by tillage treatments. p. 180. In: *Agron. Abstr.* Am. Soc. Agron., Madison WI.

Hutchinson, G.L., W.D. Guenzi, and G.P. Livingston. 1993a. Soil water controls on aerobic soil emission of gaseous N oxides. *Soil Biol. Biochem.* 25:1-9.

Hutchinson, G.L., G.P. Livingston, and E.A. Brams. 1993b. Nitric and nitrous oxide evolution from managed subtropical grassland. p. 290-316. In: R.S. Oremland (ed.). *Biogeochemistry of global change: radiatively active trace gases*. Chapman and Hall, NY.

Ingraham, J.L. 1962. Temperature relationships. p. 265-296. In: I.C. Gunsalus and R.Y. Stanier (eds.). *The bacteria*. Vol. 4. Academic Press, N.Y.

Johansson, C. and I.E. Galbally. 1984. Production of nitric oxide in loam under aerobic and anaerobic conditions. *Appl. Environ. Microbiol.* 47:1284-1289.

Johansson, C., H. Rodhe, and E. Sanhueza. 1988. Emission of NO in a tropical savanna and a cloud forest during the dry season. *J. Geophys. Res.* 93:7180-7192.

Kepros, J.G. 1992. Global warming and atmospheric altimetry. *Physics Today* 45(10):142.

Ko, M.K.W., N.D. Sze, and D.K. Weisenstein. 1991. Use of satellite data to constrain the model-calculated atmospheric lifetime for N_2O: Implications for other trace gases. *J. Geophys. Res.* 96:7547-7552.

Levine, J.S., W.S. Cofer III, D.I. Sebacher, E.L. Winstead, S. Sebacher, and P.J. Boston. 1988. The effects of fire on biogenic soil emissions of nitric oxide and nitrous oxide. *Global Biogeochem. Cycles* 2:445-449.

Li, C., S. Frolking, and T.A. Frolking. 1992a. A model of nitrous oxide evolution from soil driven by rainfall events: I. Model structure and sensitivity. *J. Geophys. Res.* 97:9759-9776.

Li, C., S. Frolking, and T.A. Frolking. 1992b. A model of nitrous oxide evolution from soil driven by rainfall events: II. Model applications. *J. Geophys. Res.* 97:9777-9784.

Linn, D.M. and J.W. Doran. 1984. Effect of water-filled pore space on carbon dioxide and nitrous oxide production in tilled and nontilled soils. *Soil Sci. Soc. Am. J.* 48:1267-1272.

Lipschultz, F., O.C. Zafiriou, S.C. Wofsy, M.B. McElroy, F.W. Valois, and S.W. Watson. 1981. Production of NO and N_2O by soil nitrifying bacteria. *Nature* 294:641-643.

Livingston, G.P. and G.L. Hutchinson. 1994. Enclosure-based measurement of trace gas exchange: applications and sources of error. Chapter 2. In: P.A. Matson and R.C. Harriss (eds.). *Methods in ecology: biogenic trace gas emissions from soil and water*. Blackwell Scientific Publ., London. In press.

Logan, J.A. 1983. Nitrogen oxides in the troposphere: Global and regional budgets. *J. Geophys. Res.* 88:10785-10807.

Luizao, F., P. Matson, G. Livingston, R. Luizao, and P. Vitousek. 1989. Nitrous oxide flux following tropical land clearing. *Global Biogeochem. Cycles* 3:281-285.

Matson, P.A. and P.M. Vitousek. 1987. Cross-system comparisons of soil nitrogen transformations and nitrous oxide flux in tropical forest ecosystems. *Global Biogeochem. Cycles* 1:163-170.

Matson, P.A. and P.M. Vitousek. 1990. Ecosystem approach to a global nitrous oxide budget. *Bioscience* 40:667-672.

Matson, P.A., P.M. Vitousek, and D.S. Schimel. 1989. Regional extrapolation of trace gas flux based on soils and ecosystems. p. 109-119. In: M.O. Andreae and D.S. Schimel (eds.). *Exchange of trace gases between terrestrial ecosystems and the atmosphere.* J. Wiley & Sons, Chichester.

McElroy, M.B. and S.C. Wofsy. 1986. Tropical forests: interactions with the atmosphere. p. 33-60. In: G. Prance (ed.). *Tropical rainforests and the world atmosphere.* Westview Press, Boulder, CO.

Melillo, J.M. and P.A. Steudler. 1989. The effect of nitrogen fertilization on the COS and CS_2 emission from temperate forest soils. *J. Atmos. Chem.* 9:411-417.

Melillo, J.M., P.A. Steudler, J.D. Aber, and R.D. Bowden. 1989. Atmospheric deposition and nutrient cycling. p. 263-280. In: M.O. Andreae and D.S. Schimel (eds.). *Exchange of trace gases between terrestrial ecosystems and the atmosphere.* J. Wiley & Sons, Chichester.

Mosier, A.R., G.L. Hutchinson, B.R. Sabey, and J. Baxter. 1982. Nitrous oxide emissions from barley plots treated with ammonium nitrate or sewage sludge. *J. Environ. Qual.* 11:78-81.

Mosier, A.R., W.J. Parton, and G.L. Hutchinson. 1983. Modelling nitrous oxide evolution from cropped and native soils. In: R. Hallberg (ed.). *Environmental Biogeochemistry.* Ecol. Bull. (Stockholm) 35:2129-2141.

Mosier, A., D. Schimel, D. Valentine, K. Bronson, and W. Parton. 1991. Methane and nitrous oxide fluxes in native, fertilized and cultivated grasslands. *Nature* 350:330-332.

Muzio, L.J. and J.C. Kramlich. 1988. An artifact in the measurement of N_2O from combustion sources. *Geophys. Res. Lett.* 15:1369-1372.

Nelson, D.W. 1982. Gaseous losses of nitrogen other than through denitrification. p. 327-363. In: F.J. Stevenson (ed.). *Nitrogen in agricultural soils.* Agron. Monogr. 22. Am. Soc. Agron., Madison, WI.

Placet, M. and D.G. Streets. 1987. Emissions of acidic deposition precursors. p. 1-78. In: *NAPAP Interim Assessment, the causes and effects of acidic deposition.* Chapter 1, Vol. II. Emissions and controls. National Acid Precipitation Assessment Program, U.S. Government Printing Office, Washington, D.C.

Poth, M. and D.D. Focht. 1985. ^{15}N kinetic analysis of N_2O production by *Nitrosomonas europaea*: An examination of nitrifier denitrification. *Appl. Environ. Microbiol.* 49:1134-1141.

Prinn, R., D. Cunnold, R. Rasmussen, P. Simmonds, F. Alyea, A. Crawford, P. Fraser, and R. Rosen. 1990. Atmospheric emissions and trends of nitrous oxide deduced from ten years of ALE-GAGE data. *J. Geophys. Res.* 95:18369-18385.

Rasmussen, R.A. and M.A.K. Khalil. 1986. Atmospheric trace gases: Trends and distributions over the last decade. *Science* 232:1623-1624.

Remde, A. and R. Conrad. 1990. Production of nitric oxide in *Nitrosomonas europaea* by reduction of nitrite. *Arch. Microbiol.* 154:187-191.

Remde, A., F. Slemr, and R. Conrad. 1989. Microbial production and uptake of nitric oxide in soil. *FEMS Microbiol. Ecol.* 62:221-230.

Robertson, G.P. and J.M. Tiedje. 1984. Denitrification and nitrous oxide production in successional and old-growth Michigan forest. *Soil Sci. Soc. Am. J.* 48:383-389.

Robertson, G.P. and J.M. Tiedje. 1988. Deforestation alters denitrification in a lowland tropical rain forest. *Nature* 336:756-759.

Rodhe, H. 1990. A comparison of the contribution of various gases to the greenhouse effect. *Science* 248:1217-1219.

Rogers, H.H., J.C. Campbell, and R.J. Volk. 1979. Nitrogen-15 dioxide uptake and incorporation by *Phaseolus vulgaris* (L.). *Science* 206:333-335.

Rowland, F.S. 1989. Chlorofluorocarbons and the depletion of stratospheric ozone. *Am. Scientist* 77:36-45.

Rudaz, A., E.A. Davidson, and M.K. Firestone. 1991. Production of nitrous oxide immediately after wetting dry soil. *FEMS Microbiol. Ecol.* 85:117-124.

Schmidt, E.L. 1982. Nitrification in soil. p. 253-288. In: F.J. Stevenson (ed.). *Nitrogen in agricultural soils.* Agron. Monogr. 22. Am. Soc. Agron., Madison, WI.

Shepherd, M.F., S. Barzetti, and D.R. Hastie. 1991. The production of atmospheric NO_x and N_2O from a fertilized agricultural soil. *Atmos. Environ.* 25A:1961-1969.

Skopp, J., M.D. Jawson, and J.W. Doran. 1990. Steady-state aerobic microbial activity as a function of soil water content. *Soil Sci. Soc. Am. J.* 54:1619-1625.

Slemr, F. and W. Seiler. 1984. Field measurements of NO and NO_2 emissions from fertilized and unfertilized soils. *J. Atmos. Chem.* 2:1-24.

Smith, M.S. and L.L. Parsons. 1985. Persistence of denitrifying enzyme activity in dried soils. *Appl. Environ. Microbiol.* 49:316-320.

Stedman, D.H. and R.E. Shetter. 1983. The global budget of atmospheric nitrogen species. p. 411-454. In: S.E. Schwartz (ed.). *Trace atmospheric constituents.* Adv. Environ. Sci. and Technol, Vol. 12. J. Wiley & Sons, NY.

Steudler, P.A., R.D. Bowden, J.M. Melillo, and J.D. Aber. 1989. Influence of nitrogen fertilization on methane uptake in temperate forest soils. *Nature* 341:314-316.

Stewart, J.W.B., I. Aselmann, A.F. Bouwman, R.L. Desjardins, B.B. Hicks, P.A. Matson, H. Rodhe, D.S. Schimel, B.H. Svensson, R. Wassmann, M.J. Whiticar, and W.-X. Yang. 1989. Extrapolation of flux measurements to regional and global scales. p. 155-174. In: M.O. Andreae and D.S. Schimel (eds.). *Exchange of trace gases between terrestrial ecosystems and the atmosphere.* J. Wiley & Sons, Chichester.

Tiedje, J.M. 1988. Ecology of denitrification and dissimilatory nitrate reduction to ammonium. p. 179-244. In: A.J.B. Zehnder (ed.). *Biology of anaerobic microorganisms.* J. Wiley & Sons, Chichester.

Tortoso, A.C. and G.L. Hutchinson. 1990. Contributions of autotrophic and heterotrophic nitrifiers to soil NO and N_2O emissions. *Appl. Environ. Microbiol.* 56:1799-1805.

Watson, R.T., H. Rodhe, H. Oeschger, and U. Siegenthaler. 1990. Greenhouse gases and aerosols. p. 1-40. In: J.T. Houghton, G.J. Jenkins, and J.J. Ephraums (eds.). *Climate change, the IPCC scientific assessment.* Cambridge Univ. Press, Cambridge.

Williams, E.J. and F.C. Fehsenfeld. 1991. Measurement of soil nitrogen oxide emissions at three North American ecosystems. *J. Geophys. Res.* 96:1033-1042.

Williams, E.J., A. Guenther, and F.C. Fehsenfeld. 1992a. An inventory of nitric oxide emissions from soils in the United States. *J. Geophys. Res.* 97:75117520.

Williams, E.J., G.L. Hutchinson, and F.C. Fehsenfeld. 1992b. NO_x and N_2O emissions from soil. *Global Biogeochem. Cycles* 6:351-388.

Williams, E.J., D.D. Parrish, and F.C. Fehsenfeld. 1987. Determination of nitrogen oxide emissions from soils: Results from a grassland site in Colorado, U.S.A. *J. Geophys. Res.* 92:2173-2179.

CHAPTER 19

Nitrous Oxide Flux from Thawing Soils in Alberta

J.W. Laidlaw, M. Nyborg, and R.C. Izaurralde

I. Introduction

Soils of north central Alberta exhibit significant denitrification activity during the spring thaw (Heaney et al., 1992; Nyborg et al., 1990) and apparently substantial gaseous losses of soil N to the atmosphere. In addition, environmental concerns and cost of commercial fertilizers have created a renewed interest in the use of animal manure as a nutrient source. Manure additions to soil, however, lead to large fluxes of N oxides (Rolston et al., 1978; Magg, 1989).

Several causes of soil N loss during soil thaw and cold temperatures have been suggested. Soil freeze/thaw increases the carbon availability from detritus or microorganisms killed by freezing (Christensen and Tiedje, 1990; Christensen and Christensen, 1991). Soil thaw can permit the physical release of subsurface produced N gas (Goodroad and Keeney, 1984). Also, the action of extra cellular denitrifying enzymes (Smith and Parsons, 1985) and chemodenitrification (Christianson and Cho, 1983) mediates N loss during soil thaw. The action of psychotrophic denitrifiers has also been suggested (Dorland and Beauchamp, 1991) and they may explain why soils from different climates respond differently to laboratory temperatures (Gamble et al., 1977; Powlson, 1988).

N_2O is usually the second greatest product of denitrification, after N_2. The ratio of N_2O to N_2 varies greatly (Gilliam et al., 1978) and is mainly affected by organic C, pH, NO_3^-, temperature, water, and redox potential. The proportion of N_2O in gaseous N emissions increased with acidity (Firestone et al. 1980; Koskinen and Keeney, 1982) and represented all of the N flux of an acid soil used by Christensen and Tiedje (1990). Denitrification at low temperature increases the ratio of N_2O to N_2 (Keeney et al., 1979). Wetter conditions increased N_2O production (Drury et al., 1992). Soil redox potential is related to many of the soil parameters which affect the $N_2O:N_2$ ratio. Some confusion may arise from consideration of the variables alone.

Because of the importance of N oxides to the environmental issues of stratospheric ozone depletion and climate warming, we conducted two laboratory experiments to assess the potential for N loss by denitrification in selected Alberta soils during the thaw. In one experiment, we incubated pre-frozen and non-frozen soils at low temperatures to find any influence of freezing on denitrification and N_2O flux. In another experiment, we measured any added effect of manure during soil thawing.

II. Methods and Materials

A. The Soils

Soil samples were taken from the cultivated horizon of four north-central Alberta soils with varying pH, C and texture (Table 1). For Experiment I, three soils of different Orders were collected during the fall of 1991, sieved through a 1 cm mesh, dried at room temperature and stored until commencement of the

Table 1. Soil properties

Soil Series	Classification	Texture	pH	Total C	Total N	Mineral N[a] NO_3^--N	NH_4^+-N
				----dag kg^{-1}----		----mg kg^{-1}----	
Experiment I							
Malmo	Typic Cryoboroll	SiCL	5.8	6.03	0.57	23.8	4.9
Breton	Typic Cryoboralf	CL	6.3	1.23	0.13	7.6	3.0
Josephine	Typic Cryaquent	CL	4.4	3.65	0.30	1.5	12.6
Experiment II							
Angus Ridge	Typic Cryoboroll						
Non-manured		SiL	6.5	3.6	0.33	0.0	2.1
Manured		SiL	7.8	6.6	0.67	0.6	3.8

[a] Values prior to initiation of experiments.

experiment. For Experiment II, samples were collected in October, 1992. Soil samples were taken from a manure treated plot (75 Mg (oven dry) ha^{-1}) and a fertilized plot (100 kg N ha^{-1} and 29 kg P ha^{-1}, spring applied). The soil samples were placed in separate pails, capped, and transported to the laboratory for overnight storage at 3 °C. The following morning, the samples were passed through a 1 cm mesh and sub sampled for moisture content.

Studies of zero-time ^{15}N recovery from soil were conducted to assess denitrification losses by difference. For the four soils used, the average zero-time recovery was 97.2%, with a range from 96.5 to 97.8%. The upper limit of the 95% confidence interval around mean recovery values reached 100% for two soils (Breton and Josephburg) and was within 0.7% of 100 in the other two (Malmo and Josephine). Consequently, labeled ^{15}N recoveries in Experiments I and II were taken as 100%.

B. Experiment I

We measured denitrification of ^{15}N labeled KNO_3 and the rate of total N_2O emissions from frozen and non-frozen soils during an 88 h incubation. The experiment consisted of three soils, four treatments (non-frozen and frozen soil, each with and without N addition) and four replications; design was randomized complete block. Air-dry soil was placed in glass jars (20 cm tall; 9 cm in diam.) to a height of 10 cm and was saturated with cold distilled water (5°C). Fertilized treatments received 80 kg N ha^{-1} as ^{15}N-labeled KNO_3 (67.7 atom % abundance) added to the surface of the soil prior to water saturation. Plexiglass lids fitted with septums and removable #12 rubber bungs were attached to the jars with silicone caulking. Jars to be frozen were placed in styro-foam containers and then maintained at -15°C (±3°C) for 14 to 21 days. Non-frozen soils were not saturated until the start of the incubation. Overhead lights with foil covered shades were placed above all jars to speed downward thaw of frozen soil and thus simulate spring thaw conditions. The air temperature 10 cm above the soil surface was maintained at 15°C. Depth of soil thaw and surface temperatures were monitored throughout the incubation (Figure 1a.).

Gas samples were collected for each eight hour period. Samples were taken 1.5 or 2 h after the jars were sealed, but otherwise the jars were left open. The gas samples were collected by inserting the needle of a 30 cc syringe, pumping the plunger several times, and then removing 24 mL of air from the head space. The air samples were stored in evacuated 25 mL Vacutainer© test tubes containing 1 g silica gel desiccant. The samples were analyzed for N_2O using a Perkin Elmer, Sigma 3 gas chromatograph with ^{63}Ni electron capture detector and a Porapak Q column. Gas concentration values were corrected for decreasing concentration gradient in the head space during trapping by multiplying by a factor of 1.47 as determined in a subsequent experiment (data not shown) using the method described by Hutchinson and Mosier (1981). Nitrous oxide data were transformed to base ten logarithms for statistical analyses. CO_2 was analyzed using a Hewlett Packard, 5890 Series II gas chromatograph with a thermal conductivity detector (TCD). After incubation, the soil was emptied into trays and quickly dried at 22°C. The soil was analyzed for ^{15}N using a Carlo Erba N Auto analyzer coupled to a VG Isogas mass spectrometer for determination of soil N mass

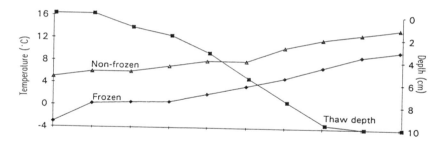
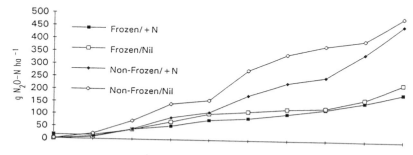
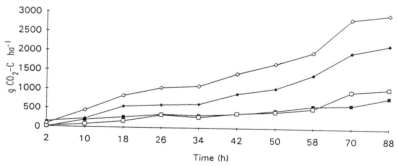

Figure 1. Experiment I: (a) Change in soil temperature of non-frozen and frozen soil and depth of thaw, (b) cumulative nitrous oxide flux, and (d) cumulative carbon dioxide emissions of Malmo soil during 88 hours.

balance of the added ^{15}N-labeled N. Mineral NO_3^- and NH_4^+ were analyzed by steam distillation (Bremner, 1965).

C. Experiment II

Fresh, cold soil was placed in glass jars as described previously. The treatments (control, control + N, manure and manure + N) were replicated four times. An additional 16 Mg ha^{-1} (oven dry) decomposed manure was mixed into the manure treated soil. The soil was brought to saturation with distilled water and ^{15}N labeled KNO_3 (5.42 atom % abundance) was added at a rate of 100 kg N ha^{-1} to +N treated soils. The soil samples were frozen for 13 days at -5°C to more realistically portray winter conditions than the -15°C used in Experiment I. Soil incubation, gas sampling and soil analyses followed the same procedure as described for Experiment I.

III. Results and Discussion

A. Experiment I

Denitrification determined by the mass balance technique showed N losses ranged from 13 to 23 kg ha^{-1} after 88 h of incubation in both frozen and non-frozen soils (Table 2). There were no differences in ^{15}N loss among soils or frozen and non-frozen treatments.

Low emissions of N$_2$O were found from all soils and treatments and were only 0.3 to 2.6% of total denitrification (Table 2). The Malmo sample had the greatest N$_2$O flux followed by Breton and then Josephine. There was also an interaction effect of freezing and the addition of N, where frozen soils with added N had significantly lower N$_2$O emissions than the non-frozen +N treatments. Christensen and Christensen (1991) and Christensen and Tiedje (1990) found that freezing of soil markedly enhanced N$_2$O emission during thaw, but our soils did not behave in that manner. N$_2$O flux remained relatively constant in all treatments throughout the incubation and increased slightly after thaw was complete (Figure 1b.). The surface temperature of non-frozen soils (Figure 1a.) was about 2°C warmer than frozen soils and may have contributed to increased N$_2$O in the Malmo and Breton series.

The low ratio of N$_2$O to N$_2$ in our work does not agree with other reports involving pH and NO$_3^-$ concentration. The brief and vigorous N emissions found from an acid soil (pH 3.8) used by Christensen and Tiedje (1990) was assumed to be exclusively N$_2$O. The Josephine soil (pH 4.4), however, had a low ratio of N$_2$O to N$_2$. Increasing NO$_3^-$ concentrations usually causes an increase in the proportion of N$_2$O as a product of denitrification (Firestone et al., 1980; Firestone et al., 1979; Letey et al., 1980; and Vermes and Myrold, 1991), but in our work N$_2$O flux did not increase with NO$_3^-$-amending the soils. Nitric oxide was not accounted for, and there is the possibility of NO being predominate to N$_2$O. Total losses of N in our experiment were similar to those found in field investigations in Alberta (Heaney et al., 1992; Nyborg et al., 1990). However, N$_2$O fluxes were about 10 times greater in the field during spring thaw than in the laboratory experiment (data not shown).

B. Experiment II

Soil ^{15}N mass balance in manure and non-manure treatments did not account for 39 and 20 kg ha^{-1} of added N, respectively (Table 3). During the 66-hour incubation, the majority of the soil volume remained frozen and the soil surface temperature remained below 5°C excepting the last 16 h (Figure 2a. and 2b.). The rate of denitrification of added NO$_3^-$-N in the manured soil was 12.7 kg ha^{-1} d^{-1} while the non-manured soil averaged 5.8 kg ha^{-1} d^{-1}. The values were consistent with the large losses of NO$_3^-$-N recorded for north-central Alberta soils in early spring when cool soil temperatures and moistures are suitable for denitrification (Malhi et al., 1990). Our measurements of denitrification loss from thawing soils were 10 to 100 times greater than the N losses reported by others (Christensen and Tiedje, 1990; Cates and Keeney, 1987; and Dorland and Beauchamp, 1991).

The estimated flux of N$_2$O-N from N-amended soil was 0.40 kg ha^{-1} day^{-1} from the manured soils and 0.15 kg ha^{-1} day^{-1} from non-manured soils. Goodroad and Keeney (1984) reported N$_2$O-N fluxes of 0.05 to 0.15 kg ha^{-1} d^{-1} in manure-amended plots during soil thaw in early to mid-April. The N$_2$O flux was relatively constant between 18 and 58 h and decreased after 58 h in manure and non-manure treated soils (Figure 2c). Our time of greatest N$_2$O production after initiation of incubation was similar to the peaking after 1 to 3 days as found by Letey et al.(1980) and Rolston et al. (1978).

The N$_2$O emissions were less than 3% of the denitrification loss from the manured soil and 2% from non-manured soil. In our work, the percent of N$_2$O-N to total N lost from soil during denitrification under cold temperatures was considerably lower than values reported elsewhere (Keeney et al., 1979; Lensi and Chalamet, 1982; Denmead et al. 1979; Christensen and Christensen, 1991; Christensen and Tiedje, 1990).

Nitrous oxide emissions of 0.25 and 0.97 kg N$_2$O-N ha^{-1} occurred from the control and No-N manure treatments, respectively, during the incubation. The concentration of NO$_3^-$-N was only 0 and 0.6 mg kg^{-1} in the control and manure treatments at the start of incubation. Therefore, immobilized N must have been mineralized to NO$_3^-$ during the incubation to supply NO$_3^-$ for denitrification on the control.

Table 2. Experiment 1: Soil and gaseous nitrogen loss during 88 h

Soil Series	Freezing[a]	^{15}N recovery (%)	Total added N emitted (kg ha^{-1})	Total N$_2$O-N emitted (kg ha^{-1})	N$_2$O/(N$_2$O + N$_2$)[b] (%)
Malmo	+	82.5	14.0	0.18	1.3
Malmo	+	NA[c]	NA	0.21	NA
Malmo	-	80.3	15.8	0.41	2.6
Malmo	-	NA	NA	0.45	NA
Breton	+	77.7	17.8	0.12	0.7
Breton	+	NA	NA	0.11	NA
Breton	-	74.0	20.8	0.16	0.8
Breton	-	NA	NA	0.22	NA
Josephine	+	71.6	22.7	0.06	0.3
Josephine	+	NA	NA	0.03	NA
Josephine	-	83.5	13.2	0.05	0.4
Josephine	-	NA	NA	0.10	NA

ANOVA

Source of Variation		Probabilities	
Soil	NS		**
Freezing	NS		**
Nitrogen			NS
Soil x freezing	NS		*
Freezing x nitrogen			NS
Soil x nitrogen			NS
Soil x freezing x nitrogen			NS

[a] Frozen (+) versus not frozen (-); [b] total gaseous N loss is assumed to be "Total added N emitted"; [c] NA = not applicable.

Table 3. Experiment II: Nitrogen after 66 hours of thaw

Treatment	Mineral N		^{15}N recovery	Total added N emitted	Total N$_2$O-N emitted	N$_2$O/(N$_2$O + N$_2$)[a]
	NO$_3^-$-N	NH$_4^+$-N	(%)	(kg kg^{-1})	(kg kg^{-1})	(%)
	(mg kg^{-1})					
Manure + ^{15}N	79.7	8.2	60.9	39.1	1.10	2.8
Manure	0.0	7.1	NA[b]	NA	0.97	NA
No manure + ^{15}N	75.9	6.5	80.0	20.0	0.42	2.1
Control	0.0	3.8	NA	NA	0.25	NA

[a] Total gaseous N loss is assumed to be "Total added N emitted"; [b] NA = not applicable.

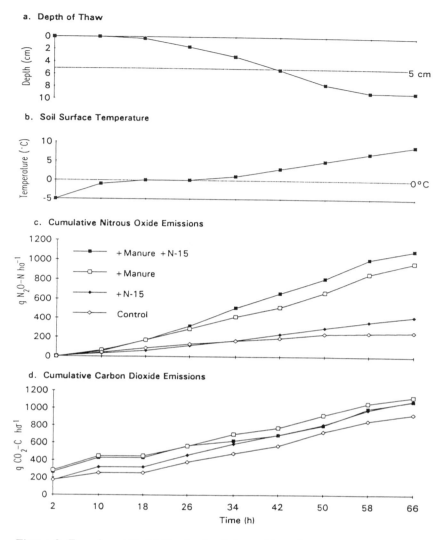

Figure 2. Experiment II: (a) Depth of soil thaw, (b) surface temperature changes, (c) cumulative nitrous oxide flux, and (d) cumulative carbon dioxide emissions during 66 hours of soil thaw.

The flux of CO_2 from the manure and non-manure treatments was much the same, with 1.1 kg CO_2-C ha^{-1} emitted during the incubation period (Figure 1d). Since CO_2 emissions were not greater in the manured soils, we suspect methane (CH_4) was a significant product during soil thaw. Future work should include an assay for CH_4 to accompany N_2O and CO_2 measurements to provide a more comprehensive view of the contribution of soil thaw to the greenhouse effect.

IV. Summary

It is apparent that large fluxes of nitrogen oxides are emitted by soil during spring thaw in north-central Alberta. This gas is important to the environmental issues of stratospheric ozone depletion and climate warming. Consequently, we investigated widely-different soils to quantify denitrification and N_2O emissions and the effect of pre-freezing under controlled conditions. We conducted an experiment with incubation of pre-frozen and non-frozen soil samples for 88 h, and with another incubation experiment with only frozen treatments for 66 h. We frequently measured N_2O emissions and measured mass balance of ^{15}N at termination of the experiments. The total denitrification losses were large among the soil-freezing and non-freezing treatments (range of 13.2 to 22.7 kg N ha^{-1}), but total N_2O emissions were consistently small (range of 0.3 to 2.6% of the denitrification loss). Pre-freezing affected neither denitrification loss nor N_2O emissions. Manure amended soil receiving labeled-fertilizer N at 100 kg ha^{-1} lost approximately twice as much N (39.1 kg ha^{-1}) as the non-manured soil. Nitrous oxide, however, remained a small part of denitrification losses.

V. Conclusions

Substantial denitrification losses occurred, in separate experiments, from the ^{15}N-labeled KNO_3 additions to thawing and non-frozen soils. Denitrification losses after 66 or 88 h ranged from 13.2 to 39.1 kg N ha^{-1} among four very different kinds of soils. However, N_2O emissions were a small proportion of the denitrification losses (<3%) of any of the soils. Whether the soil was pre-frozen, or not, had little influence on denitrification or N_2O emission. This behavior took place regardless of addition of manure.

Acknowledgments

We thank the Alberta Agriculture Research Institute (AARI) and the University of Alberta Faculty of Graduate Studies and Research for financial assistance and Ms. M. Molina-Ayala and Mr. C. Figueiredo for technical assistance.

References

Bremner, J.M. 1965. Inorganic forms of nitrogen. In: C.A. Black et al. (eds.). Methods of Soil Analysis. Part 2. *Agron.* 9:1179-1237.
Cates, R.L., Jr. and D.R. Keeney. 1987. Nitrous oxide production throughout the year from fertilized and manured maize fields. *J. Environ. Qual.* 16:443-447.
Christensen, S. and Christensen, B.T. 1991. Organic matter available for denitrification in different soil fractions: Effect of freeze/thaw cycles and straw disposal. *J. Soil Sci.* 46: 637-647.
Christensen, S. and J.M. Tiedje. 1990. Brief and vigorous N_2O production by soil at spring thaw. *J. Soil Sci.* 41: 1-4.
Christianson, C.B. and C.M. Cho. 1983. Chemical denitrification of nitrite in frozen soils. *Soil Sci. Soc. Am. J.* 47: 38-42.
Denmead, O.T., J.R. Freney, and J.R. Simpson. 1979. Studies of nitrous oxide emission from a grass sward. *Soil Sci. Soc. Am. J.* 43:726-728.
Dorland, S. and E.G. Beauchamp. 1991. Denitrification and ammonification at low soil temperatures. *Can. J. Soil Sci.* 71: 293-303.
Drury, C.F., D.J. McKenney, and W.I. Findlay. 1992. Nitric oxide and nitrous oxide production from soil: water and oxygen effects. *Soil Sci. Soc. Am. J.* 56:766-770.
Firestone, M.K., R.B. Firestone, and J.M. Tiedje. 1980. Nitrous oxide from soil denitrification: factors controlling its biological production. *Science* 208:749-751.
Firestone, M.K., M.S. Smith, R.B. Firestone, and J.M. Tiedje. 1979. The influence of nitrate, nitrite, and oxygen on the composition of the gaseous products of denitrification in soil. *Soil Sci. Soc. Am. J.* 43:1140-1144.
Gamble, T.N., M.R. Betlack, and J.M. Tiedje. 1977. Numerically dominant denitrifying bacteria from world soils. *Appl. Environ. Microbiol.* 33:926-939.
Gilliam, J.W., S. Dasberg, L.J. Lund, and D.D. Focht. 1978. Denitrification in four California soils: Effect of soil profile characteristics. *Soil Sci. Soc. Am. J.* 42:61-66.
Goodroad, L.L. and D.R. Keeney. 1984. Nitrous oxide emissions from soils during thawing. *Can. J. Soil Sci.* 64:187-194.

Heaney, D.J., M. Nyborg, E.D. Solberg, S.S. Malhi, and J. Ashworth. 1992. Overwinter nitrate loss and denitrification potential of cultivated soils in Alberta. *Soil Biol. Biochem.* 24:877-884.

Hutchinson, G.L. and A.R. Mosier. 1981. Improved soil cover method for field measurement of nitrous oxide fluxes. *Soil Sci. Soc. Am. J.* 45:311-316.

Keeney, D.R., I.R. Fillery, and G.P. Marx. 1979. Effect of temperature on the gaseous nitrogen products of denitrification in a silt loam soil. *Soil Sci. Soc. Am. J.* 43:1124-1128.

Koskinen, W.C. and D.R. Keeney. 1982. Effect of pH on the rate of gaseous products of denitrification in a silt loam soil. *Soil Sci. Soc. Am. J.* 46:165-1167.

Lensi, R. and A. Chalamet. 1982. Denitrification in water logged soils: *in situ* temperature-dependent variations. *Soil Biol. Biochem.* 14:51-55.

Letey, J., N. Valoras, Aviva Hada, and D.D. Focht. 1980. Effect of air-filled porosity, nitrate concentration, and time on the ratio of N_2O/N_2 evolution during dentirification. *J. Environ. Qual.* 9:227-231.

Maag, M. Denitrification from soil receiving pig slurry or fertilizer. p. 235-246. In: Hansen, J.A. and Henriksen, K. (eds.). Nitrogen in Organic Wastes Applied to Soils. Academic Press, London.

Malhi, S.S., Nyborg, M. and E.D. Solberg. 1990. Potential for nitrate-N loss in central Alberta soils. *Fertilizer Research* 25:175-178.

Nyborg, M., C. Figueiredo, J. Laidlaw, M. Molina-Ayala, and J. Thurston. 1990. p. 203-212. In: Emission of nitrogen gases by soil. *27th Annual Alberta Soil Science Workshop Proceedings*, University of Alberta.

Powlson, D.S., P.G. Saffigna, and M. Cattair. 1988. Denitrification at suboptimal temperatures in soil from different climatic zones. *Soil Biol. Biochem.* 20:719-723.

Rolston, D.E., D.L. Hoffman and D.W. Toy. 1978. Field measurement of denitrification: I Flux of N_2 and N_2O. *Soil Sci. Soc. Am. J.* 42:863-869.

Smith, M.S. and L.L. Parsons. 1985. Persistence of denitrifying enzyme activity in dried soils. *Appl. Environ. Microbiol.* 49:316-320.

Vermes, J.F. and D.D. Myrold. 1992. Denitrification in forest soils of Oregon. *Can. J. For. Res.* 22:504-512.

CHAPTER 20

Methane Production in Mississippi Deltaic Plain Wetland Soils As a Function of Soil Redox Species

C.R. Crozier, R.D. DeLaune, and W.H. Patrick, Jr.

I. Introduction

In recent decades, atmospheric CH_4 levels have increased (Khalil and Rasmussen, 1990; Pearman, 1991), contributing 18% to 25% of the global warming phenomenon (Ramanathan et al., 1985; Denmead, 1991). About 25% to 50% of total global CH_4 emissions are from flooded rice and natural wetland soils (Bouwman, 1990). In contrast, CH_4 oxidizing organisms can cause certain soils to serve as CH_4 sinks (Nesbit and Breitenbeck, 1992; Whalen et al., 1992). Currently, attempts are being made to correlate CH_4 emission rates with soil types, thereby refining estimates of the global impact of soils on CH_4 levels (Matthews and Fung, 1987; Van Breeman and Feijtel, 1990).

When a soil is flooded, thermodynamic principles predict that a sequential reduction of O_2, NO_3^-, Mn^{4+}, Fe^{3+}, SO_4^{2-}, and finally CO_2 (to yield CH_4) should occur (Berner, 1980). CH_4 may also be produced via the reduction of acetate (Zeikus et al., 1975; Svensson, 1984). The metabolism of labile C compounds may yield CO_2, acetate, and H_2, thereby promoting CH_4 production; while oxidized inorganic species inhibit CH_4 production (Bollag and Czlonkowski, 1973; Winfrey and Zeikus, 1979; King and Wiebe, 1980; Jakobsen et al., 1981; Inubushi et al., 1984; Yagi and Minami, 1990; Yavitt and Lang, 1990). In addition, oxidized inorganic species have been reported to exhibit a toxic effect on CH_4 formation (Jakobsen et al., 1981).

If the toxic effects of soil oxidants on methanogenesis are minimal in comparison with their role as alternate electron acceptors, then CH_4 production rates in soils within favorable temperature and pH ranges should be a function of the difference between equivalents of labile C and equivalents of inorganic oxidant species (excess soil reductant capacity, ERC). The objectives of this study were to determine if concentrations of soil redox species and excess soil reductant capacity can be used to predict net potential CH_4 production rates in freshwater wetland soils.

II. Materials and Methods

A. Sites

Three freshwater wetland soils in the Mississippi deltaic plain region of Louisiana were selected based on differences in soil type and accessibility by road. The soils selected were: 1) Allemands muck (clayey, montmorillonitic, euic, thermic Terric Medisaprist), a marsh soil with a shallow organic layer overlying a mineral layer; 2) Barbary muck (very-fine, montmorillonitic, non-acid, thermic, Typic Hydraquent), a swamp soil of predominantly mineral matter; and 3) Maurepas muck (euic, thermic, Typic Medisaprist), a marsh soil with organic material throughout the profile sampled. The Allemands muck was collected from St. Charles Parish, near the town of Boutte; the Barbary muck was collected from St. James Parish, near the town of Gramercy; and the Maurepas muck was collected from St. John the Baptist Parish, within the Manchac Wildlife Management area between Lakes Maurepas and Pontchartrain.

B. Sample Collection

Three 14.5-cm diameter cores to a depth of 30 cm were collected along randomnly placed transects at each site on October 6, 1992. Cores were divided into 0- to 10-cm, 10- to 20-cm, and 20- to 30-cm depth intervals. Core segments were placed immediately into plastic bags and sealed to minimize contact with air. Soil redox potential profiles were determined at two locations within each site using Pt electrodes inserted to depths of 5, 15, 25, and 50 cm. A calomel reference electrode was used, and a 15 min. equilibration time was allowed prior to measurment (de la Cruz et al., 1989). A limited number of electrodes were allowed to equilibrate for longer time periods for comparison.

C. Laboratory Analyses

The total core segment weight was determined. A subsample was dried at 105 °C for 24 hr to determine the initial moisture content and bulk density. Core segments were homogenized, adding distilled water as needed to form a soil paste. A subsample of the soil paste was dried as above to determine moisture content, and then ashed at 450 °C for 16 hr to determine total organic matter content. In order to convert potentially reducible Fe and Mn forms to Fe^{2+} and Mn^{2+}, a second subsample of the soil paste was flooded with a 10% dextrose solution and incubated for one month under an N_2 atmosphere. The remaining soil paste was stored in glass jars in the dark at room temperature prior to extracting NO_3^-, SO_4^{2-}, Fe, and Mn; and determining production rates of CH_4 and CO_2. Extractions and incubations were begun within 2 weeks of sample collection. Jars were filled to minimize atmospheric contact. Soil CO_2 emissions and E_h levels of wetland soils have been shown to remain stable for up to 241 days when stored at laboratory temperatures (Nyman and DeLaune, 1991). The pH of the soil paste was determined using a combination electrode. Levels of SO_4^{2-} and NO_3^- were determined from H_2O extracts using a Dionex Model 2010i Ion Chromatograph (Dionex Corp., Sunnyvale, California) equipped with a Dionex IonPak AS4A column. The eluent solution contained 2.2 mM Na_2CO_3 and 2.8 mM $NaHCO_3$, and the regenerant solution contained 0.0125 M H_2SO_4.

Iron and manganese in both wet soil pastes and soils incubated anaerobically with dextrose were extracted using 1M sodium acetate (pH 4.5) under an N_2 atmosphere (Satawathananont et al., 1991). Total extract Fe and Mn were determined using an inductively coupled argon plasma emission spectrometer (Jarrell-Ash Division, Model 855, Fisher Scientific Co., Waltham, Massachusetts). The oxidation states of Fe and Mn were not determined, presumably the soluble and exchangeable species were predominantly Fe^{2+} and Mn^{2+} (Olson and Ellis, 1982; Gambrell and Patrick, 1982). The oxidized forms of interest (Fe^{3+} and Mn^{4+}) were not determined separately due to their low solubility. Potentially reducible Fe and Mn were estimated as the difference between concentrations extracted from the dextrose-incubated soils and from wet soil pastes.

Soil respiration under aerobic conditions was used to estimate labile soil C. Aerobic conditions were selected to reduce methanogenesis and avoid the mathematical dependence of correlating (CO_2 + CH_4) emission with CH_4 emission. Respiration rates from a 10:1 sand:dry soil mix were determined using a 0.5 M KOH trap. Incubations were carried out at 23 °C in 140 ml sealed glass jars. After one week, the alkali trap was removed, 1.5 M $BaCl_2$ was added to precipitate carbonates, and samples were titrated with 0.12 M HCl using phenolphthalein indicator (Anderson, 1982).

Methane production rates were determined under anaerobic conditions. Since no attempt was made to quantify CH_4 oxidation, the values represent net potential CH_4 production. Samples were placed in glass roll tubes (Bellco Glass Co., Vineland, New Jersey), sealed with butyl rubber septa caps, and purged with N_2. Samples were incubated in the dark at 23 °C and sampled at 3, 7, and 21d. Gas samples were removed with a syringe, and CH_4 concentrations were determined with a gas chromatograph equipped with a HayeSep D polymer (100/200 mesh) column and a flame ionization detector. Sample containers were purged with N_2 following each gas sampling.

D. Calculations

Concentrations of chemical species were converted to e^- equivalents based on the following half reactions (Faust & Aly, 1981).

$1/24\ C_6H_{12}O_6 + 1/4\ H_2O = 1/4\ CO_2 + H^+ + e^-$
$1/2\ O_2 + H^+ + e^- = 1/2\ H_2O$
$1/5\ NO_3^- + 6/5\ H^+ + e^- = 1/10\ N_2 + 3/5\ H_2O$
$1/2\ Mn^{4+} + e^- = 1/2\ Mn^{2+}$
$Fe^{3+} + e^- = Fe^{2+}$
$1/8\ SO_4^{2-} + 5/4\ H^+ + e^- = 1/8\ H_2S + 1/2\ H_2O$

Methanogenesis via acetate reduction and via CO_2 reduction require different numbers of e^- equivalents (Zeikus et al., 1975; Faust & Aly, 1981).

$1/2\ CH_3COO^- + H^+ + e^- = 1/2\ CH_4 + 1/2\ HCO_3^-$
$1/8\ CO_2 + H^+ + e^- = 1/8\ CH_4 + 1/4\ H_2O$

Thus, CH_4 concentrations were not converted to e^- equivalents. Equivalents of O_2 present initially were assumed to be negligible (Inubushi et al., 1984).

E. Statistical Analyses

Correlation coefficients were calculated using the CORR procedure and mathematical regressions were performed using the REG procedure available in SAS (SAS Institute Inc., 1989).

III. Results and Discussion

A. General Soil Properties

A brief description of the soils utilized in this study is given in Table 1. The Maurepas soil (Typic Medisaprist) had very high organic matter levels (74% to 92%) throughout the 0- to 30-cm depth interval and the Allemands soil (Terric Medisaprist) had very high organic matter levels in the upper 20 cm. The Barbary soil (Typic Hydraquent) had higher organic matter levels in the deepest layer than in the surface layers due to the presence of buried log fragments. Soil pH for all samples was within the optimum range (pH 6 to 7, data not shown) for CH_4 production (Jakobsen et al., 1981). Soil redox potentials were lowest in the Maurepas soil and declined with depth for all soils within the 0- to 30-cm range. Even though the surface layers had the highest redox potentials in the field, net potential CH_4 production rates declined with depth for all three soil types. Equilibration of a limited number of electrodes for time periods up to 1 week yielded a similar pattern with depth, except that all redox potentials were approximately 250 mv lower (data not shown).

B. Soil Redox Species vs. CH_4 Production

Concentrations of soil redox species varied greatly among soil materials (Table 2). In terms of e^- equivalents, SO_4^{2-} and potentially reducible Fe were more prevalent than were NO_3^- and potentially reducible Mn. Labile C, and consequently excess reductant capacity, declined dramatically with depth in the Allemands and Maurepas soils. In the Barbary soil, labile C and excess reductant capacity did not change appreciably with depth.

Both initial (0-3 d) net potential CH_4 production rates and and total (0-21 d) net potential CH_4 production were negatively correlated with total soil oxidant concentrations and positively correlated with soil organic matter, labile C, and excess reductant capacity (Table 3). During the entire 21-day incubation period, very little CH_4 was produced by soil materials with less than 50% organic matter and the overall correlation was very weak (Fig. 1). Although labile C and excess reductant capacity were very closely correlated with each other (Table 3), regressions of these two variables with total (0-21 d) CH_4 production yielded a simple linear relationship for labile C, while the quadratic term was significant for excess reductant capacity (Figure 1). Although labile C produced a slightly better fit, the predicted relationship between excess reductant capacity

Table 1. Selected properties of the soils analyzed. Soil E_h measurements represent the means of two measurements; for other properties, values are the means (CV) of three measurements; net potential CH_4 production for each site was calculated by summing production of the three depth intervals

Soil	Depth cm	E_h mv	Bulk density g cm^{-3}	Organic matter %	CH_4 production[a] g m^{-2} d^{-1}
Allemands	0-10	+154	0.07 (16)	70 (11)	1.60 (67)
	10-20	+104	0.08 (34)	62 (20)	0.01 (135)
	20-30	+69	0.25 (44)	27 (37)	0.00 (173)
	50	+129			Σ=1.61 (67)
Barbary	0-10	+119	0.31 (81)	37 (22)	0.27 (65)
	10-20	+19	0.22 (14)	34 (12)	0.00 (62)
	20-30	+4	0.12 (21)	60 (7)	0.01 (79)
	50	+49			Σ=0.28 (61)
Maurepas	0-10	+69	0.04 (46)	92 (1)	1.00 (54)
	10-20	-16	0.05 (25)	82 (10)	0.10 (156)
	20-30	-76	0.09	74	0.01 (147)
	50	-31			Σ=1.12 (58)

[a] Net potential CH_4 production rates during the first three days of anaerobic incubation: due to rounding errors, columns may not total to the sum shown.

and CH_4 production more closely approached the origin. Thus, the excess reductant capacity provided a quantitative index separating methanogenic (ERC > 0) and non-methanogenic soils (ERC < 0), as well as an index of relative CH_4 production rates by methanogenic soils. In contrast, the use of labile C as an index provides no logical distinction between non-methanogenic soils and soils with low rates of CH_4 production.

While both labile soil C and excess reductant capacity were closely correlated with CH_4 production rates, soil oxidant concentrations were only weakly correlated with CH_4 production rates (Table 3). The three-day time lag prior to the initial CH_4 sampling date may have masked inhibitory effects of soil oxidants on CH_4 production during the initial hours of anaerobic incubation.

Net potential CH_4 production decreased with depth for all three soils. Decreases in methanogenesis with depth were also noted in a study of wetland soils in Florida (Bachoon and Jones, 1992). Since the highest rates of net potential CH_4 production were found in the surface soil layers, which are likely to be oxidized under actual field conditions (Table 1), an integration of laboratory data with actual field conditions will be required to refine estimates of CH_4 emissions. When coupled with field soil redox and temperature data, and with estimates of CH_4 oxidation rates, predictions of CH_4 emissions by different soil types and landscape regions should be possible.

IV. Summary and Conclusions

Both the amount of labile C as determined from aerobic soil respiration, and the excess reductant capacity of these soil materials were closely correlated with the net potential CH_4 production rate. Measuring soil respiration under aerobic conditions provided an estimate of soil labile C which was independent of methanogensis under the anaerobic conditions of interest. The excess reductant capacity of soils quantified both soil oxidants and soil reductants, thereby accounting for significant biochemical and geochemical variables affecting CH_4 production.

Table 2. Concentrations of soil redox species and initial (0-3 d) CH_4 production rates; values are the means (CV) of three measurements

Soil	Depth cm	NO_3^-	SO_4^{2-}	Fe_{pr}[a]	Mn_{pr}[a]	Σoxidants[b]	Labile C	ERC[c]	CH_4 production
		meq kg^{-1}							mg kg^{-1} d^{-1}
Allemands	0-10	4 (6)	30 (61)	58 (7)	0 (173)	92 (22)	1316 (44)	1223 (48)	230.2 (60)
	10-20	6 (32)	157 (33)	61 (15)	0 (100)	223 (21)	266 (74)	43 (561)	1.6 (130)
	20-30	2 (38)	256 (146)	60 (17)	0 (35)	319 (120)	12 (173)	-307 (128)	0.0 (172)
Barbary	0-10	1 (44)	20 (79)	91 (11)	2 (69)	115 (23)	364 (23)	250 (45)	15.3 (93)
	10-20	1 (25)	5 (51)	92 (7)	4 (45)	102 (9)	368 (33)	266 (42)	0.2 (72)
	20-30	1 (29)	13 (127)	71 (27)	1 (30)	86 (39)	365 (19)	279 (18)	0.6 (64)
Maurepas	0-10	7 (43)	45 (20)	5 (34)	0 (173)	57 (24)	1921 (11)	1863 (11)	239.3 (20)
	10-20	5 (9)	744 (85)	4 (66)	0 (173)	754 (84)	439 (45)	-315 (255)	24.2 (162)
	20-30	4 (30)	432 (104)	5 (46)	0 (173)	442 (102)	387 (16)	-55 (910)	1.3 (137)

[a] Potentially reducible; [b] the sum of the equivalents of NO_3^-, SO_4^{2-}, and potentially reducible Fe and Mn; [c] Excess Reductant Capacity, the difference between equivalents of labile C and equivalents of oxidants.

Table 3. Pearson product-moment correlation coefficients between selected soil parameters

	Organic matter (%)	Total oxidants (ueq g^{-1})	Labile C (ueq g^{-1})	Excess reductant capacity (ueq g^{-1})	Initial CH_4 production (mg kg^{-1} d^{-1})
Total oxidants	0.13				
Labile C	0.63**	−0.29			
Excess reductant capacity	0.44**	−0.65**	0.92**		
Initial CH_4 production[a]	0.52**	−0.26	0.89**	0.82**	
Total CH_4 production[b] (mg kg^{-1})	0.52**	−0.26	0.89**	0.81**	0.98**

** Coefficient differed significantly from zero at the 0.01 probability level; [a] net potential CH^4 production during the first 3 days of anaerobic incubation; total net potential CH_4 production during the 21 day anaerobic incubation.

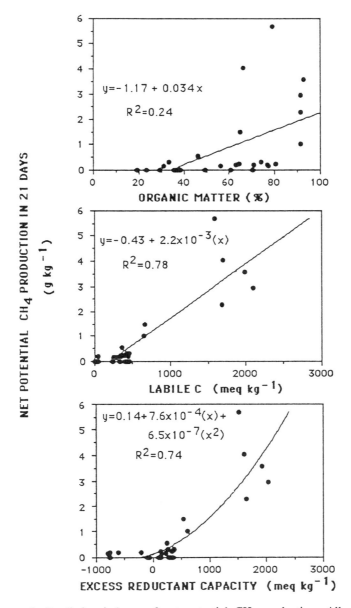

Figure 1. Predictive indexes of net potential CH_4 production. All regression coefficients, except for the y intercept in the graph depicting excess reductant capacity, differed significantly from zero at the 0.05 probability level.

V. Acknowledgements

This research was supported by the National Institute for Global Climate Change (funded by the Department of Energy) Cooperative Agreement # DE-FC03-90ER610101. W.H. Hudnall and M. Walthall helped classify soils. Field assistance was provided by D.P. Alford and T.S. Crozier. R.P. Gambrell, J. Pardue, and I. Devai assisted with planning and carrying out laboratory assays. Helpful suggestions during manuscript preparation were provided by W.J. Catallo, S.P. Faulkner, J.A. Nyman, Z. Wang, and anonymous reviewers.

VI. References

Anderson, J.P.E. 1982. Soil respiration. p. 831-871. In: A.L. Page et al. (eds.). *Methods of Soil Analysis, Part 2. Chemical and Microbiological Properties. Agronomy Monograph No. 9 (2nd Ed.).* ASA-SSSA, Madison, Wisconsin, USA.

Bachoon, D. and R.D. Jones. 1992. Potential rates of methanogenesis in sawgrass marshes with peat and marl soils in the everglades. *Soil Biol. Biochem.* 24:21-27.

Berner, R.A. 1980. *Early Diagenesis: A Theoretical Approach.* Princeton Univ. Press, Princeton, NJ. 241 pp.

Bollag, J-M. and S.T. Czlonkowski. 1973. Inhibition of methane formation in soil by various nitrogen-containing compounds. *Soil Biol. Biochem.* 5:673-678.

Bouwman, A.F. 1990. Exchange of greenhouse gases between terrestrial ecosystems and the atmosphere. p. 61-127. In: A.F. Bouwman (ed.), *Soils and the Greenhouse Effect.* John Wiley and Sons Ltd., Chichester, England.

de la Cruz, A.A., C.T. Hackney, and N. Bhardwaj. 1989. Temporal and spatial patterns of redox potential (Eh) in three tidal marsh communities. *Wetlands* 9:181-190.

Denmead, O.T. 1991. Sources and sinks of greenhouse gases in the soil-plant environment. *Vegetation* 91:73-86.

Faust, S.D. and O.M. Aly. 1981. *Chemistry of Natural Waters.* Ann Arbor Science Publishers Inc., Ann Arbor, Michigan.

Gambrell, R.P. and W.H. Patrick Jr. 1982. Manganese. p. 313-322. In: A.L. Page, et al. (eds.). *Methods of Soil Analysis, Part 2. Chemical and Microbiological Properties. Agronomy Monograph No. 9 (2nd Ed.).* ASA-SSSA, Madison, Wisconsin, USA.

Inubushi, K., H. Wada, and Y. Takai. 1984. Easily decomposable organic matter in paddy soil. IV. Relationship between reduction process and organic matter decomposition. *Soil Sci. Plant Nutr.* 30:189-198.

Jakobsen, P., W.H. Patrick Jr., and B.G. Williams. 1981. Sulfide and methane formation in soils and sediments. *Soil Sci.* 132: 279-287.

Khalil, M.A.K. and R.A. Rasmussen. 1990. Atmospheric methane: Recent global trends. *Environ. Sci. Technol.* 24:549-553.

King, G.M., and W.J. Wiebe. 1980. Regulation of sulfate concentration and methanogenesis in salt marsh soils. *Estuarine and Coastal Marine Science* 10:215-223.

Matthews, E. and I. Fung. 1987. Methane emissions from natural wetlands: Global distribution, area, and environmental characteristics of sources. *Global Biogeochem. Cycles* 1:61-86.

Nesbit, S.P. and G.A. Breitenbeck. 1992. A laboratory study of factors influencing methane uptake by soils. *Agric. Ecosys. Environ.* 41:39-54.

Nyman, J.A. and R.D. DeLaune. 1991. CO_2 emission and soil Eh responses to different hydrological conditions in fresh, brackish, and saline marsh soils. *Limnol. Oceanogr.* 36:1406-1414.

Olson, R.V. and R. Ellis Jr. 1982. Iron. p. 301-312. In: A.L. Page et al. (eds.). *Methods of Soil Analysis, Part 2: Chemical and Microbiological Properties. Agronomy Monograph No. 9 (2nd Ed.).* ASA-SSSA, Madison, Wisconsin, USA.

Pearman, G.I. 1991. Changes in atmospheric chemistry and the greenhouse effect: a southern hemisphere perspective. *Climatic Change* 18:131-146.

Ramanathan, V., R.J. Cicerone, H.B. Singh, and J.T. Kiehl. 1985. Trace gas trends and their potential role in climate change. *J. Geophys. Res.* 90:5547-5566.

SAS Institute Inc. 1989. *SAS/STAT® User's Guide, V6, 4th Ed., Vol. 2.* SAS Institute Inc., Cary, North Carolina.

Satawathananont, S., W.H. Patrick Jr., and P.A. Moore. 1991. Effect of controlled redox conditions on metal solubility in acid sulfate soils. *Plant Soil* 133:281-290.

Svensson, B.H. 1984. Different temperature optima for methane formation when enrichments from acid peat are supplemented with acetate or hydrogen. *Appl. Environ. Microbiol.* 48:389-394.

Van Breeman, N., and T.C.J. Feijtel. 1990. Soil processes and properties involved in the production of greenhouse gases with special reference to taxonomic systems. p. 195-220. In: A.F. Bouwman (ed.), *Soils and the Greenhouse Effect.* John Wiley and Sons, Chichester.

Whalen, S.C., W.S. Reeburgh, and V.A. Barber. 1992. Oxidation of methane in boreal forest soils: a comparison of seven measures. *Biogeochemistry* 16:181-211.

Winfrey, M.R. and J.G. Zeikus. 1979. Anaerobic metabolism of immediate methane precursors in Lake Mendota. *Appl. Environ. Microbiol.* 37:244-253.

Yagi, K. and K. Minami. 1990. Effect of organic matter application on methane emission from some Japanese Paddy fields. *Soil Sci. Plant Nutr.* 36:599-610.

Yavitt, J.B. and G.E.Lang. 1990. Methane production in contrasting wetland sites: response to organic-chemical components of peat and to sulfate reduction. *Geomicrobiol. J.* 8:27-46.

Zeikus, J.G., P.J. Weimer, D.R. Nelson, and L. Daniels. 1975. Bacterial methanogenesis: acetate as a methane precursor in pure culture. *Arch. Microbiol.* 104:129-134.

CHAPTER **21**

Role of Soil Survey in Obtaining a Global Carbon Budget

Richard W. Arnold

I. Introduction

Soil surveys divide the continuous surficial mantle called the pedosphere into a spectrum of recognizable segments. These segments contain less variability of most soil properties than the pedosphere as a whole. Competent soil scientists are able to delineate "areas of expectation" that are relevant to scientific understanding of soils as landscape segments. These delineated segments contribute to the practical application of such knowledge in the use and management of the pedosphere (Arnold, 1983).

The hierarchy of spatial variability that characterizes the universe is also manifest in the pedosphere. Due to the lack of a common solution to problems of scalar differences, soil maps are produced at many scales to illustrate the perceptions that exist. These scalar influenced abstractions of reality commonly take on relevance only when meaningful relationships for a specific purpose are perceived by a user. Our current inability to adequately describe and characterize scalar and temporal soil features by classical statistics, geostatistics, fractal geometry, or chaos theory techniques (Arnold and Wilding, 1991) forces us to resort to the qualitative descriptions of purposes, methods, philosophy and the resulting information embodied in the "art of soil survey".

The human mind diligently searches for and seeks out regularities, periodicities, cycles, trends, and patterns as a means of condensing and simplifying the chaotic complexities that surround us. The symmetry of classical mathematics gives way to the distraught realities of nature and stark contrasts that are ever present. The Gross National Product of a nation tells us nothing about the successes and failures of the budgets of individuals within that society; in a similar fashion a national soil map and its supporting data are a distortion of what occurs at sites within an area or at different points in time. Nevertheless, partitioning of the pedosphere has enabled sampling schemes of these strata to obtain legitimacy for spatial information about soils.

The role of the soil survey in obtaining a global carbon budget is (1) to help people make use of the stratifications unique to Pedology, (2) to provide data that may be relevant to decisions of interest, especially the stock of soil carbon, (3) to help interpret the available information that itself differs in time, space, and concept, and (4) to reduce the amount of point sampling while maintaining reliability of the information.

II. Pedological Stratifications

The paradigm of field pedology is based on relationships of "factors to processes to soil properties" which are unique to areas within landscapes. That is, the soil forming factors of climate, organisms, topography, parent material and time are distributed geographically and their interactions result in biogeochemical processes that transform parent materials into recognizable soils. Space and time variations of the soil-forming factors and soil-forming processes result in a myriad of heterogeneous mazes of juxtaposed soils.

The non-specificity of this general model of the pedosphere enables discovery and application to be employed more or less universally (Hoosbeek and Bryant, 1992).

The mapping of soil landscapes involves the use of models which link observed soil morphology and simple field measurements to specific segments of landscapes. The hypotheses of the models are tested and calibrated during the mapping procedure and represent field-scale applications of the scientific method. Soil mapping includes all of the uncertainties of applying models to reality such as incomplete data observations and biased sampling, unknown degrees of precision and accuracy of spatial extrapolation. This uncertainty is expanded by the use of some models with limited validation and verification.

Hudson (1992) refers to the science-based knowledge in soil survey as "tacit knowledge" which is learned by experience, mostly empirical, and passed on through rather elaborate schemes of self-perpetuating techniques. The potential loss of knowledge and understanding is great, in his opinion, because of the scarcity of recorded information about the individual understanding of soil-landscape functional relationships. The soil scientist mapping in an area tends to believe that the information he or she uses to make decisions in identifying soils and delineating them on a base map is rather common, simple information which is of little interest to others. This often is the nature of functionally-related knowledge gained by experience.

However, to fully appreciate soils as the surficial components of a landscape one must comprehend stratigraphy of sediments and horizons, geomorphology of events and their results, and the hydrology that integrates the segments of catchment basins (Daniels and Hammer, 1992).

In the field, pedologists learn to recognize similarities and differences of parent materials and of those properties of soil horizons that separate one kind of soil from another. Kinds and arrangement of soil horizons are used to identify soils and group them into classes. An emphasis on classification has dominated field experience rather than the quantification of soil forming processes. Classification has relied mainly on observable morphology of stable features resulting from processes of long term landscape evolution and less on the temporal aspects of biogeochemical processes themselves.

It can be concluded that the use of class-specific data has permitted fairly rapid, consistent mapping of the earth's surficial mantle of soils. The qualitative nature of the functional relationships has enabled the recognition and separation of differences that are geographically present in the pedosphere. Insofar as this basic information can be grouped and regrouped it is suitable for many other interpretations that rely on the same sets of properties.

III. Providing Relevant Data

Soil surveys at scales of about 1:20,000 are called "order 2" or detailed surveys in the United States and provide information about the geographic location and distribution of soil series. It is usual in soil survey operations to obtain and use ordinal data (Bregt et al., 1992) which consist mainly of observations placed into classes. The classes of attributes or properties can be ranked from more to less. Soil series, field estimated texture, soil drainage class and slope classes are examples of ordinal data in a soil survey. The values so obtained can be placed in an ordered list but the distance between the values is without quantitative meaning. With ordinal data the central tendency is the mode or median and dispersion is the range of the classes. It is a common mistake to treat most soil survey data as either interval or ratio data for which the mean and standard deviation are useful summary statistics (Bregt et al., 1992).

In the United States there are three main sets of geographic data for soil surveys (Bliss and Reybold, 1989). One is the county or survey area, SSURGO, which includes field surveys published at scales from 1:12,000 to 1:32,000. These are the traditional detailed soil surveys. Another set is called STATSGO which is generalized soil information recorded at a scale of 1:250,000 and soon to be available in digitized form for all states except Alaska which will be at a scale of 1:2.5 M. The composition of map units in the STATSGO database was derived from simulated transects of the SSURGO maps. A third level is a broad national perspective presented as NATSGO which outlines the major land resource areas (MLRAs). They are physiographic areas of soil associations with a database derived from a statistical sampling of about 900,000 points for the National Resource Inventory (NRI). This digitized information is commonly shown at scales of 1:7.5 M or smaller.

All of these maps show areas designated by symbols that refer to class names of soil and land components within the delineated areas. Soil properties, per se, are not listed other than slope and a few other phase criteria such as erosion and stoniness. At the SSURGO level the map units are named with phases of soil series. Soil series names are commonly place names near the location where the soils were first described and recognized (Soil Survey Staff, 1975). Soil series are classes of the lowest category in the Soil Taxonomy system of soil classification and their connection with classes in higher categories is provided in look-up tables.

Soil Taxonomy is a dynamic document whose revisions are provided in Keys to Soil Taxonomy (Soil Survey Staff, 1992) every two years. The Keys provide definitions and priority sequences for classifying soils from the soil Order category through the soil Family, but does not provide the names of soil series because they are local and usually maintained by the soil survey organization in each country that uses Soil Taxonomy. The Statistical Laboratory at Iowa State University maintains the official list of soil series for the United States.

IV. The National Cooperative Soil Survey of the United States

Federal, state and local agencies including universities work together as the National Cooperative Soil Survey (NCSS) through Memoranda of Understanding whereby they agree to follow standards and guidelines for the collection and characterization of soil data. Soil information is interpreted according to each agency's needs. For example, the Forest Service and The Bureau of Land Management use soil surveys to guide the direct management and care of much of the publicly-owned lands whereas management alternatives and technical assistance are offered to operators of privately-owned lands by the Soil Conservation Service and State agencies.

Backstopping these field survey efforts are scientists of federal and state research units who conduct field and laboratory studies associated with the management and use of soils. Many additional studies of the processes of soil and landscape development are made by scientists in other disciplines and this knowledge contributes greatly to a better understanding of how the pedosphere functions and the relationships among components of the pedosphere.

Most interpretations of the major components of soil map units are based on estimates of the common ranges of properties of the soils and of their expected behavior. Physical and chemical properties are estimated for each soil series which serve as a basis for numerous interpretations about use and management. Such information is stored in a database called the map unit interpretation record (MUIR) which currently is maintained by Iowa State University. Estimates of soil texture, organic matter, pH, and in many cases, bulk density are provided, thus it is possible to estimate carbon mass in soils.

Laboratory determined physical and chemical properties are available for many soil series of the United States. The National Soil Survey Laboratory data are housed on the State Capitol computer in Lincoln, Nebraska. Much of the information will soon be available on CD-ROM being prepared by Texas A & M University. Several thousand pedons in that data set have laboratory measured values for organic carbon and bulk density as a function of depth. The data are static in that they represent a snapshot at the time of sampling. Such characterization has traditionally been undertaken to assure the proper classification of the soil within a taxonomic system. Consequently, some pedons may not be representative of the soil series, land use or crops in the ecosystem where sampled.

Regardless of the complexity or difficulty of mapping soils, the objective of the National Cooperative Soil Survey has been to provide understandable, reasonably accurate, reasonably precise, and ultimately useful information to people who need to understand and use soils as they occur in the landscape (Brown, 1985). Statements about the precision and accuracy of mapping should provide reference as to whether or not (1) the soil map legend truly reflects the soils in the survey area, (2) the individual delineations are assigned correct names, and (3) the soil lines are drawn correctly.

V. Studies of Spatial Variability

Because of the qualitative aspects of soil survey information, researchers have examined techniques to quantify some of the spatial relationships. The more common ones are co-variance methods associated with kriging (Burgess and Webster, 1980). These techniques examine changes of variance with increasing distance between pairs of observations. Many kinds of soil data follow spherical regressions having a nugget effect which is a residual variance commonly associated with sampling and analytical errors. The nugget effect can also demonstrate inherent short-range soil variability. An increasing amount of variance with increasing distance between observational pairs indicates a spatial dependence. The distance at which the variance becomes asymptotic is the distance beyond which extrapolation is no longer related and therefore not meaningful. Each environment is different, thus it is very difficult to make simple predictions about spatial dependencies. For example, Armato (1992) measured a number of soil properties including soil organic matter in an area near Kiel, Germany and near Gunung Madu, Sumatra, Indonesia. The spatial variability range was between 70 and 120 meters in northern Germany and between 250-350 meters in South Sumatra. The nugget effects (inherent short range variations) ranged from 6-28% at the German site and from 42-76% at the Indonesian site, consequently the remaining variability from 30 to more than 90% was associated with space, primarily the elevation differences in the landscapes studied.

In a geostatistical study of soil maps in Costa Rica, the smallest distance between observations for three levels of soil maps was 250 m (Bregt et al., 1992). Bregt et al. found nugget effects of 38% for drainage class, 41% for slope class, 53% for texture class in 0-15 cm depth, and 65% for crop suitability class using spatial-difference-probability functions. This work illustrates that a lot of soil variability, even with ordinal data sets, is relatively short range in nature. This supports the notion that the geographic distribution of the soil forming factors and their interactions give rise to unique combinations of soil properties for rather small geographic areas.

Where interval and ratio data (discrete rather than by classes) are obtained within soil landscape units, means and standard deviations are appropriate measures. Based on grid samples Armato (1992) found that organic carbon means increased from interfluves to lowland depressions with coefficients of variation (CV's) from 29-74%. Thus even in detailed sampling schemes variability may limit the degree of confidence of extrapolations.

In a Massachusetts study, soil properties determined within delineations of the same map unit were appropriately analyzed with normal statistics (Mahinakbarzadeh et al., 1991). However, grids and transects crossing delineations of different map units revealed enough auto-correlation in the residuals of regression that kriging was done. When the data were transformed with logs and one axis stretched, the resulting map was similar to the "second order" soil map of the area. It is likely that soil maps inherently cluster auto-correlation of soil forming factors and processes.

In a detailed sampling study in a northern hardwood forest, Huntington et al. (Batjes and Bridges, 1992) determined that a minimum detectable change in organic carbon in the 0-10 cm soil depth was 2.4 Mg C/ha. They concluded that short range variability caused by large scale management disturbance, such as logging, could produce statistically measurable changes observable over a 5 to 20 year period making long term trends more difficult to decipher. Because standards for soil maps generally do not include information about the "recent past" land use, there is a degree of uncertainty when using a data base to assess the size of current soil carbon pools although the magnitude of effect is likely small.

VI. Pedotransfer Functions and Modeling

Some soil properties are measured or estimated in the field, some measured in the lab, and others can be derived from mathematical relationships between associated characteristics (Wagenet et al., 1991). Soil qualities are those properties that are used to assist in interpreting condition and response. Some soil qualities can be determined from basic characteristics by pedotransfer functions (algorithms) or by simulation modeling. They believe that pedotransfer functions provide only part of the solution of estimating soil qualities because they are essentially filters through which basic soil survey information passes to eventually be used in simulation models. Map unit descriptions commonly do not supply the detail necessary to understand soil behavior in a use-dependent temporal sense that is so important for simulation modeling (Wagenet et al., 1991). These authors conclude that a concerted effort should be made to further develop pedotransfer functions that link available soil data to information needed for modeling, for example, gas flux models.

There has been an increasing interest to quantify soil processes to help answer environmental questions. Many ecological models contain modules or subroutines of interest to pedologists, especially the movement of water and its dissolved constituents in the soil (Hoosbeek and Bryant, 1992). The purpose of a model commonly determines two aspects of that model, namely its scale and the fundamental unit of observation (Thompson, 1992). If we use standard order 2 soil surveys (SSURGO) as the structural suborders for genetic models that predict how particular soils might change under rapidly changing climatic conditions, we must pay considerable attention to issues of map unit variability. As previously described, scales of observation and soil variability patterns go hand in hand.

If the purpose is to estimate the gross content of carbon stored in soils globally, then the scale and the fundamental unit of observation are different than when the purpose is to determine the flux rates of carbon in the pedosphere because dynamic processes are very dependent on current conditions whereas the former rely on cumulative effects.

One approach for global inventories is to stratify the world into rather large cells, assign a dominant condition or soil class to each, estimate soil carbon for those conditions, and aggregate to the whole. In attempts to improve accuracy and precision it is common to use smaller cells or lower category classes of a classification system, assign point data or estimated representative values to components, and to aggregate this level of detail to estimate the total. Soil map units at all scales represent autocorrelation of soil forming factors and processes as perceived by pedologists. The use of soil map units is the key to maintaining the integrity of cartographic and categoric accuracy for such assessments.

Some excellent examples of using soil surveys for geographically controlled estimates of soil carbon are Franzmeier et al. (1985), Grossman et al. (1992) and Kimble et al. (1991). A number of other examples will be described at this conference by the participants.

VII. Potential Errors in Estimating Soil Carbon

Although soil surveys are the most likely basis for extrapolating point data to estimate global soil carbon values there are some cautions that should be considered (Table 1). When soils were sampled in the field, seldom were the sites selected for the purpose of soil carbon determination and extrapolation over large regions. Most soils have been sampled to characterize the major features of specific kinds of soils for soil classification which rely on the more permanent subsoil properties.

Organic matter content and distribution are important attributes of soils, however, for purposes of extrapolation it is also relevant to know if the land use associated with the sample site is representative for a much larger area. All too often a particular sample site's relation with the stratigraphy, geomorphology and hydrology of the surrounding area is not known or may be poorly documented. This reduces the reliability of interpreting and extrapolating the point data as surrogates for process to a watershed as a basis for further generalization.

The reported values for soil organic carbon are influenced by the method of analysis, by the estimation for the whole soil on a volume basis or for the fine earth fraction of the soil. Unless comparisons are made between and among methods used in different laboratories there is an uncertainty of how best to equate or correlate the numbers. Without bulk density determinations and those for coarse fragment contents, the organic carbon values cannot readily be converted into weight per unit volume for either the whole soil or for the fine earth fraction.

Extrapolating point data to areas, whether soil map polygons or grid cells imposed over such maps, involves unknown errors for the most part. Insofar as the dominant and similar soils constitute most of the area and the land uses or management practices are not strongly contrasting, the estimated values for soil organic carbon may be reasonable. A major source of error exists when small inclusions of wet soils or organic soils exist but are not adequately accounted for when assigning a value to the delineated area.

With most detailed soil maps (SSURGO) the minor wet inclusions are mentioned and can be estimated. As maps are generalized (STATSGO, NATSGO, global) the smaller inclusions become less and less and may be omitted in descriptions or definitions of the delineated area. Where these now excluded soils are wet or organic, fairly large errors in overall soil carbon may occur. Assistance from local soil scientists who are familiar with the landscapes is prudent.

VIII. Conclusions

Soil surveys stratify the earth's surface into physiographic units that correlate well with many ecosystems.

Soil surveys are prepared at different scales and soil variability is related to the scale. The larger the scale the more details can be displayed; the smaller the scale the more components become inclusions in map units.

Attribute data are usually for points and are extrapolated for map units at different scales.

Although soil properties determined for areas are usually static, the information is appropriate for carbon inventories and for locating areas having similar environmental conditions.

Table 1. Possible source and magnitude of error in estimating soil carbon from soil survey information

Sources of potential errors	Low	Medium	High
A. Site selection			
1. Criteria not soil C specific		X	
2. Unspecified relationship with local landscape		X	
B. Field sampling			
1. Small sample volume	X		
2. Horizon mixing vs. fixed depth		X	
3. Biomass contamination	X		
4. No bulk density values			X
5. Uncorrected for stone content			X
6. Improper sampling in turbated soils (e.g. permafrost)			X
C. Carbon determination			
1. Difference between methods		X	
2. Reported values; wt, wt/cc, etc.		X	
3. Estimated ranges (judgement call)		X	
D. SSURGO extrapolation - polygons			
1. Ave. data vs. dominant and components	X		
2. Ranges - dominant and components		X	
3. Use of grid cells	X		
E. STATSGO extrapolation - polygons			
1. Ave. data vs. dominant and components			X
2. Ranges - dominant and components		X	
3. Use of grid cells		X	
F. NATSGO extrapolation - polygons			
1. Ave. data vs. dominant and components		X	
2. Ranges - dominant and components		X	
3. Use of grid cells		X	

Algorithms used to predict soil property relationships should be used with caution as local physiographic conditions may alter some properties.

Soil classification provides consistent identification and naming of map units. This facilitates extrapolations of points to areas and communications about their nature.

Soil maps provide spatial extent of map units identified by classification units and represent an autocorrelation of soil forming factors and processes.

Soil scientists of the National Cooperative Soil Survey are willing to assist whenever possible.

References

Armato, M.E. 1992. Soil variability as an indicator of erosion in sloping landscapes. Schriftenreihe No. 18, Christian Albrechts Univ., Kiel, Germany.

Arnold, R.W. 1983. Concepts of soils and pedology. P. 1-21. In: L.P. Wilding, N.E. Smeck, and G.F. Hall (eds.). *Pedogenesis and Soil Taxonmy I. Concepts and Interactions*. Developments in Soils Science 11 A., Amsterdam.

Arnold, R.W. and L.P. Wilding. 1991. The need to quantify spatial variability. p. 1-8. In: M.J. Mausbach and L.P. Wilding (eds.). *Spatial variabilities of soils and landforms*. Soil Sci. Soc. Amer. Spec. Publ. No. 28, SSSA, Madison, WI.

Batjes, N.J. and E.M. Bridges. 1992. World inventory of soil emissions: identification and geographic quantification of soil factors and soil processes that control fluxes of CO_2, CH_4, and N_2O, and the heat and moisture balance. Draft background paper, ISRIC, Wageningen.

Bliss, N.B. and W.U. Reybold. 1989. Small-scale digital soil maps for interpreting natural rsources. *J. Soil Water Conserv.* 44:30-34.

Bregt, A.K., J.J. Stoorvogel, J.Bouma, and A. Stein. 1992. Mapping ordinal data in soil survey: a Costa Rican example. *Soil Sci. Soc. Am. J.* 56:525-531.

Brown, R.B. 1985. The need for continuing update of soil surveys. *Soil and Crop Sci. Soc. Florida. Proc.* 44:90-93.

Burgess, T.M. and R. Webster. 1980. Optimal interpolation and isarithmic mapping of soil properties. I. The semi-variogram and punctual kriging. *J. Soil Sci.* 31:315-331.

Daniels, R.B. and R.D. Hammer. 1992. p. 1-10. In: *Soil Geomorphology*. John Wiley & Sons, Inc. New York.

Franzmeier, D.P., G.D. Lemme, and R.J. Miles. 1985. Organic carbon in soils of North Central United States. *Soil Sci. Soc. Am. J.* 49:702-708.

Grossman, R.B., E.C. Benham, J.R. Fortner, S.W. Waltman, J.M. Kimble, and C.E. Barnham. 1992. A demonstration of the use of soil survey information to obtain areal estimates of organic carbon. ASPRS/ACSM/RT. 92. Technical Papers. Vol. 4, Remote sensing and data acquisition. p. 457-465. Am. Soc. Photogrammetry and Remote Sensing and Am. Cong. Surveying and Mapping. Bethesda, MD.

Hoosbeek, M.R. and R.B. Bryant. 1992. Toward more quantitative mechanistic modeling of pedology; a review. p. 31-46. In.: W.J. Waltman, E.R. Levine, and J.M. Kimble (eds.). Proceedings of the First Soil Genesis Modeling Conference. USDA, SCS, National Soil Survey Center, Lincoln, NE.

Hudson, B.D. 1992. The soil survey as paradigm-based science. *Soil Sci. Soc. Am. J.* 56:836-841.

Kimble, J.M., H.M. Eswaran, and T. Cook. 1991. Organic carbon on a volume basis in tropical and temperate soils. Trans. Intl. Congr. Soil Sci. Vol. V:248-253. Kyoto, Japan.

Mahinakbarzadeh, M., S. Simkins, and P.L.M. Veneman. 1991. Spatial variability of organic matter content in selected Massachusetts map units. p. 231-242. In: M.J. Mausbach and L.P. Wilding (eds.). *Spatial variabilities of soils and landforms*. Soil Sci. Soc. Amer. Spec. Publ. No. 28, SSSA, Madison, WI.

Soil Survey Staff. 1975. Soil Taxonomy. USDA Agric. Handbook 436. U.S. Govt. Print. Off., Washington, D.C.

Soil Survey Staff. 1992. *Keys to Soil Taxonomy*. SMSS Tech. Monog. No. 19., Fifth edition. Pocahontas Press, Inc., Blacksburg, VA.

Thompson, M.L. 1992. Principles for the development of quantitative soil genesis models. p. 61-71. In.: W.J. Waltman, E.R. Levine, and J.M. Kimble (eds.). Proceedings of the First Soil Genesis Modeling Conference. USDA, SCS, National Soil Survey Center, Lincoln, NE.

Wagenet, R.J., J. Bouma, and R.B. Grossman. 1991. Minimum data sets for use of soil survey information in soil interpretive models. p. 161-182. In: M.J. Mausbach and L.P. Wilding (eds.). *Spatial variabilities of soils and landforms*. Soil Sci. Soc. Amer. Spec. Publ. No. 28, SSSA, Madison, WI.

CHAPTER 22

Methods to Assess Soil Carbon Using Remote Sensing Techniques

Carolyn J. Merry and Elissa R. Levine

I. Introduction

Remote sensing provides a powerful tool for assessing the dynamics of the biosphere. With the present technology, however, chemical and physical properties of soils (e.g., organic soil carbon, nutrient content, drainage class) cannot be directly observed using satellite data. Thus, researchers attempting to draw inferences concerning soil characteristics must rely on surrogate indices, such as vegetation status or albedo, biomass estimates, or surface soil moisture or temperature, combined with modeling techniques to assess soil properties. Changes might be as subtle as small spectral changes of an individual leaf, or as obvious as the inclusion or exclusion of entire plant communities due to factors controlled by soils. The long-term goal of such research is to identify and characterize satellite measurements that may be used to infer soil characteristics beneath a vegetation canopy.

Existing data from sensors on previously launched satellites, or data from future sensors such as those onboard NASA's planned Earth Observing System (EOS), can be useful for assessing the organic soil carbon status. The following sections contain a description of these remote sensing instruments and the information that is presently available, or will be available, to assess organic soil carbon status. This is followed by an example of how remote sensing techniques were used to model organic soil carbon concentrations at an experimental site in Howland, Maine.

II. Present Remote Sensing Programs

Vegetation index mapping using the NOAA Advanced Very High Resolution Radiometer (AVHRR) data has been used extensively to monitor the status of vegetation on a global scale from the United States Corn Belt to Africa (Townshend and Justice, 1986; Achard and Blasco, 1990; Teng, 1990). NOAA global Normalized Difference Vegetation Index (NDVI) daily and weekly summaries have been produced for North America since 1982 by the NOAA Satellite Data Services Division (Hastings et al., 1989a, 1989b). Vegetation mapping from AVHRR, and more detailed mapping using Landsat Multispectral Scanner Subsystem (MSS) and Thematic Mapper (TM), and SPOT (Satellite Pour l'Observation de la Terre) High Resolution Visible (HRV) data sets are used routinely to map forested and nonforested areas (Norwine and Greegor, 1983; Tucker et al., 1985). Rates of conversion between forest and nonforest categories represent a major source of CO_2 to the atmosphere from the terrestrial biota in tropical forest regions (Nelson et al., 1987; Dale, 1990).

Some investigators indicate that a linear transformation of multispectral TM data, called the TM Tasseled Cap transformation (Crist and Cicone, 1984), can be sensitive to plant moisture stress (McDaniel and Haas, 1982; Cohen, 1991). Three separate indices or images (brightness, greenness, and wetness) are produced in this transformation. The wetness index has been found to correspond with changes in soil moisture (Crist and Cicone, 1984; Musick and Pelletier, 1988), but conclusive evidence for its relation to plant water stress has not been demonstrated (Cohen, 1991). Also, a middle infrared index (defined by Cohen, 1991 as the ratio: TM band 5/TM band 7) showed a high correlation with soil moisture (Musick and Pelletier, 1988).

Several researchers have indicted that the microwave region shows a great potential to estimate soil moisture, for a relative thin layer of soil (0 to 5 cm) or 1/10 of the wavelength (Schmugge et al., 1986; Dobson and Ulaby, 1986). Schmugge et al. (1980) and Jackson and Schmugge (1984) provide an excellent review of using passive microwave techniques to estimate soil moisture.

Infrared radiation emitted by the soil surface is a function of kinetic temperature and spectral emissivity. Temperature/emissivity algorithms have been proposed to separate these two effects. Salisbury and D'Aria (1992a, 1992b) give reflectance curves for soil spectral behavior for different soil types. They describe how their research can be used for evaluating the remote sensing data from the future EOS ASTER (Advanced Spaceborne Thermal Emission and Reflection Radiometer) sensor, described in more detail below.

III. EOS Program

Future sensors on the Earth Observing System (EOS) platforms, which are to be launched in the late 1990's, will increase our capability to monitor surface soil carbon indirectly. EOS, the centerpiece of NASA's Mission to Planet Earth, has been designed to develop an integrated scientific observing system for understanding global climate change (NASA, 1993). EOS is a data and information system that includes new sensors in a polar and low-inclination orbit to serve the needs of an integrated, multidisciplinary study of the planet. EOS is to provide systematic, continuous observations for a minimum of 15 years to study global change and other related research issues.

There will be six EOS spacecraft series to address the various environmental disciplines and include the following: EOS-AM -- characterize land and ocean surfaces, and the atmosphere using a morning crossing time; EOS-COLOR -- ocean color and productivity; EOS-AERO -- atmospheric aerosols and ozone; EOS-PM -- characterize land and ocean surfaces, and the atmosphere using an afternoon crossing time; EOS-ALT -- ocean circulation, ice sheet mass balance, and land-surface topography; and EOS-CHEM -- atmospheric chemical species and their transformation, and ocean surface stress.

The EOS data will be downlinked via the Tracking and Data Relay Satellite System (TDRSS) to the EOS Data and Information System (EOSDIS). The Direct Broadcast and Downlink systems can support transmission to ground stations for users requiring direct data reception. Presently, there are three classes of users defined by NASA: EOS team participants and scientists needing real-time data to support experiments, international meteorological and environmental agencies requiring real-time measurements of the Earth, and international partners receiving high volume data from their EOS instruments.

A description of four EOS instruments that could be used to address the soil carbon issue follows. These include the Advanced Spaceborne Thermal Emission and Reflection Radiometer (ASTER) and the Moderate-Resolution Imaging Spectroradiometer (MODIS) for mapping vegetation indices, the Multifrequency Imaging Microwave Radiometer (MIMR) for monitoring soil moisture, and the Multi-angle Imaging SpectroRadiometer (MISR) for preparing maps of vegetation albedo. At this time the EOS-AM platform, scheduled for launch in June, 1998, will include the ASTER, MISR and MODIS instruments. MIMR, and also MODIS, will be on the first EOS-PM satellite scheduled for launch in the year 2000. Descriptions of the four instrument parameters are taken from NASA (1993).

The EOS science community is presently developing a list of data products that will be available at the time of the EOS launch (Thompson, 1990). However, the final list of primary data products for each instrument is still being negotiated (Thompson, L., 1993, pers. comm.). After launch, research products will continue tp be developed by scientists analyzing the incoming EOS data stream. This research will result in an increase of available products to a level of hundreds of data products from the EOS sensors.

IV. Mapping of Vegetation Indicies

A. ASTER

The ASTER instrument is being developed by the Japanese. ASTER is an imaging radiometer that builds upon the Multispectral Electronic Self-Scanning Radiometer (MESSR), Landsat, and SPOT.

ASTER will provide high spatial resolution (15 to 90 m) images of the land surface and clouds for climatologic, hydrologic, biologic and geologic studies. There are three modes of spatial resolution for ASTER: three bands located in the visible and near-infrared region between 0.5 and 0.9 μm at a 15-m resolution, six middle infrared bands between 1.6 and 2.5 μm with 30-m resolution, and five thermal bands between 8 and 12 μm at a 90-m resolution. One of the 15 m bands (0.7-0.9 μm) will allow along-track stereo viewing with a base-to-height ratio of 0.6 that can be used to generate digital elevation models at a 15-m horizontal and 25-m vertical resolution. Products generated from ASTER will augment the present Landsat data bases by allowing coverage of areas with improved spatial resolution for the visible, near infrared and thermal wavelength regions.

The instrument will have 4% absolute radiometric accuracy in the visible, near-infrared and middle infrared regions and 1 to 3K absolute temperature accuracy (200 to 370K range). The instrument can acquire data over a 60-km swath, whose center is pointable cross-track $\pm 8.5°$ (± 106 km) in the middle and thermal infrared bands, and $\pm 24°$ (± 314 km) for the visible and near infrared bands. These pointing capabilities of ASTER will be such that any location on the Earth is accessible at least once every 16 days in all 14 bands and once every five days for the three visible and near-infrared bands.

The five thermal bands can be used to separate brightness temperature measurements into spectral emissivity and kinetic temperature. By setting emissivity in one band to a constant value and using ground truth measurements, one can solve for temperature in that band and for emissivities in the other four thermal bands. The surface kinetic temperature measurements can be related to land cover in terms of wet vs. dry conditions or vegetated vs. bare soil areas.

B. MODIS

The MODIS is designed to produce a number of global survey data products that provide a measure of biological and physical processes for the study of terrestrial, oceanic and atmospheric phenomena on a scale of 1 km^2. MODIS is a medium-resolution, cross-track, multispectral scanning radiometer that builds upon the AVHRR, High-Resolution Infrared Sounder (HIRS), Landsat TM and the Nimbus-7 Coastal Zone Color Scanner (CZCS) sensors.

MODIS is a conventional imaging radiometer, which consists of a cross-track scan mirror and collecting optics, and a set of individual detector elements. MODIS contains 36 discrete bands -- 20 bands within the spectral range 0.4 to 3 μm, and 16 within 3 to 15 μm. The spectral bands will have a spatial resolution of 250 m, 500 m and 1 km at nadir. The field of view is $\pm 55°$ across-track with a resulting image swath of 2,300 km. The instrument will have $\pm 5\%$ absolute irradiance accuracy from 0.4 to 3 μm and 1% or better in the thermal infrared region (3 to 15 μm). Continuous global coverage occurs every one to two days.

Applicable global survey data products will include vegetation (land surface) cover, conditions, and productivity. Vegetation indices will be corrected for atmosphere, soil, and directional effects.

V. Soil Moisture Monitoring Using MIMR

The MIMR is designed to obtain global observations of a variety of parameters important to the hydrologic cycle, including atmospheric water content, precipitation, soil moisture, ice and snow cover, and sea surface temperature. MIMR is a passive microwave radiometer that builds upon the Scanning Multispectral Microwave Radiometer (SMMR) and the Special Sensor Microwave/Imager (SSM/I), but has greater frequency diversity, improved spatial resolution, increased swath width and improved antenna performance, and allows for complete global coverage in less than three days.

MIMR operates at six frequencies, each with horizontal and vertical polarization, for the following wavelengths (and associated spatial resolution): 6.8 GHz (60.3 km spatial resolution), 10.65 GHz (38.6 km), 18.7 GHz (22.3 km), 23.8 GHz (22.3 km), 36.5 GHz (11.62 km), and 90 GHz (4.86 km). There are nine multiple feedhorns providing 20 channels of data. Measurements in the 20 channels have a 1 to 2 K accuracy, and 0.2 to 0.7 K radiometric stability. The channels will be converted to daily spectral and monthly average maps on a 1° grid for the variable of soil moisture index. The instrument is designed to have a cross-track swath of 1,400 km at an incidence angle of 50°. The overlap between consecutive swaths increases at latitudes greater than 45° and daily coverage is provided.

The MIMR data can be used in conjunction with data from other EOS instruments. For example, the MIMR data complement visible, infrared and active microwave observations of vegetation status, biomass and soil moisture for use in evaporation and transpiration studies.

VI. Global Maps of Vegetation Albedo Using MISR

The MISR instrument will obtain continuous, multiple-angle imagery within the 16-day orbital repeat cycle of EOS without any gaps in spatial coverage. MISR builds upon the Galileo (GLL) and the Wide-Field/Planetary Camera.

The instrument uses nine separate charge-coupled-device (CCD)-based pushbroom cameras to continuously observe the Earth at nine discrete viewing angles. These angles are at nadir and four other symmetrical fore-aft views up to $\pm 70.5°$ along-track. The cameras operate at wavelengths centered at 0.443, 0.555, 0.67, and 0.865 μm, with spectral bandwidths of 30, 20, 15, and 20 nm, respectively. Each of the 36 instrument data channels (i.e., four spectral bands for each of the nine cameras) is individually commandable to provide a spatial sampling of 240 m, 480 m, 960 m, or 1.92 km. Swath width of the MISR images is 356 km, providing multi-angle coverage of the entire Earth from nine days at the equator to two days at the poles.

MISR is designed to obtain global observations of the directional characteristics of reflected light and other information needed for studying aerosols, clouds, and the biological and geological characteristics of the land surface. In contrast to single-view direction imagers, the MISR data can be used to derive a vegetation index based on red and near-infrared fluxes, rather than radiances. Calibrated images will be provided in global (2-km grids) and local (250 m) modes. Data products of interest include global maps of spectral planetary and surface hemispherical albedo (accuracy of ± 0.03) at a 2-km grid.

VII. Deriving Soil Carbon Information from Remote Sensing: A Case Study

The capabilities of the sensors described above do not include direct measurements of organic soil carbon, but do allow assessment of many of the parameters critical to the goal of modeling soil carbon (e.g., vegetation status, soil moisture, temperature, biomass, albedo and others). The first step toward modeling soil carbon is to explore the possibility that different soils affect the spectral response of vegetation as measured by satellite.

One remote sensing technique that can be used for this purpose is to derive the NDVI. The NDVI is defined as the difference in brightness values recorded for a near infrared and a visible red wavelength band divided by the sum of the brightness values recorded in the two bands (Lillesand and Kiefer, 1987). These wavelengths are used to separate vegetation signals from variations in background brightness, and to emphasize the unique reflectance spectrum of photosynthetically active vegetation of low reflectance in the visible spectral region, and high reflectance in the near infrared spectral region (Tucker, 1979). High values of NDVI indicate the presence and condition of green vegetation, because of high near infrared reflectance and low visible reflectance. Other nonvegetated areas, such as clouds, water and snow, generally have a larger visible reflectance than a near infrared reflectance and thus yield negative NDVI values. Rocks and bare soils have similar reflectances in both wavelength bands, therefore, NDVI values are near zero. Thus, organic carbon resulting from litter and root input to the soil from vegetation should be related to greenness, as indicated by NDVI.

NDVI ratio maps can be derived from remote sensing instruments that contain these key wavelength bands. Instruments that have been used routinely to derive such maps include the Advanced Visible and Infrared Imaging Spectrometer (AVIRIS), the Landsat MSS and TM, and the NOAA AVHRR sensor.

A. Site Description

The International Paper Company's Northern Experimental Forest (NEF) at Howland, Maine, was chosen as the focus of a study to assess the relationships between soil organic carbon and NDVI. The NEF is the principal site for research within NASA's Forest Ecosystem Dynamics (FED) project, which is concerned with forest dynamics within the North American transition zone between northern hardwood forests and the boreal forest biome. To test and develop models for this biome, intensive field campaigns including an integral remote sensing component were carried out in cooperation with the University of Maine at the NEF between 1989 and 1991. Details of the design of this field experiment and preliminary results are presented in Smith et al. (1990) and Levine et al. (1994).

The NEF is a 6,300 ha tract, located in the central interior region of the state of Maine, 56.3 km north of Bangor along the Penobscot River. The climate is cold, humid, continental with occasional maritime storms from the northeast. Annual precipitation is between 80-100 cm. Winter temperatures fall below $-20°C$ with continuous snowpack between December and March, accumulating up to 2 m. Summer temperatures are generally below 30 °C. The landscape is nearly level and low-lying, increasing in elevation from east to west, with only a 68 m rise from the lowest to the highest point (U.S. Geological Survey, 1988).

In general, the NEF has always been forested, composed of mixed hardwoods, i.e. poplar/birch (*Populus* sp./*Betula* sp.), hemlock-spruce/fir (*Tsuga canadensis* (L.) Carr. - *Picea* sp./*Abies balsamea* (L.) Mill.), and hemlock-mixed hardwood stands. A fire in 1921 destroyed 1400 ha in the central section, which was reseeded to poplar and birch.

The soil parent materials vary across the site due to glacial activity during the Wisconsin glaciation. With the retreat of the glacier, approximately 12,500 years ago (Rourke et al., 1978), a mixture of coarse-loamy tills, silty lacustrine, and sandy outwash materials remained intermixed across the landscape. The terrain is hummocky and varies in drainage. In the eastern section of the study site, a gravel esker was deposited, which is much more steeply sloping and is made up of much coarser textured material than the rest of the area. Since the deglaciation, recent alluvial activities and soil-forming processes have operated on the exposed glacial material (U.S. Department of Agriculture, 1990).

B. Soil Survey and Analysis

A soil survey at a scale of 1:12,000 (minimum mapping unit size of 0.8 ha) identifying major soil associations was performed by the USDA Soil Conservation Service based in Orono, Maine, during the summer of 1990. The survey was conducted within a 3 km^2 section of the NEF, as well as on an auxiliary site on the esker located in the eastern part of the NEF. Aerial photographs and SPOT satellite imagery were used as base maps on which mapping units were drawn. Figure 1 is a map of the soil associations and Table 1 gives the taxonomic classification of each of the soil series within the associations.

Ten soil pits were dug in areas representative of 10 predominant soil series found within the soil associations at the experimental site. Standard field characterization was performed at each pit to identify the soil series and morphological properties of horizons (U.S. Department of Agriculture, 1984). During field characterization, samples for physical and chemical laboratory analysis were taken. Laboratory analyses were performed by the USDA Soil Conservation Service at the National Soil Survey Laboratory in Lincoln, Nebraska. Percent organic carbon was among the properties assessed in the lab using potassium dichromate digestion and ferrous sulfate titration (U.S. Department of Agriculture, 1984). A "cumulative" value of organic carbon (kg/m^2) in the top 50 cm was derived for each of the 10 soil series using the method given by the National Soil Survey Laboratory (1983). This information at the series level was then weighted by area to represent the mapped soil associations (Figure 2).

The completed soil map was transferred to a vector-based geographic information system (ARC/INFO, available from Environmental Systems Research Institute, Redlands, CA) by digitizing the 10 soil mapping units as they occurred across the landscape. Once digitization was complete, it was transformed to the Universal Transverse Mercator projection (UTM zone 19) using control points from a 1:24,000 U.S. Geological Survey 7-1/2-minute topographic map for the Howland, Maine quadrangle (U.S. Geological Survey, 1988). Control points were derived from the extensive and easily observed road network within the International Paper research area. Errors associated with this rectification were on the order of <2 pixels or 40 m.

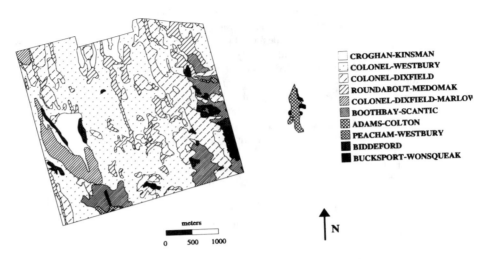

Figure 1. Soils of the Northern Experimental Forest at Howland, Maine.

Table 1. Taxonomic classification of soils at the NEF, Howland, Maine

Series	Taxonomic classification
Adams	Sandy, mixed, frigid, Typic Haplorthod
Biddeford	Fine, illitic, non-acid, frigid, Histic Humaquept
Boothbay	Fine-illitic, mixed, frigid, Andic Dystrochrept
Bucksport	Loamy, mixed, euic, terric, Borosaprist
Colonel	Coarse-loamy, mixed, frigid, Andic Dystrochrept
Colton	Sandy-skeletal, mixed, frigid, Typic Haplorthod
Croghan	Sandy, mixed, frigid, Aquic Haplorthod
Dixfield	Coarse-loamy, mixed, frigid, Andic Dystrochrept
Kinsman	Sandy, mixed, frigid, Aeric Haplaquod
Marlow	Coarse-loamy, mixed, frigid, Typic Haplorthod
Medomak	Coarse-silty, mixed, non-acid, frigid, Fluvaquentic Humaquept
Peacham	Coarse-loamy, mixed, frigid, Typic Haplohumod
Roundabout	Coarse-silty, mixed, nonacid, frigid, Aeric Haplaquept
Scantic	Fine-illitic, nonacid, frigid, Typic Haplaquept
Westbury	Coarse-loamy, mixed, frigid, Aeric Haplaquod
Wonsqueak	Loamy, mixed, euic, terric, Borosaprist

C. Image Analysis

A digital image of the Howland area from the Landsat TM was acquired for July 1, 1985. This image was used to identify heterogeneity in vegetation patterns below the scale of soil mapping. Once this predominant scale of short-range correlation was determined, an image from AVIRIS, a 224-band airborne imaging spectrometer, was used to derive NDVI. The AVIRIS scene was acquired during a September 8, 1990 overflight by the NASA ER-2 aircraft during an intensive multiaircraft field campaign (FED MAC). AVIRIS data were corrected for atmospheric effects with the technique of Gao et al. (1992) using the atmospheric optical depth measured from the ground during the overflight. Data integration was performed using several of the narrow bands (ca. 9 nm) of AVIRIS to derive broad band measurements of the red and near infrared wavelength bands (60 and 140 nm, respectively) that would be similar to data sets from other operational

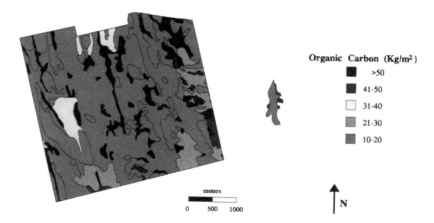

Figure 2. Organic carbon in the top 50 cm of soil for each soil association mapped at the Northern Experimental Forest at Howland, Maine. (Note: lines within areas designated with a given carbon content represent borders delineating soil association mapping units, i.e., certain soil mapping units contain the same organic carbon content. The area outside the rectangular border represents the soil associations found on the esker in the eastern part of the NEF.)

instruments (e.g., TM and AVHRR). For example, AVIRIS band numbers 25-30 (635.9 to 685.0 nm center wavelength values) covered the red TM spectral band, and AVIRIS bands 42-56 (767.0 to 901.4 nm center wavelength values) represented the near infrared. Calibrated AVIRIS radiance values were used in this integration. The NDVI map derived from the AVIRIS image was rectified to the UTM projection using the same control points as those for the soil map. The pixel size remained at 20 × 20 m before and after geometric rectification using a nearest neighbor resampling technique.

D. Results and Discussion

Figure 3 shows the strong visual relationship between carbon content of the NEF soils with NDVI patterns. Correlation of median NDVI with carbon (kg/m^2) in the top 50 cm of mapped soils was tested using Pearson's product moment correlation test (Ott, 1988). Results of the statistical analysis showed a small, but significant negative correlation (r = -0.63, p = 0.053, n = 10). Thus, the organic carbon in soils at this site increases as NDVI decreases. This relationship may be unique for the northern transition/boreal forest biome and represents slower mineralization of carbon where it is accumulating, indicating the adaptation of certain species that would tend to have lower NDVI values than more vigorously growing vegetation. Predictions of carbon status across landscapes can be estimated from NDVI values using a simple model such as this for other biomes or at other spatial scales.

Other factors that control accumulation of carbon can also be used to indirectly predict carbon status using techniques such as NDVI. Soil wetness, expressed as drainage class for example, showed a significant positive relationship with median NDVI (r = 0.75, p = 0.025, n = 10) at the same site in Maine (Levine et al., 1993a). A Spearman's rank correlation coefficient test was used to account for the ranked categories by which drainage class is described (Conover, 1980; cited in StatSci, 1991). In this work, the highest NDVI was associated with soils that were moderately well to well drained, and the lowest NDVI with poorly and very poorly drained soils where most of the organic carbon would be accumulating.

Estimates of biomass at the NEF were also derived using imagery from the NASA/JPL AIRSAR Polarimeter using a ratio of cross polarization (HV) backscattering from a longer wavelength (P or L band) to a shorter wavelength (C band radar) (Ranson and Sun, pers. comm., 1993). Future studies comparing these polarimeter results with organic carbon at the NEF soils are being planned as another surrogate for predicting soil organic carbon status.

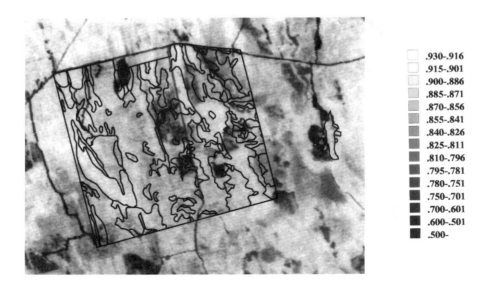

Figure 3. False color representation of broad band NDVI, calculated from atmospherically-corrected AVIRIS, subset for a 3 × 3 km study area within the NEF. Outlines of soil carbon mapping units (see Figure 2) are shown in black. (The area outside the rectangular border represents the soil associations found on the esker in the eastern part of the NEF.)

In the case study given here and in future similar studies, additional information should be included which would improve modeled relationships between organic soil carbon and remote sensing tools such as NDVI. These include properties that may be "masking" direct soil/vegetation relationships such as information about management history (e.g., clearcutting, thinning, herbicide treatment, planting, release cuts, stand age, time since disturbance, or species composition). Also, stand age or disturbance history must be included as control factors to determine their effects on NDVI. Another masking effect that may have occurred in this study is the consideration of NDVI as a single variable at the landscape level, rather than segregating NDVI by vegetation community type within soil mapping units. Optimal canopy development as indicated by high NDVI values may not be directly comparable between distinct communities (i.e., conifer vs. hardwoods).

In addition, the number of samples available to describe soils in detail is usually limited to one characterization profile per series. Without a large number of sampling points, it is difficult to statistically resolve the controls of soils on NDVI or other vegetation attributes. More intensive characterization can become extremely time consuming and costly to obtain an adequate sample. Other approaches, such as simulation modeling to predict soil properties (see Levine et al., 1993), may be a viable alternative to obtain sufficient ground data for future studies.

VIII. Conclusions

Remote sensing, with its multispectral, multiscale, and multitemporal monitoring abilities, provides a powerful tool for assessing the dynamics of the biosphere. With the present technology, however, chemical and physical properties of soils cannot be directly observed using satellite data. To draw inferences from satellite data concerning soil characteristics, surrogate indices such as vegetation status, biomass, soil moisture, or soil temperature, must be used. Future sensors on the EOS platform, which are to be launched in the late 1990's, will increase our capability to monitor soil carbon indirectly. For example, the ASTER and the MODIS offer the potential to map vegetation indices at various scales. The MIMR can be used to monitor soil moisture. The MISR can be used to prepare global maps of land surface albedo. These EOS instruments offer the potential for developing global estimates of soil carbon for various landscape environments.

An example is given of how one mapping technique, the NDVI derived from remote sensing data, can be used to assess the organic soil carbon status. This example shows that organic soil carbon can be related to NDVI data, but additional collection and manipulation of site information must be done for improved local, regional and global scale prediction.

While remote sensing technology and interpretation of imagery is in an early stage, continued research in this area is critical. The preparation for and launching of EOS is leading the way for exciting new research, different views of the Earth, and for understanding global scale processes on the Earth's surface.

Acknowledgements

This work was partially funded by the NASA Ecosystem Dynamics and Biogeochemical Processes Branch (Code SEP) at NASA Headquarters. The authors wish to acknowledge Lara Prihodko of Hughes, STX, for her assistance in image and GIS data processing.

References

Achard, F. and F. Blasco. 1990. Analysis of vegetation seasonal evolution and mapping of forest cover in West Africa with the use of NOAA AVHRR HRPT data. *Photogrammetric Engineering and Remote Sensing* 56(10):1359-1365.

Cohen, W.B. 1991. Response of vegetation indices to changes in three measures of leaf water stress. *Photogrammetric Engineering and Remote Sensing* 57(2):195-202.

Conover, W.J. 1980. *Practical Nonparametric Statistics*. 2nd Ed., John Wiley & Sons, NY.

Crist. E.P. and R.C. Cicone. 1984. A physically-based transformation of Thematic Mapper data-the TM tasseled cap. *I.E.E.E. Transactions on Geosciences and Remote Sensing* GE-22:256-263.

Dale, V.H. (ed.). 1990. Report of a workshop on using remote sensing to estimate land use change, 22-24 February 1989, Oak Ridge, Tennessee. Environmental Sciences Division, Publication no. 3397, Washington, D.C.

Dobson, M.C. and F.T. Ulaby. 1986. Active microwave soil moisture research. *I.E.E.E. Transactions on Geosciences and Remote Sensing* 24:23-35.

Gao, B., K.B. Heidebrecht, and A.F.H. Goetz. 1992. ATmosphere REMoval Program (ATREM) User's Guide. Ver. 1.1. Center for the Study of Earth From Space/CIRES, University of Colorado, Boulder, CO.

Hastings, D., M. Matson, and A.H. Horvitz. 1989a. Monitoring vegetation. *Photogrammetric Engineering and Remote Sensing* 55(1):40-41.

Hastings, D., M. Matson, and A.H. Horvitz. 1989b. Using AVHRR for early warning of famine in Africa. *Photogrammetric Engineering and Remote Sensing* 55(2):168-169.

Jackson, T.J. and T.J. Schmugge. 1984. Passive microwave remote sensing of soil moisture. *Advances in Hydroscience* 40:123-159.

Levine, E.R., R.G. Knox, and W.T. Lawrence. 1994. Relationships Between Soil Properties and Vegetation at the Northern Experimental Forest, Howland, Maine. *Remote Sensing of Environment*. In press.

Levine, E.R., K.J. Ranson, J. Smith, D. Williams, R.G. Knox, H. Shugart, D. Urban, and W. Lawrence. 1993. Forest Ecosystem Dynamics: Linking forest succession, soil process, and radiation models. *Ecological Modeling* 65:199-219.

Lillesand, T.M. and R.W. Kiefer. 1987. *Remote Sensing and Image Interpretation*, 2nd Ed., John Wiley & Sons, NY.

McDaniel, K.C. and R.H. Haas. 1982. Assessing mesquite-grass vegetation condition from Landsat. *Photogrammetric Engineering and Remote Sensing*. 48(3):441-450.

Musick, H.B. and R.E. Pelletier. 1988. Response to soil moisture of spectral indices derived from bidirectional reflectance in Thematic Mapper wavebands. *Remote Sensing of Environment* 25:167-184.

NASA. 1993. *EOS reference handbook*. (G. Asrar and D.J. Dokken, eds.). NASA Goddard Space Flight Center, Earth Science Support Office: Washington, D.C.

National Soil Survey Laboratory. 1983. Principles and Procedures for Using Soil Survey Laboratory Data. National Soil Survey Laboratory, USDA SCS, Lincoln, NE.

Nelson, R., N. Horning, and T.A. Stone. 1987. Determining the rate of forest conversion in Mato Grosso, Brazil, using Landsat MSS and AVHRR data. *International Journal of Remote Sensing* 8(12):1767-1784.

Norwine, J. and D.H. Greegor. 1983. Vegetation classification based on Advanced Very High Resolution Radiometer (AVHRR) satellite imagery. *Remote Sensing of Environment* 13:69-87.

Ott, L. 1988. *An Introduction to Statistical Methods and Data Analysis*. PWS-Kent Publishing Co., Boston, MA.

Rourke, R.V., Ferwerda, J.A., and K.J. LaFlamme. 1978. The Soils of Maine. Life Sciences and Agriculture Experiment Station, Misc. Report #203, University of Maine at Orono, Orono, ME.

Salisbury, J.W. and D.M. D'Aria. 1992a. Emissivity of terrestrial materials in the 8-14 μm atmospheric window. *Remote Sensing of Environment* 42(2):83-106.

Salisbury, J.W. and D.M. D'Aria. 1992b. Infrared (8-14 μm) remote sensing of soil particle size. *Remote Sensing of Environment*. 42(2):157-165.

Schmugge, T.J., T.J. Jackson, and H.L. McKim. 1980. Survey of methods for soil moisture determination. *Water Resources Research* 16:961-970.

Schmugge, T.J., P.E. O'Neil, and R.T. Wang. 1986. Passive microwave soil moisture research. *I.E.E.E. Transactions on Geosciences and Remote Sensing*. 24:12-22.

Smith, J.A, K.J. Ranson, D.L. Williams, E.R. Levine, M.S. Goltz, and R. Katz. 1990. A sensor fusion field optical experiment in forest ecosystem dynamics. SPIE Int. Symposium of Optical and Engineering Photonics in Aerospace Engineering. Vol. 1300. Orlando, FL.

StatSci. 1991. S-PLUS Reference Manual. Ver. 3.0, Statistical Sciences Inc., Seattle, WA.

Teng, W.L. 1990. AVHRR monitoring of U.S. crops during the 1988 drought. *Photogrammetric Engineering and Remote Sensing*. 56(8):1143-1146.

Thompson, L.L. 1990. Potential commercial uses of EOS remote sensing products. NASA Technology 2000 meeting. November 27-28, Washington, D.C.

Townshend, J.R.G. and C.O. Justice. 1986. Analysis of the dynamics of African vegetation using the normalized difference vegetation index. *International Journal of Remote Sensing* 7:1435-1445.

Tucker, C.J. 1979. Red and photographic infrared linear combinations for monitored vegetation. *Remote Sensing of Environment*. 11:171-189.

Tucker, C.J., J.R.G. Townshend, and T.E. Goff. 1985. African land-cover classification using satellite data. *Science* 227(4685):369-375.

U.S. Department of Agriculture. 1984. Procedures for collecting soil samples and methods of analysis for soil survey. Soil Survey Investigations Report No. 1., USDA Soil Conservation Service, Lincoln, NE.

U.S. Department of Agriculture. 1990. Soil Survey and Interpretations of Selected Land in Maine. USDA Soil Conservation Service, Orono, ME.

U.S. Geological Survey. 1988. Howland, Maine 1:24,000 scale quadrangle, Penobscot County. USGS, Reston, VA.

CHAPTER 23

Preparing a Soil Carbon Inventory for the United States Using Geographic Information Systems

Norman B. Bliss, Sharon W. Waltman, and Gary W. Petersen

I. Introduction

Soil and vegetation have significant roles in the carbon cycle, along with the influences of human activity. An understanding of the biogeochemical cycle of carbon is important to policy makers that are considering limitations on releases of carbon dioxide and other carbon compounds to the atmosphere. Policy makers must balance the risk of long-term environmental and economic damage due to potential global warming from increased greenhouse gases, against the near-term economic readjustments that would result from limitations on fossil fuel use.

This paper reviews the importance of soil geographic data for modeling global change, gives an overview of the structure of soil data bases, and illustrates how soil data bases can be used to analyze some aspects of the carbon cycle.

II. Role of Soil Carbon in Global Change Studies

Soil data are useful for biogeochemical and hydrologic studies at global, regional, and local scales. Total quantites of soil organic carbon, as reported in this paper, contribute to understanding the biogeochemical cycling of carbon, including future levels of carbon dioxide and other greenhouse gases in the atmosphere. Projections of the levels of greenhouse gases are used to drive the general circulation models of the atmosphere (GCM) that lead to scenarios of global warming. Soil data are also used in GCM's to characterize the land surface as part of the surface energy balance calculations.

Soil data sets for use in global models are typically derived from the Soil Map of the World (Food and Agriculture Organization of the United Nations, 1970-1978). For regional modeling of the United States, the State Soil Geographic data base (STATSGO) and the National Soil Geographic data base (NATSGO) are being developed by the U.S. Department of Agriculture (USDA), Soil Conservation Service (SCS). These data bases are being used to develop methods for improving estimates of the total amount of organic carbon in the soil, and they will contribute to an understanding of the dynamics of carbon movement into and out of the soil. Other applications of small-scale soil maps include hydrologic modeling, studies of ecosystem dynamics, and analysis of agricultural capability.

A. Global Climate Models

Moore (1992) developed a schematic diagram (Figure 1) of the interrelationships among various components of the Earth's climate system. The concept of an integrated global climate system is based on the realization that the atmosphere, biosphere, oceans, cryosphere, and lithosphere interact. The rates of mass and energy

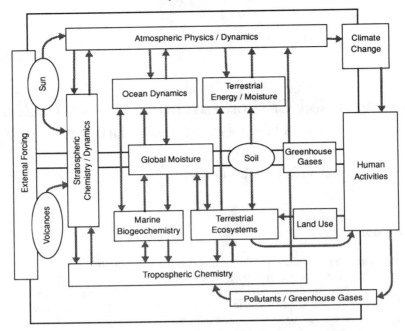

Figure 1. Interrelationships between physical and biological components of the global climate system. (From Moore, 1992.)

transfer vary within and between different subsystems. For example, the temperature of the atmosphere changes much more quickly than the temperature of the ocean, and the carbon in vegetation changes more quickly than the carbon in soil. Understanding the exchanges of moisture and carbon between soils and the atmosphere is necessary for understanding the global energy balance, the potential for global warming, and the effects of climate change on agriculture and natural ecosystems.

Oceanic and atmospheric general circulation models (GCM) are used to study the influence of changes in the concentration of greenhouse gases on future climate patterns. An evaluation of the modeling results by the Intergovernmental Panel on Climate Change (IPCC, 1990) indicates that for a business-as-usual scenario (without controlling emissions of greenhouse gases), the global mean temperature during the next century will increase 0.3 °C per decade, with an uncertainty range of 0.2 °C to 0.5 °C per decade. This will result in a likely increase in global mean temperature of about 1 °C by 2025 and 3 °C before the end of the next century.

Current research to improve the reliability of the GCM's includes efforts to improve models of cloud formation, to improve the linkages between atmospheric and oceanic models, and to improve models that simulate land surface processes. Soil data bases, along with topographic and remotely sensed data, can be used to improve the description of the land surface, and lead to better models of land surface processes at regional and global scales.

There are two principal ways in which better soil data bases can improve understanding of the physical climate system. First, research in biogeochemical cycling contributes to an understanding of the carbon cycle, so that future concentrations of greenhouse gases such as carbon dioxide (CO_2) and methane (CH_4) can be accurately modeled. Data on the physical and chemical properties of soils are used to calculate amounts of carbon in the soil, and rates of absorption and release of carbon. Second, research to improve the understanding of the hydrologic cycle will help improve the climate models that are used to predict magnitudes and patterns of climate change. Soil data are used to calculate the capacity of the soil to store water, and rates of water movement into, across, and out of the soil. The carbon cycle calculations are needed as external inputs to the climate models (to specify the amounts of greenhouse gases), and the hydrologic cycle calculations are needed to correctly simulate soil moisture within the climate models. Ultimately, the models may be able to incorporate feedback between the vegetation and atmospheric systems, so that changes in vegetation and soils as a result of changes in climate can be simulated.

Although the net increase of greenhouse gases in the atmosphere is being measured with some confidence, there is uncertainty about the magnitude of the sources and sinks. Water vapor is the dominant greenhouse gas, accounting for over half of the greenhouse effect. Carbon dioxide accounts for about 61 percent of the greenhouse effect from other greenhouse gases, with methane (15%), oxides of nitrogen (10%), and chlorofluorocarbons (13%) accounting for significant proportions on a time-averaged basis (IPCC, 1990). Both the ocean and the land surface serve as sources and sinks for carbon dioxide. The Working Group Report on Trace Gases and Nutrient Fluxes (National Academy of Sciences, 1990) states, "...it is important to know the causes and effects of the accelerated rate of increase in atmospheric CO_2, because it is a radiative active greenhouse gas that has and will continue to contribute to global warming and because it has direct effects on ecosystems." This report and others (International Geosphere-Biosphere Program, 1992; Committee on Earth and Environmental Sciences, 1992; IPCC, 1990) have clearly identified research that is needed to develop improved estimates of soil carbon and their application in various global change models.

The National Academy of Sciences (1990) identified research that is needed to better predict changes in atmospheric CO_2 concentrations, such as:
- how climatic warming and associated changes in precipitation and nutrient status will alter carbon storage in ecosystems, especially those with large amounts of soil carbon;
- how increased concentrations of atmospheric CO_2 will affect the relationships between soil nutrient status, plant litter quality, and nutrient mineralization;
- the major sources and sinks for CO_2, and the major global pattern of CO_2 transport in the atmosphere (current perceived sources and sinks for CO_2 do not match the interhemispheric CO_2 gradient);
- the global amount and distribution of biomass and soil carbon, net primary production, and rate of ecosystem respiration (considerable field data have not been adequately collated and related to vegetation and soil maps for inclusion in global climate models).

The International Geosphere-Biosphere Program (IGBP, 1992) recognizes that "...soils are of great importance in determining terrestrial carbon storage and the emissions of greenhouse gases to the atmosphere; they also influence heat exchange, and provide a key link and buffer system for the hydrological cycle. IGBP projects require the best possible global maps of soil characteristics, such as texture and organic matter content, affecting these processes." In the absence of readily available global soil maps and related attributes, "...a modelling approach is now being developed to meet that need, combining existing information on soil classification with soil property relationships, and structured to accommodate additional observations."

B. The Missing Sink in Global Carbon Balances

Sundquist (1993) indicates that,
"For more than 20 years, the principal problem in budgeting atmospheric CO_2 has been the excess of known sources over identified sinks. The amount of CO_2 produced by human activities (principally consumption of fossil fuels and destruction of forests) significantly exceeds our best estimates of the amount of CO_2 absorbed by the oceans and atmosphere (Table 1). To account for this budget imbalance (the so-called "missing" CO_2), there must be another large CO_2 sink. This sink is probably somewhere in the world's terrestrial plants and soils, but its specific identity has eluded detection. Although we can quantify the earth's major C reservoir and fluxes, balancing the anthropogenic CO_2 budget requires accounting for differences that are often small relative to the natural exchange and abundance of C (Figure 2)."

Studies of soils can contribute to understanding biogeochemical cycles in several ways. The soil acts as a "storage compartment" in the carbon cycle. Soil properties influence the form and rate of release for greenhouse gases moving from the soil to the atmosphere. Soil is a substrate for vegetation, which also acts as a "storage compartment" in the global carbon cycle, and soil fertility influences the rate of vegetation growth at a given site. Studies of the cycles of other biologically active elements, such as nitrogen, phosphorus, and sulfur are also important for assessing biological productivity.

Table 1. Intergovernmental Panel on Climate Change global CO_2 budget for 1980-1989

Reservoir	Average flux (Pg[a] of C year^{-1})
Sources	
Fossil fuels	5.4 ± 0.5
Deforestation and land use	1.6 ± 1.0
Total	7.0 ± 1.2
Sinks	
Atmosphere	3.2 ± 0.1
Oceans (modeled uptake)	2.0 ± 0.8
Total	5.2 ± 0.8
Imbalance (sources - sinks)	1.8 ± 1.4

From Sundquist (1993) as adapted from Watson et al. (1992) by Sarmiento and Sundquist (1992). [a] 1 Pg = 10^{15} g = 1 gigaton = 1 Gt.

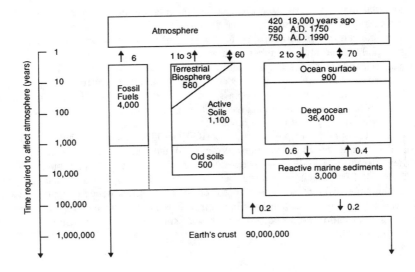

Figure 2. Principal reservoirs and fluxes in the global carbon cycle. Vertical placements relative to scale on left show approximate time scales required for reservoirs and fluxes to affect atmospheric CO_2. Double arrows represent bidirectional exchange. Single arrows to and from the atmosphere are approximate estimates of anthropogenic CO_2 fluxes for 1990. Terrestrial uptake of anthropogenic CO_2 is likely but not shown because of large uncertainties. Values in petagrams (1 Pg = 10^{15} g = 1 gigaton = 1 Gt). (From Sundquist, 1993.)

Soil represents a large reservoir of carbon. Measurements of soil carbon content in various ecosystems form a basis for making global soil carbon estimates (Zinke et. al., 1986). Changes in the rate of accumulation or decomposition of soil carbon affect the global carbon balance. These rates are sensitive to local conditions of temperature and moisture, and to soil properties such as texture, nutrient content, and organic matter content.

The form in which soil carbon is released during decomposition is important. In aerobic conditions, sufficient oxygen is available in the soil for the oxidation of carbon to CO_2. In anaerobic conditions, carbon may be released as CH_4. Methane is 32 times more effective per molecule in trapping radiation than CO_2, and accounts for 19 percent of the greenhouse effect (Bouwman, 1990, p. 27). Rice paddies, wetlands, and burning of biomass are primary sources of methane, with landfills, the intestines of ruminants, and termites also being significant sources. Improvements are needed in both field measurements of gas exchange and the mapping of soils and wetlands to reduce uncertainties in methane estimates (Bouwman, 1990, Ch. 4).

It should be cautioned that while soil data bases can be used to estimate the total amount of soil carbon, and to relate types of landscapes with their potential as sources or sinks of atmospheric carbon, they can only strengthen, not confirm, hypotheses on the annual fluxes of carbon compounds between the soil and the atmosphere. We cannot measure the amounts of carbon in global soils precisely enough to use differences in successive measurements to estimate rates of change. In Sundquist's (1993) review of the global carbon dioxide budget for conditions from the last deglaciation to the present he notes,

"The deglacial [about 10,000 to 20,000 years ago] CO_2 sink is best estimated from changes in sediment $^{13}C/^{12}C$ ratios, and the modern sink is best estimated from the imbalances calculated from atmospheric and oceanic models. Our reliance on these indirect methods is an inevitable consequence of the difficulty of assessing small differences between large and heterogeneous terrestrial assimilation and respiration fluxes. Careful analyses of atmospheric and oceanic CO_2 records may always constrain the magnitude of net terrestrial CO_2 exchange at least as accurately as direct monitoring of the land surface. Neither the deglacial nor the modern CO_2 budget will be satisfactorily balanced, however, until it can be related to specific identifiable CO_2 exchange processes. Forecasts of future CO_2 budgets cannot be based on empirical extrapolation of present unexplained imbalances but must be based on an understanding of processes that control CO_2 sources and sinks. This understanding requires intensive direct observation of terrestrial ecosystems and their record of past behavior."

C. Estimates of Global Soil Organic Carbon

Beinroth (1992) notes that the most recent estimate of total soil organic carbon, some 1,300 to 1,500 Pg (Schlesinger, 1984; Hall, 1989; Post et al., 1990) is intermediate between lower (700 Pg; Bolin 1970) and higher (3,000 Pg; Bohn, 1976) calculated values (1 Pg = 10^{15} g).

Kimble et al. (1990), considering all soil orders except Histosols (the organic soils), estimate the global total for soil organic carbon at 1,061 Pg, and calculate that half of the total (496 Pg) occurs in the tropics.

When an analysis of soil organic carbon has been completed using the methods of our study for the whole world, the totals could be used to validate or refine the estimates presented in Figure 2, of 1,600 petagrams for the sum of active soil carbon (1,100 Pg) and old soil carbon (500 Pg). There are many uncertainties in separating active from old soil carbon (Aiken et al., 1985; Stevenson, 1982). The organic matter values reported in this paper are taken predominantly from measurements of the surface layers of the soil. However, the analytical techniques used in the national data bases of the SCS measure total soil organic carbon, and do not distinguish between active and old soil carbon (Soil Survey Staff, 1992).

Soil data bases provide spatial detail to the distribution of organic carbon that will be important for understanding the processes of carbon uptake and release from the soils. The data bases include information on soil temperature and moisture regimes, landscape position, drainage, permeability, and other physical and chemical characteristics. Small areas that have high carbon content, such as peat bogs, are not necessarily mapped as separate units but can be accounted for in associated attribute files. As we use field and laboratory research to develop a better understanding of processes that exchange carbon compounds with the atmosphere, the soil data bases will provide information to spatially quantify these processes and compute totals for latitude bands, continents, or the whole world. Although knowing the spatial distribution of carbon cycling will be valuable, the caution provided by Sundquist (1993) indicates that the total fluxes calculated by these methods will need to be constrained by the results of global atmospheric and oceanic biogeochemical models.

D. Mesoscale Models

Soil data bases are also valuable for global change investigations that focus on more detailed models of processes at less than global scales. The GCMs can be used to reliably simulate the main features of the present day climate system for the whole world, such as the large-area distribution of air pressure, temperature, wind, and precipitation in both summer and winter (IPCC, 1990). They do not provide enough

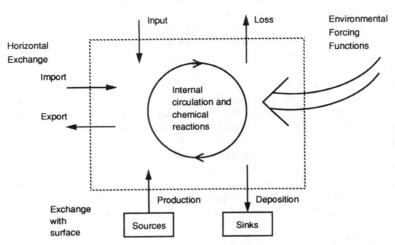

Figure 3. Essential characteristics of a mesoscale model. (From National Academy of Sciences, 1990.)

detail to confidently simulate processes that are important for understanding the effects of climate change on ecosystems and agriculture. Consequently, they cannot be used to model the feedbacks in the systems that could lead to an acceleration or a slowing of the changes being modeled.

Mesoscale climate models, suitable for areas on the order of 1 million square kilometers, are being developed to provide an essential intermediate step that links our understanding of land surface processes derived from site-specific process studies to the large grid cells used by the GCMs. Mesoscale models simulate fluxes of matter and energy for areas of similar ecosystem types. Spatial resolution on the order of 1 km is appropriate for this type of modeling. Figure 3 illustrates five essential characteristics of a mesoscale model for a gas species such as H_2O or CO_2: 1) vertical exchange with the soil, water, or vegetation that covers the earth's surface (inputs characterizing these production and deposition processes come from the microscale models, and are aggregated, if necessary, to account for the heterogeneous surface); 2) vertical exchange with the upper atmosphere; 3) horizontal exchange, via wind, with adjacent areas of the same or different ecosystem type; 4) circulation and chemical reactions within the cell that affect the concentration and flux; and 5) environmental forcing functions, such as changes in temperature, cloud cover, or the concentrations of other gases, that are likely to change the dynamics of the gas species in question (NAS, 1990).

Integrated data bases that characterize the land surface are being developed for use in mesoscale models. The overall strategy is illustrated in Figure 4 (Loveland et al., 1991). Satellite imagery, such as 1 km resolution images from the Advanced Very High Resolution Radiometer (AVHRR) sensor, may be combined with digital elevation model data, soil data, and data sets that represent land use patterns to develop a comprehensive land characteristics data base. Interpretations for a variety of purposes, including climate modeling, land management, and environmental monitoring may be derived from the land characteristics data base. The relationships between these data sources and ecological modeling at the mesoscale are explored by Burke et al. (1990) and Schimel et al. (1991).

The STATSGO and NATSGO data bases are appropriate for mesoscale modeling, and are designed so that they can be generalized to the scales of global models. The data base development is consistent with the USDA efforts in the U.S. Global Change Research Program for fiscal year 1993 (Committee on Earth and Environmental Sciences, 1992) to focus on using ground-based research to understand terrestrial systems, including soil properties such as moisture, erosion, organic matter dynamics, nutrient fluxes, and microbes.

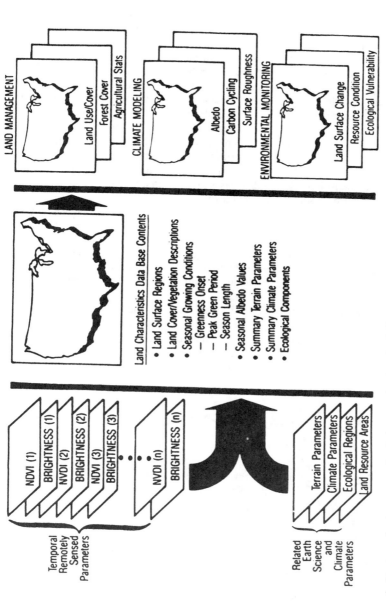

Figure 4. Strategy for large-area land characterization includes use of remote sensing and multisource data to create a spatial data base that includes seasonally distinct land-cover regions and associated attributes that can be tailored to a number of disparate applications. The normalized difference vegetation index (NDVI) is an example of a data set derived from remotely sensed data. (From Loveland et al., 1991.)

III. Representing Soils in Geographic Information Systems

The soil data bases are designed to be used in geographic information systems. A geographic information system (GIS) provides tools to link spatial data bases with attribute data bases. Geographic information systems efficiently process large amounts of spatially registered information. The systems have many functions for manipulating spatial information, including input (digitizing paper maps), registration (including changing map projections), overlay (combining information based on spatial location), analysis (of spatial relationships), and display (map products and statistical summaries).

In this research, digital soil maps are linked to data bases of soil properties. A system of soil classification, soil taxonomy (Soil Survey Staff, 1975), is used in both the digital maps and attribute data bases, and can be used as a key to link the two types of data. GIS systems that include relational data base management tools make it possible to manipulate and display information for heterogeneous map units, and account for the three dimensional structure of the soil by including information on the soil layers (horizons).

Soils are complex natural objects, and every soil map is a simplification. At all mapping scales, there may be inclusions (small areas of different soils) within a mapped soil boundary. The soil scientist that makes a soil map must make a judgement about the amount of detail to be represented on the map. Some level of soil classification or naming needs to be linked to each identifiable soil unit. A soil series is the lowest level of soil taxonomy in the U.S. system of taxonomy (Soil Survey Staff, 1975). Within a series, various phases may be distinguished according to slope, surface texture, degree of erosion, or other factors that may affect soil management. A phase of a soil series is the most detailed level of classification for which information is available in the nationwide soil data bases of the United States. We can use the phase as the most elemental unit for describing soils on national maps.

An area delineated by line segments that are stored in the GIS is called a polygon. At the mapping scales used in the national soil mapping effort, some polygons may represent the area of a single soil phase, while others may represent a soil complex or a mixture of soil phases that is too complicated to map at that scale, perhaps because data collection would be too expensive or cartographic display would be too cluttered.

The soil phases that contribute to a polygon may be referred to as the components of that polygon. A component percentage may be calculated that represents the area of the phase as a percentage of the area of the polygon.

Typically, many polygons with similar combinations of soils occur in a study area, and a soil scientist does not wish to separately describe the component percentages on a polygon-by polygon basis. A map unit may be defined, in which soil phases are components of the map unit, and the component percentages represent the area of each phase as a percentage of the area of the map unit. In Figure 5, both polygon 1 and polygon 4 are considered to be in the same map unit because they have the same proportions of phases

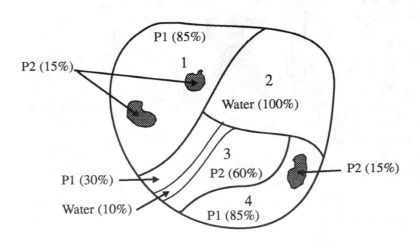

Figure 5. Schematic soil map illustrating percentages of components within map units. Each polygon has one or more components that is either a nonsoil (such as water) or a soil phase (illustrated as P1 or P2.) Polygons 1 and 4 have the same components in the same proportions, and can be mapped as the same map unit.

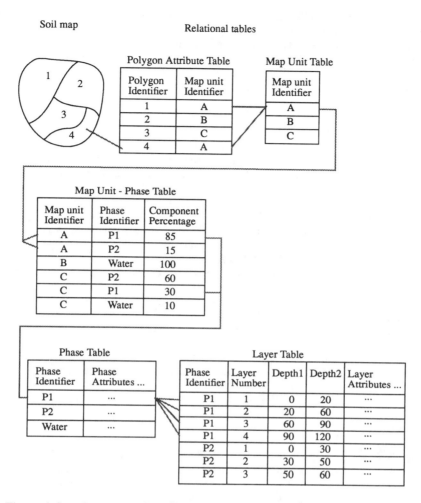

Figure 6. Sample structure for soil data in a GIS. GIS software links the digital map to the polygon attribute table. Each polygon is coded with a map unit identifier, which a relational data base management system uses to link the remaining files. The percentage of each phase within each map unit is given in the map unit-phase table. Calculations based on layer and phase attributes are carried through the intermediate tables to the polygon attribute table and can be use to make a map of the results of a query.

(P1 at 85% and P2 at 15%). Map unit identifiers are typically alphanumeric codes that are used as key fields to link the spatial data to data on soil properties. On detailed soil maps, the map unit identifier may include an abbreviation of the name of the dominant soil series with codes representing the texture of the surface horizon, the slope, and the erosion status; for example, VnC2, for Vienna silt loam, 5 to 9 percent slopes, eroded. For generalized soil maps, map units may be defined on a statewide basis; for example, WI023, for the 23rd map unit in Wisconsin. The capabilities of a GIS to quickly display information make it unnecessary to have map unit identifiers contain coded information on the soil properties.

A generic illustration of how soil data may be structured within a GIS is presented in Figure 6. Four polygons, numbered 1 to 4, are shown on the map in Figure 6. In a GIS that contains a relational data base management system, a polygon attribute table may be formed that links the polygon to the map unit identifier. In Figure 6, the map unit identifiers are labeled A, B, and C. Note that both polygons 1 and 4 are linked to map unit A.

In Figure 6, the map unit table is used to link the polygon attribute table to the rest of the data base. Additional items may be added to this table to store the results of analysis before transferring them to the polygon attribute table for display. The map unit-phase table lists the phases for each map unit and the component percentages of phases within map units. The phase table stores data on soil properties such as flooding potential and depth to bedrock--properties that characterize a whole soil. The layer table stores data on the soil properties such as texture and permeability for each soil layer.

IV. Nationwide Soil Data Bases

The SCS is developing soil data bases at three scales in support of national responsibilities for maintaining soil data (Office of Management and Budget, 1990). The most detailed mapping is being encoded into the Soil Survey Geographic Data Base (SSURGO). A soil survey area typically corresponds to a county, and is mapped at a scale between 1:15,840 and 1:31,680.

The State Soil Geographic data base (STATSGO) is developed for each State (SCS, 1991). Map units corresponding to soil associations are compiled on 1:250,000 scale quadrangle maps and are digitized. In most cases, the STATSGO data for an area are available before the SSURGO data, so an automated generalization procedure from the SSURGO data is not applicable.

The National Soil Geographic data base (NATSGO) is being developed as a set of generalizations of the STATSGO data, and will be appropriate for maps at 1:1,000,000 and smaller scales.

The attribute data for all three of these data bases are provided by the SCS's national data base of the properties of soil series, the Soil Interpretations Record (SIR). The SIR contains information on the properties of the soil phases, and these are reformatted into a set of relational tables that can be related to the map units.

Some preliminary research is also being conducted on methods to link the soil maps to the Soil Characterization Record, a data base of the detailed physical and chemical properties of soil pedons as sampled in the field.

A. NATSGO Concepts

The NATSGO data base is being developed by the SCS to provide interpretations of the nation's soils for national and regional analysis. The data base will replace the 25-year-old soils map of the U.S. National Atlas (USGS, 1987). One of the first goals for the NATSGO effort is to develop, in cooperation with the governments of Canada and Mexico, a soil organic carbon map of North America at a scale of 1:1,000,000. The NATSGO data base will be the basis of a United States contribution to improved global soil data bases. Information from the data base will be used in studies of global climate change, the balance of carbon dioxide and other compounds between the atmosphere and soils, the role of soils in ecological systems, the productivity of soils for agriculture, and rates of soil degradation.

The NATSGO data base is being formed from the data bases of the States and Territories of the United States (STATSGO). It is recognized that different users will need to emphasize different aspects of the data at the national level. The designers of the data base cannot anticipate all of these possible uses, and therefore do not wish to constrain the ways in which the STATSGO data may be aggregated for analysis and display on a nationwide basis. The STATSGO data should be generalized at the time of analysis. Thus, NATSGO consists of four parts: 1) the entire STATSGO data base for the Nation, 2) a data structure for storing rules for generalization, 3) a set of tools that allow users to define their own rules for generalization, and 4) a set of tools for creating products, such as generalized data sets or small-scale map products. Not every user of the NATSGO data will need to start with the full detail of the STATSGO data. Generalized data sets designed for particular types of analysis can be created by experts and distributed to a wider user community.

As STATSGO polygons or map units are combined to form NATSGO polygons, the NATSGO analysis tools will account for the new mixture of soil properties in each of the new polygons. Tools are being developed that allow a user to combine polygons either interactively (by pointing to the map on a computer screen) or based on the statistical similarity of soils according to particular analysis criteria. Unlike traditional map generalizations, the aggregation process is reversible; it is possible to zoom in and observe the full detail of the source data. Users can add, delete, copy, or modify the data tables that store the generalizations, according to the purpose of the analysis. The U.S. Geological Survey is cooperating with the SCS to develop and test this approach to multiscale data base development.

Why generalize? On a national or global scale, we need ways to summarize and visualize detailed information over large areas. Past efforts to do this for soil data have resulted in national and global maps that are not directly linked to the underlying field data, but are largely based on human judgement. They have also tended to emphasize those features that are visible at the mapping scales and ignore those components of the landscape that may be important but are not spatially dominant. The approach to forming generalized maps based on more detailed data bases may provide a way to overcome these limitations. A goal is to have a hierarchy of linked data bases, in which global maps are linked to national maps, which are linked, perhaps through additional levels of intermediate generalizations, to detailed maps, and ultimately to original field observations and measurements. Although this process has been present in past efforts, it has been filtered at each step of the way by a single set of judgements based on a limited set of criteria. The potential for a revised approach using high-speed computers will not eliminate the need for human judgement, but will provide a framework for documenting the choices made at each level, and will improve the speed with which alternative criteria can be used to create generalized maps. Detailed data bases that are suitable for land management can be aggregated to the level appropriate for national-level policy making. The resulting policies can then be stated in terms that are understandable to land managers who must carry out the policies.

The initial test of the method of generalization is based on generalizing the STATSGO data, which was developed for a map scale of 1:250,000, into NATSGO products such as maps at a scale of 1:1,000,000. The STATSGO data are stored as vector coverages in a geographic information system. A set of related tables defines the characteristics of each map unit, component, and soil layer, as well as various interpretations including suitability for crops, rangeland, and woodland uses.

B. The NATSGO Data Structure

A diagram of the NATSGO data structure is given in Figure 7. The polygon attribute table (PAT) is linked to the digital map by a unique identifier related to the data set name, illustrated as COVER#. A unique identifier (POLYID) is defined for each polygon for use in NATSGO. This is used to link to all other attribute tables and to return the results of analysis to the PAT for display as map products. The data tables store information about the physical and chemical properties of the soil and interpretations for soil management and use. To avoid confusion with the layer table defined in STATSGO, where layers are related to components of map units, the layer table in the NATSGO structure is called P.Layer to emphasize that the layers are related to phases.

Each polygon on the map represents a soil landscape area that is made up of one or more soil phases. A series of generalizations is shown in Figure 7, and the tables that define these generalizations store information on the proportions of the phases within the generalized units. The box labeled Full STATSGO Detail G0 represents a zero-level generalization in which there is no generalization of the data. It is structured in the same way as the other generalizations so that the same programs can be used for analysis. Generalizations (G1, G2, ..., Gn) can be created from either the Full STATSGO Detail or from other generalizations. The details of the linking tables are beyond the scope of this paper, and in the case of the full STATSGO detail used for this analysis, their structure has the same effect as the map unit and map unit-phase tables in the schematic example presented in Figure 6.

Queries such as the organic carbon analysis can be run and maps can be plotted using any generalization. The composition of the generalized polygons accounts for the area of each soil phase included in the original STATSGO data.

C. Soil Properties for Carbon Analysis

A national inventory of soil organic carbon is an example of a NATSGO product. The attribute data for STATSGO and NATSGO are extracted from the SCS's national data base of the properties of soil series, the SIR, which contains estimates of the soil organic matter content. The concept of a soil horizon, which is applicable to describing a single soil profile, has been generalized for the SIR to the concept of a soil layer, which is applicable to describing a phase of a soil series. Information on the organic O horizon is

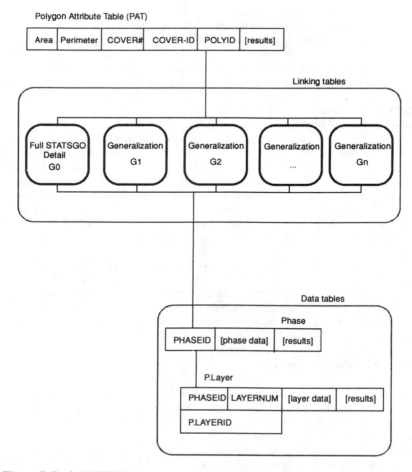

Figure 7. Basic NATSGO data structure. Alternative generalizations appropriate for mapping scales of 1:1,000,000 or 1:5,000,000 or smaller can be developed to connect the phase and P.Layer tables to the digital map. The area of each soil phase is retained in the statistics for the generalized polygons.

not included, except for Histosols. For this analysis, the organic matter values for layers contiguous to the surface are converted to soil organic carbon values. The amounts of carbon and corresponding soil volumes are accumulated across the layers within a soil profile, the components of STATSGO map units, and aggregated for each State. The results are mapped as the average soil carbon content per unit area (kg m^{-2}) and the average depth (m) of soil with carbon data.

The following data items are used in the organic carbon analysis (SCS, 1991):

Phase table
NONSOILFLAG Code indicating soil, nonsoil, or water (non-standard item)

Layer table
LAYDEPML layer depth, high (lower boundary of soil layer, m)
INCH10L   layer depth, low (upper boundary of soil layer, m)
LAYDEPMH minimum percent by weight of rock fragments > 25 mm
INCH10H   maximum percent by weight of rock fragments > 25 mm
INCH3L   minimum percent by weight of rock fragments 7.5 to 25 mm
INCH3H   maximum percent by weight of rock fragments 7.5 to 25 mm

NO10L	minimum percent by weight passing No. 10 sieve (7.5 mm)
NO10H	maximum percent by weight passing No. 10 sieve (< 7.5 mm)
BDL	bulk density, low (minimum of range, g cm^{-3})
BDH	bulk density, high (maximum of range, g cm^{-3})
OML	organic matter, low (minimum of range, percent by weight)
OMH	organic matter, high (maximum of range, percent by weight)

For the purpose of the organic carbon analysis, the Phase table has been augmented to store a code indicating whether a map unit component is a soil (phase of a soil series) or a nonsoil (water, gravel pit, rock outcrop, urban land, etc.). Some item definitions reflect a change in units (for example, inches to meters for depth) from the standard form distributed by the SCS.

Most quantitative data are reported with two values that define a range (low, high). For example, information on the carbon content of soils is stored on the layer table in terms of a range of organic matter percentages. Two values are given: OML represents the low end of the range, and OMH represents the upper end. The midpoint of the range is used to represent the organic matter content of the layer for the purpose of this paper.

V. Using NATSGO Data for Organic Carbon Analysis

The organic carbon is calculated for each layer in the P.Layer table that is not excluded as a nonsoil. Using CL to represent the organic carbon for the layer (g m^{-2}), then

CL = 5800 ODRT

where

5800 = (.01)(.58)(1,000,000) as explained below,
O = Midpoint organic matter (g/100g soil) = (OML + OMH)/2,
D = Midpoint bulk density (g soil/ cm^3 fine soil fraction) = (BDL + BDH)/2,
R = Rock fragment conversion factor (see below), and
T = Thickness of the layer (m) = LAYDEPMH - LAYDEPML.

OML and OMH are given as grams of organic matter per 100 grams of soil. This is multiplied by 0.01 to convert from a 100 g basis to a 1 g basis. The organic matter value is converted to a measure of organic carbon by multiplying by .58 (method 6A, Soil Survey Staff, 1992). The mass basis for organic carbon (appropriate for the laboratory) is converted to its equivalent volume basis (appropriate for a GIS) by multiplying by the bulk density, resulting in grams of carbon per cubic centimeter. The result is multiplied by 1,000,000 to convert from a cubic centimeter to a cubic meter basis.

A rock fragment conversion factor (R) is used to adjust for the volume of rocks in the soil sample (Method 3B2, Soil Survey Staff, 1992) according to the following relationship:

$$R = \text{Rock fragment conversion factor} = \frac{\text{Volume moist <2mm fabric}}{\text{volume moist whole soil}}$$

$$= \frac{V_{fines}}{V_{fines} + V_{rock}} = \frac{(M_f)/(D_b)}{(M_f)/(D_b)+(M_r)/(D_p)} = \frac{(R_{ft})/(D_b)}{(R_{ft})/(D_b)+(R_{rt})/(D_p)}$$

with

$R_{ft} = (R_{fs})(R_{st})$
$R_{fs} = (NO10L + NO10H)/200$
$R_{st} = 1 - R_{yt} - R_{zt}$
$R_{yt} = (INCH3L + INCH3H)/200$
$R_{zt} = (INCH10L + INCH10H)/200$
$D_b = (BDL + BDH)/2$
$R_{rt} = 1 - R_{ft}$
$D_p = 2.65 g\ cm^{-3}$

The laboratory data are reported on the basis of the mass (M_f) of the fine soil fraction (less than 2 mm diameter). The mass of rocks (M_r) represents all rocks greater than 2 mm. The conversion factor is expressed in terms of soil volumes, but the reported values in the data base are in terms of mass percentages. Dividing the masses (M_f and M_r) by the total mass of the soil allows the conversion factor to be defined in terms of R_{ft}, the ratio of the mass of soil fines to the total mass of the soil, and R_{rt}, the ratio of the mass of rock to the total mass. The bulk density (D_b) and particle density of rock (D_p) are used to convert each fraction to a volume basis. The soil less than 2 mm in diameter is reported as a percentage (mass basis) of the soil less than 3 inches in diameter in the data items NO10L and NO10H. These are used to compute the mass ratio of fines to the sampled fraction R_{fs}. The percentage of the total soil mass that is rocks between 3 and 10 inches is reported in the data items INCH3L and INCH3H. Similarly, the percentage of the total soil mass that is rocks greater than 10 inches in diameter and less than the size of the pedon is reported in the data items INCH10L and INCH10H. These are expressed as mass ratios (R_{yt} and R_{zt}), and the mass fraction of the sampled soil (R_{st}) is computed by subtraction. R_{st} is used in computing the mass fraction of fines on a total soil basis (R_{ft}). The mass fraction of rock (R_{rt}) is computed by subtraction. Tests are included in the calculation to prevent illogical values, such as more than 100% rock.

The final factor for computing C_L is the thickness (T) of the layer (in meters). Multiplying the other factors (5800 ODR) by T integrates over the depth of the soil, to give a measure of the amount of carbon per unit surface area (g m^{-2}) for the layer. This is suitable for representing the carbon values on a two-dimensional map.

The values of organic carbon (g m^{-2}) are summed over the layers, and stored on the phase table as the carbon content of the phase (C_P) per unit area (g m^{-2}), where N is the number of layers with carbon data continuous to the surface.

$$C_P = \sum_{L=1}^{N} C_L$$

For this analysis, only those layers that have organic carbon values extending continuously to the surface are included. Thus, if the surface layer has all the necessary data, the second layer is missing a carbon or bulk density value, and the third layer has all the necessary data, then only the value from the surface layer is used. If the phase does not represent water or a nonsoil, and the surface layer of the phase does not have reported carbon data, then the phase is classified as having missing data. Each phase record is classified into one of four categories: reported data, missing data, nonsoil, or water.

The information in the phase table is then summed and stored on the G0 table for the query, a linking table that has a one-to-one correspondence to records in the polygon attribute table. The areas of reported data, missing data, nonsoil, and water are summed separately within each polygon. For the phase records having reported carbon estimates, the mass of carbon (g m^{-2}) and the depth of reported carbon data (m) are weighted by the area, and summed to give polygon totals for soil organic carbon (g) and reported soil volume (m^3). For each polygon record, the total soil organic carbon (g) is divided by the reported area (m^2) to give an amount of organic carbon per unit area for the polygon (g m^{-2}). The reported volume (m^3) is divided by the reported area (m^2) to give an average depth (m).

To make a map, the polygons are classified according to carbon content per unit area (g m^{-2}), and these classes form the legend for shading the map and tabulating the carbon totals, areas, and volumes. Three maps for New Jersey are used to illustrate the results. A soil organic carbon map is shown in Figure 8, a map of the percentage of the area with reported data is given in Figure 9, and a map of the average depth is given in Figure 10. Darker tones in Figure 8 indicate higher contents of soil organic carbon per unit area. Darker tones in Figure 9 indicate a higher percentage of reported data. Darker tones in Figure 10 indicate a greater depth of soil has reported data that extends continuously to the surface.

Table 2 presents the results of carbon analysis for 40 States for which data were available at the time of writing. If a component of a map unit has both organic matter and bulk density values for the surface layer in the STATSGO data base, then the area that the component represents on the map is summed to compute the reported area for each State. The mass of soil organic carbon is computed for the reported area. A statewide average of soil organic carbon per unit area is computed by dividing the total carbon content by the reported area. The percentage of the State with reported soil organic carbon data is presented, as well as the percentage with missing data. The percentages of reported and missing data do not necessarily add to 100 percent, because areas of nonsoil (rock outcrop, gravel pits, urban land) and water are not explicitly shown in the table. The organic carbon content per unit area in this selection of States varies from 1.9 kilograms of carbon per square meter (kg C m^{-2}) for Montana to 27.5 kg C m^{-2} for Florida. Montana, Pennsylvania, and West Virginia have mountainous areas, with lower proportions of agricultural soils. Minnesota and Rhode Island have significant proportions of peat soils (Histosols), and Minnesota has extensive areas of black agricultural soils (Mollisols). Many of the other States have mixtures of agricultural and forest soils, and have statewide average carbon contents per unit area in the range 5 to 15 kg C m^{-2}.

The impact of the missing data on the totals and averages is unknown. As more soils are coded with organic matter values, the total organic carbon values may be expected to increase. It is possible, however, that a zero value (interpreted as less than 0.25 percent organic matter) is appropriate for many of the soils that are recorded as missing. It is recommended that the data structure be revised to add a data quality code for each quantitative measure, to indicate whether the quantitative measure is reported or missing. Then, if the quality code indicates reported data and the quantitative measure is zero, the zero can be used in the calculations. This coding should be applied to all layers in the soil. The same data quality code could be used to distinguish whether the quantitative value has been measured or estimated, and to indicate the analytical procedure for measurement or the method of estimation.

VI. Summary

An inventory of soil organic carbon for the United States is being developed using the soil geographic data bases of the United States Department of Agriculture, Soil Conservation Service. The soil carbon information is important for understanding the role of soils in the global carbon cycle, and will be used for climate modeling and assessing the impacts of potential global warming. The soil data bases have many other applications, and will form the basis for the United States contribution to improved global soil data bases.

The State Soil Geographic (STATSGO) data base was analyzed. Geographic information system tools are being developed so soil scientists can generalize the STATSGO data to form a new National Soil Geographic (NATSGO) data base. These data bases are designed to respond to the needs of global change modelers for current data on the Nation's soils, and to fulfill a responsibility for maintaining soil geographic data bases within the Federal Government.

Soil organic carbon has been calculated for 40 States. Among the selected States, Florida has the highest carbon per unit area, with 25.7 kg C m^{-2}, reflecting extensive wetland areas in the State. Minnesota has the second highest value, with 21.3 kg C m^{-2}, reflecting the high proportions of Histosols (organic soils) and Mollisols (mineral soils with high levels of organic matter) in the State. Montana has the lowest values, with 1.9 kg C m^{-2}, reflecting a low organic matter content of most soils. The 13 northeastern States have an average of 5.5 kg C m^{-2}, the 11 southern States have an average of 8.7 kg C m^{-2}, and the 12 midwestern States have an average of 12.2 kg C m^{-2}. The depth of soil used for the carbon calculation varies by soil type, and is determined by the available data. Preliminary analyses indicate that the 13 northeastern States have a total of 2685 teragrams (1 Tg = 10^{12} g) of soil organic carbon, 11 southern States have a total of 16,505 Tg C, and 12 midwestern States have a total of 23,568 Tg C. The analysis has identified cases of missing information on soil organic matter or bulk density in the data bases. Revisions will be made to these totals as more data are collected.

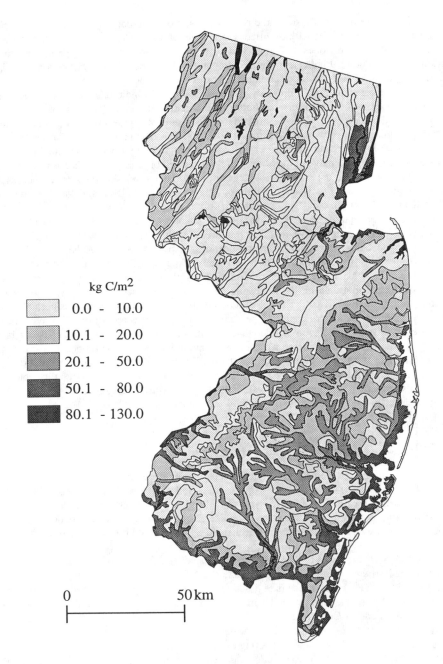

Figure 8. Soil organic carbon for New Jersey.

Preparing a Soil Carbon Inventory for the United States Using Geographic Information Systems

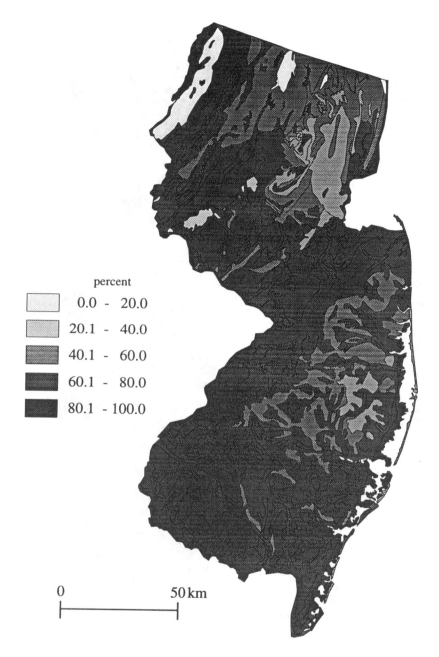

Figure 9. Percentage of area with reported carbon data.

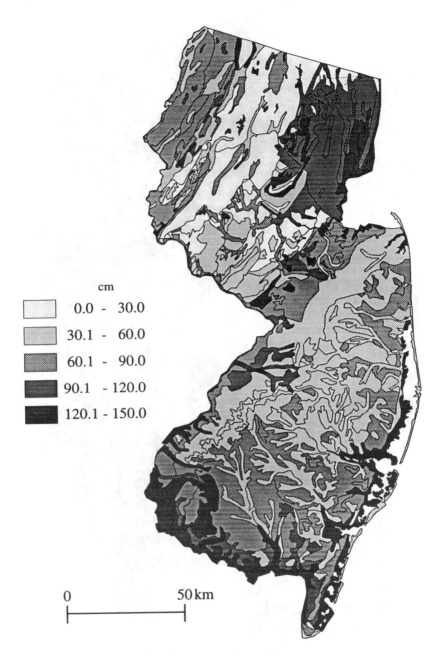

Figure 10. Depth of soil with reported carbon data.

Table 2. Soil organic carbon for selected States

State	Soil organic carbon (Tg C)	Soil organic carbon per area (kg C m^{-2})	Area with reported data (km^2)	Percent area with reported data	Percent area with missing data
Connecticut	71	10.5	6,752	53	44
Delaware	47	9.6	4,893	93	3
Maine	282	8.4	33,686	40	55
Maryland	161	6.7	23,899	87	5
Massachusetts	55	5.2	10,543	50	37
New Hampshire	74	6.6	11,087	46	48
New Jersey	229	13.9	16,503	84	8
New York	747	7.3	102,659	81	13
Pennsylvania	265	2.6	101,221	86	12
Rhode Island	22	15.8	1,390	49	41
Vermont	98	5.2	18,761	75	20
Virginia	516	5.0	102,228	97	1
West Virginia	118	2.2	53,827	86	12
Subtotal, 13 NE States	2,685	5.5	487,449	77	19
Alabama	535	4.1	128,314	96	2
Arkansas	814	6.1	132,897	96	2
Florida	3,504	25.7	136,083	89	1
Georgia	1,232	8.3	147,637	97	1
Kentucky	454	4.5	101,227	97	0
Mississippi	457	3.8	119,340	97	3
North Carolina	1,761	14.3	124,555	97	1
Oklahoma	1,132	6.5	175,226	97	1
South Carolina	888	11.3	78,406	98	0
Tennessee	408	4.1	99,678	91	4
Texas	5,320	8.2	648,449	94	2
Subtotal, 11 S. States	16,505	8.7	1,891,812	95	2
Illinois	1,653	11.5	143,837	99	1
Indiana	1,062	11.3	93,582	100	0
Iowa	2,743	19.1	143,801	99	0
Kansas	1,875	8.9	210,524	99	1
Michigan	2,726	18.5	147,101	98	1
Minnesota	4,432	21.3	207,888	95	1
Missouri	1,222	6.9	177,035	98	0
Nebraska	1,421	7.2	198,361	99	0
North Dakota	1,757	9.9	177,654	97	1
Ohio	676	6.4	105,369	99	0
South Dakota	1,590	8.3	192,544	96	1
Wisconsin	2,411	17.2	140,181	96	0
Subtotal, 12 Midwestern States	23,568	12.2	1,937,877	98	1
Colorado	747	3.1	237,499	88	6
Hawaii	186	16.1	11,491	70	1
Montana	550	1.9	286,325	75	17
Puerto Rico	44	5.6	7,902	88	2

VII. Conclusions

The analysis of soil organic carbon will be applied on a nationwide basis as soon as the STATSGO data base is complete. The STATSGO and NATSGO data bases provide better spatial and attribute detail than has been available for national soil maps in the United States. The analyses contribute to validating or refining large-area estimates of soil carbon, and will provide a prototype for mapping the spatial distribution of soil carbon globally.

The estimates of the total amount of soil carbon are not expected to have sufficient accuracy to directly assess rates of change or the magnitude of fluxes of carbon to and from the atmosphere. However, the soil data bases may contribute to understanding the spatial extent of the processes, so that reasonable hypotheses can be formulated on the mechanisms for uptake, storage, and release of carbon compounds.

Acknowledgments

Scientists in the USDA SCS Soil Survey Division, National Soil Survey Center, the U.S. Geological Survey, EROS Data Center, and universities are cooperating on NATSGO development as part of their respective contributions to the U.S. Global Change Research Program.

References

Aiken, G.R., D.M. McKnight, R.L. Wershaw, and P. MacCarthy. 1985. *Humic Substances in Soil, Sediment, and Water: Geochemistry, Isolation, and Characterization.* John Wiley & Sons, NY.

Beinroth, F.H. (ed.). 1992. *Organic carbon sequestration in the soils of Puerto Rico.* U.S. Department of Agriculture, Soil Conservation Service, World Soil Resources, Washington, D.C.

Burke, I.C., D.S. Schimel, C.M. Yonker, W.J. Parton, L.A. Joyce, and W.K. Lauenroth. 1990. Regional modeling of grassland biogeochemistry using GIS. *Landscape Ecology* 4:45-54.

Bohn, H.L. 1976. Estimate of organic carbon in world soils. *Soil Science Society of America Journal* 40: 468-470.

Bolin, B. 1970. The carbon cycle. *Scientific American* 223:124-132.

Bouwman, A.F., (ed.). 1990. *Soils and the greenhouse effect: The present status and future trends concerning the effect of soils and their cover on the fluxes of greenhouse gases, the surface energy balance and the water balance.* John Wiley & Sons, NY.

Committee on Earth and Environmental Sciences. 1992. *Our Changing Planet: The FY 1993 U.S. Global Change Research Program, A supplement to the U.S. President's 1993 Budget.* Office of Science and Technology Policy - Federal Coordinating Council for Science, Engineering, and Technology (FCCSET), National Science Foundation, Washington, D.C.

Food and Agriculture Organization of the United Nations. 1970-1978. *Soil Map of the World, Scale 1:5,000,00.*, Volumes I-X. United Nations Educational, Scientific, and Cultural Organization, Paris.

Hall, D.O. 1989. Carbon flows in the biosphere: Present and future. *Journal of the Geological Society, London* 146:175-181.

Intergovernmental Panel on Climate Change. 1990. *Climate Change: The IPCC scientific assessment.* (J.T. Houghton, G.J. Jenkins, and J.J. Ephraums, eds.). Cambridge University Press, Cambridge, England.

International Geosphere-Biosphere Program (IGBP)-Data and Information System (DIS) Standing Committee, 1992. Meeting IGBP data needs. *Global Change Newsletter.* International Geosphere-Biosphere Programme, International Council of Scientific Unions, Royal Swedish Academy of Sciences, Stockholm, December 1992.

Kimble, J.M., H. Eswaran, and T. Cook. 1990. Organic carbon on a volume basis in tropical and temperate soils. p. 248-253. In: *14th International Congress of Soil Science, Transactions.* Volume V, Commission V, Kyoto, Japan, August 1990.

Loveland, T.R., J.W. Merchant, D.O. Ohlen, and J.F. Brown. 1991. Development of a land-cover characteristics data base for the conterminous U.S. *Photogrammetric Engineering and Remote Sensing* 57:1453-1463.

Moore, Berrien III. 1992. Global Analysis, interpretation, and modelling. *Global Change Newsletter*. International Geosphere-Biosphere Programme, International Council of Scientific Unions, Royal Swedish Academy of Sciences, Stockholm, December 1992.

National Academy of Sciences. 1990. *Global Change Research Program*. Draft, Working Group Report on Trace Gases and Nutrient Fluxes, 52 pp.

Office of Management and Budget. 1990. *Coordination of Surveying, Mapping, and Related Spatial Data Activities*. Circular No. A-16 (revised), Executive Office of the President, Office of Management and Budget, Washington, D.C.

Post, W.M., T.-H. Peng, W.R. Emanuel, A.W. King, V.H. Dale, and D.L. DeAngelis. 1990. The global carbon cycle. *American Scientist* 78:310-326.

Sarmiento, J.L. and E.T. Sundquist. 1992. Revised budget for the oceanic uptake of anthropogenic carbon dioxide. *Nature* 356:589-593.

Schimel, D.S., T.G.F. Kittel, and W.J. Parton. 1991. Terrestrial biogeochemical cycles: global interactions with the atmosphere and hydrology. *Tellus* 43AB:188-203.

Schlesinger, W.H. 1984. Soil organic matter: A source of atmospheric CO_2. p. 111-127. In: G.M. Woodwell (ed.) *The role of terrestrial vegetation in the global carbon cycle: Measurement by remote sensing*. John Wiley & Sons, NY.

Soil Conservation Service. 1991. *State Soil Geographic Data Base (STATSGO): Data users guide*. U.S. Department of Agriculture, Soil Conservation Service, Miscellaneous Publication Number 1492, U.S Government Printing Office.

Soil Survey Staff. 1975. *Soil Taxonomy: A basic system of soil classification for making and interpreting soil surveys*. U.S. Department of Agriculture, Soil Conservation Service, Agriculture Handbook 436, U.S. Government Printing Office, Washington, D.C.

Soil Survey Staff. 1992. *Soil Survey Laboratory Methods Manual*. Soil Survey Investigations Report No. 42, Version 2.0. National Soil Survey Center, Soil Conservation Service, U.S. Department of Agriculture, Lincoln, NE.

Stevenson, S.J. 1982. *Humus Chemistry: Genesis, Composition, Reactions*. J. Wiley & Sons, NY.

Sundquist, E.T. 1993. The global carbon dioxide budget. *Science* 259: 934-941.

U. S. Geological Survey. 1987. *The National Atlas of the United States of America: Soils*. U.S. Geological Survey, Reston, Virginia.

Watson, R.T., L.G. Meira Filho, E. Sanhueza, A. Janetos. 1992. Sources and sinks. p. 25-46. In: J.T. Houghton, B.A. Callander, S.K. Varney (eds.). *Climate Change 1992: The Supplementary Report to the IPCC Scientific Assessment*. Cambridge University Press, Cambridge, England.

Zinke, P.J., A.G. Stangenberger, W.M. Post, W.R. Emanuel, and J.S. Olson. 1986. *Worldwide Organic Soil Carbon and Nitrogen Data*. ONRL/CDIC-18, NDP-018, Oak Ridge National Laboratory, Oak Ridge, Tennessee.

CHAPTER 24

Establishing the Pool Sizes and Fluxes in CO_2 Emissions from Soil Organic Matter Turnover

E.A. Paul, W.R. Horwath, D. Harris, R. Follett, S.W. Leavitt, B.A. Kimball, and K. Pregitzer

I. Introduction

Soil organic matter (SOM) is known to act both as a source and a sink in global CO_2 cycles. Its role in the biological productivity of managed and unmanaged systems is equally well recognized. Central to any rigorous examination of the above roles are questions relating to what is there, and what happens to it? This means that we must ask what are the pools and the fluxes involved? The degradation of the plant components and plant respiration are an integral component of these fluxes. These are difficult to separate from the soil components. Most soils contain at least 5-20% of their organic carbon (C) as partially decomposed particulate residues (Cambardella and Elliott 1992) with native grassland having >30% of its SOM in this fraction. The method of determination and the C:N ratio of this fraction indicates that it contains partly decomposed plant residues, fine roots and extensive numbers of associated microorganisms. This fraction is closely related to the short term mineralization of C and N (Janzen et al., 1992).

Most present models consider plant residues as a separate group of pools. Difficulties arise in the characterization of partially degraded residues and their associated microflora. The extent, turnover and incorporation of soil exudates into humic constituents also is poorly defined. They vary widely depending on plant age, type, and abiotic conditions. The average result from many ^{14}C studies shows that 1 to 7% of net primary plant productivity is exuded into soil. The turnover of roots is not included in this calculation. This produces much more carbon than the exudates. Studies of wheat root distribution by washing have indicated that root weights comprise approximately 12% of the stover (Wilhelm et al., 1982). Other studies have shown larger amounts, e.g. 50% of total residues (Buyanovsky and Wagner, 1986).

The most useful techniques for testing and summarizing the various soil pool sizes and their dynamics involve mathematical models. They incorporate moisture and temperature as controlling factors in both plant productivity and SOM decomposition. They also reflect the effects of the amount of clay on stabilization of SOM and soil microbial biomass (SMB). Soil depth is taken into account and the majority of the models include information on plant nutrient and water uptake. There is some but not enough integration with landscape, and plant root development, nor is there an adequate understanding of the role of soil aggregates in the stabilization of plant residues, microbial biomass, and soil humic constituents. These models have greatly helped in the general management of soils and in initial approximations of the role of SOM in global C cycles. We have to refine our models if they are going to answer the important societal questions concerning global change, manure applications, fertilizer N additions, and atmospheric pollution effects on vegetation and water quality. This refinement requires better estimates of SOM fractions and their dynamics.

II. Pool Sizes and Fluxes in Soil Organic Matter Turnover

The first research requirement in determining pool sizes and fluxes is a well-characterized site with measured abiotic characteristics, SOM levels and above and beneath ground residue inputs. Since man's activities play such a profound role, whether it be in slash/burn agriculture, clear-cut timber harvesting, or tillage, these sites should be representative of both the major climatic areas and management procedures involved. The *Global Directory of Long-Term Agronomic Experiments* (Steiner and Herdt, 1993) helps identify some of these sites. Other sites include the NSF-LTER programs and the EPA study of agronomic management in the U.S. (Elliott et al., 1994).

Identification of the various pools of SOM is the second requirement. Soil organic matter is a complex series of related but different molecules with varying physical structure, molecular weight, chemical structures and functional groups. It is associated with itself, clays, and microorganisms and is within and between aggregates as well as residing on the soil surface and having been leached to greater depth. Therefore, it is not surprising that no one physical-chemical fractionation technique will separate the components into meaningful biological fractions. Of the many that have been tried, a few can be conducted with fairly straight forward equipment in a reasonable time period. These include the measurement of microbial biomass, long-term mineralization of C and N, the use of acid hydrolysis, the separation of aggregates on a size and stability basis, the measurement of particulate SOM, the determination of the light fraction, and the measurement of organic constituents associated with soil separates. Newer techniques, such as the use of XRD-4 resin, hold promise for separating polysacharides and possibly proteinaceous materials (Wilson, 1991).

The last requirement in determining SOM dynamics relative to CO_2 fluxes is that of measuring the rates of addition of substrate and turnover of the various pools. Tracers allow one to identify specific fractions, measure their turnover rate, mineralization-immobilization processes, plant uptake, and microbial growth. The use of tracers, therefore, is vital in identifying pool sizes and determining at what speed materials flow in and out of them, e.g. the fluxes. The most useful tracers include ^{13}C in both enriched ^{13}C experiments and those that use the isotopic signal provided by the discrimination that occurs in the C_3 (Calvin cycle) and C_4 (Hatch-Slack cycle) plants grown on the same site. Alternatively, ^{14}C can be used in the enriched form or as that occurring in the atmosphere. Thermonuclear bomb testing produced a ^{14}C signal during the period from 1960 to 1985 that is useful in following medium term SOM dynamics, especially if stored reference samples are available.

The availability of accurate, automated mass spectrometers for stable isotopes, and tandem accelerator mass spectrometers for naturally occurring ^{14}C now make it possible to process numerous samples containing microgram quantities of C. The use of enriched CO_2 in experiments to measure the effects of global climate change either in enclosures or unenclosed (FACE) can provide both a ^{14}C and a ^{13}C signal.

III. The Use of Carbon Dating

Carbon dates have looked at soil development, chrono sequences, landscape positions, the effects of cultivation, and of soil forming processes (Anderson and Paul, 1984; Campbell, et al., 1967; Coleman and Fry, 1992). Soils differ greatly depending on slope, aspect, vegetation type, and position in the landscape. Related differences in parent materials and moisture availability directly affect ecosystem productivity and agronomic yield. Techniques now available for the farming of fields on a soil type-landscape basis make it possible to manage individual soil types and landscape positions. We should not expect SOM to be similar across such a range of sites. The example in Table 1 shows that the SOM content of the upper position of a catena in Saskatchewan is lower than that of the more moist toe slopes and that there are buried horizons in the bottom slope position. This shows the effects of deposition from erosion. The soils at the top of the slope have a C age of 500 yr, but the surface of the moist, lowest position in the landscape dates modern. The reason for this is not well understood. If soil is eroded from the upper to the lower slope positions, C dating should reflect this. One must conclude that the more recent C is being transported to an area that has a generally higher turnover rate.

Agriculture, and to some extent forestry research, has been dominated in the last 60 yr by the measurement of the decrease in SOM contents following disturbance. It has been easier to follow and model the drop with disturbance than it will be for us to accurately predict the buildup of SOM as we change our management techniques in an increased CO_2 environment. The example given in Table 2 shows the

Table 1. The effect of landscape on soil organic matter contents and radiocarbon age; Oxbow Catena Saskatchewan

	Top slope	Mid slope	Toe slope	Depression
Ap Horizon cm	0-10 cm	0-15 cm	0-15	0-17.5
C%	3.3	3.6	4.6	3.9
Horizon	Radiocarbon age BP			
Ap	500	270	216	Modern
Ae 17-22.5 cm			685	-
B_T 22-37 cm			930	700
B_{Tg} 60-85 cm				4870

(Adapted from Martel and Paul, 1974.)

Table 2. The effect of cultivation on the radiocarbon age; 0-10 cm Oxbow Saskatchewan

	Virgin	15 yr cultivation	60 yr cultivation
C%	5.4	3.5	2.2
N%	0.45	0.34	0.23
Radiocarbon age BP	250 ± 65	295 ± 75	170 ± 60

(Adapted from Martel and Paul, 1974.)

Table 3. The effect of bomb ^{14}C on the radiocarbon age (Indian Head, Saskatchewan).

Fraction	Proportion of total %	1963 pmC	1963 age yr	1978 pmC	1978 age yr
Soil	100	79	1900	83	1500
Nonhydrolyzable org-C	50	70	2800	77	2100

(Adapted from Anderson and Paul, 1984)

traditional 50% drop in SOM content on a weight basis. Changes in bulk density and measurement of SOM in the total profile must be taken into account, and the drop on a soil area basis is not as great as that shown when expresed as a percentage. Water erosion transports much of the soil into depressions in the same field from which it was originally displaced. Wind erosion usually is more extensive but also fills in low areas in a rolling landscape and areas with adjacent vegetation. The extent to which C deposited in the low areas is stored or lost to the atmosphere by decomposition is estimated at 10-50% of that transported.

In the example given, the radio C age of the virgin grassland soil is 250 yr. This increased through 60 yr of cultivation to 710 y showing the effects of the removal, by decomposition, of the more recently added materials. Bomb C effects are shown in Table 3. The samples collected in 1963 preceded the maximum incorporation of the ^{14}C into SOM. The overall soil at this site dated 1900 yr. This was decreased to 1500 y in 1978 as the thermonuclear-produced ^{14}C was incorporated into the SOM. Nonhydrolyzable organic C, comprising 50% of the total was very much older in the 1963 sample, it also was influenced, to some extent, by the effects of bomb C, for in 1978 this fraction had dropped an equivalent number of years. This is not surprising; the hydrolysis of modern, fairly high ligniferous plant residues also yields approximately 50% of the material as a nonhydrolyzable residue. The acid hydrolysis residue does not drop appreciably with cultivation showing that chemical recalcitrance is not the only factor causing these great ages. This led Martel and Paul in 1974 to suggest that physical protection played a major role in SOM dynamics. The role

Table 4. The effect of soil separates on radiocarbon ages; Melfort, Saskatchewan

Fraction	Proportion of organic C%	pm C	Age
Coarse silt	25	91	800 ± 50
Fine silt	29	89	965 ± 50
Coarse clay	31	86	1255 ± 60
Fine clay	8	98	170 ± 50

(Adapted from Anderson and Paul, 1984.)

Table 5. Comparison of the distribution of C and ^{14}C using radiocarbon dating and ^{14}C incorporation into microbial biomass in Western Canadian Wheat soil

	Radio-carbon age	Soil C %	Radio-carbon C %	Microbial ^{14}C %
Total soil	350 yr BP	100	100	100
Light fraction	240 pmC	6	8	5
0.5 N HCl hydrolysate	107 pmC	35	37	50
6.0 N HCl hydrolysate	161 pmC	22	24	20
6.0 N residue	1765 yr BP	37	31	25
NaOH extract	1900 yr BP	27	22	16
H$_2$O dispersion	1800 yr BP	3	3	3
Residue	1330 yr BP	7	6	6

(Adapted from Martel and Paul, 1974).

of microaggregates as outlined by Waters and Oades (1991), and Cambardella and Elliott (1992, 1993) further supports the concept of aggregation in SOM stabilization.

Numerous studies have shown that the fractionation of soil separates into sand, silt, and clay give significantly different C dates. The fine silt and course clay not only contain the largest amounts of C, they also contain the oldest C (Table 4). Fine clay is much younger. The possibility of contamination of the fine clay by microbial constituents dispersed during the fractionation procedure must be considered. The interaction of the soil separates with the soil aggregates must be further investigated, e.g. it is said that the microaggregates are the most stable. The tracer content of silt, course clay and fine clay in microaggregates and macroaggregates needs to be determined.

Table 5 shows the results of another method of fractionation. The ^{14}C distribution was measured by two techniques, by radio C dating of naturally occurring ^{14}C and by scintillation counting of the ^{14}C that had been added as glucose and was then incorporated into microbial biomass. The light fraction floated in a dense liquid represents decomposing residues and associated microorganisms. This fraction is equivalent to the particulate organic matter of other studies. In this case, the light fraction had a ^{14}C content slightly greater than twice that of the pre-bomb C standard. It represented 8% of the naturally occurring ^{14}C as determined by C dating; however, it represented 5% of the material where the ^{14}C was derived from microbial biomass, showing that significant quantities of the biomass are associated with the plant residues. Multistep acid hydrolysis has been shown to be an effective way of separating labile constituents. The high occurrence of labile C in the radio C distribution and when incorporated as microbial C shows the effectiveness of this technique; 71% of the ^{14}C introduced as microbial biomass was removed by hydrolysis. The nonhydrolyzable residue in the C dated fraction shows great age. While representing 37% of the total C, it only represents 31% of the ^{14}C. That hydrolysis is not completely effective as a separation technique

Table 6. Distribution of ^{14}C (%) in a poplar-soil system two weeks and one year after labelling in September. Standard errors of the mean shown in parenthesis

	Day 14	Day 350	From litter[a]	Total
Leaves/litter	20 (1.5)	0.12	5.5 (0.25)	
Stems/branches	28 (4.6)	13 (1.6)		
Roots < 0.5 mm	7 (1.8)	5 (0.45)		
Roots > 0.5 mm	31 (3.2)	11 (1.0)		
Root-soil respiration	12 (1.50)			50.6
Microbial biomass	1.3 (0.2)	0.5 (0.8)	0.18	
Soil	0.7 (0.5)	3.7 (1.8)	3.4	
Total	100	34.3	9.1	43.4

[a] Derived from separately placed litter after 325 days.
(From Horwath, 1993).

is indicated by the fact that 25% of the recently incorporated microbial ^{14}C is included in this fraction. Further fractionation of the soil by sodium hydroxide extraction, followed by water dispersion leaves a residue of only 7%, which upon microscopic examination was shown to be bits of charcoal, insect skeletons, etc. It is not as old as the peptizable fractions.

IV. The Use of Enriched ^{14}C in Determining Plant Inputs

Enriched ^{14}C is particularly valuable in measuring both direct and indirect plant inputs such as exudates and root turnover. The example in Table 6 shows the results of labeling 3m high hybrid poplars with ^{14}C in September. This was followed by a 14 d equilibration period before analysis. Forty-eight percent of the labelled C remained above ground in leaves, stems, and branches. Fifty-two percent was translocated underground. The majority of this was found in large roots. Respiration during the 14 d period, while the ^{14}C was being transformed to longer-life compounds within the plant accounted for 12% of the net ^{14}C fixation. The soil microbial biomass plus the associated SOM represented 2% of the net fixation. If one assumes a 35% efficiency in the incorporation of root derived substrates into microbial biomass and microbial products, 5.7% of the plant fixed ^{14}C is accounted for. Fine roots did not lose ^{14}C, but stems and branches did. We concluded that fine root growth was derived from previously stored C, rather than directly from photosynthate. In the first year after labelling, the tree components plus the leaf litter lost 56.6% of the ^{14}C. During this period the microbial biomass showed a 6-fold dilution of its ^{14}C contents, indicating the turnover rate of this component.

The leaf litter from the labelled trees was placed onto soil of unlabelled trees; the labelled trees received the unlabelled litter. This replacement experiment made it possible to differentiate the leaf derived from the root derived C. The litter placed on the surface lost 73% of its label; 17% was translocated to the soil and associated microbial biomass. At this site the leaf litter plays a significantly larger role in forming soil microbial biomass but the proportion contributed to the soil itself is approximately equal to that of roots.

The September labelling (Table 6) moved much more C beneath ground than did a July labelling (data not shown). In addition, the material from the July labelling did not decompose as rapidly as that from the September labelling. This indicates that both the proportion of above and beneath ground C and the quality of the C changes with season. Significant misinterpretations can occur if one assumes that a uniform substrate is added to the soil over the complete growing season.

V. The Use of ^{13}C

The stable isotope of C representing 1.1% of the atmosphere will receive increasing use because of safety concerns with enriched ^{14}C and the availability of accurate, sensitive, reasonably priced automatic mass spectrometers. The signal provided by the differential fractionation of $^{13}CO_2$ by C_3 or C_4 plants will be increasingly utilized. This technique is very useful for a number of reasons: 1) many fields can be sampled and multiple replicates can be utilized because of the reasonable cost of the analysis: 2) the replacement of C_4 grassland with C_3 wheat or soybeans; the growth of C_4 corn on C_3 grasslands in cooler climates and the substitution of C_4 plants onto C_3 forest vegetation in the tropics provides an extensive range of sites. 3) the length of time since cultivation of these sites ranging from 10 to 110 y is in the period of time where enriched ^{14}C experiments and naturally occurring C dates are most difficult to use.

There is an excellent range of $^{13}C:^{12}C$ ratios extending across the plains of North America where C_3 wheat is being grown on largely C_4 grasslands. On some of these sites wheat is now being replaced with maize. We thus have an identifiable ^{13}C signal moving back toward the signal provided by the original C_4 vegetation. In the eastern U.S., C_4 plants such as maize and sorghum grown on former forest vegetation provide equally good sites for analysis. The ^{13}C signal was used by Balesdent et al. in 1988 to show that in the Sanborn long-term plots at least 50% of the original tall grass prairie C remained after 100 yr of cultivation.

The ^{13}C signal has been effectively utilized in the tropics to differentiate soils from largely C_4 savannahs and the adjoining C_3 forests. Bonde et al., 1992, found that 12 yr of cropping of a forest soil to C_4 grasses resulted in a 90% turnover of the forest SOM in the sand fraction. In the clay fraction, only 40% of the C was from the C_4 grasses with the remainder being from the original forest. This represented turnover times of 4, 6, and 59 yr for the sand, silt, and clay fractions of soil, respectively. Other studies in the tropics have shown the $^{13}C:^{12}C$ ratios to change with depth. The ^{13}C signal of a podzol B horizon was from an ancient forest rather than the present savannah (Schwartz et al., 1986).

VI. Estimation of the Kinetics of Decomposition During Long-Term Incubation

The utilized chemical and physical fractionation techniques have limitations in adequately separating SOM fractions for analysis of their dynamics. An alternative, best used in conjunction with fractionation techniques, is to allow the microorganisms and their enzymes to identify the biologically active soil C. Extended incubations (100-250 d) have been found to be sensitive indicators of management effects. Table 7 illustrates that a grassland developed on a previously forested site with no cultivation during recorded time periods contained more C than cultivated sites and released significant levels of CO_2. At the time of sampling the reversion field had been allowed to revert to an old field status for 4 yr after a significant length of time in a corn-soybean rotation. Although total C contents were not yet different, the reversion treatment evolved 930 μg C g^{-1} 200 d^{-1} vs. 560 μg C g^{-1} 200 d^{-1} for the associated corn-soybean rotation. The C mineralized over the 200 d represented 7.3% of the total C in the grassland, 9.8% in the reversion and 5.9% in the corn-soybean plots, showing the rapid buildup of soil C within the reversion treatment.

Table 7 also shows that microbial biomass can be a sensitive indicator of changes, although it represents a smaller percentage of the total C than does the C mineralized in 200 d. The reversion treatment had the largest percentage of its C present as microbial biomass. The mineralization curves were fitted to the sum of two first-order equations. Carbon dating has shown that 50% of the C in most soils is very old, i.e. greater than 500 yr, therefore, it should not contribute significant amounts of C to a mineralization such as this. The definition of the second pool as $C_2 = (C_T/2)-C_1$ made it possible to obtain a good fit with the restricted number of points.

CO_2 analysis does not give an exact figure of the available C since microbial growth utilizes some of the substrate attacked during decomposition. The pool C_1 was corrected for microbial growth by utilizing growth efficiency calculations (Paul and Clark, 1989). In this study, as in many others, the decomposition rate constant is fairly consistent, but the pool size, C_1, was affected by treatment. Even though the incubation was only carried out for 200 d, kinetic analysis showed that the mean residence time of the C_2 pool ranged from 5.3 to 9.5 yr with the reversion treatment being the most active and the corn-soybean being the least active.

The use of C mineralization makes it possible to anlayse treatments from many long term experiments without the use of tracers (Elliott et al., 1993). Conducting a C mineralization study on a previously labelled sample, provides even more information. This can be either from C_3-C_4 vegetation or from enriched ^{14}C or ^{13}C studies. This type of analysis also makes it possible to further calculate pool sizes by isotope dilution calculations (Voroney et al., 1991).

Analysis by Horwath of the poplar plots (Table 6) shows that 60 d after the addition of ^{14}C the most active pool (C_1) comprised 1.4 to 3% of the total C and had a mean residence time of 25 to 50 d. Extended

Table 7. Kinetic analysis of CO_2 mineralization curves for three treatments of the KBS-LTER (Michigan State University) $C_m = C_1 (1 - e^{-k_1 t}) + C_2 (1 - e^{-k_2 t})$; $C_2 = (C_T)/2 - C_1$: standard error of the mean shown in parenthesis

	Grassland	Reversion	Corn soybeans
Total C (C_T) $\mu g\ g^{-1}$	15,000	9,500	9,500
C mineralized (C_m) $\mu g\ C\ g^{-1}\ 200\ d^{-1}$	1,100	930	560
C_m/C_T %	7.3	9.8	5.9
Microbial C $\mu g\ g^{-1}$	345 (51)	251 (38)	141 (22)
Microbial C/C_T %	2.3	2.6	1.5
Pool size (C_1) $\mu g\ g^{-1}$	623 (103)	404 (42)	170 (24)
Mineralization kinetics			
C_1^a/C_T%	4.1	4.2	1.8
$K_1\ d^{-1}$	0.022	0.029	0.024
$K_2\ d^{-1}$	0.00035	0.00052	0.00029
K_1 MRT d	45	34	41
K_2 MRT y	7.8	5.3	9.5
R_2	0.997	0.999	0.999

[a] C_1 corrected for microbial growth

mineralization studies of the SOM one year after labelling showed that the ^{14}C was now in much closer equilibration with the total soil C. The mean residence time of the ^{14}C derived CO_2 in the second pool now approximated 10 yr verses 3 yr in the sampling on 60 d after labelling.

VII. Enriched CO_2 Studies.

The use of enhanced CO_2 in experiments to measure the effect of future climate change represents a unique opportunity for determination of pool sizes and fluxes of SOM. Most of the CO_2 that is utilized in these experiments either in enclosed chambers, in enclosures open to the atmosphere at the top or completely free as in the Free Atmosphere Carbon Enrichment Experiments (FACE) comes from petroleum derived materials. The tank CO_2 has both a ^{14}C (0 PMC) and a ^{13}C signal of about -30‰. When mixed with air this becomes -10‰. The results from the FACE experiments in Arizona in which CO_2 was applied to cotton plants grown in the field for a 3 y period is shown in Figure 1. Cotton plants grown in a field with a normal atmosphere were compared to those with FACE air at 550 $\mu mol\ mol^{-1}$ of CO_2. The FACE air had a $\delta\ ^{13}C$ content of -17‰ and a ^{14}C of 75 pmC compared to the control of -7.5‰ $\delta\ ^{13}C$ and 115 pmC for ^{14}C. This resulted in control plants of -26.7‰ $\delta\ ^{13}C$ and 111.8 pmC. The FACE plants had a $\delta\ ^{13}C$ content of -38.4‰ and a ^{14}C of 72.1 pmC. The soil, after 3 y of cotton growth, was found to be labelled at -23.8‰ for the FACE compared to 22.4‰ for the control.

The associated ^{14}C signal was used to obtain further information on the dynamics of plant residues (Table 8). SOM left after flotation of the soil in the control plot at a specific gravity of 1.2 had an average age of 250 yr BP. Careful microscopic picking of the plant residues raised the total soil to a 1000 yr BP. On hydrolysis, this gave a residue of 2400 y BP and a supernatant of 250 y BP. The equivalent FACE experiments show that the incorporation of FACE residue depleted of ^{14}C changed the age to 950 yr BP. This was essentially unaltered by picking for it then approximated the picked sample from the control soil at 1000 yr BP. The HCl residue of the FACE plots was again very old, not being different than the control. This shows that after 3 yr of cotton growth most of the label was still in identifiable plant residues that could be removed by flotation and careful picking. The carbonates, at 2500 yr BP were unaffected by the increased CO_2 levels and their tracer signals. Further equilibration in the field will make it possible to follow the movement of both signals into the various SOM fractions of this site.

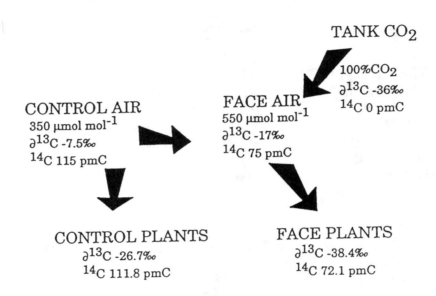

Figure 1. Summary of ^{14}C and $\partial^{13}C$ of FACE experiment air and cotton plants. (Adapted from Leavitt et al. 1992.)

Table 8. Effect of incorporation of geologically dead carbon into soil of FACE study in Arizona

	Control		FACE	
	pmC	yBP	pmC	yBP
Floated soil	0.96	250	0.89	1000
Floated &picked	0.89	1000	0.88	950
HCl-residue	0.72	2400	0.75	2300
HCl-supernatant	0.96	250	1.00	Modern
Carbonates	0.73	2500	0.73	2500

(Adapted from Leavitt et al. 1993.)

VIII. Summary and Conclusions

We have attempted to show, using examples from a number of sites, that the prerequisites for adequately determining the dynamics of SOM relative to global CO_2 fluxes are now available. Field sites with long-term management and abiotic and biological productivity measurements can provide the necessary soil samples. The use of added ^{14}C and ^{13}C either as added plant residues in enclosures or in FACE experiments gives very useful information. Modern techniques now make it possible to obtain SOM pool turnover rates without having to resort to expensive and lengthy field tracer experiments. Carbon dating using naturally occuring ^{14}C and use of the ^{13}C signal supplied by photosynthetic discrimination can be effectively used for determination of root versus soil respiration, plant residue decomposition rates, the source of mineralized CO_2, and the turnover rate of SOM fractions. Both the Saskatchewan and the Arizona data show a significant resistant fraction that is thousands of years old. More data must be obtained on intermediate sites, especially those that are affected by variations in landscapes and specific management effects.

Acknowledgements

Material present in this report was obtained from research projects supported by NSFFSA 890565 and 87-802331 as well as by USDA.

References

Anderson, D. and E.A. Paul. 1984. Organo-mineral complexes and their study by radiocarbon dating. *Soil Sci. Am. J.* 48:298-301.

Balesdent, J., G.H. Wagner, and A. Mariotti. 1988. Soil organic matter turnover in long-term field experiments as revealed by carbon-13 natural abundance. *Soil Sci. Soc. Am. J.* 52:118-124.

Bonde, T.A., B.T. Christensen, and C.C. Cerri. 1992. Dynamics of soil organic matter as reflected by natural carbon-13 abundance in particle size fractions of forested and cultivated oxisols. *Soil Biol. Biochem.* 24:275-277.

Buyanovsky, G.A. and G.H. Wagner. 1986. Post-harvest residues in cropland. *Plant and Soil* 93:57-65.

Cambardella, C.A. and E.T. Elliott. 1993. Methods of physical separation and characterization of soil organic matter fractions. *Geoderma*. In press.

Cambardella, C.A. and E.T. Elliott. 1992. Particulate soil organic-matter changes across a grassland cultivation sequence. *Soil Sci. Soc. Am. J.* 56:577-583.

Campbell, C.A., E.A. Paul, D.A. Rennie, and K.J. McCallum. 1967. Factors affecting the accuracy of the carbon dating method of soil humus studies. *Soil Sci.* 104:81-85.

Coleman, D.C. and B. Fry. 1991. *Carbon Isotope Techniques*. Academic Press, San Diego.

Elliott E.T., K. Paustian, H.P. Collins, E.A. Paul, C.V. Cole, I.C. Burke, R.L. Blevins, D.J. Monz, and S.D. Frey. Terrestrial carbon pools and dynamics: Preliminary data from the corn belt and Great Plains. ASA Special Publication, Soil Quality Symposium (in press).

Horwath, W.R. 1993. The dynamics of carbon, nitrogen and soil organic matter in populus plantations. Ph.D. dissertation, Michigan State University.

Janzen, H.H., C.A. Campbell, S.A. Brandt, G.P. Lafond, and L.Townley-Smith. 1992. Light-fraction organic matter in soils from long-term crop rotations. *Soil Sci. Soc. Am. J.* 56:1799-1806.

Leavitt, S.W., E.A. Paul, B.A. Kimball, and G.R. Hendrey. 1992. Isotopic estimation of inputs of carbon to cotton soils under FACE CO_2 enrichment. DOE Research Summary, U.S. Department of Energy Global Change Program, Carbon Dioxide Information Analysis Center, Oak Ridge National Laboratory, Oak Ridge, TN.

Leavitt, S.W., E.A. Paul, B.A. Kimball, G.R. Hendrey, J.R. Mauney, R. Rauschkolb, H. Rogers, Jr., K.F. Lewin, J. Nagy, P.J. Pinter, Jr., and H.B. Johnson. 1993. Carbon isotope systematics of FACE cotton and soils. *Agric. For. Meteorol.* Submitted.

Martel, Y.A. and E.A. Paul. 1974. The use of radiocarbon dating of organic matter in the study of soil genesis. *Soil Sci. Soc. Am. Proc.* 38:501-506.

Paul, E.A. and F.E. Clark. 1989. Soil Microbiology and Biochemistry. Academic Press, San Diego. 273 pp.

Schwartz, D., A. Mariotti, R. Lanfranchi, and B. Guillet. 1986. $^{13}C/^{12}C$ ratios of soil organic matter as indicators of vegetation changes in the Congo. *Geoderma* 39:97-103.

Steiner, R.A. and R.W. Herdt (eds.). 1993. A Global Directory of Long Term Agronomic Experiments. Vol. 1: Non-European Experiments. The Rockefeller Foundation, New York, N.Y.

Voroney, R.P., J.P. Winter, and E.G. Gregorich. 1991. Microbe/plant/soil interactions. p. 77-99. In: D.C. Coleman and B. Fry (eds.). *Carbon Isotope Techniques*, Academic Press, San Diego

Waters, A.G. and J.M. Oades. 1991. Organic matter in water stable aggregates. In: W.S. Wilson (ed.).*Advances in Soil Organic Matter Research: The Impact on Agriculture and the Environment*. The Royal Soc. Chemistry, Cambridge, United Kingdom.

Wilhelm, W.W., L.N. Mielke, and C.R. Fenster. 1982. Root development of winter wheat as related to tillage practice in western Nebraska. *Agronomy J.* 74:85-88.

Wilson, W.S. 1991. *Advances in Soil Organic Matter Research: The Impact on Agriculture and the Environment*. The Royal Society of Chemistry. Cambridge, United Kingdom.

CHAPTER 25

Fractionation and Carbon Balance of Soil Organic Matter in Selected Cryic Soils in Alaska

C.L. Ping, G.J. Michaelson, and R.L. Malcolm

I. Introduction

The recently increased interest in aquatic and terrestrial humic substances is due mainly to water quality, health-related, environmental, and global change issues. The dissolved humic and fulvic acids in soil interstitial water (soil pore water) can complex heavy metals thus increase their mobility and result in increased concentration of heavy metals in surface or stream waters (Bloom, 1979). They also can enhance the water solubility of some organic pollutants and pesticides (Chiou et al., 1987). Soil has long been speculated as one of the major contributors of humic substances to aquatic systems (Aiken et al., 1985; Malcolm, 1985). However, very few works have dealt with the capacity of particular soil to release humic substances to the surface water and the composition of these humic substances in different soil water systems. Different components of humic substances have different reactivities because of dissimilarities in molecular size as well as kinds and abundance of functional groups present.

In the upper soil profile, portions of the humic substances remain soluble in the soil interstitial water. These soluble constituents have structural and compositional differences from those humic substances precipitated around the soil mineral particles (Malcolm, 1989, 1990). These soluble humic substances may enter surface water systems through seepage, interflow, and subsurface flows (Schnitzer and Khan, 1972). There are believed to be large quantities of organic carbon, primarily humic substances, locked in the permafrost layers, i.e. arctic tundra soils. Upon warming, these humic substances would have considerable impact on the nutrient exchange/cycling, and water quality of the arctic ecosystem. Yet humic substances in the arctic tundra soils received very little attention.

Soil organic matter is an important criterion in soil classification, both in quantity expressed as % organic carbon and in quality expressed in Munsell color (Soil Survey Staff, 1992). The dominance of fulvic or humic acid fractions are important criteria to separate the Fulvic and Melanic greatgroups in Andisols, soils formed in volcanic ash (Ping et al., 1989), and the optical density of oxalate extract (ODOE) of soil material has been adapted as an index of dissolved metal-humus complexes to separate the Spodosol orders from the others (Daily, 1982; Soil Survey Staff, 1992). Soil humus, especially fulvic acids, play an important role in the formation of Spodosols (DeConinck, 1980). Spodosols are widespread throughout southern Alaska and these soils are high in humus content (Ping et al., 1989). Podzolization is the Spodosol forming process in which the organic acid complex metal ions, especially aluminum and iron, thereby enhancing weathering of minerals. As the organic acids complex metals the charges are neutralized and they may adsorb by hydrophobic interactions, or the metals may flocculate (Ugolini et al., 1988). In addition to Andisols and Spodosols, other soils derived in volcanic ash in southern Alaska exhibit characteristics of both Andisols and Spodosols, thus there are transitional problems in the classification and interpretation of these soils (Ping et al., 1990). Some dark subsurface horizons have a strong morphology of a spodic horizon but are suspected to be buried A or O horizons. The purposes of this paper are (1) to determine the organic matter

Table 1. Physical environment of the three Andic Humicryods studied

	Lat. N	Elevation m	MAP cm	MAAT °C	Slope %	aspect
Kachemak	59° 44'	455	70	2	5	S
Talkeetna	62° 44'	300	90	0	50	N
Sitka	57° 04'	470	500	6	60	N

composition of the Bhs horizons, (2) to evaluate the roles of different organic fractions in podzolization, and (3) to assess the capability of these horizons to release humic substances to soil water system.

II. Site and Soil Description

The Kachemak, Talkeetna, and Sitka soils were selected for this study because of their transitional properties in the Andisol/Spodosol classification. All three soils are formed in tephra deposit and their physiographic environment as summarized in Table 1. The Kachemak soil occurs on hilly slopes with vegetation dominated by bluejoint grass (*Calamagrostis canadensis*). The Talkeetna soil occurs on foothills with forest vegeteation dominated by white spruce (*Picea glauca*). The Sitka soil occurs on mountain slopes with forest vegetation dominated by Western hemlock (*Thuga heterophylla*). The selected morphological and chemical properties of the upper genetic horizons are given in Table 2. The Bhs and Bs horizons of all three soils have chemical properties meeting the spodic material criteria, and in addition, the Bhs horizon of the Sitka soil has cracked coatings, thus they all key out as Spodosols (Soil Survey Staff, 1992). All three soils are in the Humicryod greatgroup since they have more than 6 percent organic carbon throughout a layer of 10 cm or more within the spodic horizon, and all have a cryic soil temperature regime. They belong to the Andic subgroup because they have thickness more than 25 cm or more of andic material within 75 cm (Soil Survey Staff, 1992). The Bhs horizons of each soil were selected for this study because it is the maximum accumulation zone of translocated Al/Fe-humus complexes, and the humus is assumed to be mainly composed of fulvic acid (DeConinck, 1980).

III. Methods of Investigation

A. Soil Sampling

Thirty to fifty liters of each soil horizon were sampled for this study. The area and thickness of each horizon sampled was recorded for volumetric calculations later. All soil samples were collected with stainless steel tools and placed in Teflon lined 100-L plastic containers for shipment and storage.

B. Leachate Collection

Soil samples were stored and leached at less than 5 °C at the same temperature to minimize biological activity. A total of 70, 157, and 227 liters of water was leached through 20, 23, and 9.3 kg of Bhs soil materials from the Kachemak, Talkeetna, and Sitka Series, respectively. The soils were kept in Teflon lined containers, and intermittently saturated with deionized water, and drained through a 5-cm diameter hole was opened in the bottom of the container. The leachate was collected in a 20-L Pyrex carboy. The leachate was transferred into stainless steel filtration tanks and pressure-filtered with compressed N_2 through a 0.45-micrometer vinyl-metricel membrane filter installed in stainless steel plate filtration apparatus. The filtrate was collected in a Pyrex glass bottle in an ice-bath. The DOC of each filtrate was measured by carbon analyzer for mass balance calculations and for adjusting the DOC concentration of the filtrate to attain the column distribution coefficient (K') of 50 for maximum recovery of humic substances (Leenheer, 1981).

Table 2. Selected chemical and morphological properties of three Andic Humicryods from Alaska

Horizon	Depth cm	pH (1:1)	USDA texture	Bulk density Mg/m^3	Munsel color	O.C. %	Feo %	Alo %
Kachemak[a]								
E	0-4	4.5	fsl		2.5YR 4/2	8.5	0.14	0.15
Bs	4-9	4.5	fsl		7.5YR 4/3	9.0	0.77	1.09
Eb	9-13	4.5	sil		7.5YR 4/2	6.4	0.47	0.45
Bhs	13-20	4.7	sil	0.66	5YR 3/3	9.0	2.19	2.32
Bsl	20-29	5.1	sil	0.69	5YR 3/4	6.4	1.23	2.86
Talkeetna[a]								
Oa	0-8	3.6	peat		5YR 3/2	47.8	0.21	0.16
E	8-15	4.0	fsl		7.5YR 6/4	8.2	0.16	0.06
Bhs	15-25	4.4	sil	0.88	5YR 3/2	17.4	1.52	1.17
Bs1	25-34	4.8	sil	0.60	2.5YR 3/4	20.7	2.33	3.37
Bs2	34-42	5.0	sil		5YR 4/4	11.9	1.69	3.42
Sitka[b]								
Oa	3-7	4.2	peat		5YR 2/2	29.0	2.80	1.20
Bh	7-15	4.9	sil	0.50	5YR 2/1	12.5	8.80	3.20
Bhs	15-27	5.3	sil	0.54	5YR 3/3	10.0	6.00	3.70
Bs1	27-55	5.6	sil	0.87	5YR 3/3	8.3	5.20	5.70

[a] USDA-SCS National Soil Survey Laboratory Database; [b] Ping et al, 1989.

C. Isolation of Humic and Nonhumic Substances From Leachates

The filtrates were acidified to pH 2 with 6 M HCl, and then passed through a 4-L column of XAD-8 resin following the technique of Thurman and Malcolm (1981). The immediate measurement of DOC is necessary for quantification and for maintaining the capacity of the column within K' of 50. Effluent from the XAD-8 column was passed through a 1-L column of XAD-4 resin to adsorb the low-molecular-weight organic acids. Both the XAD-8 and XAD-4 resins were cleaned and prepared according to procedures outlined by Malcolm et al. (1977). Humic and fulvic acids adsorbed onto the XAD-8 resin were desorbed with 0.1 M NaOH; the resulting solution was acidified to pH 1 with 6 M HCl to effect a humic-fulvic separation, then centrifuged. Each humic and fulvic acid separate was desalted by readsorption onto XAD-8 resin, then desalted with deionized water, eluted with 0.1 M NaOH, hydrogen-saturated by resin exchange, evapoconcentrated and then freeze-dried. The low-molecular-weight organic acids adsorbed onto XAD-4 resin were desorbed and processed similarly. The hydrophobic neutrals retained on the XAD-8 resin after elution with 0.1 M NaOH, were desorbed with acetonitrile. The used XAD-8 and XAD-4 resins were reflux-cleaned for reuse.

D. Extraction and Isolation of Humic and Nonhumic Substances From Soils

Humic and nonhumic substances in the bulk soil (solid phase) were extracted with 0.1 M NaOH under N_2 at a solution to soil ratio about 20:1. The extraction was repeated 6 times. The extractants were separated by centrifuge and transferred to 4-L stainless steel filter tank, pressure-filtered under N_2 through a 0.45-micrometer vinyl-metricel membrane filter. The filtrates were acidified immediately with 10% HCl to pH 2 utilizing an automatic titrator and then further adjusted to pH 1 and left overnight in an ice-bath to allow precipitation of humic acids. The humic acids were then separated by centrifugation. The supernatant solutions were then processed the same as soil leachates. The humic acid fraction precipitated in the centrifuge tubes was dissolved with 0.1 M NaOH, diluted to a DOC concentration of less than 100 mg C/L, and then processed as soil leachate.

E. Other Analyses

Dissolved organic carbon (DOC) was analyzed on an OI Carbon Analyzer with a sample precision of ±0.1 mg C/L. Total soil carbon and residue soil carbon after the alkali extraction were determined by Leco carbon analyzer. The residual carbon is designated as the humin fraction. The XAD-4 resin isolates are predominantly composed of low-molecular-weight organic acids. This fraction is called XAD-4 acids.

Table 3. Carbon balance (ash free basis) in soil water extracts of Bhs horizons

Soil	Total DOC mg C	HA	FA	XAD-4 acids	Hydrophobic-neutrals	Hydrophilic-neutrals
				%		
Kachemak	1100	2.6	47.9	13.3	30.0	6.2
Talkeetna	16200	4.5	63.4	12.3	6.2	13.6
Sitka	1300	7.3	71.0	9.5	0.6	11.6

Table 4. Carbon and ash contents of organic components in soil water extracts of Bhs horizons

Soil	HA % C	HA % ash	FA % C	FA % ash	XAD-4-acids % C	XAD-4-acids % ash	Hydrophobic neutrals % C	Hydrophobic neutrals % ash
Kachemak	54.0[a]	3.0[a]	53.92	3.2	47.76	15.72	47.73	32.77
Talkeetna	54.78	1.91	53.74	0.57	48.27	3.06	--	--
Sitka	54.28	23.54	56.59	3.65	45.60	32.09	--	--

[a] Values based on best estimate.

IV. Results and Discussion

A. Soil Water Extracts

The soil water extract includes both soil interstitial water and leachates. Upon leaching, the Bhs horizons of the Kachemak, Talkeetna, and Sitka series yielded total DOC (Table 3) equivalent to 32, 624, and 69 mg C per liter of soil material, respectively. These DOC constituents present in the soil water extract are water soluble; they are percolated from the overlying horizons and solibulized from the bulk soils upon leaching. The total DOC constituents collected in the soil water extracts correspond to 0.01%, 0.07%, and 0.14% of the total soil carbon of the Kachemak, Talkeetna, and Sitka Bhs horizons, respectively. The total DOC concentration in the leachate are low at any given time, but they are regenerated yearly and the long term effect registered in the genetic soil horizons. All three soils are on either hilly or mountainous terrain; therefore it is speculated that the DOC constituents would be potential contributor to water constituents depending on transport mechanisms. On a volume basis, the forested soils, such as Sitka and Talkeetna series, generate 2 to 20 times more DOC constituents, respectively, than the grassland soils such as the Kachemak series. The possible reason for Talkeetna soil having such a high DOC concentration is that the Bhz horizon may be a buried surface horizon which releases decomposed organic constituents upon burial by tephra (E horizon).

The carbon balance in the soil water extracts is calculated based on the carbon content and ash contents of the organic components fractionated and isolated from the water extracts of the 3 soils (Table 4). The fulvic and humic acids have carbon contents ranging from 53 to 56%. The XAD-4 acids and hydrophobic neutrals have carbon contents ranging from 45 to 48%. The same tandem XAD-8/XAD-4 resin isolation procedure was applied to all soil interstitial waters and the alkali extracts of bulk soils. However, the fractionated organic components vary widely in their purity, i.e. ash contents. Generally the humic substances have lower ash contents than those of the nonhumic fractions. This may be due to the higher P and S contents in those fractions as pointed out by Malcolm et al. (1994). But for those humic fractions with ash content over 1%, incomplete removal of ash by filtration is likely the cause. High ash content would interfere with characterization analysis such as titration, elemental and CPMAS ^{13}CNMR spectral analyses (Malcolm, 1976). Those samples with high ash content need to be dissolved and refiltered.

The carbon balance of the soil water extracts of all three soils is presented in Table 4. The hydrophilic neutral fractions are the DOC which were not adsorbed by either of the two XAD-resins. Thus the %C of hydrophilic neutrals is calculated from the breakthrough curve by dividing the DOC concentration from the XAD-4 resin column by the initial DOC concentration. The high ash contents of the hydrophobic neutrals preclude accurate elemental analysis, therefore the %C of hydrophobic neutrals was calculated from the difference between total organic carbon measured in the soil water extract and the sum of all isolated fractions plus hydrophilic neutrals. The fulvic acids account for 47% to 71% of the total DOC in soil water extracts. The XAD-4 acids are slightly higher in the Kachemak and Talkeetna soils than that in the Sitka soils. The combined hydrophobic neutrals and hydrophilic neutrals were highest (36%) in the Kachemak soil and lowest in the Sitka soils (12%). The higher amount of these nonhumic organic components in the grassland soil (Kachemak) may be due to its faster biological turnover rate under grass than that of the forest soils (Sitka). The humic acids content is very low in Kachemak but over 7% in Sitka soils.

The method of using macroporous resins and multiple cycles of adsorption and desorption developed by Thurman and Malcolm (1981) is novel not only to concentrate humic substance from aqueous solution but also to separate them from inorganic solutes and nonacidic organic components. The tandem XAD-8/XAD-4 resin procedure developed by Malcolm (1991) provides a breakthrough to obtain three fractions of organic compounds (fulvic acids, humic acids, and hydrophobic neutrals) from XAD-8 resin and low-molecular-weight acids from XAD-4 resin. The nonhumic fractions (neutrals and XAD-4 acids) account for from 20% to 50% in the total DOC of soil extracts of the Bhs horizons of Sitka and Kachemak soils, respectively. Without going through the tandem XAD-8/XAD-4 resin isolation procedure, some fulvic acid would coprecipitate with humic acid, and the nonhumic fractions would be included as fulvic acid fractions. The results would lead to incorrect proportions of the fulvic acids to humic acids, and/or to the overestimation of the proportion of fulvic acids, and the potential role of XAD-4 acids in the podzolization process would be overlooked.

B. Alkali extracts of Bulk Soil Material

The total carbon, DOC recovery, and ash contents of organic components in bulk soil and alkali extracts of the bulk soils are given in Tables 5 and 6.

The commonly reported higher carbon content of the humic acids is observed here (55-56%). The fulvic acids and XAD-4 acids have carbon contents of 50 to 52 and 42% to 45% respectively. The elemental composition of the hydrophobic neutrals was precluded due to high ash content. The ash content of all the humic and fulvic acids in the alkali extracts are reasonably low enough to allow characterization analysis. The trend of elevated ash content of XAD-4 acids in the soil water extracts is also observed in the bulk soil alkali extracts, and the ash content of the XAD-4 acids in alkali extracts of Sitka soil exceeded 34%.

The carbon balance of the bulk soils and the alkali soil extracts are given in Table 7, and it was constructed on ash-free basis and on recovered DOC in the alkali extracts (Table 5 and 6). The residual carbon in the bulk soil unextractable by alkali solution is referred to as humin. The amount of hydrophilic neutrals were not determined by direct DOC analysis, but were estimated by difference measurement. Hydrophobic neutrals were determined by resin elution with acetonitrile. The total neutral content of the alkali extracts of the Kachemak, Talkeetna, and Sitka Bhs soil materials was estimated to be 20, 17, and 12%, respectively.

It is postulated that XAD-4 acids, mainly low-molecular-weight organic acids, and neutrals are biologically more active and thus have higher turnover rate and have less resident time than fulvic and humic acids. This is supported by the fact that in the alkali extracts, the proportions of XAD-4 acids and neutrals are relatively lower than that in the soil interstitial water.

Table 5. Total carbon balance in bulk soil and soil alkali extracts of Bhs horizons

Soils	Soil extracted O.D. wt. g	Total C %	Alkali solution L	% C extracted %
Kachemak	85.2	10.8	14.4	68.4
Talkeetna	55.0	18.5	15.1	85.7
Sitka	42.5	11.1	14.7	77.0

Table 6. Carbon and ash contents of organic components in bulk soil alkali extracts

Soil	HA % C	HA % ash	FA % C	FA % ash	XAD-4-acids % C	XAD-4-acids % ash	Neutrals[a] % C	Neutrals[a] % ash
Kachemak	54.97	4.23	50.62	3.60	45.12	12.92	55	30
Talkeetna	55.31	1.80	51.61	2.39	45.28	10.11	55	30
Sitka	56.27	3.70	52.91	3.59	42.02	34.53	55	30

[a] Estimated by difference.

Table 7. Total carbon balance (ash free) in Bhs horizon

Soils	Humin	Humic acids	Fulvic acids	XAD-4 acids	Neutrals[a]
			---%---		
Kachemak	31.3	13.0	33.3	2.4	20.0
Talkeetna	12.9	26.2	38.3	1.6	17.0
Sitka	23.7	39.0	24.0	1.3	12.0

[a] Estimated by difference.

As commonly observed, the humus of forest soils (Spodosols) are characterized by a high content of fulvic acids, and grassland soils (Mollisols) contain a high amount of humic acid (Stevenson, 1985). The results in this study indicated otherwise. The grassland Kachemak soil has high content of fulvic acids (33%), and has all the morphological and chemical criteria of a Spodosol (Ping et al., 1990; Soil Survey Staff, 1992). As pointed out by Tsutsuki et al. (1988), higher bases such as Ca or Mg are required to stabilize the humus to form Mollisols; yet the Kachemak soil is formed in felsic tephra which has very low base status (Ping et al., 1989). The forested Sitka soil has low content of fulvic acids (24%) but high content of humic acids (39%). The forest biomass has higher lignin content which favors the formation of humic acids (Tate III, 1987). This trend is also noted in the soil water extracts. In the Sitka soil, humic acids probably play a significant role in the podzolization process because of the higher proportions of humic acids in soil water extract.

V. Summary and Conclusions

The total carbon balance of soil water extracts and alkali extracts of the bulk soils were achieved by applying the tandem XAD-8/XAD-4 resin isolation procedure. Fulvic acids are the dominant fraction of the DOC in soil water extracts ranging from 48 to 71%. This is in agreement with the theory of podzolization process in which fulvic acids fraction is the primary complexing agent and translocating agent for weathered products such as Al and Fe to the underlying soil horizons (DeConinck, 1980). The XAD-4 acids or the low-molecular-weight organic acids account for 10 to 13%. In their soil solution analysis, Ugolini and Sletten (1991) showed a variety of low-molecular-weight organic acids. However, the concentrations are low reflecting a relatively short turnover time because of their use as energy supplies by soil microorganisms (Tan, 1986). The hydrophobic neutrals and hydrophilic neutrals account for 12 to 36%. The Bhs horizon of the Kachemak soil, a grassland Spodosol has only trace amount of humic acids but more than 50% nonhumic substances in the soil water extract suggesting most likely the products of the active biological activities in the overlying horizons.

In the bulk soils, there is a marked accumulation of humic acids with decreased fulvic acids. Even though fulvic acids are more active in podzolization processes, they are biologically less stable than humic acids. The XAD-4 acids are chemically more active than fulvic acids and are speculated to carry an important role in complexing and transporting metal ions. The greatly reduced proportion of XAD-4 acids in the bulk soil also suggest this fractions are readily available for biological decomposition thus accumulate less readily than fulvic acids in the soil.

The proportions of humic and fulvic acids in the bulk soils vary widely among the three soils. This wide variability reflects the soil-forming processes or the genetic environments of the three soils. Further investigations are needed to compare the composition of humic and nonhumic substances in the overlying and underlying horizons to access the carbon transformation in the soil system.

References

Aiken, G.R., D.M. McKnight, R.L. Wershaw, and P. MacCarthy. 1985. An introduction to humic substances in soil, sediment, and water. p. 1-9. In: Aiken et al. (eds.). *Humic substances in soil, sediment, and water*. John Wiley & Sons, New York.

Bloom, P.R. 1979. Metal-organic matter interactions in soils. p. 129-150. In: R.H. Dowdy et al. (eds.). *Chemistry in the soil environment*. ASA Spec. Publ. 40. ASA and SSSA, Madison, WI.

Chiou, C.T., D.E. Kile, R.L. Malcolm, and J.A. Leenheer. 1987. A comparison of water solubility enhancement of organic solutes by aquatic humic substances and commercial humic acids. *Environ. Sci. and Technol.* 21:1231-1234.

Daily, B.K. 1982. Identification of podzols and podzolized soils in New Zealand by relative absorbance of oxalate extracts of A and B horizons. *Geoderma* 28:29-38.

DeConinck, F. 1980. Major mechanisms in formation of spodic horizons. *Geoderma* 24:101-128.

Leenheer, J.A. 1981. Comprehensive approach to preparative isolation and fractionation of dissolved organic carbon from natural waters and wastewaters. *Environ. Sci. Technol.* 15:578-587.

Malcolm, R.L. 1976. Method and importance of obtaining humic and fulvic acids of high purity: *Journal of Research*, U.S. Geological Survey, 4:37-40.

Malcolm, R.L. 1985. Geochemistry of stream fulvic and humic substances. p. 181-209. In G.R. Aiken et al. (eds.) *Humic substances in soil, sediment, and water*. John Wiley & Sons, New York.

Malcolm, R.L. 1989. Applications of solid-state ^{13}CNMR spectroscopy to geochemical studies of humic substances. p. 339-372. In: M.H.B. Hayes, P. MacCarthy, R.L. Malcolm, and R.S. Swift (eds.). *Humic substances II. In search of structure*. John Wiley & Sons, Chichester, England.

Malcolm, R.L. 1990. Evaluation of humic substance from Spodosols. In: J. Kimble and R.D. Yeck (eds.) *Proceedings of the Fifth International Soil Correlation Meeting (ISCOM V). Characterization, Classification, and Utilization of Spodosols*. USDA, Soil Conservation Service, Lincoln, NE.

Malcolm, R.L. 1991. Factors to be considered in the isolation and characterization of aquatic humic substances. p. 369-391. In: H. Boren and B. Allard (eds.). *Humic substances in the aquatic and terrestrial environment*. John Wiley and Sons, London.

Malcolm, R.L., K. Kennedy, C.L. Ping, and G.J. Michaelson. 1994. Fractionation, characterization, and comparison of bulk soil organic substances and water-soluble soil interstitial organic constituents in selected Cryosols in Alaska. In *Proceedings, International Symposium on Carbon Sequestration*. April 5-9, 1993. Columbus, OH. (This volume.)

Malcolm, R.L., E.M. Thurman, and G.R. Aiken. 1977. The concentration and fractionation of trace organic solutes from natural and polluted waters using XAD-8, a methylmethacrylate resin. 11th Annual Conference on Trace Substances in Environmental Health, Columbia, Missouri, p. 307-314.

Ping, C.L., S. Shoji, T. Ito, J.M. Kimble, and F. DeConinck. 1990. Andisol/Spodosol transition problems. p. 252-266. In: J.M. Kimble and R.D. Yeck. (eds.). *Proceedings of the 5th International Soil Correlation Meeting (ISCOM V)*. USDA-Soil Conservation Service, Washington, D.C.

Ping, C.L., S. Shoji, T. Ito, and J.P. Moore. 1989. Characteristics and classification of volcanic ash-derived soils in Alaska. *Soil Sci.* 148:8-28.

Schnitzer, M. and S.U. Khan. 1972. *Humic substances in the environment*. Marcel Dekker, New York. 327 pp.

Stevenson, F.J. 1985. *Cycles of soil: C, N, P, S, micronutrients*. John Wiley & Sons, New York. 380 pp.

Soil Survey Staff. 1992. *Keys to Soil Taxonomy*. 5th Ed. SMSS Tech. Monogr. No. 7. Virginia Polytech and State University, Blacksburg, VA.

Tan, K.H. 1986. Degradation of soil minerals by organic acids. p. 1-27. In: P.M Huang, and M. Schnitzer (eds.). *Interaction of soil minerals with natural organics and microbes*. SSSA Special Pub. 17. Soil Society of America. Madison, WI.

Tate III, R.L. 1987. *Soil organic matter*. John Wiley & Sons, New York. 291 pp.

Thurman, E.M. and R.L. Malcolm. 1981. Preparative isolation of aquatic humic substances. *Environ. Sci. Technol.* 15:463-466.

Tsutsuki, K., C. Suzuki, S. Kuwutsuka, P. Becker-Heidmann, and H.W. Scharpenseel. 1988. Investigation on the stabilization of the humus in Mollisols. *Z. Pflanzenernahr. Bodenk.* 151:87-90.

Ugolini, F.C., R. Dahlgren, S. Shoji, and T. Ito. 1988. An example of Andisolization and Podzolization as revealed by soil solution studies, southern Hokkaido, northeastern Japan. *Soil Sci.* 145:111-125.

Ugolini, F.C. and R.S. Sletten. 1991. The role of proton donors in pedogenesis as revealed by soil solution studies. *Soil Sci.* 151:59-75.

CHAPTER 26

Fractionation, Characterization, and Comparison of Bulk Soil Organic Substances and Water-Soluble Soil Interstitial Organic Constituents in Selected Cryosols of Alaska

R.L. Malcolm, K. Kennedy, C.L. Ping, and G.J. Michaelson

I. Introduction

The organic content of soil interstitial waters has been studied sparsely during the past decade. Dissolved organic carbon constituents in soils, such as fulvic acids, have been postulated for many years to contribute to podzolization processes. Recently, dissolved interstitial soil organic constituents have also been linked to the accelerated movement of certain pesticides, insecticides, and other anthropogenic organic contaminants through soils (Chiou et al., 1986; Gerstl et al., 1989; Kogel-Knabner et al., 1991; Dunnivant et al., 1992; Malcolm, 1993). Changes in the concentration and composition of interstitial waters in cryosols may be sensitive and critical factors for evaluating climate change, CO_2 and CH_4 gas fluxes, and pedogenic processes associated with global warming. XAD resin methods, which have been successful for the concentration and fractionation of dissolved stream organic substances, may be directly applicable to the study of soil interstitial waters and the fractionation of bulk soil alkali extracts.

The objectives of this report are threefold: 1) to characterize the classes of organic compounds isolated by the tandem XAD-8/ XAD-4 resin procedure from interstitial waters of B_{hs} horizons of three Alaskan Spodosols, 2) to characterize the same classes of organic compounds isolated by the resin procedure from alkali extracts of bulk B_{hs} soil samples from the three same soils, and 3) to determine the extent of possible differences in the organic constituents between interstitial waters and bulk soil alkali extracts of the same B_{hs} horizon for these three Alaskan soils.

II. Background for this Study

This study is the first of a series on the genesis and composition of specific organic constituents in Cryosols and their relative importance as precursors of CO_2 and methane. It is hypothesized that current traditional approaches which relate soil fluxes of these gases to total soil organic carbon are imprecise and inappropriate. More than 80 percent of soil organic constituents (humic acids, fulvic acids, and humin) are decomposed so slowly that many investigators have considered them to be recalcitrant. The active constituents in CO_2 and methane production from soils are postulated to be saccharidic compounds, low-molecular-weight acids, and various other hydrophilic soil components which generally account for 20 percent or less of the total soil organic carbon. These postulated active and recalcitrant fractions can now be separated and isolated by procedures presented in this paper. These fractions will be tested for gas production in future experiments.

Another hypothesis in the long-term research perspective is that gas production in soils may be determined more precisely by evaluating soil interstitial organic constituents than by the whole bulk soil organic matter. To test this hypothesis, it is necessary to fractionate and isolate the respective organic fractions from both bulk soil and soil interstitial waters. Water soluble leachates from the soils studied are assumed to approximate soil interstitial water because, 1) leachates were collected from large volumes of soil which were leached for a relatively long period of time (several months); 2) leachates were collected at leaching rates and intervals simulating natural rainfall; 3) the soils were periodically saturated with water but were never water-logged or anaerobic; and 4) the soils were maintained at temperatures similar to natural soil conditions.

III. Methods and Materials

A. Soil Sampling

The Bhs horizon of the Talkeetna soil series near Talkeetna, Alaska, the Sitka soil series near Sitka, Alaska, and the Kechemak soil series near Homer, Alaska, were sampled during the summer of 1991. A sufficient amount of each B_{hs} horizon was collected to fill a Teflon-lined, 100-L plastic container. Complete site characterization and classification of these soils are given in the paper by Ping et al. in these proceedings.

B. Soil Interstitial Water Sampling

After collection, the soil samples were kept cold and moist in a dark, temperature-controlled room at 5 °C to maintain natural soil conditions and to minimize biological activity during storage and subsequent leaching. A hole was made in the bottom of each container for percolating interstitial waters to flow into a glass collection vessel. During the fall of 1991, while the soil remained in the dark temperature-controlled room, 5 to 10 L of distilled water was added to the top of the soil every two to three days. This slow addition of water was intended to simulate natural rainfall events, to simulate natural percolation of water through soil, and to prevent water-logging, and to maintain natural soil aeration. The production of soil interstitial waters in this way enabled the collection of moderate amounts of sample sufficient for organic characterization. This technique was also believed to closely approximate natural soil interstitial waters without contamination or problems associated with Teflon or ceramic, porous cups.

C. Tandem XAD-8/XAD-4 Resin Isolation Procedure

The vessel for collection of soil interstitial water was kept in an ice bath at 1 °C. Each day the collected soil interstitial water was pressure filtered through a 0.45 mm Gelman[1] vinyl metricel membrane filter which was free of any detergent wetting agents. The filtered interstitial water sample was kept in an ice bath for 5 to 10 days until a sufficient amount of water was collected for processing on the resin columns. The isolation procedure used two types of XAD resin in series: XAD-8 resin first, followed by XAD-4 resin. The acidified, filtered interstitial water was first passed through a 4-L column of XAD-8 resin. The effluent from the XAD-8 column was then passed through a 4-L column of XAD-4 resin. Three classes or fractions of organic compounds (fulvic acids, humic acids, and hydrophobic neutrals) were obtained from the XAD-8 resin and a fourth fraction (XAD-4 acids) was obtained from the XAD-4 resin. Detailed procedures and discussion of resin methods are presented in Malcolm (1991) and Malcolm and MacCarthy (1992). The weights and proportions of each fraction from the soils studied are presented by Ping et al. in another paper in these proceedings.

To check the reproducibility of the sampling technique for interstitial water, the Talkeetna soil was treated for a second series of the interstitial water collection. The two series were referred to as Talkeetna I and Talkeetna II samples.

---

[1] Use of the Gelman trade name in this report is for identification purposes only and does not constitute endorsement by the U.S. Geological Survey.

D. Alkali Soil Extraction Procedure

After the interstitial water collections, each B_{hs} soil sample was subsampled for alkali extraction of organic substances. Each of the soil samples was extracted with a 1:20 ratio (one gram of soil to 20 ml of alkali) in 0.1 M NaOH under a nitrogen atmosphere with constant shaking for a 24-hour period. The extractant mixture was centrifuged, the liquid decanted, and then pressure filtered through a 0.45 mm Gelman vinyl metricel membrane filter. The alkali extract was filtered directly into a container where the pH was held constant by an automatic titrator at pH 4.0 with HCl. The extraction procedure was repeated five times with each sample. The alkali extracts were combined and the fulvic/humic separation accomplished by humic precipitation at pH 1. The fulvic acids were diluted with distilled water to a DOC concentration of less than 250 mg C/L and processed by the tandem resin procedure in the same manner as used for as water samples. The precipitated humic acids were dissolved in dilute NaOH, and diluted to less than 100 mg C/L; the pH was slowly reduced to pH 2, and the sample was processed on the tandem resin procedure as for a water sample.

E. DOC Analysis

Dissolved organic carbon was analyzed on an Oceanographics International Carbon Analyzer[2] with a sample precision of ± 0.1 mg C/L. Direct organic carbon analysis was accomplished after acidification and nitrogen purging to remove dissolved inorganic carbon.

F. Elemental Analyses

Elemental analyses for C, H, O, N, P, and S were performed by Huffman Laboratories, Inc., Golden, Colorado. Moisture and ash determinations were conducted in order to calculate elemental contents on an ash-free and moisture-free basis.

G. Amino Acid Analyses

The amino acid composition after hydrolysis for 16 amino acids was determined by Dr. J.E. Fox, Department of Chemistry, University of Birmingham, United Kingdom, using the procedure for amino acid analysis according to Spackman et al. (1958).

H. Titration Analyses for Carboxyl, Phenolic, and Ester Functional Groups.

Titration analyses for carboxyl, phenolic, and ester functional groups were performed on a 10 mg hydrogen-saturated sample in a of 4 ml volume of water. Carboxyl and phenolic acidity was determined from the titer up to pH 10 in a rapid 8-minute time period; ester content was determined from a prolonged 20-hour period where the pH was held constant at 10 with an automatic titrator. This titration procedure, a modification of the method of Bowles et al. (1989), corrects for ester hydrolysis titer, which elevates phenolic contents during slow titrations. Carboxyl equivalents are computed up to pH 8; phenolic equivalents are calculated as twice the titer from pH 8-10.

---

[2]Use of Oceanographics International Carbon Analyzer trade name in this report is for identification purposes only and does not constitute endorsement by the U.S. Geological Survey.

I. Cross Polarization Magic Angle Spinning ^{13}C-Nuclear Magnetic Resonance Spectroscopy (CPMAS ^{13}C-NMR)

The ^{13}C-NMR solid-state spectra were obtained by cross-polarization magic angle spinning (CPMAS) at 4 kHz. As a result of a variable contact time experiment, the T_{CH} and T_{1P} cross-polarization parameters for the most quantitative spectra were obtained at a 1 ms contact time and a repeat time of 1 s. The CPMAS ^{13}C-NMR spectra were recorded on a custom-made spectrometer at the laboratory of Dr. G. Maciel, Regional NMR Facility, Colorado State University. The ^{1}H frequency was 90.1 MHz, the ^{13}C frequency 22.6 MHz, acquisition time 1024 ms, sweep width 531.11 ppm, and line broadening 39.999 Hz. The number of scans normally varied from 3500 to 10,000.

IV. Results and Discussion

Because of the page limitations of papers in these proceedings, the major results of this paper will focus on the results from the B_{hs} horizon of the Talkeetna soil. These data are generally representative for the Sitka and Kechemach soil series. It is also customary to refer to alkali extracts of soils merely as fulvic or humic acids, but due to the inclusion of soil interstitial water fulvic or humic acids, it is necessary to designate the alkali extracts of bulk soil material as alkali fulvic or alkali humic acids to differentiate them from their soil interstitial water counterparts.

A. Elemental Content

The elemental contents of fulvic acids, humic acids, and XAD-4 acids of the alkali extracts and interstitial waters from the Talkeetna B_{hs} horizon are given in Table 1. The elemental contents are generally normal for alkali extractable bulk soil fulvic and humic acids with the exception of the high oxygen contents. The oxygen contents in excess of 40% is not high for B_{hs} horizon organic matter, but should be considered as a good marker or index as being water mobilized B_{hs} organic material. The humic substances are also normal in that they contain higher contents of C and N, and a lower content of O than fulvic acids. The relatively low ash content enables the elemental composition to be determined with little interference. The H, S, and P contents of both samples are low and essentially the same for both humic and fulvic acids.

The alkali soluble bulk soil XAD-4 acids are similar to those normally found in fresh waters with the highest O content and lowest C content of all the isolated organic fractions. The high ash content (10.11%) may be due to the elevated S and P contents of this fraction.

The elemental contents of the interstitial water fulvic acids, humic acids, and XAD-4 acids, are somewhat different from their traditional alkali extractable bulk counterparts, but yet have many similarities, especially between humic acids. The traditionally higher C and N contents of humic acids over fulvic acids are not observed and, in fact, are reversed with fulvic acids containing the highest N content (2.79%). The carbon contents of interstitial water fulvic acids and XAD-4 acids are higher than bulk soil counterparts as well as having consistently lower O contents for the same comparisons. The interstitial water XAD-4 acids are also similar to stream water XAD-4 acids in that they are much lower in C content and much higher in O content than both interstitial water fulvic or humic acids.

B. Carboxyl, Phenolic, and Ester Functional Groups by Titration Analysis

The carboxyl, phenolic, and ester functional group content of XAD-8/XAD-4 alkali extracts and interstitial waters from the Talkeetna B_{hs} horizon are given in Table 2. The carboxyl and phenolic contents of soil fulvic acids are slightly higher than common, but normal for illuvial organic matter in B_{hs} horizons. The lower carboxyl content of humic acids (4.75 meq/g) than for fulvic acids (6.54 meq/g) is also common, but the phenolic content of humic acids is low, only slightly higher than for fulvic acids. The carboxyl and phenolic contents of the XAD-4 acids are very similar to those of fulvic acids. The ester contents of all soil and interstitial water fractions are nearly the same (0.7 meq/g) with the exception of the soil XAD-4 acids (1.20 meq/g).

Table 1. Elemental content (in percent by weight on a moisture-free and ash-free basis) of XAD-8 and XAD-4 fractions of alkali extracts and interstitaial waters from the Talkeetna B_{hs} soil horizon

Fraction	C	H	O	N	S	P	Total %	% Ash
Interstitial water fulvic acids	55.24	3.98	39.21	2.79	0.30	0.01	101.53	0.57
Alkali fulvic acids	51.61	3.28	44.51	1.17	0.31	0.01	100.89	2.39
Interstitial water humic acids	55.30	3.92	37.70	2.54	0.33	0.01	99.80	1.91
Alkali humic acids	55.31	3.91	39.68	2.62	0.35	0.13	102.00	1.81
Interstitial water XAD-4 acids	48.27	3.43	45.40	1.23	0.52	0.02	98.87	3.06
Alkali XAD-4 acids	45.28	4.42	48.34	1.81	0.92	0.35	101.12	10.11

Table 2. Carboxyl, phenolic, and ester functional group content (meq/g) of XAD-8 and XAD-4 fractions of alkali extracts and interstitial waters from the Talkeetna B_{hs} soil horizon

Fraction	COOH	Θ-OH	R-COOR'
Interstitial water fulvic acids	6.57	1.34	0.75
Alkali fulvic acids	6.54	1.60	0.69
Interstitial water humic acids	5.10	1.88	0.75
Alkali humic acids	4.75	1.73	0.61
Interstitial water XAD-4 acids	7.12	1.40	0.79
Alkali XAD-4 acids	6.28	1.75	1.20

The carboxyl, phenolic, and ester contents of the interstitial water fulvic and humic acids are approximately the same as for the bulk soil alkali extracts of fulvic and humic acids, respectively. There are large differences between bulk soil alkali extracts and interstitial water XAD-4 acids for all three types of functional groups. The limited data base for XAD-4 acids of bulk soil extracts or interstitial waters are insufficient to discuss the general magnitude of these differences.

C. CPMAS ^{13}C-NMR Spectroscopy

The CPMAS ^{13}C-NMR spectra for XAD-8 and XAD-4 fractions of alkali extracts and interstitial waters from the B_{hs} horizon of the Talkeetna soil are shown in Figures 1 and 2. The types of carbon and the carbon distribution in percent from these spectra are given in Table 3. The large differences in composition among the four isolates (fulvic acids, humic acids, XAD-4 acids, and hydrophobic neutrals) are readily determined by observing Figure 1. The changes in chemical shift and the intensity differences at various chemical shifts give each fraction individuality.

The Talkeetna soil interstitial water fulvic acids have two large intense aliphatic carbon peaks, a moderate aromatic peak, and a very intense carboxyl peak. The interstitial water humic acids have three moderate aliphatic peaks, a broad and intense aromatic carbon peak, and a moderate carboxyl peak. The interstitial XAD-4 acids have two intense aliphatic peaks, weak anomeric and aromatic carbon peaks, and a very intense carboxyl peak. The hydrophobic neutrals exhibit a very intense and moderately narrow aliphatic peak at a chemical shift of 25 ppm, three additional, weak aliphatic peaks, a weak aromatic peak, and intense carboxyl and ester peaks. The intense and narrow aliphatic peak may partially represent an unknown organic contaminant in the hydrophobic neutrals fraction. Such narrow intense peaks are suspect in the ^{13}C-NMR spectra, but no verification of contamination in this fraction has been shown.

As shown in Figure 2 and Table 3, the CPMAS ^{13}C-NMR spectra show small to moderate differences between the composition of interstitial water and bulk soil alkali extracts of the same Talkeetna B_{hs} horizon. The Talkeetna interstitial water humic acids and the bulk soil alkali humic acids are the most similar, with only a 10% average difference for the 8 carbon assignments. The highest average difference of 20% is for the interstitial water XAD-4 acids and alkali soil extract XAD-4 acids. Major differences between the interstitial water XAD-4 acids and the bulk soil alkali XAD-4 extract are in the aliphatic carbon region, as evident in Figure 2, spectra c and d. The 0-50 ppm peak of bulk soil alkali XAD-4 acids is 13 relative percent less and the 64-85 ppm peak is 25 relative percent greater than the corresponding peak areas for the interstitial water XAD-4 acids. The intense peak in the 64-95 ppm area of the bulk soil alkali XAD-4 acids fraction also has a different peak maximum which is near 75 ppm, whereas the peak maximum is near 85 ppm for the interstitial water XAD-4 acids.

A comparison of the CPMAS ^{13}C-NMR spectra of the Talkeetna B_{hs} interstitial water fulvic acids with the bulk soil alkali fulvic acids is also shown in Figure 2, spectra a and b. The spectra are somewhat similar, with only minor differences in peak intensities. The bulk soil alkali fulvic acids are less aliphatic (0-108 ppm) and more aromatic (108-160 ppm) than the interstitial water fulvic acids.

The variation in composition within each of the various organic fractions from the three different B_{hs} soil horizons is small to moderate (an average of 10-20 percent for the eight carbon chemical shift areas). The variation within each fraction is approximately the same as the differences between interstitial water and

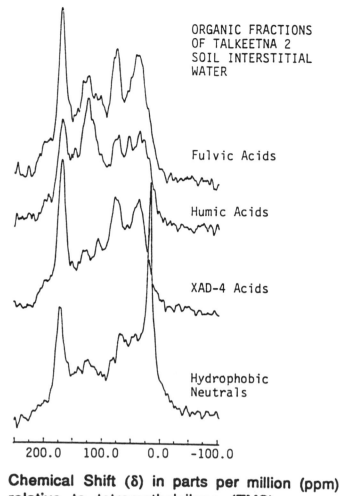

Figure 1. CPMAS ^{13}C-NMR spectra for XAD-8 and XAD-4 fractions of interstitial water from the B_{hs} horizon of the Talkeetna soil series.

bulk soil alkali isolates for each organic fraction. The variation among soil interstitial water fulvic acids from the three B_{hs} soil horizons are given in Figure 3. The soil interstitial waters from the same Talkeetna B_{hs} soil horizon were collected and fractionated twice before aliquots of the soil were extracted with alkali. The respective soil interstitial water fractions from the two experiments were quite similar, as shown from the interstitial fulvic acids in Figure 3. The CPMAS ^{13}C-NMR spectra for the three B_{hs} soil horizons and the two leaching experiments from the same Talkeetna B_{hs} soil horizon are all very similar on the number of chemical shift peaks and the maximum for each peak (type of carbon). The peaks at each chemical shift vary 10-20 percent only in peak area (intensity) relative to the total spectra (0-230 ppm chemical shift). The intense carboxyl (160-190 ppm) and C-O chemical shift regions (64-95 ppm) are very similar; the major differences are in the magnitude and ratios of the 0-50 ppm unsaturated aliphatic peaks with the 108-160 ppm aromatic carbon peaks. The within fraction variation for the other three fractions are similar to those shown in Figure 3 for fulvic acids. Space limitations do not permit the presentation of the data for other fractions.

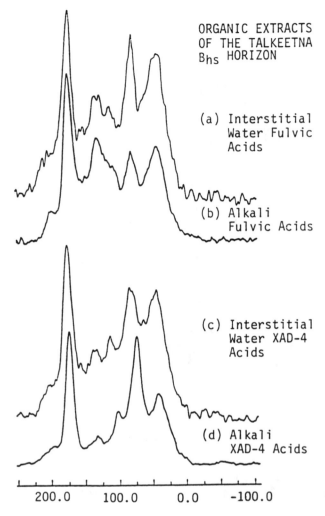

Figure 2. CPMAS ^{13}C-NMR spectra for alkali fulvic acids, interstitial water fulvic acids, alkali XAD-4 acids, and interstitial water XAD-4 acids from the B_{hs} horizon of the Talkeetna soil series.

D. Amino Acid Analyses

The amino acid content for 16 amino acids (in ηhmole/mg) for XAD-8/XAD-4 fractions of alkali extracts and interstitial waters from the Talkeetna B_{hs} horizon are given in Table 4. The most striking result of the amino acid data is that the alkali soil extracts are all much higher in amino acid content than the respective soil interstitial water fraction. The most pronounced difference is for humic acids where the average concentration of all amino acids is five times higher for alkali soil extracts as compared to soil interstitial water humic acids. The average concentration for alkali extracts of XAD-4 acids and fulvic acids are three and two times higher, respectively, than their soil interstitial water counterparts. Among interstitial water fractions, the fulvic acids are lowest in amino acid content (64.4 ηmole/mg), the humic acids highest (122.8 ηmole/mg), and the XAD-4 acids intermediate (72.5 ηmole/mg). Among the alkali soil extract fractions,

Table 3. Carbon distribution (in percent) by CPMAS ^{13}C-NMR Spectroscopy for XAD-8 and XAD-4 fractions of alkali extracts and interstitial waters from the Talkeetna B_{hs} soil horizon

Fraction	Unsubstituted aliphatic carbon 0-50 ppm	Singly substituted O or N-aliphatic carbon 50-64 ppm	Singly substituted O or N-aliphatic carbon 64-95 ppm	Anomeric carbon O-C-O 95-108 ppm	Aromatic carbon 108-143 ppm	Aromatic carbon 143-160 ppm	Carboxyl and ester carbon 160-190 ppm	Ketonic carbon C=O 190-220 ppm
Interstitial water fulvic acids	21.5	6.5	18.0	5.3	18.0	6.3	19.0	5.3
Alkali fulvic acids	17.6	6.9	15.0	6.8	19.9	6.8	22.5	4.5
Interstitial water humic acids	20.8	7.5	15.4	6.1	24.0	8.0	14.8	3.5
Alkali humic acids	21.5	7.3	12.3	6.9	23.4	6.6	18.1	3.7
Interstitial water XAD-4 acids	23.1	6.0	22.6	6.9	11.4	5.6	19.8	4.7
Alkali XAD-4 acids	20.1	5.0	30.1	7.6	8.0	2.8	22.1	4.1
Interstitial water hydrophobic neutrals	35.6	6.8	15.7	3.6	12.4	6.0	17.5	2.4
Alkali hydrophobic neutrals	32.9	6.4	13.9	3.0	16.3	4.3	19.3	3.9

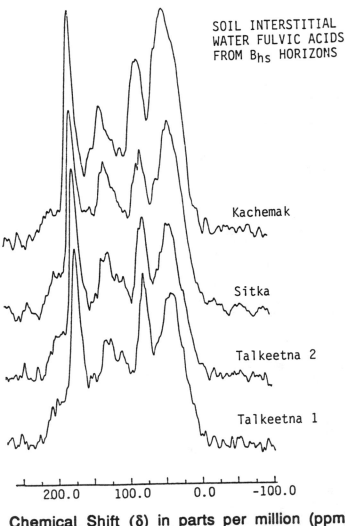

Figure 3. CPMAS ^{13}C-NMR spectra of soil interstitial water fulvic acids from B_{hs} horizons of the Talkeetna, Sitka, and Kachemak soil series.

the order is the same, with fulvic acids lowest (113.2 ηhmole/mg), humic acids highest (558.8 ηmole/mg), and XAD-4 acid intermediate (196.4 ηmole/mg).

Glycine is the amino acid in highest concentration for the three soil interstitial water fractions, whereas aspartic acid was the highest in concentration for two of the three alkali extract fractions. Glycine, aspartic acid, glutamic acid, proline, and alanine were the five most abundant amino acids in all three of the soil interstitial water fractions, whereas aspartic acid, glycine, glutamic acid, alanine, and serine were the five most abundant in all three alkali extract fractions. Proline was noticeably absent from the five most abundant amino acids in all of the three alkaline extract fractions, whereas serine was noticeably absent from the top five amino acids in soil interstitial water fractions. The five most abundant amino acids accounted for 81% of the total amino acid content in soil interstitial XAD-4 acids, but the top five only accounted for 53% of the soil alkali extract for humic acids.

Table 4. Amino acid content (in nanno mole/mg) for XAD-8 and XAD-4 fractions of alkali extracts and interstitial waters from Talkeetna B_{hs} soil horizon

Amino acid	Interstitial water fulvic acids	Alkali fulvic acids	Interstitial water humic acids	Alkali humic acids	Interstitial water XAD-4 acids	Alkali XAD-4 acids
Alanine	5.0	12	11	57	6.3	25
Arginine	0.7	1.1	1.5	16	0.3	0.5
Aspartic acid	10	21	17	72	13	36
Glutamic acid	6.7	17	12	59	8.2	34
Glycine	16	20	24	68	22	45
Histodine	1.7	3	3.2	23	2.7	8
Iso-leucine	1.4	2.9	3.6	21	0.8	5
Leucine	1.9	4.9	5.4	41	0.9	6
Lysine	1.7	3.2	3.9	19	1.5	2
Methionine	0.3	0.2	0.6	5.8	0.3	1.5
Phenylalanine	0.7	1.9	2.4	2.3	0.2	--
Proline	7.3	4.6	11	31	5.3	5.6
Serine	4.1	7.7	9.5	40	4.5	13
Threonine	3.7	7.2	8.9	39	3.8	10
Tyrosine	0.5	0.6	1.5	10	0.2	--
Valine	2.9	5.9	7.3	34	2.5	4.8
Total	64.6	113.2	122.8	558.8	72.5	196.4

V. Summary and Conclusions

The compositions of fulvic acids, humic acids, and XAD-4 acids which were isolated from soil interstitial waters and bulk soil alkali extracts of the B_{hs} horizons of the Talkeetna, Sitka, and Kechemak soils of Alaska were investigated by elemental analyses, titration analyses for carboxyl, phenolic, and ester functional groups, CPMAS ^{13}C-NMR spectroscopy, and amino acid analyses. The composition of the hydrophobic neutrals fraction isolated from the same soil samples was also analyzed by CPMAS ^{13}C-NMR spectroscopy. All the test analyses found large differences among the various fractions (fulvic acids, humic acids, XAD-4 acids, and hydrophobic neutrals) within the soil interstitial water series and the soil alkali extract series. All test analyses indicated differences between bulk soil alkali extracts and soil interstitial waters for all four respective organic fractions. Elemental analyses, titration data, and CPMAS ^{13}C-NMR spectral data indicate small to moderate differences in composition between bulk soil and interstitial water organic fractions of the three spodic soil horizons. Elemental analyses and CPMAS ^{13}C-NMR spectroscopy were used to analyze the entire carbon content of the fractions. The summation of carboxyl, ester, and phenolic moities in titration analyses represents approximately 25% of the total carbon population of each fraction.

Amino acid analyses of the fractions indicate large differences in amino acid content among the fractions of each series and between the respective fractions of soil interstitial water and alkali soil extracts. The amino acid analyses appear to be the most definitive probe to evaluate similarities and differences between soil interstitial water and bulk soil alkali extractable organic fractions, but the data must be interpreted with some reservations because the amino acid content of the various organic fractions represents only from 1-10% of the total carbon in various fractions.

Large compositional differences are known to exist between fulvic acid and humic acid components in soils, and the four XAD-8/XAD-4 fractions isolated from water. The application of this tandem resin isolation scheme appears to be useful for the study of alkali soil extracts and soil interstitial waters, with the result of large compositional differences among the four isolated fractions (fulvic acids, humic acids, XAD-4 acids, and hydrophobic neutrals).

Small to moderate differences were postulated to exist between the compositions of soil interstitial water and bulk soil and alkali extracts because the small amount of organic substances in soil solution should be more water soluble than organic constituents in the bulk soil. These differences between the organic composition in soil solution and the bulk soil should be minimal in B_{hs} horizons because the majority of B_{hs} organic soil components were mobilized in moving water from more surficial horizons. After immobilization in the B_{hs} horizon, the organic constituents should remain similar to soil interstitial waters from which they were derived. The results of this study appear to support this hypothesis.

The organic fractions isolated by the tandem resin procedure from both the bulk soil and soil interstitial water sources will be evaluated for gas production in future decomposition experiments. The method presented appears to be a useful technique to divide soil organic constituents into fractions whose activity for production of CO_2 and methane can subsequently be evaluated.

References

Chiou, C.T., R.L. Malcolm, T.I. Brinton, and D.E. Kile. 1986. Water solubility enhancement of some organic pollutants and pesticides by dissolved humic and fulvic acids. *Env. Sci. & Tech.* 20:354-366.

Bowles, E.C., R.C. Antweiler, and P. MacCarthy. 1989. Acid-base titration and hydrolysis of fulvic acid from the Suwannee River. p. 205-229. In: R.C. Averett, J.A. Leenheer, D.M. McKnight, and K.A. Thorn (eds.). *Humic substances in the Suwannee River, Georgia: interactions, properties, and proposed structures*. U.S. Geological Survey Open-File Report 87-557, Washington, D.C.

Dunnivant, F.M., P.M. Jardine, D.L. Taylor, and J.F. McCarthy. 1992. Cotransport of cadmium and hexichlorobiphenol by dissolved organic carbon through columns containing aquifer material. *Env. Sci. Tech.* 26:360-368.

Gerstl, Z., Y. Chen, U. Mingelgrin, and B. Yaron. 1989. *Toxic organic chemicals in porous media.* Springer-Verlag, NY.

Kogel-Knabner, I., P. Knabner, and H. Deschauer. 1991. Dissolved organic matter as carrier for exogenous organic chemicals in soils. p. 121-128. In: W.S. Wilson (ed.). *Advances in soil organic matter research: the impact on agriculture and the environment.* The Royal Society of Chemistry, Cambridge, England.

Malcolm, R.L. 1991. Factors to be considered in the isolation and characterization of aquatic humic substances. p. 369-391. In: H. Boren and B. Allard (eds.) *Humic substances in the aquatic and terrestrial environment*. John Wiley and Sons, London.

Malcolm, R.L. 1993. The concentration and composition of DOC in soils, streams, and groundwaters. p. 3-11. In: K.C. Jones and A.J. Beck (eds.) *Organic substances in soil and water*. University of Lancaster, England.

Malcolm, R.L. and P. MacCarthy. 1992. Quantitative evaluation of XAD-8 and XAD-4 resins used in tandem for removing organic solutes from water. *Environment International* 18:597-607.

Spackman, D.H., W.H. Stein, and S. Moore. 1958. Automatic recording apparatus for use in the chromatography of amino acids. *Anal. Chem.* 30:1190-1205.

CHAPTER 27

CO_2 Efflux from Coniferous Forest Soils: Comparison of Measurement Methods and Effects of Added Nitrogen

Kim G. Mattson

I. Introduction

Soil CO_2 efflux results from the production and evolution of CO_2 from the forest floor and mineral soil. It reflects decomposer and root activities in the soil and conversion of carbon from organic to inorganic forms as an energy yielding process (Singh and Gupta, 1977). Soil CO_2 efflux is an extremely useful measure of ecosystem behavior because it integrates all below-ground carbon transformations. For example, by assuming the total belowground carbon content is at steady state, CO_2 efflux equals carbon inputs to the belowground system from detritus and from root respiration. Under this assumption, comparison of CO_2 efflux among sites reveals differences in below-ground carbon flows.

Accurate measurement of soil CO_2 efflux for a site is difficult. Simple techniques, such as static chambers, disrupt the turbulent flux during the measurement. Non-disrupting methods, such as micrometerological techniques require more highly sophisticated equipment and greater constraints on the choice of field site. In addition, fluxes are highly variable spatially and temporally and accurate quantification requires intensive sampling. Several comparisons of methods have concluded that static chamber based measures are typically the best choice when relative comparisons are desired (deJong et al., 1979; Cropper et al., 1985). However, in order to construct carbon budgets or quantify carbon fluxes from landscapes, absolute or unbiased measures of carbon flux are required. To this end, knowledge of potential biases in measurement techniques is needed.

Knowledge of changes in carbon cycling due to management and disturbance is required for predicting effects on carbon balance and for developing for managing the sink/source characteristics of ecosystems. For example, nitrogen fertilization may be one of the most cost-effective means by which to increase above-ground carbon gain in terrestrial systems (Entry et al., 1993a). However, the effects of added nitrogen (N) on below-ground carbon turnover are complex and not well understood (Fog, 1988; Nohrstedt et al., 1989). Without a better understanding of the below-ground response, one cannot predict effects of such management options on the overall ecosystem carbon balance.

My primary objective in this study was to quantify and compare CO_2 fluxes from the forest floor using a static chamber method and a soil profile method (deJong and Schappert, 1972). My secondary objective was to examine CO_2 efflux from three sites that had received different levels of added N.

II. Methods

A. Site Description

This research was performed in conjunction with the Oregon Transect of Ecosystem Research (OTTER) Project whose goal was to estimate major fluxes of carbon, nitrogen, and water in coniferous forests using an ecosystem process model (Peterson and Waring, 1994). One of the OTTER sites, a Douglas-fir (*Pseudotsuga menziesii*) and western hemlock (*Tsuga heterophylla*) forest on the west slope of the Cascade Range that had received nitrogen-fertilizer applications, was chosen to for the tests of CO_2 efflux measures.

The area was located approximately 20 km east of Scio, Oregon (45 °45' N, 122 °35' W) at 650 to 800 m elevation on the northwest slopes of Snow Peak. The sampling areas were composed of 40-year old Douglas-fir and western hemlock stands under natural regeneration following logging of older-growth forests. Soils of the area are classed as clayey-skeletal, mixed, frigid Umbric Dystrochrepts and Medial, frigid Andic Haplumbrepts (Flane and Moe gravelly loams, respectively). These soils are deep, well drained, high in organic matter, low in bulk density, and were formed in colluvium derived from basic igneous and tuffaceous rock and from breccia (Soil Survey of Linn County Area, Oregon 1983). Forest stands were pre-commercially thinned approximately five years prior to our field measures.

In the spring of 1990, three forested stands were chosen for measures based on past nitrogen histories. These sites are referred to as: the control (CO), the single fertilized site (1X site), and the double fertilized site (2X site). Both the 1X and 2X sites were part of an aerial application of nitrogen in 1988 (45 g N m^{-2} as urea). The 2X site received additional hand applications of urea (30 g N m^{-2}) in the spring of 1990 and 1991 and of ammonium forms (15 g N m^{-2}) in the fall of 1990 and 1991. The control received no additions of nitrogen.

B. Dates of Field Collections

Data used to calculate CO_2 efflux by two methods, together with moisture and temperature were collected on all three sites on the same day for 15 dates over a 277 day period. Additional soil and forest floor physical measures were made at the end of the period.

C. Static Chamber Method

CO_2 was measured by the static method (Anderson, 1982) at 12 locations at each site. Chambers consisted of cylinders 15 cm diam, 20 cm long, made of PVC sewer pipe pressed through the forest floor and into the top 2 cm of the mineral soil. An alkali absorbent (20 ml of $1N$ NaOH) contained in jars of 6 cm diam and 8 cm height was periodically placed directly on the ground inside the cylinder and the cylinder was sealed for 24 or 48 hours. The dimensions of the jar are nearly the same as those recommended by Anderson (1982). Jars were not placed on a tripod because sticks and irregularities in the forest floor inside the cylinders provided the same effect as a tripod would have on level soils. The diameter of the cylinders used here are smaller than those recommended by Anderson (1982) and had the effect of decreasing the ratio of soil surface area to absorbent surface area. Decreasing this ratio was reasoned to increased the efficiency of the absorbent and reduce underestimation bias of the chamber. CO_2 absorbed by the alkali was measured by titration with HCl using $BaCl_2$ to first precipitate the carbonate. The CO_2 trapped by the NaOH is divided by the cross sectional surface area of the cylinder and by the incubation time to provide a estimate of carbon flux from the forest floor to the atmosphere. Repeatability of replicate titrations of field blanks was generally ±0.05 g C m^{-2} d^{-1}.

As a check of the efficiency of the NaOH to absorb the CO_2, the headspace concentrations inside several chambers were monitored during an incubation with NaOH. The chambers were placed over soils of varying efflux rate at one of the sites and headspace concentrations were collected using 1 ml plastic tuberculin syringes every 3 hours for 30 hours. Analysis was via gas chromatography as described below.

D. Soil Profile Method

At four of the static chamber locations in each site, CO_2 efflux was also measured by the soil profile method (deJong and Schappert, 1972). The location for each soil profile was within 0.5 m of the static chamber. CO_2 efflux was calculated as the product of CO_2 gradients and diffusivities of CO_2 within defined soil horizons. CO_2 concentrations were measured at 5 vertical points: in the air just above the forest floor (depth 1), at the forest floor-mineral soil interface (depth 2), and at the following depths in the underlying mineral soil: 10 cm (depth 3), 30 cm (depth 4), and 40 cm (depth 5). These depths were chosen because CO_2 production in each intervening layer could be calculated and soil moistures were already being monitored at three of these layers.

Gas samples were collected directly from the air above the forest floor and from soil gas wells permanently installed at the four depths in the soil. The wells consisted of steel tubing 3 mm i.d. and 45 cm long with one tip crimped shut and 1 mm diam holes drilled just behind the tip. The opposite end was fitted with a 3.2 mm thick butyl rubber septum using Swagelock compression fittings (Crawford Fitting, Solon, OH). During installation, the tips of the wells were pressed into the soil to the desired depth. During sampling, gases were extracted from the well using 1.0 ml plastic tuburculin syringes fitted with 2.5 cm long needles. Syringes were first purged with 1 ml of gas from a well and a second sample taken for analysis. The tips of the needles were imbedded into rubber stoppers during transport back to the laboratory.

CO_2 and O_2 were analyzed within 4 to 9 hours after sampling via gas chromatography using either a 5730-A Hewlett Packard or Shimadzu GC-8A. Both chromatographs used a thermal conductivity detector, He as the carrier gas, Alltech CRT I dual columns (Alltech Associates, Inc., Deerfield, IL) and were standardized using a 9900 ppm CO_2 standard. I found that a one-point calibration was insufficient for accurate detection of CO_2 over a wide range of concentrations and at least one additional standard at concentration < 1000 ppm CO_2 was also used. Repeatability of replicate samples of standard gases was generally $\pm$ 5 %. Repeatability of replicate samples of gases drawn from wells was generally $\pm$ 10 %. Time of sampling and time of analyses were recorded for individual samples in order to estimate storage time in the syringes.

I found that the plastic tuburculin syringes could not hold CO_2 at high concentrations for several hours without loss. The rate of CO_2 loss a predictable ($r^2 = 0.96$) function of storage time and CO_2 concentration. Measured CO_2 concentrations > 6000 ppm were corrected using $e^{(0.0216 \text{ hr} + 0.0682)}$ and measured CO_2 concentrations between 3000 and 6000 ppm were corrected using $e^{(0.00930 \text{ hr})}$, (hr = hours of storage time). No correction factors were used for CO_2 concentrations less than 3000 ppm as tests showed no losses over 5 to 6 hours.

Percent moisture (gravimetric, Gardner, 1986) of the forest floor and soil at 0 to 10 cm and at 30 to 40 cm depths was measured at four representative locations in each site. Soil moisture at 20-30 cm depths was estimated as the mean of soil moisture values of the 0-10 cm depth and the 30-40 cm depth. At each of the 12 static cylinders in each site, soil temperature at 10-12 cm depth was measured using the mean of three probe-type digital thermometers inserted at locations within 0.3 m of the cylinder. The digital thermometers were calibrated against mercury thermometers to $\pm 0.1°$ C.

E. Forest Floor and Soil Measures

Forest floor depth and bulk density (measured volume of excavated sample/sample mass) and mineral soil bulk density, organic matter content, rock volume and root volumes for the three mineral soil layers were measured at each soil CO_2 profile station. Forest floor was collected within an area of 625 cm^2; soil samples were collected using a 5.3 cm diam steel tube and pressing into the soil to the desired depth. Organic matter particle density (pychnometer method, Blake and Hartge, 1986) was estimated from a mean of six random samples of forest floor material collected at the 2X site. Soil bulk density was calculated for soil portion with rocks, roots, and other debris > 2 mm Oremoved. Soil organic matter was determined by loss on ignition at 435 °C for 24 hours (Davies, 1974).

F. Calculations for Soil Profile Method

CO_2 efflux calculations followed deJong and Schappert (1972) and assume that gases move between the soil and the atmosphere by diffusion, driven by concentration gradients and modified by available air-filled pore spaces. The following calculations were performed for each soil layer and collection date:

$$Q_{n\,n+1} = Ds \cdot (C_{n+1} - C_n)/(Z_{n+1} - Z_n) \cdot 126.3/T_n \tag{1}$$

where:

$Q_{n\,n+1}$ = efflux to depth n from depth $n + 1$ (g C·m^{-2}·day^{-1})
Ds = soil diffusivity of CO_2 between depth n and $n + 1$ (cm^2·sec^{-1})
C_n = CO_2 concentration at depth n (ppm)
Z_n = depth of collection C_n (cm)
T_n = temperature at depth n (°K)
126.3 is a unit conversion factor and was derived as the product of:
(10^{-6} atm·ppm^{-1} ·10^{-3} L·cm^{-3})/(0.0821 L·atm·mole^{-1}·deg^{-1}) · 12 g·mole^{-1}
and (10000 cm^2·m^{-2} · 8.64*10^4 sec·day^{-1}).

C and Z were measured for each depth; representative values of C by sites are given in Table 1. T was measured at 10 cm soil depth and assumed constant for all depths. This assumption was reasonable because expected variation in soil temperature with depth (2 °C) would produce less than 2% change in the calculation of Ds. Ds was calculated as a function of diffusivity in air corrected by the proportion of soil pores that are air filled. Among several equations to calculate Ds, the Millington and Quirk equation was chosen (W.A. Jury, University of California, Riverside, pers. comm.):

$$Ds/Do = (a^{3.33}/p^2) \tag{2}$$

where:

Do = diffusivity of CO_2 in air = 0.139 cm^2·s^{-1} · (air temp °K/273)^2
a = air filled porosity in layer
p = total porosity in layer.

Calculation of p was based on bulk density and particle density (Danielson and Sutherland, 1986):

$$p = 1 - BD/PD. \tag{3}$$

where:

BD = bulk density (g·cm^{-3})
PD = particle density (g·cm^{-3}).

For the forest floor, BD and was measured for forest floor samples; PD was assumed to be constant and equal to 1.54 g·cm^{-3} based measures of six representative samples.

For mineral soil, p was calculated with a modification of equation (3):

$$p = 1 - 0.377\,BD - 0.272\,BD \cdot POM - PVR \tag{4}$$

where:

POM = proportion organic matter in soil (g·g^{-1}),
PVR = proportion rocks and roots in soil (cm·cm^{-1}).

Equation (4) was used because Scio soils are high in organic matter content and organic matter has lower values of both BD and PD than the mineral component of soils. Equation (4) is the same as equation (3) except that it uses the sum of volume-weighted PD for the organic component and volume-weighted PD for the mineral component and the volume of rocks and roots are subtracted out. The proportional volume of organic component was calculated as $BD \cdot POM/0.327$. The divisor represents bulk density of organic

Table 1. Seasonal CO_2 concentrations (ppm) by depths for the three sites

Depth	July-August			December-February			April-May		
	2X[a]	1X	CO	2X	1X	CO	2X	1X	CO
1	478	465	509	464	501	448	432	472	438
2	885	621	1052	799	643	629	1077	862	766
3	1630	1606	1947	375	1511	877	371	1688	1140
4	4488	2797	3810	2256	2144	2170	3445	2750	2728
5	4957	3484	3917	2215	2068	2590	3838	3936	3278

[a] One profile at the 2X site had CO_2 concentrations at depth 4 and 5 as high as 30,000 ppm during July and August. The profile was located on an old skid trail in highly compacted soils and was not thought to be representative of the site. These data were not used to calculate mean concentrations here.

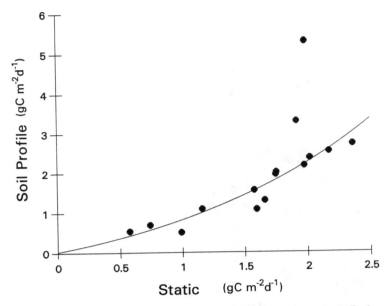

Figure 1. CO_2 efflux by the soil profile method versus the static method. Each point represents the mean of 12 estimates made on a single date. The line represents best fit of a single exponent model: $y = e^{0.576 x + 0.025}$, $r^2 = 0.84$ with the highest y value deleted.

matter in the soil and was based on the mean of 14 samples of wood, dead roots, and uniformly shaped debris found in the soil. The proportional volume of the mineral component was equal to *1 - (proportional volume of organic matter)*. *PD* for the organic component was estimated to be 1.54 (see above), *PD* for the mineral component of the soil was assumed to be a constant and equal to 2.65 (Danielson and Sutherland, 1986). The portion of the total soil column occupied by gas impermeable objects averaged 13 % of belowground volume.

a for specific layer was calculated as:

$$a = p - BD \cdot \%M/100, \tag{5}$$

where:
$\%M = \%$ moisture by weight.

III. Results

A. Comparison of Methods

An exponential model fit to the a plot CO_2 efflux from the soil profile method versus the static method suggests that the soil profile method was 10% lower than the static method when effluxes were less than 2 g C m^{-2} d^{-1}, and up to 40% higher when effluxes were greater than 2 g C m^{-2} d^{-1} (Figure 1). Fluxes greater than 2 g C m^{-2} d^{-1} typically occur during the warmer summer months from June through September when soil temperatures exceed 10 °C. Data in Figure 1 are means across sites for specific dates. Measures of CO_2 efflux at a single soil profile location did not agree as well with the co-located measures by the static method for a specific date. This is to be expected since the two methods measured points that were 0.5 m distant and soils have shown no or little spatial structure with regard to CO_2 efflux over distances of less

Table 2. Annual CO_2 efflux (g C m^{-2} yr^{-1}) by sites estimated by the static and soil profile methods

	Site		
	2X	1X	Control
Static[a]	425	610	600
Soil Profile	540	469	692

[a] only those cylinders that were co-located with the permanent gas sampling wells used in this comparison

than a meter (Robertson et al., 1988; Aiken et al., 1991). Such high variation in CO_2 efflux makes it difficult to match side by side comparisons.

Annual fluxes estimated by soil profile method were 15% greater than the static method at the control, 27% greater at the 2X site, but 23% lower at the 1X site (Table 2). Given the relationship shown in Figure 1, it would be expected that the soil profile method would produce higher annual fluxes than the static method, particularly if most of the flux occurs during the warmer seasons at rates greater than 2 g C m^{-2} d^{-1}. The concentration gradient between depth 1 and depth 2 together with the diffusivity determines the CO_2 efflux at the surface. This gradient was typically lower at the 1X site than the other two sites (Table 1), while diffusivities were either similar or also lower (data not shown). The observation that the bias between the two methods was not consistent among the three sites suggests that either method has a degree of error due to some combination of measurement errors, method bias, and random errors due to inability to sample the same soil. (Sources of error are discussed in detail in the Discussion.)

During a static chamber incubation with NaOH, the CO_2 concentration inside the headspace was observed to vary considerably from ambient. In chambers placed over soils producing high rates of CO_2 efflux, the headspace concentration of CO_2 increased above the ambient concentrations while in chambers placed over soils of low CO_2 efflux the headspace concentrations fell below ambient concentrations (Figure 2). Ambient CO_2 concentrations (i.e., in the absence of chambers) in the air just above the forest floor ranged from 400 to 650 ppm while concentrations at the bottom of the forest floor ranged from 600 to 1800 ppm. With the dimensions of the chamber and jars used in this study, CO_2 effluxes from the soil in the range of 1.6 g C m^{-2} d^{-1} were approximately balanced by the rate of absorption by NaOH and is consistent with the pattern shown in Figure 1.

B. Effects of Added Nitrogen

Additions of N to the 2X site were associated with 29% lower annual CO_2 efflux when compared to the control site using the static method and with 22% lower efflux when using the soil profile method (Table 2). A more thorough examination of site differences was made using all 12 of the static chambers at a site over the entire sampling period available (June, 1990 through December, 1992). These data indicated that CO_2 efflux from the 1X site was similar to the control and the 2X site was still about 25% less than the control during the first year of N additions (Figure 3).

After September, 1991, the site differences diminished. However, summer rates of CO_2 efflux from the control, the 1X site, or a second control established in 1992, never quite approached the high efflux rates observed during the summer of 1990. These observations indicate that the N effect on the 2X site was temporary or that added N inhibits CO_2 effluxes only when effluxes exceed 2.0 g C m^{-2} d^{-1}, or both. In addition, the differences in CO_2 efflux between the 2X site and the control were greatest in the forest floor and 0 to 10 cm mineral soil layer during the summer and fall of 1990 (Table 3).

C. CO_2 Efflux by Layers

Seasonal CO_2 efflux for each of the 4 layers averaged across all sites indicated that the forest floor and top soil layers produce most of the soil CO_2 efflux (Figure 4). However, during December through April, CO_2 efflux from the top mineral soil layer was quite low and actually went negative in March. This is a result of the particularly low CO_2 concentrations measured at depth 3 at the 2X site when soils were very wet

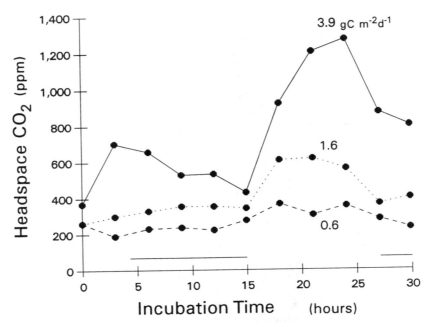

Figure 2. CO_2 concentrations inside static chambers during a 30-hour measurement in the field. The three lines represent three chambers, each with their respective efflux determinations via titration of the NaOH. The horizontal lines indicate nighttimes.

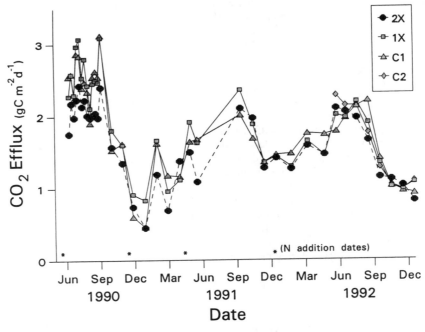

Figure 3. CO_2 efflux over time from sites with varying N additions. The 2X site received N on dates indicated. No recent additions of N were made to the 1X site, or the controls C1 or C2. CO_2 measured by static method with 12 cylinders per site. Coefficient of variation was 25 to 35%.

Table 3. Difference in seasonal mean CO_2 efflux rates (g C m^{-2} d^{-1}) between 2X and control by horizons as estimated by the soil profile method

	2X	Control	Difference[a]
		July-September	
Q_{12}	2.2	3.9	-1.7
Q_{23}	1.0	2.0	-1.0
Q_{34}	0.6	0.7	-0.1
Q_{45}	0.1	0.2	-0.1
		October-December	
Q_{12}	0.8	1.1	-0.3
Q_{23}	0.3	0.7	-0.4
Q_{34}	0.2	0.2	0.0
Q_{45}	0.0	0.3	-0.3
		January-March	
Q_{12}	0.8	0.9	-0.1
Q_{23}	0.0	0.2	-0.2
Q_{34}	0.1	0.2	-0.1
Q_{45}	0.1	0.0	0.1
		April-May	
Q_{12}	2.0	1.5	0.5
Q_{23}	0.0	0.3	-0.3
Q_{34}	0.2	0.2	0.0
Q_{45}	0.1	0.0	0.1

[a] Difference = 2X - control

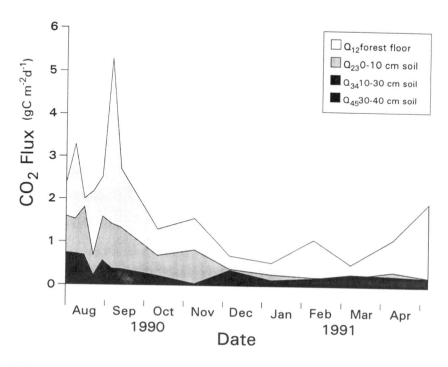

Figure 4. CO_2 flux by layers as measured by the soil profile method. Data are means for all three sites.

(Table 1). The 1X site and the control did not behave in the same manner as the 2X site which suggests a local or site effect. The fall measures may better reflect the relative carbon turnover by depths in these forests. The high CO_2 production in the these layers indicate greater carbon cycling as a result of inputs of litter from above-ground sources and the higher root densities and therefore increased root respiration and root turnover in the upper soil layers.

IV. Discussion

A. Soil Profile

The two main sources of error associated with the soil profile method are measurement of CO_2 concentration and calculation of Ds. Accurate measurement of CO_2 concentration in the field requires several precautions. Since the most important flux is that to the atmosphere, the measures for the forest floor layer are the most important. Within this layer, CO_2 concentration typically varies from 1000 ppm at the bottom to 500 ppm at the top, and Ds's are relatively high. An error of only 100 ppm in CO_2 concentration gradient could alter the calculated flux by 20%. Calculation of Ds of the forest floor and soil were prone to measurement errors mostly in moisture, somewhat in bulk density and only slightly in particle density. Percent moisture was not measured for the same volume in which the CO_2 concentrations were made because sampling would have disturbed the profile; estimates had to be taken from another representative location. Percent moisture in the top 10 cm of soil had a 23% coefficient of variation for 16 samples taken within a 4 by 4 m grid on a dry day. Percent moisture in the forest floor samples taken at locations more than 10 m apart had a coefficient of variation of over 30% during dry periods and approximately 20% during wet periods. An error in moisture of 25% could result in an 15 to 30 % error in CO_2 fluxes in the forest floor and 20 to 60% errors in the fluxes in the soil. Errors in bulk density contributed similar errors in CO_2 fluxes as percent moisture, but bulk density could be measured with higher confidence than moisture, particularly if measured at the end of the experiment using the same location as that used for the profiles. Errors in particle density of 25% contributed only 5% errors in Ds. These calculations suggest that a single measure of flux occasionally could have a random error as high as 50%. Finally, it was important to correct total porosity by the volume of soil containing rocks and large roots since CO_2 cannot diffuse through these substances. Without the correction, the estimated soil fluxes would have been approximately 13% higher.

In 25 instances, measured CO_2 concentration in the mineral soil was less than 300 ppm. In many of these instances, water was also collected out of the soil wells, low concentrations of O_2 were measured, and a methane peak on the chromatograms was observed. Soil CO_2 concentrations less than those found in the air are not expected in the soil unless there is a CO_2 sink. Microbial consumption of CO_2 such as during methanogenesis is possible.

B. Static Chamber

The static method is relatively easy to use and should give reliable effluxes estimates as long as headspace CO_2 concentrations do not vary widely from ambient concentrations during the incubation. The greatest potential bias occurs under conditions of high CO_2 efflux when CO_2 concentrations can build up inside the chamber. To a lesser extent, CO_2 can be removed too rapidly from the headspace when soil effluxes are low. Under these conditions, the vertical concentration gradient from the soil into the chamber headspace will be altered resulting in some soil CO_2 at depths below the chamber bottom either diffusing away from or toward the chamber, following the steepest concentration gradients (provided that diffusivity of the soil is uniform).

Refinements to reduce chamber bias might involve increasing the height of the chambers, reducing the height to diameter ratio of the jars containing the absorbent, and varying the height of the absorbent above the soil surface. Taller chambers would provide a larger headspace per surface area of soil and would act to buffer against changes in headspace CO_2 concentrations. Increasing the jar diameter or decreasing the height of the jar above the absorbent surface would act to reduce the boundary layer over the absorbent and its resistance to diffusion. This technique would enhance the efficiency of the absorbent and reduce the problem of headspace CO_2 increases but would increase the problem of CO_2 decreases under the conditions of low soil efflux. To alleviate the problem of depression of headspace CO_2 concentration caused by too rapid of CO_2 absorption, the position of the jar above the soil surface might be varied so as to keep the jar close to the soil surface when effluxes are expected to be high and far from the soil surface when effluxes are expected to be low.

The static chamber method of measuring CO_2 efflux has been compared with other methods with somewhat mixed results. Static chambers have been reported to underestimate CO_2 effluxes when compared to "dynamic" methods (those that utilize flow-through chambers and infrared gas analyzers) by at least 1/3 (Edwards and Sollins, 1973; Cropper et al., 1985; Freijer and Bouten, 1991). Van Cleve et al. (1979) compared four methods under laboratory conditions and found the static method to give the highest effluxes. deJong et al. (1979) observed close agreement between the soil profile and the static method at one site, but at another site the soil profile gave much higher estimates of CO_2 efflux. It has been recommended to use sufficiently large volumes of alkali so that 80% is still unreacted at the end of an incubation to avoid underestimation of CO_2 effluxes (Kirita and Hozumi, 1966). However, this concept is not supported by the data as rates of absorption of CO_2 by alkali appear to be linear irrespective of the degree to which the alkali is saturated (Table 3 in Kirita and Hozumi, 1966; Figure 6 in Van Cleve et al., 1979; unpublished data of the author). Thorough comparison of methods may have to wait for the development of a means to produce a uniform and controllable CO_2 emission surface instead of using soil.

C. Effects of Nitrogen

Depression of soil CO_2 efflux with additions of nitrogen has also been reported by others (deJong et al., 1974; Robarge, 1976; Kowalenko et al., 1978; Nohrstedt et al., 1989). The mechanism of suppression is not known, however there is evidence in the literature that increased inputs of exogenous nitrogen may reduce the flux of carbon through the belowground subsystem. Additions of N have been associated with increase shoot:root ratios of laboratory-grown plants (Vessey and Layzell, 1987) and reduced root production in forests (Vogt et al., 1985 and Ahlstrom et al., 1988). One explanation is that less carbon is allocated to the root and mycorrhizal system since nitrogen is relatively easier to obtain. It is a working hypothesis in mycorrhizae research that at either high or low ranges of N availability, colonization of roots by ectomycorrhizal fungi is reduced (Harley and Smith, 1983). At high nitrogen availability, sufficient nitrogen can be acquired by the host without the aid of the symbiosis. While at very low nitrogen availability, the overall health and ability to fix carbon of the host is impaired such that the host can not maintain the symbiosis. Changes in the carbon allocation to root systems as a function of added nitrogen may result in greater carbon accretion aboveground at the expense of carbon flow belowground. Another mechanism by which N could suppress CO_2 efflux from soils is through suppression of decomposition of resistant organic matter. In a review of 60 reports, Fog (1988) concluded that nitrogen additions enhanced decomposition in cases where the carbon source was relatively labile. But in cases where the carbon source was resistant, added nitrogen either had no effect or acted to reduce decomposition processes. Added N has been shown to suppress the oxidation of aromatic herbicides in soil (Entry et al., 1993b) and thought to suppress the lignolytic enzyme system of white rot fungi (Kirk and Farrel, 1987).

Considering that many forest ecosystems now receive additions of nitrogen via atmospheric sources or intensive fertilization, and that purposeful nitrogen fertilizations is one management option under consideration to mitigate global climate change, it is important to understand the nature of interactions between nitrogen availability and the belowground carbon cycle of forests.

V. Summary and Conclusions

The soil profile method shows promise as a reliable means to measure CO_2 efflux. If only efflux from the forest floor surface is desired, CO_2 concentrations at the top of the forest floor and at some depth in the forest floor need only be made. These samples could even be collected without the need for permanent gas wells if a proper syringe were used. To obtain a better estimate of the true CO_2 gradient, analysis should be performed on a bulked sample composed of at least three gas samples collected from an area of 1 to several square meters. Forest floor measures, particularly moisture, could also be better monitored within the area without disrupting the area for future determinations of the CO_2 profile.

There were several sources of measurement error associated with the soil profile method. Differences in replicate samples of CO_2 concentrations from the same well occasionally exceeded 10%, therefore several profiles need to be developed. A one-point calibration of the gas chromatograph with a standard gas of 9900 ppm CO_2 was found to underestimate CO_2 concentrations of 400 ppm by 100 ppm, therefore a two-point calibration is recommended for analyses of wide ranges of CO_2 concentrations. In the deeper or compacted soil layers, CO_2 concentrations as high as 30,000 ppm were observed. Under these conditions, it is difficult to store the gas during transport back to the laboratory without some loss. At least an estimate of the loss should be performed. Accurate estimation of soil diffusivity is difficult and is dependent on accurate measures of soil moisture, bulk density and rock contents.

Static chambers can underestimate and overestimate CO_2 efflux depending on the dimensions of the chamber, the jars containing the absorbent, and the rate of soil CO_2 efflux. The bias comes from alterations in the chamber headspace CO_2 concentration caused by flux rates into the absorbent not being equal to flux rates from the soil. Refinements to the chambers to reduce their bias include taller chambers that buffer against CO_2 concentrations changes, jars that reduce the boundary layer above the absorbent, and varying the height of the absorbent above the soil surface as a function of expected soil efflux rates.

CO_2 effluxes from three forested sites with differing histories of N additions suggests that added N acts to suppress soil CO_2 effluxes either during the first year after additions and/or during periods of high soil effluxes. The suppression appeared to occur mostly in the forest floor and upper soil layer which were the most active sources of CO_2 production. The mechanism of suppression is not known but the literature suggests N may suppress carbon allocation to root systems and/or suppress heterotrophic oxidation of resistant carbon compounds.

Acknowledgments

This study was supported in part by funds from the Environmental Protection Agency and USDA's National Research Inititative Competitive Grants Program.

References

Aiken, R.M., M.D. Jawson, K. Grahammer, and A.D. Polymenopoulos. 1991. Positional, spatially correlated and random components of variability in carbon dioxide efflux. *J. Environ. Qual.* 20:301-308

Ahlstrom, K., H. Perrson, and I. Borjesson. 1988. Fertilization in a mature Scots pine (*Pinus sylvestris* L.) stand: effects on fine roots. *Plant Soil* 106:179-190.

Anderson, J.P.E. 1982. Soil respiration. p. 831-871. In: A.L. Page, R.H. Miller, and D.R. Keeney (eds.). *Methods of Soil Analysis.* Part 2, 2nd Ed. American Society of Agronomy, Madison, WI.

Blake, G.R. and K.H. Hartge. 1986. Particle density. p. 377-382. In: A. Klute, (ed.). *Methods of Soil Analysis.* Part 1, 2nd Ed. American Society of Agronomy, Madison, WI.

Cropper, W.P., Jr., K.C. Ewel, and J.W. Raiche. 1985. The measurement of soil CO_2 evolution *in situ*. *Pedobiologia* 28:35-40.

Danielson, R.E. and P.L. Sutherland. 1986. Porosity. p. 443-462. In: A. Klute, (ed.). *Methods of Soil Analysis.* Part 1, 2nd Ed. American Society of Agronomy, Madison, WI.

Davies, B.E. 1974. Loss-on-ignition as an estimate of soil organic matter. *Soil Sci. Soc. Amer. Proc.* 38:150-151.

deJong, E. and H.J.V. Schappert. 1972. Calculation of soil respiration and activity from CO_2 profiles in the soil. *Soil Science* 113:328-333.

deJong, E., H.J.V. Schappert, and K.B. MacDonald. 1974. Carbon dioxide evolution from virgin and cultivated soil as affected by management practices and climate. *Can. J. Soil Sci.* 54:299-307.

deJong, E., R.E. Redmond, and E.A. Ripley. 1979. A comparison of methods to measure soil respiration. *Soil Science* 127:300-306.

Edwards, N.T. and P. Sollins. 1973. Continuous measurements of carbon dioxide evolution from partitioned forest floor components. *Ecology* 54:406-412.

Entry, J.A., K.G. Mattson, and M.B. Adams. 1993a. Carbon balance in forests ecosystems: response to nitrogen. p. 155-165. In: T.S. Vinson and T.P. Kolchugina, (eds.). International Workshop on Carbon Cycling in Boreal Forests and Sub-Arctic Ecosystems: Biospheric Responses and Feedbacks to Global Climate Change. U.S. Environmental Protection Agency. EPA/600R-93/084.

Entry, J.A., K.G. Mattson, and W.H. Emmingham. 1993b. The influence of nitrogen on atrazine and 2,4-D mineralization in grassland soils. *Biol. Fert. Soils* 16:179-182.

Fog, K. 1988. The effect of added nitrogen on the rate of decomposition of organic matter. *Biol. Rev.* 63:433-462.

Freijer, J.I. and W. Bouten. 1991. A comparison of field methods for measuring soil carbon dioxide evolution: Experiments and simulation. *Plant Soil* 135:133-142.

Gardner, W.H. 1986. Water content. p. 493-544. In: A. Klute (ed.). *Methods of Soil Analysis*. Part 1, 2nd Ed. American Society of Agronomy, Madison, WI.

Harley, J. and S.A. Smith. 1983. *Mycorrhizae Symbiosis*. Academic Press, NY.

Kirita, H. and K. Hozumi. 1966. Re-examination of the absorption method of measuring soil respiration under field conditions. I. Effect of the amount of KOH on observed values. *Physiol. and Ecol.* 14:23-31.

Kirk, K.T. and R.L. Farrel. 1987. Enzymatic "combustion": the microbial degradation of lignin. *Ann. Rev. Microbiol.* 41:465-505.

Kowalenko, C.G., K.C. Ivarson, and D.R. Cameron. 1978. Effect of moisture content, temperature, and nitrogen fertilization on carbon dioxide evolution from field soils. *Soil Biol. Biochem.* 19:417-423.

Nohrstedt, H.-O., K. Arnebrant, E. Baath, and B. Soderstrom. 1989. Changes in carbon content, respiration rate, ATP content, and microbial biomass in nitrogen-fertilized pine forest soils in Sweden. *Can. J. For. Res.* 9:323-328.

Peterson, D.L. and R.H. Waring. 1994. Overview of the Oregon Transect Ecosystem Research Project. *Ecological Applic.* (In press)

Singh, J.S. and S.R. Gupta. 1977. Plant decomposition and soil respiration in terrestrial ecosystems. *Bot. Rev.* 43:449-528.

Robarge, M.R. 1976. Respiration rates for determining the effects of urea on the soil surface organic horizon of a black spruce stand. *Can. J. Microbiol.* 22:1328-1335.

Robertson, G.P., M.A. Huston, F.C. Evans, and J.M Tiedje. 1988. Spatial variability in a successional plant community: Patterns of nitrogen availability. *Ecology* 69:1517-1524.

Van Cleve, K., P.I. Coyne, E. Goodwin, C. Johnson, and M. Kelly. 1979. A comparison of four methods for measuring respiration in organic material. *Soil Biol. Biochem.* 11:237-246.

Vessey, J.K. and D.B. Layzell. 1987. Regulation of assimilate partitioning in soybean. *Plant Physiol.* 83:341-348.

Vogt, K.A., D.J.Vogt, E.E. Moore, W. Littke, C.C. Grier, and L. Leney. 1985. Estimating Douglas-fir fine root biomass and production from living bark and starch. Canadian Journal of Forest Research 15:177-179.

CHAPTER **28**

In Search of the Bioreactive Soil Organic Carbon: The Fractionation Approaches

H.H. Cheng and J.A.E. Molina

I. Introduction

A major issue in the current concern on the impact of global climate change is the potential effect of such change on the present status of the earth's environment. The warming trends observed globally in recent years have been postulated to be related to the increase in the levels of certain "greenhouse" gases, especially CO_2, in the atmosphere. However, the effects of increasing greenhouse gas concentrations cannot be assessed until the magnitude of the net concentration increase can be ascertained. This increase is not only affected by the quantities of these gases produced, especially from such human activities as fossil fuel burning and forest destruction, it is also governed by the modifying effects of the environment which can serve as a sink of these gases. Unfortunately, the identity and nature of the sinks are far from being elucidated. As pointed out by Sundquist (1993):

> "For more than 20 years, the principal problem in budgeting atmospheric CO_2 has been the excess of known sources over identified sinks. The amount of CO_2 produced by human activities ... significantly exceeds our best estimates of the amount of CO_2 absorbed by the oceans and atmosphere. To account for this budget imbalance (the so-called "missing" CO_2) there must be another large CO_2 sink. This sink is probably somewhere in the world's terrestrial plants and soils, but its specific identity has eluded detection." (Sundquist, 1993. p. 934)

Although terrestrial plant-soil systems are recognized as potential sinks for atmospheric CO_2, much of the current effort to account for the missing CO_2, including that of Sundquist himself, is still only focused on the oceans and forests as sinks. It is imperative that the carbon budget of the vast land areas covered by prairies and agricultural crops and the dynamics of carbon cycling processes affecting the amount of carbon stored in these systems be also accurately assessed so that the actual role of terrestrial plant-soil systems in the global carbon budget can be realistically determined.

A. Significance of Terrestrial Carbon

Soil is a vast depository for carbon, water, and heat in the environment and plays a crucial role in the carbon cycle, the hydrologic cycle, and general climate. Soil, climate, and vegetation vary over a landscape and change with time according to season and land use practices. Most estimates have indicated that the carbon stored in terrestrial systems is two or more times greater than that in the atmosphere (e.g. Bohn, 1976; Sundquist, 1993). The annual exchange of carbon between the atmosphere and terrestrial systems represents more than one tenth of the carbon in the atmosphere. Thus any change in terrestrial carbon pools should affect the level of atmospheric CO_2 significantly. In comparison with terrestrial systems, even though the ocean is by far the larger reservoir for carbon, the kinetics of exchange between the carbon species in the ocean and those in the atmosphere is extremely slow, and biological carbon fixation is limited to the ocean surface (Sarmiento, 1993).

B. Sensitivity of Total Soil Organic Carbon to Climate Changes

There is certainly no lack of soil carbon data in journal literature and in soil survey reports (Grossman et al., 1992). Yet, there are few valid estimates of the total carbon stored in the earth's surface soil. Most global estimates are based on data obtained from very few samples, each of which represents a very large geographic area (e.g., Bohn, 1976; Kimble et al., 1990). These estimates could not account for the local influence of different soil formation factors in different regions that affect soil organic carbon (SOC) contents as well as the natural variability of soils within a landscape. Moreover, most studies on the effect of environmental changes on the status of soil organic matter have invariably attempted to assess changes in **total** SOC. Since only a small portion of the terrestrial carbon is actively involved in the biological carbon transformation processes, total SOC is not always a sensitive or accurate indicator of climate changes, especially in the short term, although it can be an indicator of change over a lengthy period of time (Odell et al., 1984). To study the influence of climate changes on the dynamics of SOC transformations, it will be necessary to identify the specific soil carbon components which are actively involved in the carbon cycling processes.

C. Constituents and Cycling of Soil Organic Carbon

Soil organic matter is a composite of carbon-rich organic materials derived from plant and animal residues that have been variably degraded by soil microorganisms, mixed with microbial remains and organic complexes that have reacted with or been stabilized by soil minerals. The relationship among the various components and their transformations are often depicted by a carbon cycle (Stevenson, 1986). The terrestrial carbon cycle involves fixation of atmospheric CO_2 by plants through photosynthesis, partitioning of photoassimilated carbon within plants and into animal pools, incorporation of plant, animal, and other carbon sources into soil organic matter through residue decay and resynthesis, and the return of carbon to the atmsophere by mineralization of soil carbon. This depiction provides a conceptual framework that broadly identifies the major carbon pools in the total environment. However, it provides little information on the dynamics of carbon transformation processes within the soil, nor identifies the controlling mechanisms for carbon to move from one pool to the next. Since soil carbon exists in many different forms and stability or bioavailability states, a general description of the carbon cycle is not helpful in identifying the nature of specific organic fractions in the soil pools involved in the transformation and cycling processes.

D. Need for Identifying Bioreactive Soil Organic Carbon

Since soil carbon includes a multitude of different organic components ranging from plant residues and microbial biomass to highly stable humus, not all SOC participates equally in the transformation processes. Some of the components are formed and degraded more rapidly than others. The more bioreactive components are influenced more readily by any change to the soil system. To assess the effects of climate changes on the terrestrial carbon cycle, it is necessary to differentiate the carbon associated with bioreactive soil organic components subject to rapid turnover from the carbon which is tied up in more stable forms subject only to long-term turnovers. Although some soil carbon components, such as the microbial biomass and the water-soluble organic carbon, are readily identified as bioreactive SOC, there is little information on the magnitude or the total nature of this pool.

The purpose of this paper is to assess the applicability of various current SOC characterization methods for identifying the bioreactive SOC pool and to evaluate other approaches which may be more fruitful for characterizing and quantifying the magnitude of this pool. This will not be a comprehensive review of the literature on soil organic matter extraction, fractionation, or characterization. The emphasis will be to point out the problems and constraints of existing SOC fractionation methods and to suggest means for developing more appropriate analytical approaches for isolation and characterization of the bioreactive SOC.

II. Approaches to Fractionation of Soil Organic Carbon

A. Importance of Total Soil Organic Carbon Determination

Before we consider the approaches to define and identify the bioreactive SOC, it will be necessary to consider how total SOC is determined, since the carbon fractions are usually expressed as a percentage of the total SOC. If the total SOC cannot be determined accurately and precisely, the magnitude of the carbon fractions isolated may not be correctly estimated and their significance may not be appreciated. The importance of accurate total SOC determination can be illustrated by the following calculations. For instance, it is known that the earth's land surface area is approximately 1.48×10^8 km^2, and 1 ha of surface soil (15 cm in depth) contains approximately 2.5×10^6 kg of soil. If the soil carbon content of just one-tenth of the earth's land is increased by 0.1%, this is equivalent to adding some 3.7 gT of carbon to the soil. This amount is more than double the amount needed to account for the "missing carbon" or imbalance between the global carbon source and sink estimated by Sundquist (1993). Few of the commonly used methods for total SOC determination have either the accuracy or precision to detect a difference of 0.1% carbon. If the method used does not have the sensitivity of detecting such small differences, they should not be used for these studies. Therefore, it is essential to assess the applicability of an analytical method critically before accepting the data generated using this method.

All current methods for determination of SOC are based on the conversion or combustion of soil carbon to CO_2. Their differences are in the oxidation agent and procedure, the quantitativeness of carbon conversion, and the method for assessing the product of conversion. Oxidation can be either by wet or dry combustion. Although such oxidants as hydrogen peroxide and alkaline hypobromide have been used to remove soil organic matter, they have not been used for soil carbon analysis. Wet combustion methods are most commonly based on oxidation by dichromate dissolved in concentrated sulfuric acid or a sulfuric and phosphoric acid mixture. Completeness of carbon conversion depends on the strength of the oxidant used and whether external heat is applied to assist in the combustion process. Dry combustion methods use either oxygen or metal oxides under high temperature to convert organic carbon. The method of determination can be based either on a measurement of the CO_2 produced, or on the loss in oxidative power of the oxidant used. The accuracy and precision of the procedure for final assessment depends upon the precautions taken in eliminating interferences before measuring the final product, and the sensitivity of the measurement mode to interference.

Common methods for soil carbon determination have been thoroughly examined and reviewed in the literature (e.g., Bremner and Jenkinson, 1960; Allison et al., 1965; Nelson and Sommers, 1982). The most frequently used method for total soil carbon determination found in the literature is probably the rapid dichromate oxidation method, commonly known as the Walkley-Black method (see Nelson and Sommers, 1982). Because many versions or modifications of this method have been proposed and many soil carbon values reported in the literature do not specify which version was actually used, it would be impossible to evaluate the extent of error in the carbon values reported without knowing the specific analytical procedure used. A mere citation of the Walkley-Black method as reference is not sufficient to ascertain which version of the method was used.

Regardless the version used, it is not likely that many of the reported carbon values are sufficiently quantitative or accurate to differentiate carbon content at the 0.1% level. Nelson and Sommers (1982) showed that the total SOC recovered by the Walkley-Black method ranged from 45 to 85%, averaging at best around 80%. Incomplete recovery results in an underestimation of soil carbon content. Some versions of the Walkley-Black method use a constant factor to compensate for the incomplete recovery. This correction may bring the analytical results closer to an accurate estimation, but the precision of such an estimate is no more satisfactory. Other versions improve the carbon recovery by supplying external heat during the combustion step (Mebius, 1960; Allison et al., 1965). These modifications can usually yield more accurate and reproducible results. Instrumental methods using the dry combustion approaches have also been shown to be more accurate. If appropriate precautions are taken, these modified procedures have the sensitivity and accuracy to differentiate carbon contents at the 0.1% level. To be useful in evaluating the impact of global climate change on terrestrial carbon cycling processes, analytical methods must be able to detect or differentiate small changes in the actual carbon content.

B. Chemical Fractionation Approaches

Numerous references in the literature have reported on results of fractionation of soil organic matter or SOC. The general approach to characterizing the nature of SOC is first to bring the organic matter into solution by an extractant, and then to fractionate the contents of the extract. Most of the methods for fractionation favor those extractants that are the most efficient in solubilizing organic matter. The more organic matter is solubilized, the more effective the extraction is thought to be. For characterization of the chemical structure of the extracted organic matter, such preference is desirable. However, for extracting the bioreactive SOC pool, such approaches may not be appropriate.

1. Classical Fractionation

Among the commonly known fractionation methods is the classical scheme of using NaOH as an extractant at concentrations varying from 0.1 to 0.5 M, and subsequent acidification of the extract to separate fulvic acid, humic acid, and humins (Stevenson, 1965). Although NaOH has been used for over 200 years as an extractant for soil organic matter, it is still the most efficient extractant for dissolving SOC. The amount of SOC solubilized usually ranges from 50 to 70% of the total in the soil. Many other extractants have also been proposed and tested. Among the more successful ones are those which are capable of forming complexes with polyvalent metals, such as oxalate, citrate, and sodium pyrophosphate (Bremner and Lees, 1949). Sodium pyrophosphate, for instance, has an extraction efficiency of around 35%. Most of the studies using these fractionation schemes were aimed to examine the structure or composition of soil organic matter. Such schemes are not designed to separate or differentiate the bioreactive SOC from other SOC. Even though humic acid has sometimes been assumed to be a resynthesized product of fulvic acid, these chemical fractions cannot be correlated to soil bioreactivity. Studies of soil incubated with ^{14}C-labeled glucose have shown that even after a brief incubation period, ^{14}C could already be found in all such chemical fractions (e.g., Simonart and Mayaudon, 1958).

2. Isolation of Structually-Known Components

Occasionally, such methods as the proximate analysis procedure for plant materials has been used for soil fractionation (Stevenson, 1965). Soil samples are first extracted with a sequence of organic solvents and dilute acids to remove fats, waxes, oils, resins, and water-soluble polysaccharides, then hydrolyzed with increasing strengths of acid to remove celluose and hemicelluloses, and finally acid-digested to estimate protein and lignin-humus (Waksman and Stevens, 1930). The aim of this type of approach is to obtain structurally-known or chemically-identifiable fractions from soil. Proximate analysis tends to identify components associated with plant residues rather than with the humified portion. The readily extractable fractions are likely more bioreactive. However, in proximate analysis of soils, over 80% of the carbon would remain in the final acid-digested fraction or in the undigested residue. Whether any of the carbon not extracted in the initial steps is also bioreactive is not known.

A number of analytical methods are available to isolate and identify specific organic chemicals or components in the soil. Whether these components constitute the bioreactive SOC have not been evaluated. A major obstacle to developing reliable methods for estimating bioreactive SOC is the lack of knowledge on the nature of this SOC. In the 1940-50s, there was a great effort in developing methods for characterizing soil polysaccharides (see Greenland and Oades, 1975). These chemicals serve a variety of functions in the soil, but cannot be identified as the sole bioreactive components. Similarly, since the pioneering studies by Jenkinson and Powlson (1976), there have been a proliferation of studies on estimation of soil microbial biomass. A number of modifications of the fumigation method have been suggested (e.g., Vance et al., 1987). Other approaches, such as by the respiration method (Anderson and Domsch, 1978; Smith et al., 1985), ATP or dehydrogenase assays, etc., are various indicators of soil microbial population (Jenkinson and Ladd, 1981; Kaiser et al., 1992). Soil microbial biomass is certainly a significant fraction of bioreactive SOC. It is by no means representative of the total bioreactive SOC. Thus, in spite of the extensive research efforts on soil organic matter worldwide over the past century, no specific methods have been developed to account for all the bioreactive SOC or to differentiate it from the more stable carbon in soil organic matter.

3. Lessons from Nitrogen Fractionation Studies

Although carbon cycling has not received needed attention until recently, the nitrogen cycling processes have been studied extensively. The intimate relationship between soil organic carbon and soil nitrogen (commonly expressed as the C/N ratio) and its importance in nitrogen transformations is common knowledge. Jansson (1958) has clearly pointed to soil carbon as the driving force for the nitrogen cycle. Thus one should be able to gain insights on the nature of the bioreactive carbon from nitrogen transformation studies. However, nitrogen fractionation studies have traditionally not given much attention to the nature of associated carbon, except as part of the nitrogenous components, such as amino acids and hexosamines (e.g., Cheng and Kurtz, 1963; Bremner, 1965). Even though fractionation schemes have been developed for soil nitrogen characterization, the fractions isolated cannot be specifically identified as bioreactive. As pointed out by Keeney and Bremner (1964), it may not be possible to isolate specific chemical fractions from soils that is the bioreactive or "available" fraction of soil nitrogen. This could also be the case for soil carbon.

C. Modeling Approaches

Models per se are not useful as a soil fractionation tool for bioreactive SOC, but they could provide valuable information on the nature and magnitude of carbon pool representative of the bioractive fractions. For the past half of a century, there has been a mushrooming of models depicting the carbon and nitrogen transformation processes in the soil. Early modeling efforts to estimate soil carbon losses or accumulation have been based on extrapolation of experimental data for documenting the dynamics of total soil carbon (Jenny, 1941; Henin and Dupuis, 1975). More recently, models have been constructed to include pools of SOC grouped in terms of their decay rate constants, indicative of their stability. Molina et al. (1994) recently surveyed the range of stability constants used in over 20 models and summarized the number of pools identified at each decay rate constant in a histogram of dynamic pool stability. Soil organic carbon depicted by these models could be seen as a continuum of soil organic pools with stability ranging from decay rates of transient states to those of infinite stability.

On the other hand, studies measuring isotopic carbon variations caused by such occurrences as isotopic fractionations during chemical and physical transformations, radioactivity decay, or thermonuclear testing, do not support the concept of a continuum of stability which would have resulted from the gradual stabilization of in situ synthesized soil organic matter (Campbell et al., 1967; Paul and Van Veen, 1978). Instead, experimental evidence tends to support the existence of separate soil organic matter pools representing dynamic fractions of modern origin and stable fractions of much older origin. A number of studies have identified the retention time of modern fractions ranging from a few years to 20 or 40 years; whereas the older fractions are shown to be from over 500 to thousands of years in age (Martel and Paul, 1974; O'Brien and Stout, 1978; Jenkinson et al., 1987; Schiff et al., 1990; and Hsieh, 1992). These studies have given support to a bimodal distribution of soil organic matter components based on their stability. The experimental evidence has now led to modifications of carbon and nitrogen transformation models so that the dynamics of younger and older organic matter in the soil could be depicted separately. Such an attempt has recently been validated with long-term laboratory and field data (Nicolardot et al., 1994; Nicolardot and Molina, 1994).

D. Radioisotope Fractionation Approaches

A promising approach to estimate the stability of SOC on the basis of its age is the carbon isotopic fractionation technique. This technique is based on the differential discrimination of carbon isotopes in the metabolites of C3 and C4 plants, and it has proved to be also a powerful tool to distinguish new from old soil organic matter (Martin et al., 1990). Using this technique, Cerri et al. (1985) demonstrated that the organic matter newly formed from sugar-cane grown on soil developed under forest cover exhibited a turnover of 40 years. Balesdent et al. (1987) showed that the photoassimilated carbon from corn grown on pine forest exhibited a SOC turnover rate of about 36 years. The same dichotomy between the new and old organic matter was found when the prairie soil of the Sanborn field, Missouri, was broken and cultivated with wheat. The old prairie soil exhibited a turnover time of 1400 years or more while the organic matter synthesized from wheat had a half-life of 10 to 15 years (Balesdent et al., 1988).

The bimodal distribution of the soil organic matter age has been shown to be likely associated with soil particle size fractions (Anderson et al., 1981; Elliott and Cambardella, 1991). The older, stable organic matter tended to be associated with silt fractions while the rapidly turned-over organic matter appeared to be in the coarse and very fine fractions (Tiessen and Stewart, 1983; Anderson and Paul, 1984; Christensen, 1987). Newly synthesized organic matter with a short half-life was found to be associated with fine sand particles and fine organo-mineral complex (Cerri et al., 1985; Balesdent et al., 1987). Bioresistant aromatic chemicals and compounds of microbial origin were mostly localized in the fine silt fraction (Turchenek and Oades, 1979). Stable old prairie organic carbon was also observed in ferromanganic concretions, an indication that carbon stability could be linked in part to co-precipitation with Fe and Mn (Balesdent et al., 1988). By separating the soil organic materials into mineral-free "light" organic fraction and organo-mineral "heavy" fraction (Strickland and Sollins, 1987), Janzen et al. (1992) showed that the light fraction of SOC was more transient and more sensitive to cropping and management changes. These evidences all indicate that certain components of SOC are more bioreactive than others. It should then be possible to develop an appropriate analytical method to isolate those chemical fractions which are identified by models as the bioreactive SOC, sensitive to climate changes.

III. Conclusion: Prospect Brightening

Although at present there does not exist a readily available method for identifying and characterizing the bioreactive organic carbon components in soils, this brief review has brought out several possible approaches toward the development of such methodology. Attention needs to be shifted away from the strictly chemical fractionation approaches to methods which can be related to the availability of those fractions of SOC to the biological activities in the soil. Methods for such specific bioreactive components as polysaccharides and microbial biomass will continue to be important. Methods developed for nitrogen availability estimations (e.g., Keeney, 1982; Gianello and Bremner, 1986) can also be useful guides. Novel extraction and fractionation methods, such as using the supercritical fluid extraction technique (Cheng et al., 1993) should be further explored. Most importantly, reliable carbon transformation models should be validated for estimating the size of the bioreactive SOC pool. Radioisotope fractionation technique should be used as a guide to identify the soil fractions which contain the bioreactive SOC components. Appropriate chemical fractionation methods may then be developed for estimating the magnitude and characterizing the nature of the bioreactive SOC.

Acknowledgement

Minnesota Agricultural Experiment Station Publication No. 20,837. Contribution from the Department of Soil Science, University of Minnesota, St. Paul, MN.

References

Allison, L.E., W.B. Bollen, and C.D. Moodie. 1965. Total carbon. p.1346-1366 In: C.A. Black (ed.). *Methods of soil analysis.* Agronomy 9.

Anderson, D.W. and E.A. Paul. 1984. Organo-mineral complexes and their study by radiocarbon dating. *Soil Sci. Soc. Am. J.* 48:298-301.

Anderson, D.W., S. Saggar, J.R. Bettany, and J.W.B. Stewart. 1981. Particle size fractions and their use in studies of soil organic matter: I. The nature and distribution of forms of carbon, nitrogen, and sulfur. *Soil Sci. Soc. Am. J.* 45:767-772.

Anderson, J.P.E. and K.H. Domsch. 1978. A physiological method for the quantitative measurement of microbial biomass in soils. *Soil Biol. Biochem.* 10:215-221.

Balesdent, J., A. Mariotti, and B. Guillet. 1987. Natural abundance as a tracer for studies of soil organic matter dynamics. *Soil Biol. Biochem.* 19:25-30.

Balesdent, J., G.H. Wagner, and A. Mariotti. 1988. Soil organic matter turnover in long-term field experiments as revealed by carbon-13 natural abundance. *Soil Sci. Soc. Am. J.* 52:118-124.

Bohn, H.L. 1976. Estimate of organic carbon in world soils. *Soil Sci. Soc. Am. J.* 40:468-470.

Bremner, J.M. 1965. Organic forms of nitrogen. p. 1238-1255. In: C.A. Black, (ed.). *Methods of soil analysis.* Agronomy 9.

Bremner, J.M. and D.S. Jenkinson. 1960. Determination of organic carbon in soil. I. Oxidation by dichromate of organic matter in soil and plant materials. *J. Soil Sci.* 11:394-402.

Bremner, J.M. and H. Lees. 1949. Studies on soil organic matter: II. The extraction of organic matter from soil by neutral reagents. *J. Agric. Sci.* 39:274-279.

Campbell, C.A., E.A. Paul, D.A. Rennie, and K.J. McCallum. 1967. Applicability of the carbon-dating method of analysis to soil humus studies. *Soil Sci.* 104:217-224.

Cerri, C., C. Feller, J. Balesdent, R. Victoria, and A. Plenecassagne. 1985. Application du traçage isotopique naturel en ^{13}C, á l'étude de la dynamique de la matiére organique dans les sols. *Compt. Rend. Acad. Sci. Ser. II*, 300:423-428.

Cheng, H.H. and L.T. Kurtz. 1963. Chemical distribution of added nitrogen in soils. *Soil Sci. Soc. Am. Proc.* 27:312-316.

Cheng, H.H., J. Gan, W.C. Koskinen, and L.J. Jarvis. 1993. Potential of the supercritical fluid extraction technique for characterizing organic-inorganic interactions in soils. Proc. Intern. Soc. Soil Sci. Working Gr. MO First Workshop on Impact of Inorganic, Organic, and Microbiological Soil Components on Environmental Quality, Edmonton, Canada, August 1992. (In press)

Christensen, B.T. 1987. Decomposability of organic matter in particle size fractions from field soils with straw incorporation. *Soil Biol. Biochem.* 19:429-435.

Elliott, E.T. and C.A. Cambardella. 1991. Physical separation of soil organic matter. *Agric. Ecosyst. Environ.* 34:407-419.

Gianello, C. and J.M. Bremner. 1986. Comparison of chemical methods of assessing potentially available organic nitrogen in soil. *Commun. Soil Sci. Plant Anal.* 17:215-236.

Greenland, D.J. and J.M. Oades. 1975. Saccharides. p. 213-261. In: J.E. Gieseking (ed.). *Soil components Volume 1 Organic components.* Springer Verlag, New York.

Grossman, R.B., E.C. Benham, J.R. Fortner, S.W. Waltman, J.M. Kimble, and C.E. Branham. 1992. A demonstration of the use of soil survey information to obtain areal estimates of organic carbon. Am. Soc. Photogram. Remote Sens./Am. Congr. Survey. Map. Tech. Papers 4:457-465.

Henin, S. and M. Dupuis. 1945. Essai de bilan de la matiére organique du sol. *Ann. Agron.* 15:17-29.

Hsieh, Y.P. 1992. Pool size and mean age of stable organic carbon in cropland. *Soil Sci. Soc. Am. J.* 56:460-464.

Jansson, S.L. 1958. Tracer studies on nitrogen transformations in soil with special attention to mineralisation-immobilization relationships. *Kungl. Lantbrukshogsk. Ann.* 24:101-361.

Janzen, H.H., C.A. Campbell, S.A. Brandt, G.P. Lafond, and L. TownleySmith. 1992. Light-fraction organic matter in soils from long-term crop rotations. *Soil Sci. Soc. Am. J.* 56:1799-1806.

Jenkinson, J.S. and J.N. Ladd. 1981. Microbial biomass in soil: Measurement and turnover. p. 415-471. In: E.A. Paul and J.N. Ladd (ed.). *Soil biochemistry. Vol. 5.* Marcell Dekker, New York.

Jenkinson, D.S. and D.S. Powlson. 1976. The effects of biocidal treatments on metabolism in soil. V. A method for measuring soil biomass. *Soil Biol. Biochem.* 8:209-213.

Jenkinson, D.S., B.S. Hart, J.H. Rayner, and L.C. Parry. 1987. Modelling the turnover of organic matter in long-term experiments at Rothamsted. *INTECOL Bull.* 15:1-8.

Jenny, H. 1941. *Factors of soil formation - System of quantitative pedology.* McGraw-Hill, New York.

Kaiser, E.-A., T. Mueller, R.G. Joergenson, H. Insam, and O. Heinemeyer. 1992. Evaluation of methods to estimate the soil microbial biomass and the relationship with soil texture and organic matter. *Soil Biol. Biochem.* 24:675-683.

Keeney, D.R. 1982. Nitrogen-availability indices. p. 711-733. In: A.L. Page, et al. (ed.). *Methods of soil analysis. Part 2: Chemical and microbiological properties.* 2nd Ed. Agronomy 9

Keeney, D.R. and J.M. Bremner. 1964. Effect of cultivation on the nitrogen distribution in soils. *Soil Sci. Soc. Am. Proc.* 28:653-656.

Kimble, J.M., H. Eswaran, and T. Cook. 1990. Organic carbon on a volume basis in tropical and temperate soils. *Trans. 14th Intern. Congr. Soil Sci.* V:248-253.

Martel, Y.A. and E.A. Paul. 1974. The use of radiocarbon dating of organic matter in the study of soil genesis. *Soil Sci. Soc. Am. Proc.* 38:501-506.

Martin, A., A. Mariotti, J. Balesdent, P. Lavelle, and R. Vuattoux. 1990. Estimate of organic matter tourover rate in a savanna soil by ^{13}C natural abundance measurements. *Soil Biol. Biochem.* 22:517-523.

Mebius, L.J. 1960. A rapid method for the determination of organic carbon in soil. *Anal. Chim. Acta* 22:120-124.

Molina, J.A.E., H.H. Cheng, B. Nicolardot, R. Chaussod, and S. Houot. 1994. Biologically active soil organics: A case of double identity. In: J.W. Doran (ed.). *Defining Soil Quality for a Sustainable Environment*. SSSA Special Publ. No. 34.

Nelson, D.W. and L.E. Sommers. 1982. Total carbon, organic carbon, and organic matter. p. 539-579. In: A. L. Page et al. (eds.). M*ethods of soil analysis. Part 2: Chemical and microbiological properties*. 2nd Ed. Agronomy 9.

Nicolardot, B. and J.A.E. Molina. 1994. C and N fluxes between rapid and slow turnover pools of soil organic matter: Model calibration with long-term field experimental data. *Soil Biol. Biochem.* 26. (In press)

Nicolardot, B., J.A.E. Molina, and M.R. Allard. 1994. C and N fluxes between rapid and slow turnover pools of soil organic matter: Model calibration with long-term incubation data. *Soil Biol. Biochem.* 26. (In press)

O'Brien, B.J. and J.D. Stout. 1978. Movement and turnover of soil organic matter as indicated by carbon isotope measurements. *Soil Biol. Biochem.* 10:309-317.

Odell, R.T., S.W. Melsted, and W.M. Walker. 1984. Changes to soil organic carbon and nitrogen of Morrow Plots soils under different temperatures, 1904-1973. *Soil Sci.* 137:160-171.

Paul, E.A. and J.A. Van Veen. 1978. The use of tracers to determine the dynamic nature of organic matter. *Trans. 11th Intern. Congr. Soil Sci.* 3:61-102.

Sarmiento, J.L. 1993. Ocean carbon cycle. *Chem. Engin. News* 71(22):30-45.

Schiff, S.L., R. Aravena, S.E. Trumbore, and P.J. Dillon. 1990. Dissolved organic carbon cycling in forested watersheds: A carbon isotope approach. *Water Resour. Res.* 26:2949-2957.

Simonart, P. and J. Mayaudon. 1958. Étude de la décomposition de la matière organique dans le sol, au moyen de carbone radioactif. II. Décomposition du glucose radioactif dans le sol. A. Répartition de la radioactivité dans les fractions humiques du sol. *Plant Soil* 9:376-380.

Smith, J.L., B.L. McNeal, and H.H. Cheng. 1985. Estimation of soil microbial biomass: An analysis of the respiratory response of soils. *Soil Biol. Biochem.* 17:11-16.

Stevenson, F.J. 1965. Gross chemical fractionation of organic matter. p. 1409-1421. In: C.A. Black (ed.). *Methods of soil analysis*. Agronomy 9.

Stevenson, F.J. 1986. *Cycles of soil: Carbon, nitrogen, phosphorus, sulfur, micronutrients*. John Wiley & Sons, New York. 380p.

Strickland, T.C. and P. Sollins. 1987. Improved method for separating light- and heavy- fraction organic material from soil. *Soil Sci. Soc. Am. J.* 51:1390-1393.

Sundquist, E.T. 1993. The global carbon dioxide budget. *Science* 259:934-941.

Tiessen, H. and J.W.B. Stewart. 1983. Particle-size fractions and their use in studies of soil organic matter: II. Cultivation effects on organic matter composition in size fractions. *Soil Sci. Soc. Am. J.* 47:509-514.

Turchenek, L.W. and J.M. Oades. 1979. Fractionation of organo-mineral complexes by sedimentation and density techniques. *Geoderma* 21:311-343.

Vance, E.D., P.C. Brookes, and D.S. Jenkinson. 1987. An extraction method for measuring soil microbial biomass. *Soil Biol. Biochem.* 19:703-707.

Waksman, S.A. and K.R. Stevens. 1930. A critical study of the methods for determining the nature and abundance of soil organic matter. *Soil Sci.* 30:97-116.

CHAPTER **29**

The Use of ^{13}C Natural Abundance to Investigate the Turnover of the Microbial Biomass and Active Fractions of Soil Organic Matter under Two Tillage Treatments

M.C. Ryan, R. Aravena, and R.W. Gillham

I. Introduction

Considerable effort is being spent on the modelling of soil organic matter (SOM) dynamics as part of a larger effort to understand the sources and sinks of atmospheric CO_2. The soil microbial biomass is the living part of the SOM, whose size is less than 0.45 μm^3, excluding plant material. An understanding of the cycling of organic carbon through the microbial biomass is of particular interest in SOM cycling for numerous reasons: 1) the microbial biomass is a 'transformation station' for all SOM processes, where materials are taken up, converted into new compounds and actively or passively released to various other carbon pools, 2) the microbial biomass is a functionally relevant carbon pool (as opposed to those defined by various operational methods of separation), 3) an understanding of carbon turnover in the microbial biomass provides a rigorous test of model data and concepts, and suggests where further research might be focused, and 4) the microbial biomass is sensitive to changes in soil management, and provides an indicator of long term effects on SOM as a whole.

Previous investigators have taken advantage of the *in-situ* labelling of plant carbon resulting from a vegetation change from one of the two major photosynthetic pathways (C_4 and C_3) to the other to study the dynamics of SOM (Martin et al., 1990; Balesdent et al., 1990; Balesdent et al., 1987). This approach is extended in the present case to investigate the carbon dynamics of the microbial biomass and two active fractions of the SOM in a soil where a C_4 plant, maize, was cultivated under two tillage systems which had previously supported C_3 vegetation. The results presented in this paper are part of an on-going investigation whose main objective is the understanding of carbon and nitrogen cycling in agricultural soils and their impacts on shallow groundwater (Ryan, 1993).

II. Materials and Methods

A. Study Site

The study site is located at the Agriculture Canada Research Station near Delhi, Ontario. The soil is a well-drained member of the Fox series, and is classified as a Brunisolic Grey Brown Luvisol. Soil samples were taken from the A_k horizon, which extends to an average depth of 26 cm. While the site has historically supported only C_3 vegetation, including native forest, and a tobacco (*nicotina tobacum*) and rye (*secale*) rotation, an ongoing study at the site is evaluating maize (*zea mays*) as a replacement crop in the area (Burton et al., 1993).

As part of the replacement crop study, four randomized plots were available for sampling during the fourth and fifth year maize crops under each of conventional tillage (CT, with ploughing to 30 cm depth) and no-tillage (NT, where planting is carried out with a seed drill) management. In addition, soil was sampled from adjacent plots carrying their first year of maize (C1), and from soils undergoing tobacco-rye rotations (TOB). In the tobacco-rye rotation, tobacco plants are grown in every second year, with winter rye sowed immediately after the tobacco is harvested.

B. Soil Sampling

For analyses of the microbial biomass and active fractions of the SOM, a bucket auger was used to take twelve bulk soil samples from each soil type (3 samples from each random block in the alternative crop experiment) from 0 to 20 cm depth, from each soil management type. The twelve samples were then composited into a single sample, from which three duplicates were taken. Thus the values reported are mean values and standard deviations for the composite samples.

Soil for bulk SOM measurements was sampled at various depths (including the litter layer) using a 2.5 cm soil probe, and composited in a similar manner to the soil used for microbial biomass and active SOM fraction analyses.

The sampling dates were chosen to try and elucidate seasonal changes in the incorporation of C_4 carbon into the microbial biomass and active carbon pools. Samples were taken when the greatest fluctuations were expected, for example, immediately before planting in the spring (92.04.29), when the lowest annual amount of "young", or C_4, carbon was expected, and immediately before and after maize silking when abundant recent root exudates and senescing plant material might result in the highest amounts of annual C_4 incorporation into the more rapidly cycling SOM pools (91.08.28 and 92.07.28).

C. Bulk Soil Organic Matter and Plant $\delta^{13}C$ Analyses

SOM samples were air-dried and sieved to 2 mm prior to being crushed. Inorganic carbon was removed by heating the samples to 60 °C in 10% HCl. The samples were subsequently rinsed until the pH was neutral, with supernatant discarded between rinses, and freeze dried for analysis.

The fraction organic carbon (f_{OC}) was estimated by infrared measurement of the CO_2 evolved on combustion of dry samples at 800 °C. All results are expressed as fraction organic carbon, based on the initial dry weight of the acidified sample. Duplicate samples were within 10% of one another, unless otherwise noted.

Whole plants, including roots, stems, and leaves (except for tobacco, where only leaves were available for analysis) were oven dried, homogenized and ground to prior to determination of their ^{13}C composition.

D. Measurement of Active Fractions of SOM

Soil samples for analysis of the microbial biomass, extractable carbon, and incubation CO_2 were sieved at field moisture content to 2 mm diameter to remove large roots. The three composite samples, comprised of equal portions of the twelve soil samples taken from each soil type, were processed as follows.

The microbial biomass was estimated using the chloroform fumigation extraction method (Vance et al., 1987) with extractions carried out on "control" and "fumigated" samples. The carbon pool extracted from the control samples is referred to as 'extractable carbon'. Subsamples of the control and fumigated extracts were frozen and submitted for dissolved organic carbon analysis and larger samples were freeze-dried for ^{13}C analysis.

The amount of microbial biomass carbon was estimated using the relationship $C_{BM} = F_C/k_C$, where F_C is the difference in the amount of carbon extracted from the fumigated and control samples (ie. $C_{FUM} - C_{CONT}$), and k_C is the proportion of microbial carbon extracted from the fumigated sample. A value of $k_C = 0.45$ was used (Vance et al., 1987).

For soil CO_2 incubations, soil samples were pre-incubated for a 10 day period. After a subsequent 10 day incubation, gas samples were collected directly from the incubation jars by means of a septa incorporated into the lid. The evolved CO_2 was analyzed for ^{13}C to estimate the fraction of C_4 carbon in the

SOM mineralized during the incubation. This carbon pool is referred to as incubation CO_2, and is by definition the most labile, or most easily mineralizable, carbon pool in the SOM.

E. ^{13}C Analyses

Stable carbon isotope ratios were measured on the bulk SOM and extracted carbon by combustion at 550°C with CuO and silver wire in sealed pyrex tubes (Boutton et al., 1983). The evolved CO_2 from combustion and soil incubation was purified cryogenically and analyzed using VG Micromass PRISM and 903 isotope ratio mass spectrometers at the Environmental Isotope Laboratory, University of Waterloo. The ^{13}C analyses are expressed in the usual delta per mille (‰) notation, defined as

$$\delta^{13}C = \left(\frac{R_{SAMPLE}}{R_{STANDARD}} - 1\right) * 1000 \tag{1}$$

where and R_{SAMPLE} and $R_{STANDARD}$ are the $^{13}C/^{12}C$ ratio of the sample and international standard (PeeDEE belemnite), respectively. The analytical reproducibility is better than 0.2 ‰.

F. Calculation of the Fraction of Carbon Derived From Maize

The $\delta^{13}C$ of the microbial biomass is estimated as the $\delta^{13}C$ of the carbon in the fumigated extraction ($^{13}C_{FUM}$) in excess of that in the control extraction ($^{13}C_{CONT}$), and is calculated as

$$\delta^{13}C = \frac{(\delta^{13}C_{FUM} * C_{FUM}) - (\delta^{13}C_{CONT} * C_{CONT})}{C_{FUM} - C_{CONT}}. \tag{2}$$

where C_{FUM} and C_{CONT} are the amount of carbon extracted from the fumigated and control samples respectively.

The $\delta^{13}C$ calculated for the microbial biomass represents the portion of biomass that is rendered extractable by chloroform fumigation. According to the K_c value cited above, this portion represents only 45% of the biomass. Strictly speaking, the values of $\delta^{13}C$ refer to the chloroform-labile fraction of the microbial biomass, although they are referred to herein simply as microbial biomass $\delta^{13}C$ values.

The fraction of carbon originating from C_4 plants, x, is calculated using a two end-member mixing model,

$$\delta_S = x\delta_{C4} + (1-x)\delta_{C3}, \quad or \quad x = \frac{\delta_S - \delta_{C3}}{\delta_{C4} - \delta_{C3}} \tag{3}$$

where δ_S is the $\delta^{13}C$ of the sample, and δ_{C3} and δ_{C4} are the $\delta^{13}C$ of the C_3 and C_4 carbon sources (Cerri et al., 1985). If the total amount of carbon, C, is known, the amounts of carbon from C_4 and C_3 sources are

$$C_{C4} = xC \quad and, \quad C_{C3} = (1-x)C. \tag{4}$$

III. Results

A. ^{13}C Composition of Vegetation and SOM

The ^{13}C value of the three vegetation types are included in Table 1. The two C_3 'background', or reference, vegetation types, tobacco and rye, have isotopic ratios that vary by about 3.5‰. The difference between the average $\delta^{13}C$ value of the reference vegetation and maize is about 15‰, which is similar to the difference in ^{13}C often cited between C_3 and C_4 vegetation. This shift is assumed to be maintained constant during the decomposition of the different types of material (Balesdent et al., 1990), and is used as the denominator in equation 3.

The ^{13}C and f_{OC} of the SOM are presented in Table 2. The most depleted $\delta^{13}C$ value, -29.6‰, was observed in the litter layer in the tobacco-rye rotation, and is characteristic of the rye plant material which was most recently planted in the soil prior to sampling. The average $\delta^{13}C$ value of the reference soil,

Table 1. $\delta^{13}C$ of Plant Materials (in ‰)

	Tobacco	Rye	Maize
		(‰)	
	-26.6	-29.2	-13.2
	-25.8	-30.0	-12.8
		-29.8	-12.8
			-12.6
Average	-26.2	-29.6	-12.8

Table 2. $\delta^{13}C$ and f_{OC} of bulk soil organic matter ; soil sampled April 29, 1992

Depth (cm)	f_{OC} (%)	$\delta^{13}C$ (‰)
Tob/rye		
Leaf Litter	--	(-29.6)
0-2	.40	-26.9
2-5	.48	-26.8
5-10	.52[b]	-26.7
15-20	.42	-26.7
15-20	.39	-26.3
Average[a]	.44	-26.6
(Std. dev.)	(.05)	(.2)
NT corn		
0-2	.36	-24.0
2-5	.41	-25.2
5-10	.52[c]	-25.2
15-20	.34	-25.9
10-20	.37	-25.6
Average	.41	-25.4
(Std. dev.)	(.06)	(.6)
CT Corn		
Leaf	--	(-13.8)
0-2	.39	-25.5
2-5	.37	-26.7
5-10	.42	-25.1
15-20	.31	-25.1
10-20	.34	-25.2
Average	.36	-25.4
(Std. Dev.)	(0.04)	(.6)

[a]Averages weighted for different depth intervals; [c]four samples analyzed, std. dev. = 0.03; [d]four samples analyzed, std. dev. = 0.08.

-26.65‰, is enriched with respect to the average value for tobacco and rye plants of 27.9‰. This profile also shows a slight ^{13}C enrichment with depth. This isotopic difference between plant materials and SOM is attributed mainly to isotopic effects occurring during decomposition (Nadelhoffer and Fry, 1988; O'Brien and Stout, 1978).

The shallow samples from the NT soils are more enriched in ^{13}C than the CT, indicating a greater amount of C_4 carbon. Although both the NT and CT soils show a decrease in f_{OC} values relative to the tobacco-rye soil, the decrease is more marked in the CT soil. Average f_{OC} values for the NT and CT soils after four years of maize were 0.41 and 0.36, relative to a reference value of 0.44 for the tobacco-rye soil (Table 2). The average $\delta^{13}C$ value for the bulk SOM in both the NT and CT soils was -25.4‰, which according to equations 3 and 4 corresponds to the incorporation of 8% of C_4 carbon into the bulk SOM pools. The

Table 3. $\delta^{13}C$ of various organic matter fractions in tobacco-rye rotation; standard deviation of three replicate samples in parentheses

Sample type	Sampling date	Biomass carbon ($\mu gC/gm$ dry soil)	Control extraction $\delta^{13}C$ (‰)	Incubation $\delta^{13}CO_2$ (‰)	Biomass $\delta^{13}C$ (‰)
Tob (plant)	08/28/91	--	--	-25.7 (.1)	--
Rye (sprouting)	04/29/92	95 (3)	-24.6 (.1)	--	-24.9 (.3)
Rye (senescent)	07/28/92	70 (1)	-24.8 (.1)	-25.7 (.1)	-23.6 (.1)
Average reference value			-24.7	-25.7	-24.2

remaining C_3 carbon in the NT and CT soils (corresponding to f_{oc} values of 0.38 and 0.33, respectively) correspond to half-lives of 10 and 18 years, if degradation of the carbon in the bulk SOM is assumed to be a first order process.

B. ^{13}C Composition of Microbial Biomass and Labile Fractions of SOM in Tobacco-Rye Soil

The ^{13}C values obtained for the extractable carbon, incubation CO_2 and microbial biomass from the reference tobacco-rye soil are all enriched relative to the plant material and SOM (Table 3). The reference $\delta^{13}C$ of each carbon pool is taken as an average of the values measured from the tobacco-rye soil. While the extractable and incubation carbon do not show significant variation between the two sampling events, the ^{13}C of the microbial biomass shifts by about 1.3‰. This can be explained by increased incorporation of depleted carbon due to active rye growth during the spring sampling period. Further investigations are being carried out to evaluate seasonal changes in the ^{13}C composition of the reference microbial biomass. For the purposes of this paper, the average $\delta^{13}C$ value will be used to determine the fraction of C_4 carbon incorporated into the microbial biomass of soils under maize crops.

C. ^{13}C Composition of Microbial Biomass, Extractable Carbon, and Incubation CO_2 of SOM Under Maize.

The ^{13}C composition, and the fraction, x, of C_4 carbon incorporated into each of the three labile pools of SOM measured are included in Table 4. Generally, the soils planted to maize show a noticeable enrichment in ^{13}C in the SOM fractions, compared to the reference soil.

The results for the first year corn indicate that at the onset of silking (3 months after the vegetation change), the microbial biomass had incorporated about 10% of C_4 carbon. The fraction of C_4 carbon then decreased by about one-half immediately before planting in the second year, or 1 year after the vegetation change. A similar pattern of increased C_4 incorporation during the growing season seems to be exhibited by both the NT and CT treatments. The fraction of C_4 carbon in the microbial biomass increased by a factor of two between samples taken immediately before planting of the fifth year of maize, and those taken immediately before silking, three months later. During this period, the fraction of C_4 carbon in the microbial biomass increased from 16 to 40% in the NT soils, and from 27 to 47% in the CT soils.

The fraction of C_4 carbon in the extractable carbon pool exhibits less variation over the growing season, particularly in the NT treatment. Except for samples taken from the NT treatment immediately before planting, there is a consistently greater fraction of C_4 carbon incorporated into the microbial biomass pool.

Table 4. $\delta^{13}C$ and fraction C_4 carbon (x) of various organic matter fractions after the introduction of maize; standard deviation of three replicate samples in parentheses

	Years since maize intr'd	Sample date	Biomass (μgC/gm dry soil)	Control extraction $\delta^{13}C$ (‰)	x (%)	Incubation CO_2 $\delta^{13}C$ (‰)	x (%)	Biomass C $\delta^{13}C$ (‰)	x (%)
First year maize	.25	92.07.28	59 (1)	-24.0 (.1)	4	--	--	-22.8 (.2)	10
First year maize	.3	91.08.28	--	--	--	-25.7 (.1)	<1	--	--
First year maize	1	92.04.29	50 (8)	-24.3 (.1)	3	--	--	-23.4 (.2)	5
No-tillage	3.3	91.08.28	--	--	--	-17.6 (.1)	54	--	--
No-tillage	4	92.04.29	57 (2)	-21.9 (.1)	19	--	--	-21.7 (.5)	16
No-tillage	4.25	92.07.28	61 (5)	-21.5 (.1)	21	--	--	-17.5 (.4)	40
Conventional tillage	3.3	91.08.28	--	--	--	-19.6 (.5)	41	--	--
Conventional tillage	4	92.04.29	62 (23)	-22.4 (.2)	15	--	--	-20.2 (.1)	27
Conventional tillage	4.25	92.07.28	74 (2)	-21.1 (.1)	24	-18.4 (.3)	48	-17.2 (.3)	47

The ^{13}C composition of the CO_2 evolved on soil incubation indicates a substantially greater fraction of C_4 carbon is available for mineralization in the NT and CT soils after 3 years of maize, compared to the first year. In the only sampling event where all three carbon pools were measured in the maize soil, the microbial biomass carbon and incubation CO_2 had incorporated similar fractions of C_4 carbon, equal to roughly double the fraction of C_4 carbon that was in the extractable carbon pool.

IV. Discussion

The $\delta^{13}C$ depth profiles in SOM can be attributed to changes in the source vegetation, and to processes that fractionate the carbon during decomposition (e.g. Balesdent et al., 1990). Assuming that the processes responsible for ^{13}C fractionation during decomposition are similar in all soils studied, the differences in the ^{13}C composition between the reference and maize soils in the present study can only be attributed to vegetation change.

A. Turnover of Bulk SOM

Decomposition rates calculated from the amount of C_3 carbon remaining in the bulk SOM after 4 years indicate that overall, the C_3 carbon is being lost from the CT treatment at a greater rate than from the NT soils. The decomposition rates, which assume that the decay process is first order, correspond to average half lives for the C_3 carbon of 18 years for NT and 10 years for CT. These rates of degradation are similar to those determined in other ^{13}C natural abundance studies (e.g. 25 years, Balesdent et al., 1987, and 10 to 15 years for an 'easily mineralizable component', Balesdent et al., 1988). Accelerated decomposition in CT relative to NT management was also reported by Balesdent et al. (1990).

B. Incorporation of C_4 into Microbial Biomass, Extractable and Incubation Carbon Pools

Seasonal changes in the ^{13}C composition are considerably more pronounced in the microbial biomass carbon than in the extractable carbon pool. Except for the NT soil sampled immediately before planting, the microbial biomass consistently incorporated a greater fraction of C_4 carbon than did the extractable carbon pool, demonstrating the effectiveness of the microbial biomass in sequestering labile carbon. This is also indicated by the incorporation of a similar fraction of C_4 carbon into the microbial biomass and into the most labile pool, the incubation carbon.

To some degree the ^{13}C of the SOM fractions measured can be correlated with the mean age of the carbon in the pools. The age of the C_4 carbon in each pool is less than the time since maize was first introduced to the soil. Thus, along with seasonal fluctuation in the ^{13}C of each fraction, the mean age of the carbon in the pool must also fluctuate. For instance, after 4 years of CT management, 20% of the carbon has a mean age less than 4 years. Three months later, 47% of the carbon has a mean age of less than 4.25 years.

C. Turnover of Microbial Biomass

The results in Table 4 suggest that the incorporation of C_4 carbon into the microbial biomass varies seasonally according to crop growth and senescence. The fraction of C_4 carbon in the microbial biomass immediately before planting may be an annual minima, with rapid incorporation of C_4 carbon into the microbial biomass occurring during the growing season, when abundant C_4 substrate is available. After crop death and senescence, the fraction of C_4 carbon presumably decreases again, until the following crop is planted. A conceptual model of this process is postulated in Figure 1. The model is represented by the incorporation of C_4 carbon into two pools of the microbial biomass; one that continually cycles at a relatively slow rate (perhaps corresponding to a protected, or fungal fraction of the biomass), and a second that responds rapidly to short term supplies of labile substrate, sometimes referred to as an opportunistic, or bacterial fraction of the biomass (Schnurer et al., 1986; Bonde and Rosswall, 1987).

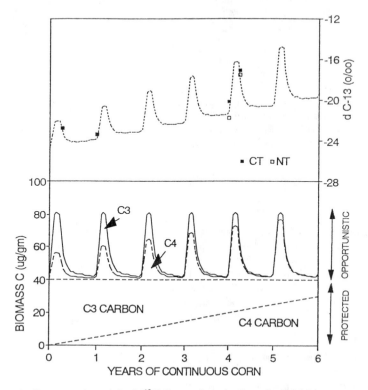

Figure 1. Conceptual model of $\delta^{13}C$ fluctuations in the microbial biomass.

Soil organic models in general have treated the microbial biomass as a "black box"; or a single fraction of SOM cycling at a constant rate throughout the year. In part this has been done in order to simplify an already complex system, and in part because 'great difficulties exist in establishing the basic parameters governing the kinetics of microbial turnover' (VanVeen et al., 1985). Calculated microbial biomass half-lives have ranged from 0.2 to 4.2 years (e.g. McGill et al., 1986; Jenkinson and Rayner, 1977). Some investigators have invoked half-lives for two compartments of the microbial biomass that cycle at different rates. Ladd et al. (1981) concluded from radiocarbon labelling studies that microbial biomass ^{14}C formed from decomposing residue was lost in fast and slow cycling compartment with half-lives of 0.24 and 4.2 years. More complex models of SOM turnover (e.g. McGill et al., 1981) have also included both bacterial and fungal biomass pools that cycle at different rates.

The calculation of half-lives from the data in Table 4 allows further insight into the turnover of the microbial biomass. After 4 years of continuous corn under conventional till management, the fraction of C_3 carbon remaining is 0.73, which corresponds to a half-life of $t_{1/2} = (\ln(2) \cdot 4 \text{ yrs})/\ln(.73) \approx 9$ years (assuming first order degradation is occurring and a constant amount of biomass carbon exists in the soil immediately before planting each year). A similar calculation for the NT treatment yields a half-life of 16 years. These half-lives are calculated for the degradation of C_3 carbon in the microbial biomass in soil samples taken immediately before planting, when the amount of labile carbon in the soil is likely at an annual minima, and might reflect the decomposition of carbon in the protected pool of the microbial biomass. The calculated half-lives are considerably longer than those published in the literature, and may reflect the nature of the microbial population associated with the very low organic matter in the Delhi soil.

A similar calculation for the change in C_3 carbon in the microbial biomass pools between the samples taken immediately before planting, and at silking (3 months later) yields half-lives of about 0.5 years for both CT and NT soils supporting their fifth year of maize. These rates, which are much faster than those calculated above, are based on the loss of C_3 carbon during the period in which abundant C_4 substrate is available for the microbial biomass. Over the same period between planting and silking, the amount of C_4 carbon in the microbial biomass increases, as abundant C_4 substrate becomes available. The doubling or generation time for the fraction of C_4 carbon in the microbial biomass are 0.2 years for the NT soil, and 0.3 years for the CT treatment. These parameters might be characteristic of the active fraction of the microbial biomass.

V. Summary

The natural abundance of ^{13}C has been used in combination with the chloroform fumigation extraction method to elucidate information about the turnover of carbon in the microbial biomass and other soil organic matter pools, in a soil undergoing a change from C_3 to C_4 vegetation. The results suggest that the incorporation of 'young' (or C_4) carbon occurs in a cyclical pattern, with an annual maxima during the active growing season, superimposed on a more constant gradual increase. This pattern may be an expression of the dynamics of two fractions of the microbial biomass. The more protected fraction, which turns over slowly and consistently, may be responsible for the steady, gradual incorporation of C_4 carbon into the biomass pool. In addition, a more opportunistic population, responding to short term increases in labile substrate during the growing season, may be responsible for seasonal maxima in the C_4 composition of the microbial biomass. First order degradation half-lives estimated for the degradation of C_3 carbon in the microbial biomass were 16 and 9 years for the protected fraction under NT and CT management, respectively, and 0.5 years for the opportunistic fraction under both management systems. These results suggest that carbon turnover in the microbial biomass is more rapid under conventional till, than no-till management system.

VI. Conclusions

This study has shown that ^{13}C natural abundance signals in SOM can provide valuable information about carbon cycling through the microbial biomass. The distinct and well documented ^{14}C input to SOM attributed to thermonuclear explosions during the last 35 years can be an additional tool to evaluate the two carbon pools, fast and slow, that are involved in cycling of the microbial biomass. Ongoing research is being carried out to improve our understanding of seasonal fluctuations of the incorporation of C_4 carbon into the microbial biomass at our site, and to obtain more statistically rigorous data. It is anticipated that the data generated in this type of study will prove valuable in modelling microbial biomass dynamics in soil.

Acknowledgments

Funding for the study was provided by R.W. Gillham's National Science and Engineering Research Council of Canada Research Grant. Thanks are extended to Richard Elgood for his invaluable assistance in the laboratory, and to Ron Beyaert and David Burton for facilitating the use of the Delhi Research Station plots.

References

Balesdent, J., A. Mariotti, and B. Guillet. 1987. Natural ^{13}C abundance as a tracer for studies of soil organic matter dynamics. *Soil Biol. Biochem.* 19(1):25-30.
Balesdent, J., G.H. Wagner, and A. Marriotti. 1988. Soil organic matter turnover in long-term field experiments as revealed by the carbon-13 natural abundance. *Soil Science Society of America Journal* 52:118-124.
Balesdent, J., A. Mariotti, and D. Boisgontier. 1990. Effect of tillage on soil organic carbon mineralization estimated from ^{13}C abundance in maize fields. *J. of Soil Science* 41: 587-596.
Bonde, T.A. and T. Rosswall. 1987. Seasonal variation of potentially mineralizable nitrogen in four cropping systems. *Soil Science Soc. Am. J.* 51:1508-1514.

Boutton, T.W., W.W. Wong, D.L. Hachey, L.S. Lee, M.P. Cabrera, and P.D. Klein. 1983. Comparison of quartz and pyrex tubes for combustion of organic samples for stable carbon isotope analysis. *Anal. Chem.* 55:1832-1833.

Burton, D.L., M.F. Younie, E.G. Beauchamp, R.G. Kachanoski, D.M. Brown, and D.E. Elrick. 1993. Alternative crop management practices and nitrate contamination of groundwater with sandy soils used for tobacco production. Final Report prepared for Tobacco Diversification/Alternate Enterprise Initiative Development Branch, Agriculture Canada and Supplies and Services Canada. File Number 07SE.01687-7-0278.

Cerri, C., C. Feeler, J. Balesdent, R. Victoria, and A. Plenecassagne. 1985. Application du traçage isotopique naturel en ^{13}C á l'étude de la dynamique de la dynamique del la matière organique dans les sols. *Comptes Rendus del l'Académie des Sciences de Paris* T.300,II,9,423-428.

Jenkinson, D.S. and J.H. Rayner. 1977. The turnover of soil organic matter in some of the Rothamsted classical experiments. *Soil Sci.* 123(5):298-305.

Ladd, J.N., J.M. Oades, and M. Amato. 1981. Microbial biomass formed from ^{14}C, ^{15}N-labelled plant material decomposing in soils in the field. *Soil Biol. Biochem.* 13:119-126.

Martin, A., A. Mariotti, J. Balesdent, P. Lavelle, and R. Vuattoux. 1990. Estimate of organic matter turnover rate in a savanna soil by ^{13}C natural abundance measurements. *Soil. Biol Biochem.* 22(4):517-523.

McGill, W.B., K.R. Cannon, J.A. Robertson, and F.D.Cook. 1986. Dynamics of soil microbial biomass in Breton L after 50 years of cropping to two rotations. *Can. J. Soil Sci.* 66:1-19.

McGill, W.G., H.W. Hunt, R.G. Woodmansee, and J.O. Reuss. 1981. Phoenix, a model of the dynamics of carbon and nitrogen in grassland soils. *Ecol. Bull.* 33:49-115.

Nadelhoffer, K.J. and B. Fry. 1988. Controls on natural nitrogen-15 and carbon-13 abundances in forest soil organic matter. *Soil Science Soc. Am. J.* 52:1633-1640.

O'Brien, B.J. and J.D. Stout. 1978. Movement and turnover of soil organic matter as indicated by carbon isotope measurements. *Soil Biol. Biochem.* 10:309-317.

Ryan, M.C., 1993. Personal communication.

Schnurer, J., M. Clarholm, and T. Rosswall. 1986. Fungi, bacteria and protozoa in soil from four arable cropping systems. *Biol. Fertil. Soils* 2:119-126.

Vance, E.D., P.C. Brookes, and D.S. Jenkinson. 1987. An extraction method for measuring soil microbial biomass. *Soil Biol. Biochem.* 19(6):703-707.

van Veen, J.A., J.N. Ladd, and M. Amato. 1985. Turnover of carbon and nitrogen through the microbial biomass in a sandy loan and a clay soil incubated with [^{14}C(U)] glucose and [^{15}N(NH$_4$)$_2$SO$_4$] under different moisture regimes. *Soil Biol. Biochem.* 17:747-756.

CHAPTER 30

Trace Gas and Energy Fluxes: Micrometeorological Perspectives

S.B. Verma, J. Kim, R.J. Clement, N.J. Shurpali, and D.P. Billesbach

I. Introduction

To assess accurately the possible causes and potential effects of increasing atmospheric trace gas concentrations requires a quantitative understanding of the surface fluxes of these gases (e.g., carbon dioxide, methane). Traditionally, trace gas fluxes have been measured using chambers (enclosures). The fluxes so obtained are limited to small areas. The application of the micrometeorological eddy correlation technique is recommended for larger scale, "areally-integrated" measurement of trace gas flux (e.g., National Research Council, 1984; University Corporation for Atmospheric Research, 1986; National Center for Atmospheric Research, 1989). This technique allows continuous flux measurements with minimal disturbance to the microenvironment. In the past 15 years, we have employed this technique to measure fluxes of carbon dioxide in a variety of ecosystems (e.g., agricultural crops, a temperate grassland, a deciduous forest and an open peatland). Results from these and other relevant studies are summarized here. Recently, we have initiated a program of research to measure methane fluxes using the eddy correlation technique. A tunable diode laser spectrometer is employed to measure rapid response fluctuations in methane concentrations. Methane fluxes measured in our study in a Minnesota peatland and other selected studies are briefly described.

Surface exchange of these trace gases in terrestrial ecosystems is closely linked with hydrology, of which evapotranspiration (water vapor flux) is a fundamental component. Knowledge of water vapor and heat fluxes is needed to develop a more complete understanding of trace gas dynamics. Data on these fluxes are also needed for computation of density effects on the measured fluxes of trace gases (e.g., Webb et al., 1980). Thus, results on evapotranspiration and energy partitioning in several ecosystems are briefly reviewed here.

II. Micrometeorological Eddy Correlation Technique

The eddy correlation technique provides a measure of the vertical flux (F_s) of a transported entity at a point in the atmosphere by correlating the fluctuations (s') in the concentration of that entity with the fluctuations (w') in the vertical wind speed (e.g., Businger, 1986):

$$F_s = \overline{w's'}. \tag{1}$$

The overbars indicate time averages and the primes denote deviations from the mean. Further details of this technique, including information on sensor and site requirements and data acquisition procedures, can be found in previous publications (e.g., Baldocchi et al., 1988; Verma, 1990).

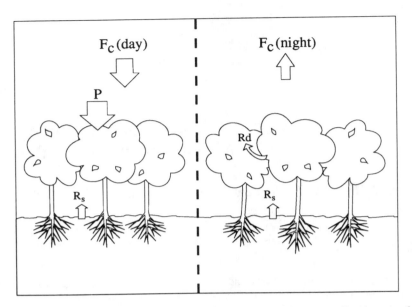

Figure 1. Schematic representation of CO_2 exchange between soil, plant, and atmosphere during the day and night.

Rapid response sensors for eddy correlation measurements of fluxes of momentum, sensible heat, and water vapor have been available for some time. Recently, rapid response sensors have also been developed and employed in various field experiments for measuring fluxes of carbon dioxide (e.g., Bingham et al., 1978; Brach et al., 1981; Ohtaki and Matsui, 1982; Anderson et al., 1984), methane (e.g., Verma et al., 1992a) and other entities such as ozone, sulfur dioxide, and nitrogen oxides.

III. Carbon Dioxide Flux

A. Background Information

Atmospheric CO_2 flux (F_c) obtained from the micrometeorological technique represents the net ecosystem CO_2 uptake or release (Figure 1). The daily net ecosystem CO_2 exchange (F) can be estimated by integrating daytime and nocturnal F_c over a period of 24 hours. During daytime, F_c is the sum of the net uptake of CO_2 by plants (P) and the CO_2 emanating from the soil surface (R_s). The term R_s includes contributions from respiration of soil organisms and root respiration (the contributions of roots range from 10% to 90%, depending on factors such as soil water availability and belowground productivity). At night, F_c represents the sum of R_s and R_d (plant respiration of the aboveground portion of the plants). Note that the fluxes are expressed on per unit ground area basis.

Typical growing season values of F_c, F_c/P, and R_s in selected ecosystems in temperate climates are presented in Table 1. In general, values of $F_c(day)$ in agricultural crops are among the highest, while fairly low values are observed in a peatland. Variation in $F_c(day)$ values from different ecosystems is primarily because of differences in plant photosynthesis and respiration rates and soil biological activities (e.g., root respiration, decomposition of organic matter). Factors controlling these processes are briefly discussed in the next subsection. As is indicated by $F_c(day)/P$ values in Table 1, about 60-90% of the net CO_2 uptake by vegetation in most ecosystems comes from the atmosphere during the day. The remainder is contributed by the CO_2 emanating from the soil surface. Compared to $F_c(day)$, the magnitude of $F_c(night)$ is small, ranging from 0.1 to 0.5 mg m^{-2} s^{-1}. The magnitude of $R_s(day)$ ranges from 0.1 to 0.6 mg m^{-2} s^{-1}. Generally, these values appear to be of similar order as those of $F_c(night)$. This may result because the reduction in R_s (from day to night) due to lower nocturnal soil temperature, may be of approximately similar magnitude as the contribution of R_d in $F_c(night)$.

Table 1. Typical midday values during the growing season of the atmospheric CO_2 flux (F_c), the ratio of atmospheric CO_2 flux to net CO_2 uptake by plants (F_c/P), and the soil CO_2 efflux (R_s) in various ecosystems in temperate climates

Temperate ecosystems	F_c(day) mg m^{-2} s^{-1}	F_c(day)/P mg m^{-2} s^{-1}	F_c(night) mg m^{-2} s^{-1}	R_s(day) mg m^{-2} s^{-1}
Agricultural crops	1.0 ~ 2.0	0.7 ~ 0.9	-0.5 ~ -0.1	-0.6 ~ -0.1
Grassland	0.6 ~ 1.3	0.7 ~ 0.9	-0.4 ~ -0.2	-0.6 ~ -0.1
Deciduous & coniferous forest	0.4 ~ 1.0	0.6 ~ 0.9	-0.5 ~ -0.2	-0.5 ~ -0.2
Open peatland	0.1 ~ 0.4	0.6 ~ 0.9	-0.2 ~ -0.1	-0.2 ~ -0.1

Data from Anderson and Verma (1986), Baldocchi (1982), Baldocchi et al. (1986), Beadle et al. (1985), Clement (1988), Jarvis et al. (1976), Kim and Verma (1990), Kim and Verma (1992), Kim et al. (1992), Monteith et al. (1964), Monteith (1976), Norman et al. (1992), Shurpali et al. (unpublished data), Verma et al. (1986), Verma et al. (1989), Verma et al. (1992b).

B. Primary Controlling Factors

Light, temperature, and moisture are considered major factors which control the exchange of CO_2. In the short-term, these environmental factors directly influence processes such as plant photosynthesis and respiration. In the long-term, these factors can affect the length of the growing season by influencing the timing of plant development and senescence, thereby affecting the long-term carbon budget of ecosystems. Also, temperature and moisture influence[1] the CO_2 emanating from the soil.

1. Light

The relationships between P and photosynthetically active radiation (Q_p) observed (under nonlimiting soil water conditions) over a variety of canopies (e.g., wheat [*Triticum aestivum* (L.)] field in Nebraska, grassland in Kansas, deciduous forest in Tennessee, coniferous forest in eastern England, and open peatland in north central Minnesota) are schematically shown in Figure 2. The shapes of P-Q_p relationship in a wheat field, a grassland, and deciduous and coniferous forests are generally similar and do not indicate light-saturation (up to Q_p level of 2000 μmol m^{-2} s^{-1}). This is because the available Q_p deep inside these canopies (with leaf area index of ≥ 3) increases with increasing solar elevation, providing more energy to the lower light-unsaturated leaves. Vegetation in the above mentioned peatland, however, is dominated by *Sphagnum* moss and short vascular plants. The peatland vegetation seems to attain light saturation at Q_p of approximately 600 μmol m^{-2} s^{-1}. The influence of vertically distributed leaves of vascular plants on light interception may also have contributed. Prior chamber measurements of photosynthesis in various *Sphagnum* species indicate the occurrence of light saturation at Q_p ranging from 250 to 620 μmol m^{-2} s^{-1} (e.g., Harley et al., 1989).

2. Temperature

The response of P to temperature can be considered similar to that of overall plant growth. With increasing soil temperatures production initially increases rapidly and then slows down as the optimum temperature is

---

[1] In wetlands, water table elevation can also influence the soil CO_2 flux.

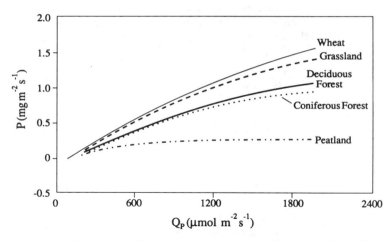

Figure 2. Net CO_2 uptake (P) by plants as a function of photosynthetically active radiation (Q_p) under well watered conditions for wheat grown in Nebraska (Clement, 1988), a grassland in Kansas (Verma et al., 1992b), a deciduous forest in Tennessee (Baldocchi et al., 1987), a coniferous forest in eastern England (Beadle et al., 1985), and an open peatland in Minnesota (Shurpali et al., unpublished data).

approached, beyond which decline in production is observed (e.g., Parton et al., 1992). For example, under nonlimiting soil moisture conditions, the response of P to air temperature in a grassland dominated by C_4 vegetation showed an optimum (at temperature $\approx$ 24 °C), whereas R_s increased continuously with increasing soil temperature (Kim et al., 1992). The difference in temperature response of P and R_s has important implications concerning the impact of the predicted global warming on net ecosystem CO_2 exchange.

3. Moisture

Net uptake of CO_2 by plants can be affected significantly by a reduction in soil water or an increase in atmospheric dryness (e.g., Schulze, 1986; Ball et al., 1987; Chaves, 1991). With increasing soil moisture stress, photosynthesis is suppressed significantly and its diurnal peak is shifted from midday to morning hours. High or extreme atmospheric dryness also reduces photosynthesis. Soil CO_2 flux is also affected by the availability of soil moisture (e.g., Singh and Gupta, 1977).

Seasonal courses of P, R_s, and extractable soil water (W_E) in a temperate grassland are illustrated in Figure 3. The P and R_s (midday values) showed significant seasonal variations which were quite similar to that in W_E. When W_E dropped below 30%, P was almost zero (during this period, the occurrence of high vapor pressure deficit also affected P) and the magnitude of R_s also reduced. When soil moisture conditions improved (in mid August), the magnitudes of P and R_s increased accordingly. During the senescence period (late September to mid October) the magnitudes of both P and R_s decreased.

In Figure 4, P is plotted against Q_p for a grassland and open peatland when the soil water was not limiting. The data are grouped into two intervals of vapor pressure deficit (D). These results indicate a reduction in P with an increase in D. Similar results have been reported for agricultural crops (e.g., soybean [*Glycine max* (L.)], Baldocchi, 1982; sorghum [*Sorghum bicolor* (L.)], Anderson and Verma, 1986; cereal crops, Clement, 1988) and coniferous forest (e.g., Jarvis et al., 1976; Jarvis, 1981). Reductions in P resulting from increases in D is generally attributed to a decrease in stomatal conductance (e.g., Ball et al., 1987).

Trace Gas and Energy Fluxes: Micrometeorological Perspectives 365

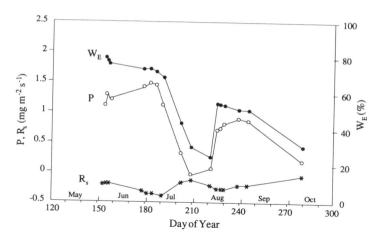

Figure 3. Seasonal variations in net CO_2 uptake (P) by plants, soil CO_2 flux (R_s) and extractable soil water (W_E) in a temperate grassland (adapted from Verma et al., 1990). Values of W_E were computed as the ratio of actual to total soil moisture held with a water potential between -0.03 and -1.50 MPa obtained over the primary root zone (0-1.4 m).

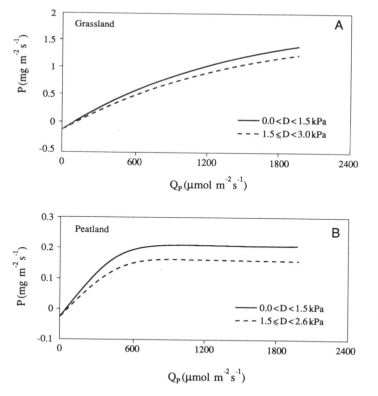

Figure 4. Relationship between net CO_2 uptake (P) by plants and light (Q_p) under different ranges of vapor pressure deficit (D) in: (A) a temperate grassland and (B) an open peatland in Minnesota. Note that the scales of P used for the ordinate in the two figures are different.

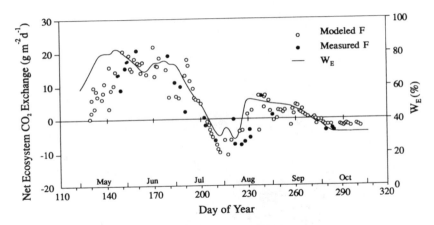

Figure 5. Seasonal distribution of the (daily) net ecosystem CO_2 exchange (F) in a temperate grassland in 1987 (adapted from Kim et al., 1992). Both measured and modeled values are shown. Values of W_E (extractable soil water) are also shown.

C. Net Ecosystem CO_2 Exchange

In order to compute the seasonal distribution of the (daily) net ecosystem carbon dioxide exchange (F), we need frequent (preferably continuous) micrometeorological measurements of F_c throughout the season. Supplementary measurements of soil CO_2 flux and dark respiration of key plant species are also desirable for a thorough analysis of the data. Figure 5 shows an example of net ecosystem CO_2 uptake for a temperate grassland. As expected, the grassland was a sink for atmospheric CO_2 with an uptake rate of 5-20 g m^{-2} d^{-1} when soil water was not limiting (May-mid July). However, the 3 weeks of a dry spell (late July-early August) changed this ecosystem from a sink to a source for atmospheric CO_2 in the middle of the growing season. Even under improved moisture conditions after early August with frequent rainfall, the ecosystem released CO_2 at an average rate of 5 g m^{-2} d^{-1} for about 2 weeks. The degree to which this ecosystem can become a source of CO_2 depends on the severity, duration and the timing of the drought event in relation to the growth stage (Kim et al., 1992). Significant day-to-day variation in the net ecosystem CO_2 exchange (particularly with rainfall events) indicates that caution should be exercised in the broad interpretation of CO_2 exchange when results are based on short-term measurements.

IV. Methane Flux

A. Micrometeorological Measurements

We have recently begun a research program on micrometeorological measurements of methane fluxes. Our first study site, referred to as the Bog Lake Peatland, is located in the Chippewa National Forest, adjacent to the Marcell Experimental Forest (47° 32' N, 93° 28' W) in north central Minnesota. Bog Lake Peatland is transitional between poor-fen and bog. The methane sensor used in this study is a prototype (fast response) tunable diode laser spectrometer (TDLS). A brief description of our TDLS sensor can be found in Verma et al. (1992a).

A pilot study was conducted in the summer of 1990 to establish the feasibility of using the prototype TDLS in the field (Verma et al., 1992a). Detailed studies were conducted over a five-month (mid May-mid October) period in 1991. The seasonal distribution of measured methane flux is shown in Figure 6. We interpret these observations to indicate two levels of variability in methane fluxes in this ecosystem. The first level represents a "gradual" seasonal pattern of methane flux (indicated by the eye fit curve). The flux ranged from 35 to 160 mg m^{-2} d^{-1} and showed a similar overall seasonal trend as the peat temperature. Superimposed on this seasonal pattern were several episodic emissions of methane. These events generally spanned over 4-5 days, and the methane flux was 2-3 times as large as the values in the "gradual" seasonal

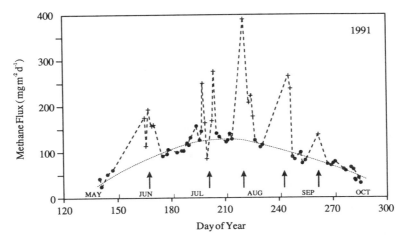

Figure 6. Seasonal distribution of methane flux (F_m) in a Minnesota peatland in 1991 (adapted from Shurpali et al., 1993). Data during episodic emissions are denoted by +. (Also, vertical arrows at the bottom of the figure are provided to help clearly identify the periods with episodic emissions.) Data during non-episodic periods are denoted by solid circles.

trend, discussed above. These episodic emissions were associated with substantial drops in atmospheric pressure and a declining water table (for details, see Shurpali et al., 1993). Episodic releases of methane have been reported in several previous studies (e.g., Bartlett et al., 1988; Devol et al., 1988; Moore et al., 1990; Dise, 1991).

B. Primary Controlling Factors

Micrometeorologically measured methane flux (F_m) represents the net ecosystem release (or uptake) of methane, which is the result of different processes such as methane production, oxidation of the produced methane, and transport of methane to the atmosphere. In general, factors such as soil (peat) temperature, water table position, and substrate availability have been reported to influence F_m. For peatlands in northern Minnesota, Dise (1991) reported that peat temperature largely controlled flux within an individual location, but the depth of water table beneath the peat surface distinguished the flux among locations. Water table position, peat temperature, and (less importantly) degree of peat humification (or type of vegetation) explained 90% or more of variance in log CH_4 flux over two years of measurements.

In our study at the Bog Lake peatland, we examined the combined effect of peat temperature and water table depth on F_m using a multiplicative equation: $F_m = \left[\dfrac{1}{1 + bD_1}\right] a\, c^{(T_p - 10)/10}$, where F_m is the methane flux (mg m^{-2} d^{-1}, averaged between 1000 and 1700 hrs.); D_1 is the distance (m) of the water table from an average hollow surface; T_p is the mean peat temperature (°C) at 0.1 m depth; and a, b, and c are coefficients (Shurpali et al., 1993). This equation accounted for approximately 70% of the variability in the methane flux data (excluding those during episodic emissions). A review of the literature indicates that the mechanisms controlling these individual processes are not yet fully known (e.g., Aselmann and Crutzen, 1989; Schutz et al., 1989; Dise, 1991). Detailed process-level studies are needed to quantify factors limiting the production, oxidation, and emission of methane. The episodic emissions deserve further investigation. Integrated efforts on such studies along with methane emission measurements (preferably, using micrometeorological techniques) are recommended to advance our overall understanding in this area.

Table 2. Mean summertime methane fluxes from different wetland types; values in parentheses represent the maximum methane fluxes

Wetland type	Methane flux (mg m^{-2} d^{-1})	Location	Reference
Bogs	21-356	Minnesota, U.S.A.	3, 5, 7
	5 (21)	Stordalen, Sweden	10
	2 (18)	Alaska, U.S.A.	2b
	<5	Quebec, Canada	8
Fens	121-402 (727)	Minnesota, U.S.A.	3, 5, 7, 9
	39 (52)	Alaska, U.S.A.	11
	20-200	Quebec, Canada	8
	95 (350)	Stordalen, Sweden	10
Swamps	23 (73)	South Carolina, Georgia, Florida, U.S.A.	6
	108-192 (219)	Amazon, Brazil	2a, 4
Marshes	572 (664)	Minnesota, U.S.A	3, 7
	304 (550)	Michigan, U.S.A.	1
	230-590 (885)	Amazon, Brazil	2a, 4

[1]Baker-Blocker et al. (1977), [2a]Bartlett et al. (1988), [2b]Bartlett et al. (1992), [3]Crill et al. (1988), [4]Devol et al. (1988), [5]Dise (1991), [6]Harriss and Sebacher (1981), [7]Harriss et al. (1985), [8]Moore and Knowles (1990), [9]Shurpali et al. (1993), [10]Svensson and Rosswall (1984), [11]Whalen and Reeburgh (1988).

C. Methane Fluxes from Other Studies in Natural Wetlands

Prior (chamber) measurements made at the Marcell Experimental Forest (MEF) in Minnesota include those by Crill et al. (1988) and Dise (1991). Crill et al. (1988) observed a seasonal average of 77-294 mg CH_4 m^{-2} d^{-1} in different bogs and fens at MEF during May-August. Dise (1991) also reported methane fluxes varying from 21 to 402 mg m^{-2} d^{-1} during summer months. The results (seasonal mean of about 125 mg m^{-2} d^{-1}) presented in Figure 6 are within the range of values obtained in these studies at MEF. Generally, the flux values measured in Minnesota peatlands are higher than those observed in subarctic peatlands of Canada (Moore and Knowles, 1990), Sweden (Svensson and Rosswall, 1984), or Alaska (Whalen and Reeburgh, 1988; Bartlett et al., 1992) (Table 2). These results may reflect the relative effects of cooler temperatures and nutrient limitations on methane production in boreal and subarctic zones.

As indicated in Table 2 and also in the survey by Aselmann and Crutzen (1989), methane fluxes generally range in an increasing order from bogs, fens, swamps, and marshes (the relative magnitudes of fluxes are likely associated with factors such as availability of nutrients and water). Yet, due to the large variability of fluxes, Aselmann and Crutzen (1989) concluded that the data base is still rather small to accurately assess emission characteristics of various wetlands.

V. Water Vapor and Heat Fluxes

A. Background Information

Information on the partitioning of incident solar radiation into water vapor flux (λE) and sensible heat (H) is essential to develop a better understanding of the land surface exchange processes. The term H contributes to the heating of the atmosphere locally, whereas λE affects the surface temperature by evaporative cooling. The latter also contributes to the formation of clouds (affecting surface-atmosphere radiative exchange) and helps redistribute the energy by releasing heat to the atmosphere upon condensation. The partitioning of net radiation (R_n) into H and λE is controlled by the rate of evapotranspiration (ET), which depends on several soil, plant and atmospheric factors (discussed below).

B. Energy Partitioning and Controlling Factors

The surface energy balance is expressed as:

$$A + H + \lambda E \approx 0 \tag{2}$$

where A ($= R_n + G$) is the available energy, G the soil heat flux. Micrometeorological measurements of these energy balance components provide direct information on surface energy partitioning. The most commonly used parameter to describe the energy partition is the Bowen ratio ($\beta = H/\lambda E$). When the soil water is not limiting, daytime β typically ranges from 0.1 to 0.5 for agricultural crops, grassland and open peatland (Table 3). On the other hand, for deciduous and coniferous forests, values of β are larger and range from 0.2 to 5.0.

Figure 7A shows typical diurnal patterns of β on clear days over a temperate grassland canopy under two different soil moisture regimes. On a day with favorable soil moisture, β remained low and relatively constant (≈ 0.3) during the day. With severe moisture stress, however, β increased rapidly from morning to late afternoon hours (midday mean ≈ 1.3) as an increasing proportion of the available energy was partitioned to H.

A seasonal picture of energy partitioning over the same grassland canopy is given in Figure 7B. A general decrease in seasonal β occurs as vegetation develops with increasing water use. Midday values of β declined until mid June when leaf area index reached near maximum. Later in the growing season (late July-early August), progressive reduction in W_E caused significant increase in β. A brief recovery in mid August-September was due to replenished soil water by ample rainfall. Similarity between the seasonal patterns of F and β (Figs. 5 and 7) indicates the strong influence of soil water availability on the mass exchange (CO_2 and water vapor) in this ecosystem.

How effectively the evapotranspiration is controlled by atmospheric and biological factors can be examined in terms of the "coupling concept" of McNaughton and Jarvis (1983). They expressed evapotranspiration of well watered vegetated surfaces in the form:

$$\lambda E = \Omega (\lambda E_{eq}) + (1 - \Omega)(\lambda E_{imp}) \tag{3}$$

where Ω ($= [1 + \gamma(g_a/g_c)/(s + \gamma)]^{-1}$) is a "decoupling factor", s is the slope of the saturation vapor pressure-temperature curve, γ is the psychrometric constant, g_a is aerodynamic conductance, and g_c is canopy conductance. The term, λE_{eq} [$= sA/(s + \gamma)$], is the equilibrium latent heat flux which would be achieved if the surface were decoupled from the regional conditions above due to inefficient turbulent mixing. The term, λE_{imp} ($= \rho C_p g_c D/\gamma$, where D is the vapor pressure deficit of air), is the imposed latent heat flux which would occur if regional vapor pressure deficit were imposed on the vegetation by efficient turbulent transport. The value of Ω sets the relative importance of these two terms. The two extreme cases (Ω approaching either 0 or 1) can now be examined.

When stomata are widely open or when g_a is small so that $g_c \gg g_a$ (e.g., in case of unstressed short vegetation), Ω approaches 1 and Eq. (3) reduces to $\lambda E \approx \lambda E_{eq}$. Under such conditions, λE is mainly a function of A (hence, R_n and G) and air temperature. The surface evapotranspiration is less dependent on λE_{imp}, and is said to be weakly coupled to the atmospheric driving potential (i.e., D) and biological control (i.e., g_c). As indicated in Table 3, the mean value of Ω for unstressed short vegetation (e.g., agricultural crops, grassland, open peatland) approaches 1.

At the other extreme (e.g., when g_a is large and g_c relatively small), as would be the case for forests (Table 3), Ω approaches zero. Then, Eq. (3) reduces to $\lambda E \approx \lambda E_{imp}$. In such a case, the λE becomes more sensitive to changes in g_c and D, and is less dependent on R_n. Large values of g_a for these tall canopies indicate that the vapor pressure deficit difference (between the canopy and the ambient air) is small. This has led to the notion of a "strong coupling" between tree canopies and regional atmospheric conditions, in

Table 3. Typical growing season values of the Bowen ratio (β), decoupling factor (Ω), canopy conductance (g_c), aerodynamic conductance (g_a) and evapotranspiration (ET) for various vegetation canopies in temperate climates under nonlimiting soil moisture conditions

Vegetation	β	Ω	g_c (mm s^{-1})	g_a (mm s^{-1})	ET (mm d^{-1})
Agricultural crops	0.1 - 0.5	0.6 - 0.9	20 - 50	20 - 50	3 - 10
Grassland	0.2 - 0.5	0.5 - 0.9	10 - 25	5 - 55	3 - 6.5
Open peatland	0.2 - 0.5	0.6 - 0.9	10 - 30	5 - 30	2 - 6
Deciduous forest	0.2 - 1.2	0.3 - 0.6	5 - 20	45 - 70	3 - 4.5
Coniferous forest	0.5 - 5	0.1 - 0.2	5 - 20	100 - 330	1.5 - 4

Data are from Brutsaert (1982), Jarvis (1981), Jarvis and McNaughton (1986), Jones (1983), Kelliher et al. (1990), Kim et al. (unpublished data), McNaughton and Jarvis (1983), Monteith (1976), Rosenberg et al. (1983), Stewart and Thom (1973), Stewart (1988), Verma et al. (1986), Verma et al. (1989), Verma et al. (1992b).

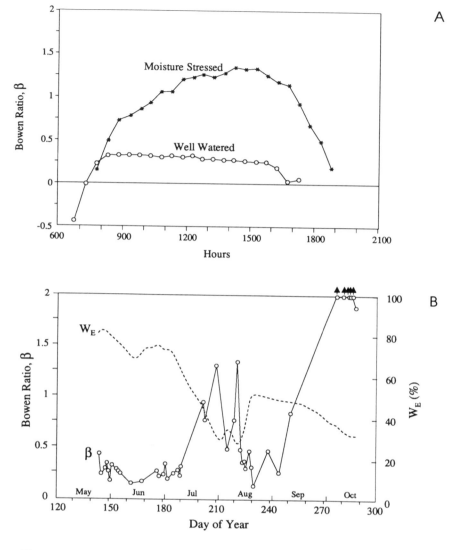

Figure 7. A. Diurnal patterns of the Bowen ratio (β) on a well watered day (11 July, 1987) and a moisture stressed day (30 July, 1987) in a temperate grassland (adapted from Kim and Verma, 1990). B. Seasonal variations in midday (1230-1430 hrs, local time) values of Bowen ratio (β) and extractable soil water (W_E) in a temperate grassland (adapted from Kim and Verma, 1990).

contrast to short canopies where larger gradients (differences) in D between the canopy air and regional conditions prevail.

In reality, conditions in most canopies do not exactly correspond to one of the two extreme cases mentioned above. Even in the case of well watered short vegetation, equilibrium conditions are encountered only rarely. The second term on the right of Eq. (3) is not completely negligible and has been found typically about one-fourth the size of the first term (McNaughton and Jarvis, 1983). The idea underlying equilibrium conditions has led to the development of equations such as that given by Priestley and Taylor (1972):

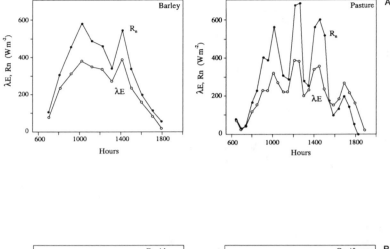

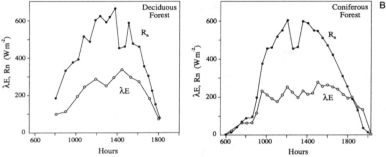

Figure 8. Diurnal variation in latent heat flux (λE) and net radiation (R_n) for (A) a barley crop at Mead, Nebraska (Kim, 1986) and a pasture in New Zealand (McNaughton and Jarvis, 1983) and (B) a deciduous forest at Oak Ridge, Tennessee (Verma et al, 1986) and a coniferous forest in British Columbia (McNaughton and Jarvis, 1983).

$$\lambda E = \alpha_{P-T}\, \lambda E_{eq}, \qquad (4)$$

where α_{P-T} is the Priestley-Taylor coefficient, which is incorporated to account for some contribution from λE_{imp}. With a proper selection of α_{P-T} value, this approximation can often yield reasonable estimates of λE. Under limiting soil moisture conditions, however, other factors need to be considered. The value of α_{P-T} can vary significantly (e.g., 0.1-1.5) depending on the factors such as soil moisture availability (e.g., Brutsaert, 1982) and canopy conductance (e.g., McNaughton and Spriggs, 1989; Verma et al., 1992b).

C. Evapotranspiration Rates in Selected Ecosystems

Hourly or half-hourly micrometeorological measurements of λE are integrated on a 24-hour basis to obtain daily ET rates. Typical growing season ET rates from a variety of vegetation types (well supplied with water) in temperate climates are summarized in Table 3. The ET rates for agricultural crops, grassland, and open peatland range as high as 6-10 mm d^{-1}. On the other hand, rates for deciduous and coniferous forests are smaller (<5 mm d^{-1}). As has been discussed earlier, these differences arise from the fact that λE of well watered short vegetation (crops or grassland) is determined mainly by R_n whereas λE of forests is not. This is evident in Figure 8A, which shows that λE of barley and pasture canopy closely follows R_n. In

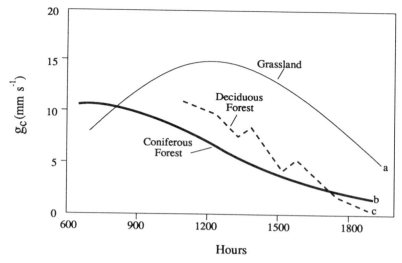

Figure 9. Diurnal variation in canopy conductance (g_c) for three different canopies: (a) a temperate grassland (Kim and Verma, 1991; seasonal average); (b) coniferous forest (Shuttleworth, 1989; seasonal average); and (c) a deciduous forest (Verma et al., 1986; six-day average).

Table 4. Typical midday values for $\Omega \lambda E_{eq}$, $(1 - \Omega) \lambda E_{imp}$, and λE with $A = R_n + G = 500$ W m^{-2}, $T = 20$ °C, and $D = 1.5$ kPa

Canopy	$\Omega \lambda E_{eq}$ (W m^{-2})	$(1 - \Omega) \lambda E_{imp}$ (W m^{-2})	λE (W m^{-2})
Forest: $g_c \approx 7$ mm s^{-1}, $\Omega \approx 0.2$	70	160	230
Grasslands and Crops: $g_c \approx 14$ mm s^{-1}, $\Omega \approx 0.8$	275	75	350

contrast, λE of deciduous and coniferous forest canopies does not follow changes in R_n as closely (Figure 8B). Figure 9 illustrates typical diurnal variation in g_c on clear days for a grassland, deciduous and coniferous forests under well watered conditions. During the day, g_c in forest canopies (which are strongly coupled with the air above) falls rapidly due to stomatal closure with increasing D. However, g_c in grassland canopy (which is weakly coupled with the air above) is more sensitive to irradiance than to ambient D, and follows the diurnal pattern of irradiance. Let us, for example, consider a clear day with midday values of $A = 500$ W m^{-2}, $T = 20$ °C, and $D = 1.5$ kPa. Use of Eq. (3) would yield midday λE of 230 W m^{-2} for a forest canopy ($g_c \approx 7$ mm s^{-1}; $\Omega \approx 0.2$) and 350 W m^{-2} for a grassland ($g_c \approx 14$ mm s^{-1}; $\Omega \approx 0.8$). As indicated in Table 4, λE from a forest canopy is largely determined by λE_{imp}, which is limited by the small magnitudes of g_c during midday. For grasslands and crops, on the other hand, because of greater dependence of λE on λE_{eq} and higher values of g_c, the ET rates are higher.

VI. Summary and Conclusions

This paper concentrates on micrometeorological measurements of fluxes of carbon dioxide, methane, water vapor, and sensible heat. The micrometeorological eddy correlation technique provides continuous, areally-integrated measurement of trace gas and energy fluxes. It requires rapid response instrumentation and extensive sites with relatively homogeneous terrain.

The data on net ecosystem exchange of trace gases show significant day-to-day variations, indicating that interpretations solely based on short-term flux measurements can be misleading. Information on seasonal distributions of trace gas fluxes from key ecosystems is needed to rationally resolve the future effects on the climate and the biosphere. Also needed are process-level studies to establish a better understanding of the factors controlling the emission of these gases. Coordinated efforts on such studies along with energy flux measurements are recommended to further our overall understanding of the source-sink strengths of such trace gases.

Acknowledgements

This study was supported by the National Science Foundation under Grant ATM-9006327 and the Midwestern Regional Center of the National Institute for Global Environmental Change. We thank Mrs. Sharon Kelly for the stenographic work, Ms. Sheila Smith for preparing the figures, and Drs. T. J. Arkebauer, J. M. Norman, and E. S. Verry for their review of this paper. Published as Paper No. 10300, Agricultural Research Division, University of Nebraska, Lincoln.

References

Anderson, D.E. and S.B. Verma. 1986. Carbon dioxide, water vapor and sensible heat exchanges of a grain sorghum canopy. *Boundary-Layer Meteorol.* 34:317-331.

Anderson, D.E., S.B. Verma, and N.J. Rosenberg. 1984. Eddy correlation measurements of CO_2, latent heat and sensible heat over a crop surface. *Boundary-Layer Meteorol.* 29:263-272.

Aselmann, I. and P.J. Crutzen. 1989. Global distribution of natural fresh water wetlands and rice paddies, their net primary productivity, seasonality and possible methane emissions. *J. Atmos. Chem.* 8:307-358.

Baker-Blocker, A., T.M. Denahue, and K.H. Mancy. 1977. Methane flux from wetland areas. *Tellus* 29:245-250.

Baldocchi, D.D. 1982. Mass and energy exchanges of soybeans: microclimate-plant architectural interactions. Ph.D. dissertation. University of Nebraska, Lincoln. 265 pp.

Baldocchi, D.D., S.B. Verma, and D.E. Anderson. 1987. Canopy photosynthesis and water-use efficiency in a deciduous forest.. *J. Appl. Ecol.* 24:251-260.

Baldocchi, D.D., B.B. Hicks, and T.P. Meyers. 1988. Measuring biosphere-atmosphere exchanges of biologically related gases with micrometeorological methods. *Ecology* 69(5):1331-1340.

Ball, J.T., I.E. Woodrow, and J.A. Berry. 1987. A model predicting stomatal conductance and its contribution to the control of photosynthesis under different environmental conditions. p. 221-224. In: J. Biggins (ed.). *Progress in Photosynthesis Research*, Vol. IV. Martinus-Nijhoff, Dordrecht.

Bartlett, K.B., P.M. Crill, D.I. Sebacher, R.C. Harriss, J.O. Wilson, and J.M. Melack. 1988. Methane flux from the Central Amazonian Floodplain. *J. Geophys. Res.* 93(D2):1571-1582.

Bartlett, K.B., P.M. Crill, R.L. Ross, R.C. Harriss, and N.B. Dise. 1992. Methane emissions from tundra environments in the Yukon-Kuskokwin Delta, Alaska. *J. Geophys. Res.* 97(D15):16645-16660.

Beadle, C.L., H. Talbot, R.E. Neison, and P.G. Jarvis. 1985. Stomatal conductance and photosynthesis in a mature Scots pine forest. III. Variation in canopy conductance and canopy photosynthesis. *J. Appl. Ecol.* 22:587-595.

Bingham, B.E., C.H. Gillespie, and J.H. McQuaid. 1978. Development of a miniature, rapid response carbon dioxide sensor. Lawrence Livermore Laboratory, Rept. UCRL-52. Livermore, CA.

Brach, E.J., R.L. Desjardins, and G.T. St. Amour. 1981. Open path CO_2 analyzer. *Phys. Elec. Sci. Instrum.* 14:1415-1419.

Brutsaert, W. 1982. *Evaporation into the atmosphere: theory, history and application*. D. Reidel Publishing Company, Dordrecht, Holland. 299 pp.

Businger, J.A. 1986. Evaluation of the accuracy with which dry deposition can be measured with current micrometeorological techniques. *J. Clim and Appl. Meteorol.* 25:1100-1124.

Chaves, M.M. 1991. Effects of water deficits on carbon assimilation. *J. Experimental Botany* 42:1-16.

Clement, R.J. 1988. Carbon dioxide and energy exchanges of cereal crops. M.S. thesis. University of Nebraska, Lincoln. 112 pp.

Crill, P.M., K.B. Bartlett, R.C. Harriss, E. Gorham, E.S. Verry, D.I. Sebacher, L. Madzar, and W. Sanner. 1988. Methane flux from Minnesota peatlands. *Global Biogeochem. Cycles* 2:371-384.

Devol, A.H., J.E. Richey, W.A. Clark, S.L. King, and L.A. Martinelli. 1988. Methane emissions to the troposphere from the Amazon floodplain. *J. Geophys. Res.* 93:1583-1592.

Dise, N.B. 1991. Methane emission from peatlands in northern Minnesota. Ph.D. dissertation. University of Minnesota. 130 pp.

Harley, P.C., J.D. Tenhunen, K.J. Murray, and J. Beyers. 1989. Irradiance and temperature effects on photosynthesis of tussock tundra *Sphagnum* mosses from the foothills of the Philip Smith Mountains, Alaska. *Oecologia* 79:251-259.

Harriss, R.C. and D.J. Sebacher. 1981. Methane flux in forested freshwater swamps of the southeastern United States. *Geophys. Res. Lett.* 8:1002-1004.

Harriss, R.C., E. Gorham, D.I. Sebacher, K.B. Bartlett, and P.A. Flebbe. 1985. Methane flux from northern peatlands. *Nature* 315:652-654.

Jarvis, P.G. 1981. Stomatal conductance, gaseous exchange, and transpiration. p. 175-204. In: J. Grace, E.D. Ford, and P.G. Jarvis (eds.). *Plants and their atmospheric environment*. Blackwell Scientific, Oxford.

Jarvis, P.G. and K.G. McNaughton. 1986. Stomatal control of transpiration: scaling up from leaf to region. *Adv. Ecol. Res.* 15:1-48.

Jarvis, P.G., G.B. James, and J.J. Landsberg. 1976. Coniferous forest. p. 171-240. In: J.L. Monteith (ed.). *Vegetation and the atmosphere. Vol. 2, Case Studies*. Academic Press, New York. 439 pp.

Jones, H.G. 1983. *Plants and microclimate*. Cambridge University Press, Cambridge. 323 pp.

Kelliher, F.M., D. Whitehead, K.J. McAneney, and M.J. Judd. 1990. Partitioning evapotranspiration into tree and understory components in two young *Pinus Radiata* D. Don stands. *Agric. For. Meteorol.* 50:211-227.

Kim, J. 1986. Environmental and physiological effects on water use of cereal crops. M.S. thesis. University of Nebraska, Lincoln. 119 pp.

Kim, J. and S.B. Verma. 1990. Components of surface energy balance in a temperate grassland ecosystem. *Boundary-Layer Meteorol.* 51:401-417.

Kim, J. and S.B. Verma. 1991. Modeling canopy stomatal conductance in a temperate grassland ecosystem. *Agric. For. Meteorol.* 57:55:149-166.

Kim, J. and S.B. Verma. 1992. Soil surface CO_2 flux from a Minnesota peatland. *Biogeochem.* 18:37-51.

Kim, J., S.B. Verma, and R.J. Clement. 1992. Carbon dioxide budget in a temperate grassland ecosystem. *J. Geophys. Res.* 97(D5):6057-6063.

McNaughton, K.G. and P.G. Jarvis. 1983. Predicting effects of vegetation changes on transpiration and evaporation. p. 1-47. In: T.T. Kozlowski (ed.). *Water deficits and plant growth*, Vol. VII. Academic Press, New York, NY.

McNaughton, K.G. and T.W. Spriggs. 1989. An evaluation of the Priestley and Taylor equation and the complementary relationship using results from a mixed-layer model of the convective boundary layer. p. 86-101. In: T.A. Black, D.L. Spittlehouse, M.D. Novak, and D.T. Price (eds.). *Estimation of aerial evapotranspiration*, IAHS Publ., 177.

Monteith, J.L. 1976. *Vegetation and the atmosphere. Vol. 2, Case studies*. Academic Press, New York. 439 pp.

Monteith, J.L., G. Szeicz, and K. Yabuki. 1964. Crop photosynthesis and the flux of carbon dioxide below the canopy. *J. Appl. Ecol.* 1:321-337.

Moore, T. and R. Knowles. 1990. Methane emissions from fen, bog and swamp peatlands in Quebec. *Biogeochem.* 11:45-61.

Moore, T., N. Roulet, and R. Knowles. 1990. Spatial and temporal variations of methane flux from subarctic/northern boreal fens. *Global Biogeochem. Cycles* 4:29-46.

National Center for Atmospheric Research. 1989. Global tropospheric chemistry: chemical fluxes in the global atmosphere. In: D.H. Lenschow and B.B. Hicks (eds.). Report of the Workshop on Measurements of Surface Exchange and Flux Divergence of Chemical Species in the Global Atmosphere. Boulder, CO, October 1987. 107 pp.

National Research Council. 1984. *Global tropospheric chemistry: a plan for action*. National Academy Press, Washington, D.C. 194 pp.

Norman, J.M., R. Garcia, and S.B. Verma. 1992. Soil surface CO_2 fluxes on the Konza Prairie. *J. Geophys. Res.* 97(D17):18845-18853.

Ohtaki, E. and T. Matsui. 1982. Infrared device for simultaneous measurement of atmospheric carbon dioxide and water vapor. *Boundary-Layer Meteorol.* 24:109-119.

Parton, W., S. Running, and B. Walker. 1992. Report: a toy terrestrial carbon flow model. p. 281-302. In: D. Ojima (ed.). *Modeling the earth system*. UCAR, Office for Interdisciplinary Earth Studies Global Change Institute Volume 3. 488 pp.

Priestley, C.H.B. and R.J. Taylor. 1972. On the assessment of surface heat flux and evaporation using large-scale parameters. *Mon. Weather Rev.* 100:81-92.

Rosenberg, N.J., B.L. Blad, and S.B. Verma. 1983. *Microclimate: the biological environment*, 2nd Ed., John Wiley & Sons, New York. 495 pp.

Schulze, E.D. 1986. Carbon dioxide and water vapor exchange in response to drought in the soil. *Annu. Rev. Plant Physiol.* 37:247-274.

Schutz, H., W. Seiler, and R. Conrad. 1989. Processes involved in formation and emission of methane in rice paddies. *Biogeochem.* 7:33-53.

Shurpali, N.J., S.B. Verma, R.J. Clement, and D. Billesbach. 1993. Seasonal distribution of methane flux in a Minnesota peatland measured by eddy correlation. *J. Geophys. Res.* 98:20,649-20,655.

Shuttleworth, W.J. 1989. Micrometeorology of temperate and tropical forest. *Phil. Trans. R. Soc. Lond.* B 324:299-334.

Singh, J.S. and S.R. Gupta. 1977. Plant decomposition and soil respiration in terrestrial ecosystem. *The Botanical Review* 43:449-528.

Stewart, J.B. 1988. Modelling surface conductance of pine forest. *Agric. For. Meteorol.* 43:19-35.

Stewart, J.B. and A.S. Thom. 1973. Energy budget in pine forest. *Quart. J. R. Met. Soc.* 99:154-170.

Svensson, B.H. and T. Rosswall. 1984. In situ methane production from acid peat in plant communities with different moisture regimes in a subarctic mire. *Oikos* 43:341-350.

University Corporation for Atmospheric Research. 1986. Global tropospheric chemistry/plans for the U.S. research effort. OIES Report 3, University Corporation for Atmospheric Research, Boulder, CO. 110 pp.

Verma, S.B. 1990. Micrometeorological methods for measuring surface fluxes of mass and energy. *Remote Sens. Rev.* 5:99-115.

Verma, S.B., D.D. Baldocchi, D.E. Anderson, D.R. Matt, and R.J. Clement. 1986. Eddy fluxes of CO_2, water vapor and sensible heat over a deciduous forest. *Boundary-Layer Meteorol.* 36:71-91.

Verma, S.B., J. Kim, and R.J. Clement. 1989. Carbon dioxide, water vapor and sensible heat fluxes over a tallgrass prairie. *Boundary-Layer Meteorol.* 46:53-67.

Verma, S.B., F.G. Ullman, D. Billesbach, R.J. Clement, J. Kim, and E.S. Verry. 1992a. Eddy correlation measurements of methane flux in a northern peatland ecosystem. *Boundary-Layer Meteorol.* 58:289-304.

Verma, S.B., J. Kim, and R.J. Clement. 1992b. Momentum, water vapor, and carbon dioxide exchange at a centrally located prairie site during FIFE. *J. Geophys. Res.* 97(D17):18629-18639.

Webb, E.K., G.I. Pearman, and R. Leuning. 1980. Correction of flux measurements for density effects due to heat and water vapor transfer. *Quart. J. R. Met. Soc.* 106:85-100.

Whalen, S.C. and W.S. Reeburgh. 1988. A methane flux time series for tundra environments. *Global Biogeochem. Cycles* 2:399-409.

CHAPTER **31**

A Micrometeorological Technique for Methane Flux Determination from a Field Treated with Swine Manure

J.H. Prueger, T.B. Parkin, and J.L. Hatfield

I. Introduction

Methane is recognized as an important greenhouse gas with potentially far reaching implications for global climate change. Methane effectively absorbs in the near and far infrared ranges (peaks at 3 and 7.5 μm, Rosenberg et al., 1983). Rodhe (1990) reported that on a mole basis, and in conjunction with the decay time of gases resident in the atmosphere, CH_4 has approximately 5 times greater absorbing capability than CO_2 and can account for approximately 15 to 33 percent of the anticipated contribution of atmospheric warming. Houghton et al. (1990) reported that the 20 year global warming potential for methane is about 63 times that of CO_2, on a per-molecule basis. The annual rate of increase in atmospheric CH_4 is currently estimated at 1.1% (Cicerone and Oremland, 1988).

Whalen et al. (1989) estimated that sources of atmospheric methane originated primarily from biological sources such as peat bogs, wetlands, rice fields, and tundra that utilize recently fixed carbon (approximately 80%) and the production and combustion of fossil fuels (approximately 20%).

Global estimates of CH_4 emissions reported by Baker-Blocker (1977), indicated the contribution to methane flux from four different wetlands was about 6.8×10^{13} g or about 2% annually. Estimates from Harriss et al. (1985) and Matthews and Fung (1987) indicated approximately 66% of total global CH_4 emissions came from northern (>40°N) wetlands. Bartlett et al. (1988) calculated a 12% global methane contribution from the Amazonian floodplain. While marsh-like ecosystems are significant sources of methane, other ecosystems such as semi-arid native grasslands can be sinks for atmospheric methane (Mosier et al., 1991). Whereas CH_4 flux from natural ecosystems have been intensively studied, virtually no data exists for non-flooded cultivated agricultural systems.

Understanding complex interactions of CH_4 transport between the biosphere and atmosphere is important to understanding global climatic changes. This understanding should include local, regional and global scales as well as varying temporal scales. Large scale estimates reported in the current literature are largely based on extrapolations from numerous ground-based chamber studies that have been found to contain considerable spatial and temporal variability (Matson et al., 1989). Cicerone and Shetter (1981) compared their methane flux results from marshes and lakes to similar earlier data, and concluded that existing data and flux-measurement methods were insufficient for reliable global extrapolations.

Part of the spatial variability can be attributed to techniques involving field CH_4 flux measurements. Many of these measurements have been made *in situ* using closed static or open/dynamic chambers placed over a variety of different surfaces. Closed static chambers are commonly used for flux measurements of atmospheric trace gases and generally employ an open bottom chamber which is placed over soil/vegetative sites in which air samples are drawn from the chamber at discrete time intervals and then analyzed for the concentration of trace gas in question. The ease and simplicity of the system as well as the low cost of constructing and maintaining such a system are the main advantages. A key disadvantage of the closed chamber method in a field situation is the disruption and isolation from local microclimatic parameters. Wind, solar radiation, temperature, and gas concentration gradients as well as turbulence exchange phenomena can significantly affect gas exchange between a surface and the atmosphere above it. Anderson (1973), Gupta and Singh 1(977), and Sharkov 1984 identified many factors that can lead to under or over estimation of CO_2 fluxes under static chambers. It can be logically expected that the factors which affect CO_2 flux from under a static chamber can also affect the flux of other trace gases. Dynamic chambers are normally equipped with a mechanism such as a pump or fan in which air within a chamber can be circulated and exchanged at a constant flow rate during the sampling period. The general design of this type of system is intended to reduce increases of air temperature, relative humidity, and pressure within the chamber by circulating the air within the chamber by mechanical turbulence. Rochette et al. (1992) compared static and dynamic chambers for soil respiration under field conditions and found consistently lower soil respiration values than did the dynamic system. Disadvantages of this system are essentially the same as for the closed static system.

Baldocchi et al. (1988) proposed alternative methods for evaluating trace gas fluxes between the biosphere and atmosphere under natural conditions using micrometeorological techniques used for measuring and estimating water vapor flux (latent heat, LE) and sensible heat flux, H. Micrometeorological methods provide several advantages for measuring CH_4 flux from a surface compared to the chamber methods. The techniques are non-disruptive in terms of radiant energy transfer and turbulence exchange between a surface and the atmosphere. A micrometeorological technique for measuring CH_4 flux into the atmosphere allows for "natural" *in situ* measurements. Flux values are time averaged point measurements in which if sufficient fetch (100:1 as a minimum) is available can represent temporal and spatially integrated values. Fetch is defined as upwind distance of a uniform surface roughness. The ratio 100:1 represents 100 m of surface distance for every 1 m height of a measurement instrument.

A. Bowen-Ratio

The Bowen-ratio approach is a flux gradient technique that assumes turbulent transport of an atmospheric gas to be analogous to molecular diffusion. In this case the turbulent flux of a gas is proportional to the product of the mean vertical gas concentration gradient, Δc, between two heights and a turbulent-transfer coefficient k which relates the turbulent flux to the gradient of the associated mean variable concentration, Baldocchi et al. (1988). This relationship is expressed as;

$$F = k\left(\frac{\partial c}{\partial z}\right) \tag{1}$$

where F is the flux density ($\mu g\ m^{-2}\ hr^{-1}$) of the gas in question, $\partial c/\partial z$ is the measured vertical gas concentration gradient between two heights ($\mu g\ m^{-3}$) and k is the turbulent-transfer coefficient for the gas in question with units of $m^{-2}\ s^{-1}$. In this study, transfer of CH_4 away from a surface to the atmosphere is considered positive. The turbulent-transfer coefficient can be calculated using a Bowen-ratio energy balance technique described by Baldocchi et al. (1988). The Bowen-ratio (Bowen 1926), is a ratio between sensible and latent heat flux energy at an evaporating surface and can be expressed as;

$$\beta = \gamma\ \Delta T\ /\ \Delta e \tag{2}$$

where β is the Bowen-ratio which is dimensionless, γ is the psychrometric constant, ΔT is the temperature difference between two heights and Δe is the difference in vapor pressure between two heights.

The energy balance at a surface is defined as the difference between energy sources and sinks. Baldocchi et al. (1988) expressed surface energy balance as

$$R_n = H + S + LE + P_s + G \tag{3}$$

where R_n is net radiation, H is sensible heat flux, S is soil heat flux, LE is latent heat flux, P_s is energy consumed by photosynthesis, and G is the canopy heat storage, all are expressed in units of $W\ m^{-2}$. In most micrometeorological cases P_s and G are considered negligible relative to R_n, H, S, and LE and are normally ignored. Based on the assumption of equality of turbulent-transfer coefficients for mass and energy, Baldocchi et al., (1988) expressed the turbulent-transfer coefficient for a trace gas as

$$k = -\frac{(R_n - S - G)}{\rho_a(C_p\frac{\partial T}{\partial z} + L\frac{\partial v}{\partial z} + \psi\frac{\partial C}{\partial z})} \quad (4)$$

where R_n, S, and G have been previously defined, v and C are the mixing ratios of water vapor and CO_2 respectively, ρ_a is the density of air, C_p is the specific heat of air, T is potential air temperature and ψ is energy equivalent of CO_2 fixation. In cases where CO_2 is not routinely measured, an assumption of equality of exchange coefficients is made for momentum and energy.

Results for LE and H flux calculations using the Bowen-ratio energy balance technique have been found to be reasonably reliable. Verma and Rosenberg (1978) reported that errors using the energy balance technique were between 10-5%.

The objective of this study was to evaluate the flux-gradient technique using a Bowen-ratio energy balance technique as described by Baldocchi et al., (1988) to measure CH_4 flux over a corn field that was treated with injected swine manure during the 1992 growing season. Swine production in the midwest an important agricultural activity which produces a considerable amount of animal waste. Typically swine waste serves as a soil amendment and when spread on to agricultural soils provides a means of disposing the waste.

II. Materials and Methods

The field was located in the Walnut Creek watershed approximately 10 km south of Ames, Iowa with geographic coordinates of 41° 57′ 53″ N latitude and 93° 39′ 8″ W longitude. This field is approximately 237 ha and was planted to corn using conventional tillage and agronomic practices. The soil type was a Webster loam, which had an average organic material content between 6-7% and a 0-2% slope.

A micrometeorological station was erected 150 meters from the southern edge of the field to measure surface energy balance components of net radiation, soil heat flux, sensible heat flux, and latent heat flux densities of water. Sensible and latent heat flux were calculated using the Bowen-ratio technique (Bowen, 1926). Anemometers and aspirated wet and dry bulb psychrometer profile masts were constructed near the Bowen-ratio station to evaluate wind speed and water vapor pressure profiles over the surface and to compare temperature, wind speed, and vapor pressure values to those from the Bowen-ration station. The anemometers and psychrometers were placed at heights of 30, 55, 80, 130, 230, and 280 cm above the corn crop surface.

A Q*6 REBS[1] net radiometer and soil heat flux plate were used to measure net radiation and soil heat flux. Net radiation was measured at 2 m above the top of the soil and subsequently adjusted to measure 2 m above the corn canopy as the season progressed. The psychrometer and anemometer instruments were also adjusted periodically throughout the growing season to maintain the same heights over the corn canopy as had been maintained over the bare soil. Soil heat flux was measured at depth of 10 cm and soil thermocouples were placed at 1, 2, 5, 10, 25, and 50 cm depths to measure soil temperature gradients to evaluate the storage component for soil heat flux.

The Bowen-ratio was calculated as a ratio of the difference between the dry bulb temperatures measured at 25 and 125 cm above the surface to the difference between vapor pressures (e) calculated from the wet and dry bulb temperatures measured at the two heights. Latent heat flux was calculated as

---

[1]The USDA-ARS does not endorse any product or company, trade/product names are only for facilitation purposes for the reader.

$$L_e = \frac{(R_n - G)}{(1 + \beta)} \tag{5}$$

where L_e is latent heat flux in W m^{-2}, R_n and G are net radiation and soil heat flux in W m^{-2} and β is the Bowen-ratio calculated as

$$\beta = \frac{pC_p(T_1 - T_2)}{\lambda \epsilon (e_1 - e_2)} \tag{6}$$

where p is atmospheric pressure (kPa), C_p is specific heat of air (J kg^{-1} K^{-1}), T is air temperature at two different points above a surface, and e is vapor pressure from the same heights as air temperature (kPa), λ is the latent heat of vaporization (2.45 x 10^6 J kg^{-1}), and ϵ is the ratio of the mole weight of water vapor and air (0.622). Sensible heat flux (H) was calculated as a residual from equation 3. Transport coefficients for water vapor and heat, k_v and k_h were calculated using equation 4.

A. Methane Sampling Mast

The methane sampling mast erected next to the Bowen-ratio station consisted of 1 L evacuated metal canisters logarithmically spaced at four heights above the surface (30, 56, 100, 200 cm). The canisters were purged with He and evacuated to a vacuum of 78 kPa. Each canister was fitted with a fine metering valve calibrated to deliver an air flow intake rate of 333 ml ± 5 ml hr^{-1}. The air was sampled for three hours during the afternoon from 1300 to 1600 hours. At the end of the three hour sampling period, the canisters were in equilibrium with atmospheric pressure. Latent and sensible heat values were calculated for 30 minute averages and then averaged for the three hour interval.

B. Swine Manure Application and Air Sampling

Several days prior to the swine manure injection, ambient methane concentrations were measured in the field using the methane sampling mast. An area of 200 x 200 m (4 ha) was marked within the 237 ha field for swine manure treatment. The methane sampling mast was located in the center of this plot next to the Bowen-ratio station providing a minimum fetch of 100 meters in all directions. On April 3, 1992 swine manure was injected approximately 10 cm below the soil surface using a dual knife applicator with a knife spacing of 1.5 m at an application rate of 25,000 L ha^{-1}.

After each sampling event the sampling canisters were transported to the laboratory where a 30 ml aliquot from each canister was drawn for analysis using a 50 ml syringe. From each 30 ml aliquot three 0.5 ml samples were drawn and analyzed for methane concentration by gas chromatography. Results were used to calculate methane flux using the Bowen-ratio method described by Baldocchi (1988).

III. Results

Temporal variation of CH$_4$ flux could be observed during the sampling period from day of year (DOY) 85 to DOY 230 (Figure 1a). DOY 98 showed an increase of methane flux to approximately 28 μg m^{-2} hr^{-1} from a previous average of about 1.8 μg m^{-2} hr^{-1}. The following days 99 and 100 showed further increases to 107 and 62 μg m^{-2} hr^{-1} respectively. This first plume of CH$_4$ flux was preceded by a 15 mm precipitation event on DOY 90 followed by two smaller precipitation events (4 mm each) on DOY 98 and 99, Figure 1b. The next plume event was observed on DOY 119, after two precipitation events (20.1 and 12.7 mm) which occurred on DOY 111 and 114 respectively. After DOY 119 problems were encountered with the sampling

A Micrometeorological Technique for Methane Flux Determination

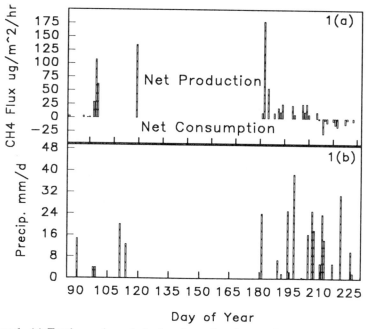

Figure 1. (a) Total sample period of methane flux from a field treated with swine manure. (b) Precipitation totals for same period of time.

canisters. The next period in which the data were collected began on DOY 180 and continued through DOY 231. The only large flux event that appeared during this period occurred on DOY 182 and 184 which was preceded by a 24.1 mm precipitation event. Days subsequent from DOY 184 to DOY 211 showed positive fluxes to the atmosphere which ranged between 7 to 28 μg m^{-2} hr^{-1} indicating low methane transport to the atmosphere from the soil. After DOY 211, CH_4 fluxes became negative suggesting that consumption of CH_4 had exceeded production of CH_4. These values remained negative despite several large precipitation events in which the soil became saturated for a period of 24-48 hours as evidenced by water ponding on the soil surface. Figures 2(a) through 2(c) show an enhanced view of the sampling period from DOY 85 to DOY 110 with the addition of average soil temperature during the 3 hour sampling period at a depth of 10 cm, the approximate depth at which the swine manure was applied. Methane flux can be seen to be related to soil temperature. The daily trend observed in Figure 2(c) is similar to the monthly trend of CH_4 production as a function of sediment temperature presented by Thebrath et al. (1993). Figures 3(a) through 3(c) show results from DOY 201 to DOY 230. These results show CH_4 production relative to CH_4 consumption to steadily decrease towards the end of the sampling period. Methane production and consumption can be seen to correlate well with soil temperature on some days but not on others.

In general, gradients of CH_4 above the soil surface fluctuated throughout the sampling period (Figure 1a). Net positive gradients indicate greater CH_4 production than CH_4 consumption while negative gradients indicate increased CH_4 consumption over CH_4 production. The positive CH_4 gradients observed in the spring suggest increased active CH_4 production by soil microbes in response to injected swine manure that was applied on DOY 93. Methane production, an anaerobic process, was likely occurring in microsites in the soil. Anaerobic microsites supporting high denitrification activity have been observed to be associated with particulate organic matter (Parkin, 1987), and with zones in which organic waste has been injected into the soil (Rice et al.,1988). The apparent trend indicated by these data suggest that early in the season the soil treated with injected swine manure is a source for CH_4, while in the latter part of the season (DOY 209-219), net consumption of CH_4 exceeds production which results in the soil becoming a sink for CH_4. The gradients observed in June, July, and August were smaller than those observed in April, and in many cases, may have been due to analytical variation. Net methane production ranged between 2 μg m^{-2} hr^{-1} to 177 μg

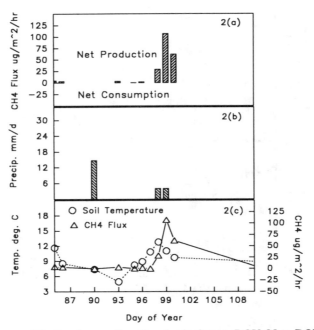

Figure 2. (a) CH$_4$ flux from soil surface beginning on DOY 85 to DOY 110. (b) Precipitation totals for the same time period as 2(a). (c) Average soil temperature at 10 cm for the time corresponding to the sampling period of CH$_4$.

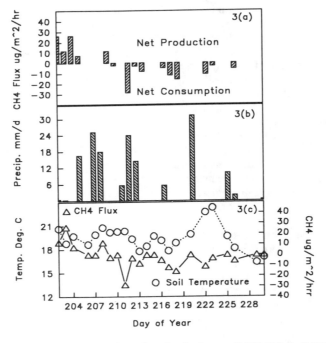

Figure 3. (a) CH$_4$ flux from soil surface beginning on DOY 201 to DOY 230. (b) Precipitation totals for the same time period as 2(a). (c) Average soil temperature at 10 cm for the time corresponding to the sampling period of CH$_4$.

m^{-2} hr^{-1} with an average value of about 42 μg m^{-2} hr^{-1} during periods after a precipitation event when soil conditions were favorable for anaerobic decomposition, Whalen et al. 1992. During the latter part of the season net methane consumption was approximately -7 μg m^{-2} hr^{-1}.

IV. Conclusions

This preliminary study for evaluating a micrometeorological technique for measuring methane flux suggests that a Bowen-ratio energy balance technique offers an alternative to chamber based techniques for measuring CH_4 transport from the soil into the atmosphere. Temporal variability will continue to be associated with CH_4 flux measurements due to variations of biological and micrometeorological parameters controlling CH_4 at the soil surface. Modifications of sampling protocol will allow more accurate determinations of small gradients of CH_4 as well as shorter sampling intervals (30-60 min.). These improvements should allow for greater accuracy of field measurements, as well as for accounting for diurnal temporal variation normally associated with CH_4 transport into the atmosphere.

References

Anderson, J.M. 1973. Carbon dioxide evolution from two temperate, deciduous woodland soils. *J. Appl. Ecol.* 10:361-378.

Baker-Blocker, A., T.M. Donahue, and K.H. Mancy, 1977. Methane flux from wetland areas. *Tellus* 29:245-250.

Baldocchi, D.D., B.B. Hicks, and T.P. Meyers. 1988. Measuring biosphere-atmosphere exchanges of biologically related gases with micrometeorological methods. *Ecology* 69:1331-1340.

Bartlett, K.B., P.M. Crill, D.I. Sebacher, R.C. Harriss, J.O. Wilson, and J.M. Melack. 1988. Methane flux from the Amazon River floodplain. *J. Geophys. Res.* 93:1571-1582.

Bowen, I.S. 1926. The ratio of heat losses by conduction and by evaporation from any water surface. *Physical Review* 27:779-787.

Circerone, R.J. and R.S. Oremland. 1988. Biogeochemical aspects of atmos-pheric methane. *Global Biogeochem. Cycles* 2:299-327.

Circerone, R.J. and J.D. Shetter 1981. Sources of atmospheric methane:measurements in rice paddies and a discussion. *J. Geophy. Res.* 86:7203-7209.

Gupta, S.R. and J.S. Singh. 1977. Effect of alkali, volume and absorption area on the measurement of soil respiration in a tropical sward. *Pedobiologia* 17:223-239.

Harriss, R.C., E.Gorham, D.I. Sebacher, K.B. Bartlett, and P.A. Flebbe, 1985. Methane flux from northern peatlands. *Nature* 315:652-653.

Houghton, J.T., G.J. Jenkins, J.J. Ephramus, (eds.). 1990. *Climate Change: The IPCC Scientific Assessment*. Cambridge University Press, Cambridge.

Matson, P.A., P.M. Vitousek, and D.S. Schimel, 1989. Exchange of Trace Gases between Terrestrial Ecosystems and the Atmosphere. p. 97-108. M.O. Andreae and D.S. Schimel (eds.). John Wiley & Sons. Chichester, New York.

Matthews, E.O. and I. Fung, 1987. Methane emission from natural wetlands: Global distribution, area, and environmental characteristics of sources. *Global Biogeochem. Cycles* 1:61-86.

Mosier, A., D. Schmiel, D. Valentine, K. Bronson and W. Parton. 1991. Methane and nitrous oxide fluxes in native, fertilized and cultivated grasslands. *Nature* 350:330-332.

Parkin, T.B. 1987. Soil microsites as a source of denitrification variability. *Soil Sci. Soc. Am. J.* 51:1194-1199.

Rice, C.W., P.E. Sierzega, J.M. Tiedje, and L.W. Jacobs. 1988. Stimulated denitrification in the microenvironment of a biodegradable organic waste injected into soil. *Soil Sci. Soc. Am. J.* 52:102-108.

Rochette, P., E.G. Gregorich, and R.L. Desjardins. 1992. Comparison of static and dynamic closed chambers for measurement of soil respiration under field conditions. *Can. J. Soil Sci.* 72:605-609.

Rodhe, H. 1990. A comparison of the contribution of various gases to the greenhouse effect. *Science* 248:1217-1219.

Rosenberg, N.J., B.L. Blad, and S.B. Verma. 1983. *Microclimate: The biological environment.* John Wiley & Sons, Chichester, New York. 495 pp.

Sharkov, I.N. 1984. Determination of the rate of CO_2 production by the absorption method. *Sov. Soil Sci.* 16:102-111.

Thebrath, B., F. Rothfuss, M.J. Whiticar, and R. Conrad. 1993. Methane production in littoral sediment of Lake Constance. FEMS Microbiology *Ecology* 102:279-289.

Verma, S.B., N.J. Rosenberg, and B.L. Blad. 1978. Turbulent exchange coefficients for sensible heat and water vapor under advective conditions. *J. Appl. Meteorol.* 17:330-338.

Whalen, M., N. Tanaka, R. Henry, B. Deck, J. Zeglen, J.S. Vogel, J. Southon, A. Shemesh, R. Fairbanks, and W. Broecker. 1989. Carbon-14 in methane sources in atmospheric methane: The contribution from fossil carbon. *Science* 245:286-290.

Whalen, S.C., W.S., Reeburgh, and V.A., Barber. 1992. Oxidation of methane in boreal forest soils: A comparison of seven measures. *Biogeochemistry* 16:181-211.

CHAPTER 32

Application of the CENTURY Soil Organic Matter Model to a Field Site in Lexington, KY.

A.S. Patwardhan, R.V. Chinnaswamy, A.S. Donigian, Jr., A.K. Metherell, R.L. Blevins, W.W. Frye, and K. Paustian

I. Introduction

The CENTURY model developed by Parton et al. (1987, 1988) simulates the dynamics of carbon, nitrogen, phosphorus, and sulphur in cultivated and grassland soils. The model has been previously used for various field applications. Most of these studies, however, were limited to grassland conditions, because the model lacked the capability of simulating complex cropping practices. Recently researchers at NREL at Colorado State University (Metherell, 1992) have added algorithms and incorporated crop parameters to CENTURY which allow for the simulation of crop growth, crop rotations, and tillage practices. CENTURY model can now simulate commercial crops such as corn, soybean, wheat, and hay.

The Environmental Research Laboratory of the Environmental Protection Agency (EPA) at Athens initiated a study in 1991 to estimate greenhouse gas emissions and carbon sequestration potential of agricultural production systems (Barnwell et al., 1992). The EPA Climate Change study region consists of the Corn Belt, the Great Lakes area, and a portion of the Great Plains. Corn, soybean, and wheat are the major crops grown in this region. Since CENTURY was being used for assessing the carbon sequestration potential of agroecosystems in the study region, there was a need to conduct field testing of the model to better assess its predictions of soil organic carbon (SOC). In this paper we describe the application of the CENTURY model to a field site in Lexington, KY.

Soil carbon is a direct function of the carbon inputs to the soil derived from the crop, both as roots and crop residues. Thus the impacts of tillage and harvest practices that control the disposition of the crop biomass, and associated carbon, must be accurately represented to mimic the carbon balance of the soil. The approach used in this study to simulate soil organic carbon is based on the assumption that if the CENTURY model simulates corn grain production with a reasonable degree of accuracy, thus simulating appropriate soil carbon input due to grain production which is a major factor in determining soil carbon, then soil organic carbon dynamics will be adequately simulated. The objective of our study was to evaluate the capability of CENTURY for simulating soil carbon and nitrogen under varying agricultural management systems. Complete details of the model application and results are described by Chinnaswamy et al. (1993).

II. Century Model Overview

CENTURY simulates dynamics of C, N, P and S in cultivated and grassland soils using a *monthly* time step for model simulations in the top 20 cm of the soil profile. The model was not developed with an intent to simulate detailed crop physiology and thus crop yield, however, the purpose of the model is to simulate *long term dynamics* of soil organic carbon, nitrogen, phosphorus, and sulphur. The model allocates organic matter into five pools: two of which represent litter or crop residues and the remaining three representing soil organic matter. The soil organic matter is divided into three fractions: 1) an "active" fraction which has rapid turnover rate and consists of microbial biomass and metabolites, 2) a "slow" fraction with an intermediate turnover time which represents stabilized decomposition products, and 3) a "passive" fraction which represents the highly stabilized, recalcitrant organic matter. The model is described in detail by Parton et al. (1987, 1988).

III. Lexington Field Experiment

A field experiment was initiated in the spring of 1970 (Blevins et al., 1977, 1983a) at the Kentucky Agricultural Experiment Station Farm in Lexington. The soil at the site is a Maury silt loam (fine, mixed, mesic Typic Paleudalf) with slopes ranging from 1 to 3%. The Maury soil is deep, well-drained and originated in residuum of phosphatic limestone. The experimental area, a broad ridgetop, that was under bluegrass pasture for approximately 50 years until the initiation of the experiment (Blevins 1992, personal communication). Management practices on the pasture area consisted of low intensity grazing with cattle and horses, and occasional mowing. The ecosystem was altered from the permanent bluegrass pasture to continuous corn production in April 1970.

The objective of the experiment was to study the effects of conventional tillage and no-till systems on chemical and physical properties of soils. The cropping practices consisted of corn followed by a winter cover crop of rye. The experimental design was split block with four replications. The experimental area (46.1 x 48.4 m^2) was divided laterally into four strip blocks receiving four levels of ammonium nitrate fertilizer (0, 84, 168, and 336 kg N ha^{-1}). In the text that follows below, the zero fertilizer treatment will be referred to as the control treatment.

Plots under conventional tillage system were plowed in late April, about one to two weeks before corn planting. The tillage implements consisted of a four-bottom 16-inch-trailing moldboard plow and a tandem disk harrow. All plots were sprayed with herbicide just before or immediately after planting corn. Corn was planted in May using a conventional planter, and rye was broadcast on all plots in mid September, two weeks prior to harvesting corn, so as to produce a cover crop and additional mulch. No cultivation practices were used for weed control, and no irrigation practices were followed. For the no-till plots herbicide was sprayed in April before planting corn. A no-till planter was used for planting corn, rye was broadcast on all plots in mid-September, two weeks prior to harvesting corn.

IV. CENTURY Model Application

The CENTURY model was used first to simulate the historic grassland conditions and then the long-term continuous corn-tillage experiments. The simulation of grassland conditions provided initial values for various organic matter pools as they might have existed at the beginning of the experiment in 1970. The data available for CENTURY simulations, calibration, and validation consisted of meteorologic data (monthly precipitation, and maximum and minimum temperatures), soil physical properties, soil carbon and nitrogen measurements, atmospheric nitrogen deposition, and annual crop yield.

The following assumptions were made to simulate the historic grassland conditions:

a. The site was under grassland conditions.
b. The period between April and October was the growing season for grass and November was the senescence month.
c. Grazing events occurred throughout the year with winter grazing during winter months of November, December, and January through March.
d. A fire event occurred once every ten years in July. The fire intensity was assumed to be moderate.

The model was run for a period of 5000 years using the option for stochastic generation of climate data in the model. The initial soil carbon levels in the various pools at the beginning of the historic run were established using the IVAUTO option in the CENTURY model. The IVAUTO option allows the user to estimate the beginning SOC conditions based on regression models developed by Burke et al. (1989) that estimate SOC as a function of soil texture and regional climate. The state variables such as total soil organic carbon and nitrogen from this run were then compared with the field data observed in 1970. The observed soil carbon level at the beginning of the experiment was reported as 5210 g C m^{-2}, whereas the CENTURY simulated value was 4995 g C m^{-2}, which is an under simulation by only 4% with respect to the observed value. Soil organic nitrogen at the beginning of the experiment was reported to be 468.9 g N m^{-2}, while CENTURY simulated soil organic nitrogen was 469.1 g N m^{-2}, which is an over estimation by 0.04%.

For both the conventional tillage and no-till experiments conducted at the site between 1970 and 1991, the CENTURY model was run with four input scenarios consisting of four fertilizer application rates (0, 84, 168, 336 kg N ha^{-1}). Output variables analyzed were grain yield, soil organic carbon, grain nitrogen concentration, plant nitrogen uptake, and net carbon input into soil. The analysis of model behavior consisted of a comparison of simulated vs. observed data.

V. Results and Discussions

The results obtained from the simulations are explained in the following order below: 1) corn grain production, 2) SOC, 3) grain N concentration and N uptake by plants, and 4) net carbon input to soils. First the conventional tillage experiment results are presented and they are followed by the no-till results. In the final sub-section the conventional tillage and no-till results are compared.

A. Corn Grain Production

The crop production parameters (crop production and harvest index) were calibrated to observed crop yields under the conventional tillage system, for the highest fertilizer application rate of 336 kg N ha^{-1}. These same parameters were then used for both conventional tillage and no-till and for all fertilization treatments, including the control treatment. This approach was adopted to maintain consistency between tillage levels and to approximate the nitrogen stress induced at the lower nitrogen fertilizer applications. These procedures resulted in generally good agreement between the observed and simulated crop yields for most all of the tillage and fertilizer treatments.

The model over estimated average corn grain yield by approximately 6 to 10% for corn grown under conventional tillage for all N-fertilization levels (Table 1). Simulated grain yields increased with the increasing N fertilizer rates as expected. From Table 1, it is observed that there was a greater variation in annual yield under the control treatment than under the treatments where fertilizer was applied (higher standard deviation from the mean under control treatment). This variation in corn yield for the control treatment suggests that corn production was limited by nitrogen availability for crop growth. For the treatments where fertilizer was applied corn production was not limited by nitrogen availability, however, it may have been limited by the maximum crop production parameter as specified in the crop input file.

Table 1. Average simulated and observed corn yields (kg ha^{-1}, 21 year average, 1970-1991)

Fertilizer rates	----Simulated----		----Observed----		----% Difference----	
	Average	S.D.[a]	Average	S.D.[a]	Average	S.D.[a]
Conventional tillage						
0	4947	924	4619	2004	9.2	35.0
84	7156	679	6752	1991	6.7	27.3
168	7780	699	7103	1856	9.3	24.9
336	7803	714	7316	1891	7.1	24.4
No-Till						
0	5306	979	4247	1279	19.9	26.3
84	6900	744	6762	1808	2.9	26.6
168	6953	782	7553	1688	-8.1	26.8
336	6962	783	7638	1820	-8.8	27.5

[a] S.D. stands for Standard Deviation

Crop production parameters for the conventional tillage system were also used for simulating the no-till experiment. For the no-till experiment the model overestimated the average grain yield by approximately 20% and 3% for the control treatment and for N rate of 84 kg N ha^{-1} respectively (Table 1). The average grain yields corresponding to the high fertilizer applications were underestimated by the model by approximately 8-9%. The simulated grain yields increased with increasing N fertilizer rates, as did the observed data. Additionally, the model showed little or no significant difference between the average grain yields corresponding to the three fertilizer rates of 84, 168 and 336 kg N ha^{-1}.

B. Soil Organic Carbon

In general, for the conventional tillage experiment, simulated soil carbon values for all the fertilizer rates followed the trend of the observed SOC (Figure 1). However, the simulated carbon values corresponding to low fertilizer rates were more in line with the observed data than those at the higher rates, indicating that CENTURY has a greater SOC response for higher fertilization levels (Figure 1). Nearly identical carbon inputs were simulated for the fertilization levels of 168 and 336 kg N ha^{-1} (Table 2) resulting in identical SOC for these two fertilizer rates.

Soil organic carbon values increased with increasing N fertilizer rates. This is due to the resulting increases in crop production which in turn increases the surface residue after harvest. Incorporation of this residue into the soil leads to increasing soil organic matter. This conclusion is justified by the carbon inputs simulated (Table 2), which increase with increasing fertilizer rates. Blevins et al. (1977, 1983b) reported similar observations in their conventional tillage experiments. Organic carbon corresponding to all the fertilizer rates indicated a net SOC loss at the end of the simulation compared with the original pasture soil, with the minimum loss for the highest fertilizer application and vice versa.

A decrease in soil organic matter is generally found when grassland or permanent pasture is cultivated (Allison, 1973). Plowing may increase decomposition rates in the disturbed plow layer and conversion of grassland to annual crops may also result in lower net carbon inputs of carbon to soil due to the export of organic matter in the harvested products. In simulations of long-term field experiments in Sweden, Paustian et al. (1992) also reported that simulated and measured SOM decreased in treatments receiving only crop residues as carbon inputs. The net SOC loss could be due to the fact that plowing increases oxidation and thus the decomposition of organic matter at least in the top 20 cm. Under conventional tillage systems, for the three fertilizer N rates, 84, 168, and 336 kg N ha^{-1}, the model overestimated the carbon values at the end of the simulation by 5 to 10%.

For the no-till experiment for N fertilizer rates, 0, 84 and 168 kg N ha^{-1}, the simulated values appeared to follow the trend of the observed data (Figure 2). Also, the model predictions compared reasonably well with the observed data for 1980 and 1992. There is a concern about the 1990 observed SOC for fertilization levels of 0, 84, and 168 kg N ha^{-1}, as it does not coincide with the field observations made in 1992; the validity of the 1990 data point is being investigated. For the highest N fertilizer rate, the simulated values neither followed the trend of the observed data nor compared well. In addition, the observed data showed an increase in the organic C values from 1980 for the highest fertilizer rate, and the model did not show this response. Soil organic carbon levels corresponding to the three fertilizer rates were higher than those of the control treatment. However, there was not much difference between the simulated C values corresponding to the three N fertilizer rates (84, 168, and 336 kg N ha^{-1}). In fact, they appear to be reaching an equilibrium after 1980.

For no-till systems, simulations predicted a net loss in SOC levels for all the fertilizer rates compared to the original pasture soil; this is due to the fact that cultivation of grassland or permanent pasture results in a loss of soil organic matter (Allison, 1973). The observed data support the above findings indicating a decrease in SOC for the 0, 84, and 168 kg N ha^{-1} fertilizer rates; however, the model results for the 336 kg N ha^{-1} fertilizer level were in the opposite direction. Blevins et al. (1983b) reported that under the no-till system the organic C in the surface increased at the end of a 10-year field experiment. This suggests that the model could be modified to better represent the changes in the organic matter dynamics, especially for the no-till system, possibly by extending the model to a depth of 30 cm and adding a layered soil profile representation. The issues and problems, along with recommended research to better represent no-till systems with the CENTURY model are presented in detail by Donigian et al. (1993).

Application of the CENTURY Soil Organic Matter Model to a Field Site in Lexington, KY.

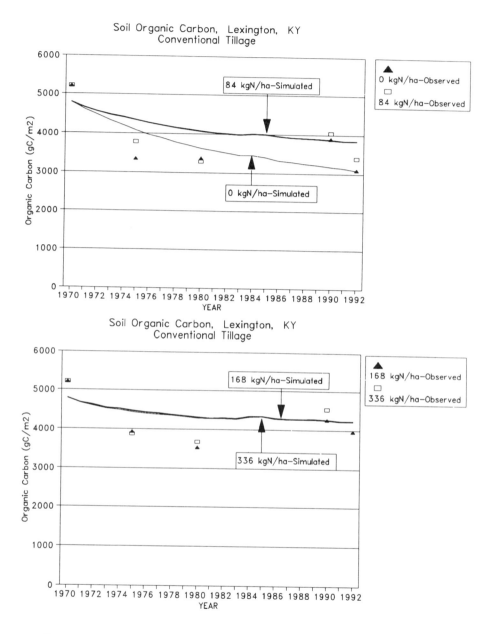

Figure 1. Simulated and observed soil organic carbon in the top 20 cm of the soil profile under conventional tillage.

Table 2. Average carbon inputs and plant nitrogen uptake, predicted by CENTURY

Fertilizer rates Kg N ha^{-1}	Conventional Tillage		No-Till	
	Carbon inputs Kg N ha^{-1}	Plant N uptake Kg N ha^{-1}	Carbon inputs Kg N ha^{-1}	Plant N uptake Kg N ha^{-1}
0	2570	68	2590	65
84	4100	118	3740	110
168	4680	136	3790	120
336	4710	143	3800	121

C. Grain Nitrogen Concentration and N Uptake by Plants

Observed grain nitrogen concentrations data were only available for fertilization levels of 84 and 168 kg N ha^{-1} and for the years 1980 and 1981 only. For the conventional tillage experiment the model underestimated grain nitrogen concentration for 1981; however, there is an excellent agreement between observed and simulated values for 1980 (Figure 3a). Simulated grain nitrogen concentration compared reasonably well with the observed data for the no-till case (Figure 3b). In comparison to the grain nitrogen concentration range of 1.35 to 1.75 reported in the literature (Follet et al., 1991), the model underestimated the grain nitrogen concentrations for all fertilization levels except for a few years of simulation.

Simulated nitrogen uptake by plants increased with increasing N fertilizer rates with the highest uptake corresponding to the highest fertilizer rate. The average plant N uptake and carbon inputs for both tillage systems and for all fertilizer rates are presented in Table 2. Typical corn plant N uptake as reported in the literature ranges from 135-168 kg N ha^{-1} (Tisdale et al., 1985). The observed plant N uptake was reported to be 100-160 kg N ha^{-1}. The plant N uptake under the control treatment for both the tillage experiments was under simulated; however, the model did a reasonably good job for all the other fertilizer rates (Table 2).

D. Net Carbon Input into the Soil

The net carbon input into the soil increased with increasing N fertilizer rates (Table 2). The carbon inputs were approximately one and half times that of the control treatment for the case when fertilizer was applied. This is due to higher N fertilizer rates leading to greater crop growth, which in turn results in higher crop residue after harvest and greater carbon input into the soil upon incorporation. The CENTURY predictions in Table 2 indicate that the carbon inputs under the no-till system were lower than that of conventional tillage systems for the cases when fertilizer was applied. It appears that the no-till model parameters such as those which transfer the crop litter after harvest into various soil pools, need to be reevaluated.

E. Comparison of Conventional Tillage and No-Till Simulations

The simulated average grain yields corresponding to no-till were less than those of conventional tillage for all the fertilizer rates except for the control treatment (Table 1), which is due to the simulation of lower soil temperatures under the no-till system resulting in retarding crop germination and hence crop yield. This result was not evident for the control treatment because the higher crop production parameters (crop production parameters of the highest fertilizer application which resulted in highest yield were used) offset the effect of soil temperature on crop germination. These simulation predictions are in contrast to the field data. Blevins et al. (1986) reported that at low N fertilizer rates, the 10-year average corn yield was higher for conventional tillage than for no-till. However, at moderate to high rates of N fertilization, average corn yield was equal or higher under no-till. As reported by Hill et al. (1973), no-tillage can cause an increase in corn yields by making water more readily available to the plants due to a lower evaporation rate from the soil surface. On the other hand, in a conventional tillage system, the water stress resulting from high evaporation rate in the early growing period could reduce the yields. The model behavior, which does not

Application of the CENTURY Soil Organic Matter Model to a Field Site in Lexington, KY.

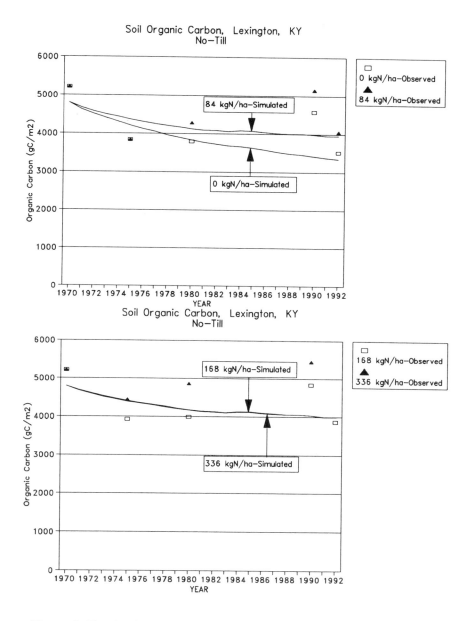

Figure 2. Simulated and observed soil organic carbon in the top 20 cm of the soil profile under no-till.

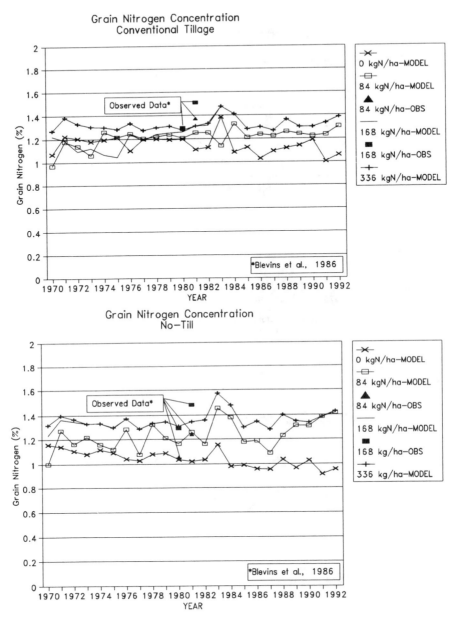

Figure 3. Simulated and observed grain nitrogen concentrations at different fertilizer additions a) conventional tillage, and b) no-till.

coincide with Hill's observations suggests that nutrients and other factors could be impacting the yield levels apart from available water. However, since the predictions generally deviated by less than 10% (mostly) from the respective observed data, for both tillage types, the model is performing adequately in predicting the yield levels, and one must again note that CENTURY model was not developed to predict yearly crop yields.

The organic carbon values for the no-till simulation were greater than those of conventional tillage for the fertilizer rates of zero and 84 kg N ha^{-1}. For the high fertilizer rates of 168 and 336 kg N ha^{-1}, the reverse was true. This is in contrast to literature reports which indicate that tillage operations associated with the conventional (tillage) treatment usually promote more rapid oxidation of organic matter (Blevins et al., 1977) thus resulting in lower organic carbon. As discussed before, lower soil temperature can limit crop production under a no-till system. Since the residue inputs, shoot and root, contribute directly to the surface and soil litter, respectively, a reduction in their quantity would have reduced the SOC levels at high fertilizer rates in the no-till treatment. The tillage parameters used for no-till result in lower transfer of root and shoot litter thus producing lower carbon inputs, and resulting in lower SOC than for conventional tillage.

VI. Summary and Conclusions

The CENTURY model predicted the corn grain yields reasonably well. At high fertilizer rates, the crop production parameter used for CENTURY simulations limits the crop yield and at low fertilizer rates the nutrients (nitrogen) limit crop production. In spite of nutrient limitations on crop production, the CENTURY model predictions, especially the 21-year average yields matched reasonably well with the field data.

In general, the organic C values predicted by the model adequately represented the dynamics of the organic matter in the soil. The field data indicated higher SOC under no-till systems than under conventional tillage systems. The model simulated this response reasonably well for the lower fertilization rates. However, the model simulations for the no-till system when 336 kg N ha^{-1} fertilizer was applied resulted in a decreased SOC, and the field observations contradict this result. From the soil carbon inputs simulated it can be concluded that for both tillage systems, there is a need to evaluate the cultivation parameter values in order to better represent the impacts of tillage types on soil organic matter dynamics.

The grain nitrogen concentrations simulated by the model adequately matched with the observed values. In general, the predicted values were in the range reported by the researchers. The nitrogen uptake by the plants was adequately predicted by the model for the conventional tillage and no-till systems. However, the model under estimated the N uptake for the control treatment.

In conclusion, the CENTURY model can simulate soil organic matter dynamics under various crop-tillage-management scenarios, but needs further improvements and modifications as reported by Donigian et al. (1993). Additionally, the testing results further confirm that the CENTURY model can be an effective tool for assessing carbon sequestration potential of agroecosystems in the central United States.

Acknowledgements

This study was funded by the Environmental Research Laboratory in Athens, GA under EPA Contract No. 68-CO-0019 with AQUA TERRA Consultants, Mountain View, CA. Mr. Tom Barnwell was the Project Officer from the EPA Athens ERL. His support and assistance was critical to the success of the project.

References

Allison, F.E. 1973. *Soil organic matter and its role in crop production*. Developments in Soil Science 3. Elsevier, Amsterdam-London-New York.
Barnwell, T.O.,Jr., R.B. Jackson, IV, E.T. Elliott, I.C. Burke, C.V. Cole, K. Paustian, E.A. Paul, A.S. Donigian, Jr., A.S. Patwardhan, A. Rowell, and K. Weinrich. 1992. An approach to assessment of management impacts on agricultural soil carbon. *Water, Air, and Soil Pollution*. 64:423-435.
Blevins, R.L., G.W. Thomas, and P.L. Cornelius. 1977. Influence of no-tillage and nitrogen fertilization on certain soil properties after 5 years of continuous corn. *Agron. J.* 69:383-386.

Blevins, R.L., M.S. Smith, G.W. Thomas, and W.W. Frye. 1983a. Influence of conservation tillage on soil properties. *J. Soil and Water Conserv.* 38(3):301-305.

Blevins, R.L., G.W. Thomas, M.S. Smith, W.W. Frye, and P.L. Cornelius. 1983b. Changes in soil properties after 10 years continuous no-tilled and conventionally tilled corn. *Soil and Tillage Research.* 3 (1983) 135-146.

Blevins, R.L., J.H. Grove, and B.K. Kitur. 1986. Nutrient uptake of corn grown using moldboard plow or no-tillage soil management. *Commun. Soil Sci. Plant Anal.* 17(4):401-417.

Burke, I.C., C.M. Yonker, W.J. Parton, C.V. Cole, K. Flach, and D.S. Schimel. 1989. Texture, climate, and cultivation effects on soil organic matter contents in U.S. grassland soils. *Soil Sci. Soc. Am. J.* 53:800-805.

Chinnaswamy, R.V., A.S. Patwardhan, and A.S. Donigian, Jr. 1993. Application of the CENTURY model to the Lexington, KY Site. Appendix B of Draft Report Prepared for U.S. EPA, Environmental Research Laboratory, Athens, GA.

Donigian, A.S., Jr., T.O. Barnwell, R.B. Jackson, A.S. Patwardhan, K.B. Weinrich, A.L. Rowell, R.V. Chinnaswamy, and C.V. Cole. 1993. Assessment of alternative management practices and policies on soil carbon in agroecosystems of the central U.S. Draft Report Prepared for U.S. EPA, Environmental Research Laboratory, Athens, GA.

Follett, R.F., D.R. Keeney, and R.M. Cruse. 1991. *Managing nitrogen for groundwater quality and farm profitability.* Soil Science Society of America, Inc., Madison, WI.

Hill, J.D. and R.L. Blevins. 1973. Quantitative soil moisture use in corn grown under conventional and no-tillage methods. *Agron. J.*:945-949.

Metherell, A.K. 1992. Simulation of soil organic matter dynamics and nutrient cycling in agroecosystems. Ph.D. Dissertation. Colorado State University, Fort Collins, Colorado.

Parton, W.J., D.S. Schimel, C.V. Cole, and D.S. Ojima. 1987. Analysis of factors controlling soil organic matter levels in great plains grasslands. *Soil Sci. Am. J.* 51:1173-1179.

Parton, W.J., J.W.B. Stewart, and C.V. Cole. 1988. Dynamics of C, N, P and S in grassland soils: A model. *Biogeochemistry* 5:109-131.

Paustian, K., W.J. Parton, and J. Persson. 1992. Influence of organic amendments and N-fertilization on soil organic matter in long-term plots: model analyses. *Soil Sci. Soc. Am. J.* 56(2):476-488.

Tisdale, S.L., W.L. Nelson, and J.D. Beaton. 1985. *Soil fertility and fertilizers.* 4th Ed. Macmillan Publishing Company, New York, NY. 754 pp.

CHAPTER **33**

The Exchange of Carbon Dioxide between the Atmosphere and the Terrestrial Biosphere in Latin America

C.G.M. Klein Goldewijk and M. Vloedbeld

I. Introduction

Houghton (1991) estimated a global release of carbon to the atmosphere from land use change in 1990 of 1.1-3.6 Pg C yr^{-1}. This number has been adopted by the IPCC for the 1990 estimate (Houghton et al, 1990). Compared with a global carbon flux estimate of 5.5-6.5 Pg C yr^{-1} resulting from fossil fuel combustion, this flux from land conversion is not negligible. This paper describes the development of a global terrestrial carbon model which quantifies the different fluxes of carbon dioxide between the atmosphere and terrestrial biosphere. This model is embedded in an integrated modelling framework of global change, which takes into account the most important land use conversions. The conversions are crucial for accurately assessing terrestrial carbon fluxes. The advantage of such integrated approach is that the consequences of climatic change, changing atmospheric composition and changing land cover are comprehensively taken into account. The implementation of the modelling framework on a high-resolution grid further allows for a regional explicit assessment of these consequences and their linkages.

The objective of this paper is to illustrate some recent developments in carbon cycle modelling (c.f. Esser (1991), Melillo (1993), Vloedbeld and Leemans (1993)). For this paper we have used a preliminary version of our carbon cycle model. We give a detailed model description and several simulations of the carbon budget for Latin America. This region is chosen, because of the relevance of its recent, well documented, land cover changes for the carbon budget. We further estimate future exchanges of carbon dioxide under a reference scenario with different atmosphere-biosphere feedback assumptions.

II. The Terrestrial Environment System

The IMAGE 2.0 model is a multi-disciplinary, integrated model of global climate change, designed to provide a scientifically-based overview of climate change issues to support the evaluation of policies. It consists of three fully linked model sub-systems: industry-energy, terrestrial environment and atmosphere-ocean (Figure 1). Dynamic calculations are performed for a hundred year time horizon. Some calculations are grid-based and others are on world regional level, depending on the sub-model. IMAGE 2.0 has evolved from earlier versions of the IMAGE model (Rotmans, 1990) and from the ESCAPE model (CEC, 1992). These models made progress in linking socio-economic, biogeochemical and climatological factors together in a single model for evaluating scenarios for greenhouse gas emissions and climate change.

The Terrestrial Carbon model is part of the TES of the IMAGE 2.0 model. The TES consists of a linked set of five models for *Agricultural Demand, Terrestrial Vegetation, Land Use Change, Land Use Emissions, and Terrestrial Carbon.* The core of the system is the Land Use Change model, which takes into account two important influences on future land cover: (1) future *demands* for all kinds of land, coming from demographic/economic trends, and (2) future *potential* for land, resulting from changes in local climate and

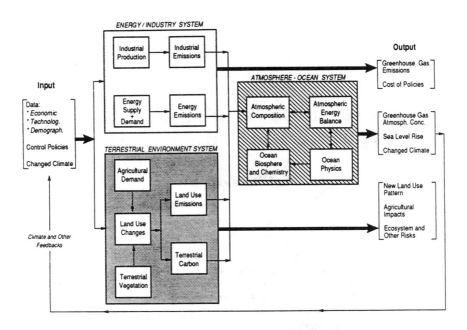

Figure 1. Framework of models and linkages of IMAGE 2.0.

atmospheric composition. Demand for land is calculated by the Agricultural Demand model which uses socio-economic variables as input and calculates the demands for agricultural products (including lumber and livestock) on a world regional level. Potential for land is computed on a 0.5° latitude by 0.5° longitude grid, on the basis of local climate, topography, and soil conditions. This is done by the Terrestrial Vegetation model and is expressed as *potential* vegetation and crop productivity (Leemans and Solomon, 1993). The demand and potential for land are combined in a regionally explicit, rule-based Land Use Change model. The rules specified in this model, attempt to capture current understanding and important driving factors of global land use change. For the purpose of the global land cover simulations, initial land cover is divided into categories based on the global database of Olson et al (1983). Olson's 47 categories were aggregated into 17 new categories, in order to make dynamic calculations more tractable.

The computed future land cover pattern is then used by the Land Cover Emissions model and the Terrestrial Carbon model to compute emissions to the atmosphere of gases that contribute to radiative forcing such as CO_2, CH_4, NO_x and other indirect greenhouse gases like CO and VOC's. The future land cover patterns are also used by the Terrestrial Carbon model to compute carbon fluxes between the biosphere and the atmosphere.

III. The Terrestrial Carbon Model

A. Overview

The Terrestrial Carbon model computes the carbon fluxes of the terrestrial environment. The model is partly a grid-scale adaption of the terrestrial part of the global carbon cycle model proposed by Goudriaan and Ketner (Goudriaan and Ketner, 1984) with many modifications, as will be explained. The driving force of the terrestrial biosphere is the net primary production (NPP), which results in a net uptake of carbon dioxide from the atmosphere. It is the result of gross primary production (radiant energy that is transformed by photosynthesis into organic compounds), minus the losses of autotrophic (plant) respiration (Waring and

Schlesinger, 1985). The global carbon cycle can be conceptually divided into different compartments, each consisting of carbon pools that store carbon at different rates and quantities, and for different residence times. The most important carbon pools in the terrestrial biosphere are phytomass (=living biomass), litter, humus and stable humus or charcoal. Total phytomass estimates range from 400 Pg C to 830 Pg C (Atjay et al., 1979; King et al., 1992; Schlesinger, 1977; Whittaker and Likens, 1975). The litter pool is estimated to range from 50 Pg C to 100 Pg C (Esser, 1991; Schlesinger, 1977). Humus and charcoal are often taken together as soil carbon; this large carbon pool ranges from 1390 Pg C to 1700 Pg C (Atjay et al., 1979; Bouwman, 1989; Esser, 1991; Post et al., 1982). Carbon is released to the atmosphere when carbon pools decrease. This is represented by the net ecosystem production (NEP), which takes into account natural processes like biomass decomposition and soil respiration. Another source of carbon dioxide is human-induced activities such as land cover changes, which are explained in more detail in part C.

B. Net Ecosystem Production

Net ecosystem production is the result of the net primary production minus the natural decay of all carbon pools. There is not much information about net primary production of natural vegetation types. Data given by e.g. Whittaker and Likens (1975), Larcher (1980); (Olson et al., 1983; Raich et al., 1991) show considerable differences (Table 1), some of which can be explained by the use of different classifications. The initial net primary production (NPPI) for 1970 used in IMAGE 2.0 is characteristic for each land cover type. NPPI is corrected at a later time period for temperature, precipitation, CO_2-concentration, altitude, and soil characteristics.

Net primary productivity is partitioned among four main plant components (leaves, branches, stemwood, and roots), according to fixed distribution coefficients p_{lk}, where l is landcover type (1-17) and k is the plant component number (1-4). Initial parameters are listed in Table 2. The increase of biomass in ecosystems is represented by a simple logistic function after (Cooper, 1983) and (Dewar, 1990; Dewar, 1991; Dewar and Cannell, 1992). This function is used here to describe the increase in net primary production, and consequently the increase in biomass, given by:

$$y_{g,t} = \frac{1}{1 + b \cdot e^{-r(t-t_0)}} \qquad (1)$$

with $y_{g,t}$ the fractional increase in net primary production in grid cell g at time t. The parameter b is a scaling factor for carbon storage, depending on the initial carbon content of a land cover type (b_o) and the carbon content of old-growth land cover types (b_m); it is calculated as $(b_m-b_o)/b_o$. The relative growth rate of carbon storage in trees is represented by r.

Initial net primary production is reduced or enhanced by a so-called CO_2 fertilization factor (β) (Bastacow and Keeling, 1973; Goudriaan and Ketner, 1984). This factor is computed by combining several coefficients related to temperature, water availability, nutrient availability, plant species, and altitude (Fitter and Hay, 1981; Parton et al., 1987; Vloedbeld and Leemans, 1993), for each grid cell and is given by:

$$\beta_{g,t} = \gamma_{g,t}^{tot} \cdot \beta_i \qquad (2)$$

with β_i the initial value of the beta factor for C_3 crops (C_3 is defined below) at 25°C, under good nutrient supply. The reference value of beta (0.7) was based on CO_2 fertilization experiments (Goudriaan and de Ruiter, 1983; Luxmoore and Baldocchi, 1993)). γ^{tot} is expressed as:

$$\gamma_{g,t}^{tot} = \gamma_{g,t}^{tmp} \cdot \gamma_{g,t}^{wat} \cdot \gamma_g^{alt} \cdot \gamma_{g,t}^{spe} \cdot \gamma_g^{nut} \qquad (3)$$

The mean β values for each land cover type as calculated by IMAGE 2.0 are given in Table 3. The values correspond well with data (ranges) given by other authors, based on many experiments or modelling efforts (Bazzaz, 1990; Kienast and Luxmoore, 1988; Luxmoore and Baldocchi, 1993; Polglase and Wang, 1992).

Table 1. Net primary production estimates of different authors, and IMAGE 2.0 estimate (g C/m² yr)

	Lieth (1973)	Whittaker & Likens (1975)	Atjay et al. (1979)	Larcher (1980)	Olson et al. (1983)	Waring & Schlesinger (1985)	Goudriaan & Ketner (1984)	Raich et al. (1991)
Agricultural land		290	430	290	550	325	430	
Ice, Polar desert		0		1	0		0	
Cool (semi-)desert	30	30	60	40	70	35	100	
Hot desert		0	5	1	25	1	70	
Tundra	30	65	90	65	130	70	70	
Cool grass shrub	280	225	350	205	390	250	570	470
Warm grass shrub	280	315	350	405	430	350	570	470
Xerophytic woods	450	270	490	405	360	300	510	110
Boreal forest	450	360	380	360	490	400	510	
Cool conifer forest	450	585	380	360	490	400	510	725
Cool mixed forest	450	540	660	540	600	650	510	725
Deciduous forest	450	540	660	540	600	650	510	725
Warm mixed forest	450	540	660	585	600	650	510	435
Trop. dry forest/savanna	280	315	790	405	430	350	770	435
Trop. seasonal forest	675	675	710	720	550	1000	770	700
Trop. rain forest	675	900	1020	720	780	1000	770	1190
Wetlands		1125	470		125-1440	1000	770	

Table 1. continued--- Net primary production estimates of different authors, and IMAGE 2.0 estimate (g C/m² yr)

	Mc Guire et al. (1992)	King et al. (1992)	Melillo et al. (1993)	Minimum	Maximum	Mean	IMAGE 2.0 (1994)
Agricultural land		290		290	550	372	400
Ice, Polar desert		10		0	10	2	0
Cool (semi-)desert	65	30	90	30	100	55	50
Hot desert		30	50	0	70	23	50
Tundra	120	65	120	30	270	83	100
Cool grass shrub	200	225	215	205	570	307	300
Warm grass shrub	425	315	335	280	570	386	400
Xerophytic woods	110		130	110	510	314	400
Boreal forest	535	360	240	240	535	409	500
Cool conifer forest	535	360	240	240	585	431	500
Cool mixed forest	650	540	670	450	725	594	600
Deciduous forest	650	540	620	450	725	590	600
Warm mixed forest	725	585	740	450	740	615	600
Trop. dry forest/savanna	435	315	390	280	790	447	400
Trop. seasonal forest	700	675	870	550	1000	731	800
Trop. rain forest	1050		1100	675	1100	921	900
Wetlands				125	1440	673	500

Table 2. Parameters used in the Terrestrial Carbon submodel of IMAGE 2.0

Nr. landcover type	Partitioning coefficients (unitless)				Lifetime (yr)						
	Leaf	Branch	Stem	Root	Leaf	Branch	Stem	Root	Litter	Humus	Charcoal	H.F.[a]	C.F.[b]
Agricultural land	0.8	0.0	0.0	0.2	1	1	1	1	1	20	1000	0.2	0.05
Ice, Polar desert	0.5	0.1	0.1	0.3	1	10	50	10	5	100	1000	0.4	0.05
Cool (semi-)desert	0.5	0.1	0.1	0.3	1	10	50	10	5	100	1000	0.4	0.05
Hot desert	0.5	0.1	0.1	0.3	1	10	50	10	3	50	1000	0.4	0.05
Tundra	0.5	0.1	0.1	0.3	1	10	50	10	5	100	1000	0.4	0.05
Cool grass shrub	0.6	0.0	0.0	0.4	1	10	50	3	1	60	1000	0.4	0.05
Warm grass shrub	0.6	0.0	0.0	0.4	1	10	50	3	1	20	1000	0.4	0.05
Xerophytic woods	0.3	0.2	0.3	0.2	1	10	50	10	1	50	1000	0.4	0.05
Boreal forest	0.3	0.2	0.3	0.2	3	30	50	10	5	60	1000	0.4	0.05
Cool conifer forest	0.3	0.2	0.3	0.2	3	20	50	10	3	50	1000	0.4	0.02
Cool mixed forest	0.3	0.2	0.3	0.2	1	20	50	10	1	40	1000	0.4	0.05
Deciduous forest	0.3	0.2	0.3	0.2	1	20	50	10	1	40	1000	0.4	0.05
Warm mixed forest	0.3	0.2	0.3	0.2	1	20	50	10	1	40	1000	0.4	0.05
Trop. dry forest/savanna	0.3	0.2	0.3	0.2	2	10	30	5	1	20	1000	0.4	0.05
Trop. seasonal forest	0.3	0.2	0.3	0.2	2	10	30	8	1	20	1000	0.4	0.05
Trop. rain forest	0.3	0.2	0.3	0.2	2	10	30	8	1	20	1000	0.4	0.05
Wetlands	0.3	0.2	0.3	0.2	2	10	30	8	1	60	1000	0.4	0.05

[a]Humification factor (unitless); [b]Carbonization factor (unitless).

Table 3. Mean calculated CO_2 fertilization factors

Land cover type	β 1970	β 2050
Agricultural land	0.49	0.64
Ice	0.03	0.04
Cool semidesert	0.32	0.39
Hot desert	0.05	0.09
Tundra	0.21	0.28
Cool grass	0.44	0.53
Warm grass	0.09	0.11
Xerophytic woods	0.26	0.32
Taiga	0.24	0.30
Cool conifer	0.31	0.38
Cool mixed	0.28	0.35
Temperate deciduous	0.29	0.37
Warm mixed	0.44	0.54
Tropical dry/savanna	0.35	0.41
Tropical seasonal	0.50	0.58
Tropical rain	0.54	0.60
Wetlands	0.44	0.53
World	0.28	0.35

The effect of temperature on CO_2 fertilization rate (γ^{tmp}) is shown by several experiments (Idso et al, 1987; Sage and Sharkey, 1987). Depending on plant species, CO_2 fertilization may not be apparent below a certain minimum temperature.

The effect of water availability on the CO_2 fertilization rate (γ^{wat}) is related to the water use efficiency of plants. This effect is described using a Priestly-Taylor coefficient, which is the ratio of actual evapotranspiration to equilibrium evapotranspiration, assessed over a full year (Prentice et al, 1992).

The response of plant growth to CO_2 fertilization is also influenced by altitude (γ^{alt}). Roughly, the optimum temperature for net photosynthesis decreases by 5 °C if altitude rises with 2000 m (Fitter and Hay, 1981). Because atmospheric CO_2 pressure decreases with altitude, plants at high altitude are more CO_2 limited. Therefore they can profit more from CO_2 fertilization than low altitude plants. For a doubling of CO_2, Körner and Diemer (1987) found that plants growing at 2600 m had more than twice the photosynthetic response of plants at 600 m.

The role of CO_2 fertilization is most strongly related to the photosynthetical pathway of plants; C_3 or C_4 (γ^{spc}). In general, C_3 plants show a greater increase in growth with increasing CO_2 concentration than C_4 plants (Körner and Diemer, 1987; Larcher, 1980). In this study a simple approach has been taken, by dividing grasses into C_3 and C_4 species, and distinguishing evergreen trees from deciduous trees.

The effect of nutrient availability on CO_2 fertilization is expressed by γ_{nut}. Goudriaan and de Ruiter (1983) found that the biotic growth factor decreased to nearly half its original value if plants suffered from a nitrogen deficiency and dropped to zero when phosphorus was also limiting (Goudriaan and de Ruiter, 1983). Nutrient availability in soils is taken into account by assigning values "high", "medium", or "low" to each grid cell, based on the FAO/UNESCO Soil Map of the World (FAO, 1987).

Temperature also has a direct influence on the processes of photosynthesis and respiration. Both are stimulated by temperature increase, but their response varies within different temperature ranges (Fitter and Hay, 1981; Larcher, 1980) (Figure 2 + 3). Photosynthesis begins at a minimum temperature just below 0 °C for most mid-latitude plants, and its response increases rapidly until an optimum is reached (which is specific for each land cover type, assuming no other factor is limiting). Beyond this point, gross photosynthesis declines rapidly. Photorespiration begins at a low rate, but increases exponentially with temperature. At a certain maximum temperature, gross photosynthesis equals photorespiration, so that the CO_2 uptake equals CO_2 respiration. Further temperature increase would result in a net release of CO_2. The feedbacks used here are described in more detail in Vloedbeld and Leemans (1993). The above-described direct influence of temperature on photosynthetic rates is given by the coefficient $\sigma_{g,t}$.

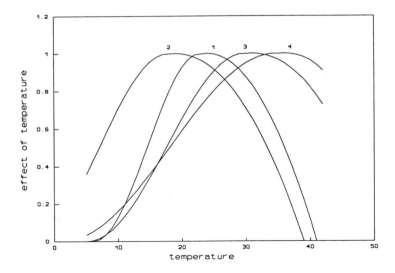

1 = agricultural land
2 = boreal forest
3 = tropical rain forest
4 = warm grassland

Figure 2. Assumed effect of temperature on plant growth for some land cover types in IMAGE 2.0. (After Fitter and Hay, 1981.)

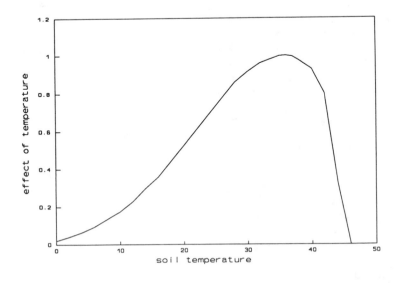

Figure 3. Assumed effect of soil temperature on soil decomposition rates. (After Parton et al., 1987.)

Following the formulation of Goudriaan and Ketner (1984) we propose:

$$NPP_{g,t} = NPPI_{g,l} \cdot y_{g,t} \cdot \sigma_{g,t}^{npp} \cdot (1 + \beta_{g,t} \cdot \ln(C_t/C_i)) \tag{4}$$

with $NPP_{g,t}$ the corrected net primary production in grid cell g at time t and $NPPI_{g,l}$ is the initial NPP which is characteristic for land cover type l in grid cell g. C_t is the actual carbon dioxide concentration at time t and C_i the initial carbon dioxide concentration.

In natural undisturbed ecosystems the carbon flux from each plant component equals its content divided by its average lifespan τ. Leaves, branches, and stemwood decay normally to the litter pool. The root biomass is transferred gradually to the humus pool. The resulting equation for the change of carbon of plant biomass component B is:

$$\frac{dB_{g,k}}{dt} = p_{l,k} \cdot NPP_g - \frac{B_{g,k}}{\tau(B_{g,k})} \tag{5}$$

For litter (L) the equation is:

$$\frac{dL_{g,k}}{dt} = \sum_{k=1}^{3} \frac{B_{g,k}}{\tau(B_{g,k})} - \frac{L_g}{\tau(L_g)} \cdot \phi_{g,t} \tag{6}$$

In this equation, $\phi_{g,t}$ represents the effect of temperature and water availability on soil respiration. Decomposition rates of litter, humus and charcoal in soil increase nearly linearly with water availability up to a threshold value. The threshold value is reached if the ratio between monthly precipitation and potential evapotranspiration is about 1.25 (Parton et al., 1987). If precipitation exceeds this value, soil moisture is no longer a limiting factor for decomposition. The Priestly-Taylor coefficient is used to describe the relationship between soil decomposition rates and water availability (Parton, 1987) (Figure 4).

In boreal and arctic regions the lifetime of litter is about 1-5 years, in temperate areas 1-3 years and in the tropics 1 year. Only a fraction λ_l is transferred to the humus pool. Humus (H) decays much slower than litter, and a fraction δ enters a charcoal pool of relatively stable carbon (K) that contains recalcitrant humus, charcoal, and other forms of elementary carbon (Bouwman, 1989; Kortleven, 1963). The equation for humus is:

$$\frac{dH_g}{dt} = \lambda_l \cdot \left(\frac{B_{g4}}{\tau(B_{g4})} + \frac{L_g}{\tau(L_g)} \cdot \phi_{g,t} \right) - \frac{H_g}{\tau(H_g)} \cdot \phi_{g,t} \tag{7}$$

The equation for charcoal is:

$$\frac{dK_g}{dt} = \delta \cdot \left(\frac{H_g}{\tau(H_g)} \cdot \phi_{g,t} \right) - \frac{K_g}{\tau(K_g)} \cdot \phi_{g,t} \tag{8}$$

Following the definition given at the beginning of this section, the net ecosystem production is represented by:

$$NEP_{g,t} = NPP_{g,t} - (1-\lambda_l) \cdot \left(\frac{B_{g4}}{\tau(B_{g4})} + \frac{L_g}{\tau(L_g)} \cdot \phi_{g,t} \right) - (1-\delta) \cdot \left(\frac{H_g}{\tau(H_g)} \right) \cdot \phi_{g,t} - \frac{K_g}{\tau(K_g)} \cdot \phi_{g,t} \tag{9}$$

where the terms to the right of NPP represent the carbon flux due to soil respiration. Making the common assumption that NEP from undisturbed ecosystems with relatively stable climates is zero, we can compute

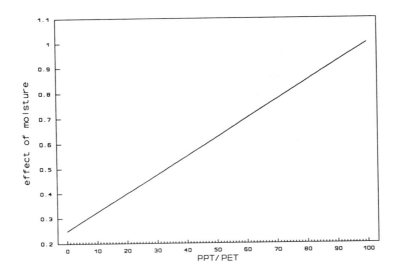

Figure 4. Assumed effect of water availability on soil decomposition rates (approximated from Parton et al., 1987). PPT/PET is the ratio of actual to equilibrium evapotranspiration, assessed over a full year.

initial values of B_{gk}, L_g, H_g, and K_g. However, changing climate and land conversion disrupts the balance between NPP and soil respiration, and results in a net carbon flux to the atmosphere. Land use conversion may result in an additional carbon flux due to biomass burning. The following section describes how these processes are included in the model.

C. Land Conversions.

1. Forest to Agricultural Land, Forest to Grassland

Deforestation is known to have a great impact on the carbon cycle of deforested land. For example, carbon is released to the atmosphere when forests are cut down and the unharvested remainder is burnt at the site. The process of shifting cultivation, by which areas are used for agriculture for a short period and then abandoned and left for regrowth, causes deforestation in the tropics. Important factors are the expansion of crop producing areas (Africa, Asia) and rangeland (Latin America). When a forest is cut down, the Terrestrial Carbon model assumes that a fraction of the aboveground biomass goes to the charcoal pool, a fraction to the wood products pool, another fraction to the humus pool, and the remainder is emitted to the atmosphere. The fraction of biomass going to the charcoal pool is 0.01 for leaves, branches, and litter, and 0.02 for stems (pers. comm. J.M. Robinson). The root biomass and the humus pool of the old land cover type are converted into the humus pool, of the new land cover type. Also the charcoal pool of the old land cover type enters the same pool of the new land cover type. All carbon pools will then have the pre-assigned lifetimes of the new land cover types, which leads to changes in the rate of decomposition and release of CO_2. When forest is converted into agricultural land or grassland, the model computes that NEP will be strongly negative just after the conversion, because of the accelerated rate of soil decomposition, and after some years will decrease in magnitude but remain negative. Converted land will then act as a source of CO_2. In tropical regions the flux will be smaller than in temperate regions because of the smaller amount of soil carbon present in the converted land, and the longer lifetime of humus.

2. Grassland to Agricultural Land.

In warm grasslands all the leaves are presumed to be burnt, with a small fraction ending up in the charcoal pool. In cool grasslands, all the leaf biomass is assumed to enter the litter pool. In both grassland types all the root biomass enters the humus pool. The model computes that net primary production in grasslands is more than 95% of its maximum in the first year after the conversion. Litter decomposition progresses somewhat slower so that the net ecosystem production is positive for a short time period. The NEP will be negative in the following years after the conversion, but gradually decrease to near zero.

3. Agricultural Land to Forest

The computed net ecosystem production of a forest is negative in the first years after the conversion. The net primary production of a new forest is still very low, while the decomposition of the soil carbon present still continues. When the increase in plant biomass accelerates, NEP becomes positive, resulting in an uptake of CO_2. The converted land acts as a sink for CO_2 and will remain doing so for a long time, despite the fact that after many years the NPP is at its steady state level. This is because soil decomposition rates tend to adjust slowly to the aboveground changes, and it can take many years before the decomposition rates are adjusted to the new environmental circumstances.

4. Agricultural Land to Grassland

The model computes that net primary productivity is at its maximum in one year, while the litter decomposition starts rapidly afterwards. Therefore, the NEP is positive in the beginning, becomes negative after some years and will progress eventually to zero in the long term.

IV. Preliminary Results and Discussion

For the input variables of the TES simulations we made the following assumptions. We used actual population and personal income data for Latin America for the period 1970-1990, and assumed a 82% increase in population and a increase of 242% in personal income for the period 1990-2050. These increases correspond to the lower growth scenario of the World Bank (World Bank, 1987). We also assumed an increase of CO_2 from pre-industrial levels to 570 ppm in 2050. Regional temperature, precipitation, and other climatic parameters were based on the IIASA climate database (Leemans and Cramer, 1991) and a climate change scenario generated with the Geophysical Fluid Dynamics Laboratory General Circulation Model (GFDL-GCM; (Manabe and Wetherald, 1987)). The original GFDL-GCM results were corrected for the actual atmospheric CO_2 concentration to obtain a transient climate change. This corresponds to a calculated global mean Δt of 4.0 °C. The regional temperature increase in Latin America was 0.1-1.0 °C.

In Table 4, the IMAGE 2.0 land cover types are listed with their number of grid cells (base year 1970), areas, total net primary production and living biomass. Calculations for Latin America gave an initial total NPP of 10.8 Pg C yr^{-1} in 1970, which is close to the 1990 estimate of Raich et al (1991) of 12.5 Pg C yr^{-1}. Differences can be explained by the use of a different classification and area distribution. Especially the agricultural area used in IMAGE 2.0 with its relative low NPP contributes already 10% of the total NPP.

In Figure 5, the net carbon flux to the atmosphere is given as a result of land cover changes in Latin America. The period 1970-1975 is used for initialization, starting with the assumption that the ecosystems were in a steady state in 1970. Of course, this is not the case for most of the ecosystems involved. Currently, work is being done to initialize the model from 1900 until now, in order to take the history of the land cover changes into account.

Table 4. Assumed initial state for Latin America in 1970

IMAGE classes	Nr. grid cells	Area M ha	Total NPP Pg C yr^{-1}	Phytomass Pg C
Agricultural land	1017	287.6	1.15	1.15
Ice, polar desert	23	4.7	0.00	0.00
Cool (semi-)desert	321	70.8	0.04	0.26
Hot desert	191	53.3	0.03	0.20
Tundra	102	28.2	0.03	0.27
Cool grass shrub	2	0.6	0.00	0.00
Warm grass shrub	1599	457.7	1.83	3.28
Xerophytic woods	519	149.4	0.60	11.08
Boreal forest	0	0.0	0.00	0.00
Cool conifer forest	0	0.0	0.00	0.00
Cool mixed forest	24	4.7	0.03	0.58
Deciduous forest	26	5.0	0.03	0.61
Warm mixed forest	378	106.1	0.64	13.04
Tropical dry/savanna	859	259.2	0.91	5.83
Tropical seasonal forest	981	298.2	2.39	30.78
Tropical rain forest	871	269.1	2.42	31.25
Natural wetland	274	78.2	0.70	9.29
Total	7187	2072.8	10.80	107.62

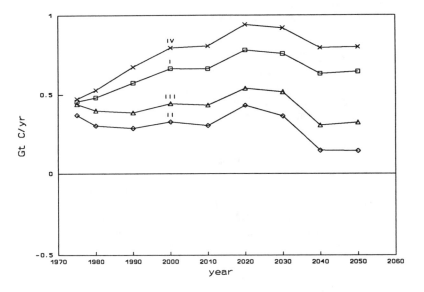

I   = no feedbacks, only land cover changes

I   = land cover changes with CO_2 fertilization effect

III   = land cover changes with direct temperature effect on plant growth

IV   = land cover changes with effect of temperature and water availability on soil respiration

Figure 5. Net carbon flux to the atmosphere from Latin America, as result of land cover changes, combined with individual feedback mechanisms.

The fluctuations in the carbon fluxes due to human disturbances (e.g. deforestation) show a rather discontinuous pattern. This may be partly caused by the calculation routine of the Land Use Change model. Another reason for the discontinuity could be the assumptions made in the Agricultural Demand model. When, according to the scenario, the increase in population slows down next century, or the amount of agricultural land eventually is enough to satisfy the food demand, the rates of land conversion also decrease, and this would result in lower carbon fluxes.

To examine the effect of only land cover changes the model has been run with no feedbacks included (Run 1). Other runs have been made with direct temperature effect on plant growth, temperature effect on CO_2 fertilization, and a combined effect of temperature and water availability on soil respiration.

A. Run 1: Only Land Cover Changes

The total net primary productivity decreased from 10.5 Pg C yr^{-1} in 1975 to 9.2 Pg C yr^{-1} in 2050, which is a loss of 12%. This loss is probably due to lower productivity of land cover types that replace forests. After deforestation soil respiration continues, while the above-ground productivity of the new land cover type is much smaller than the previous one. This results in a negative net ecosystem production, meaning a loss of carbon to the atmosphere ranging from 0.1 Pg C yr^{-1} to 0.3 Pg C yr^{-1} during the simulation period. Combined with the carbon released by burning immediately after conversion, the total flux to the atmosphere ranges from 0.4 Pg C yr^{-1} to 0.8 Pg C yr^{-1} during the simulation period, with a maximum around 2020.

B. Run 2: Land Cover Changes with only CO_2 Fertilization (β)

As result of the CO_2 fertilization effect the net primary production increased from 10.7 Pg C yr^{-1} in 1975 to 12.0 Pg C yr^{-1} in 2050, a rise of 12%. Although this increase is reduced every time forests are converted into grasslands or agricultural land, the CO_2 fertilization effect is still sufficient enough to stimulate the biomass production. The total carbon flux to the atmosphere is 0.3 Pg C yr^{-1} in the 70s, rises to 0.4 Pg C yr^{-1} until 2020 and decreases again to almost zero in 2050. This means that in spite of the CO_2 fertilization effect, the immediate flux resulting from land cover changes tends to exceed the enhanced carbon uptake by the ecosystems.

C. Run 3: Land Cover Changes with only Direct Temperature Effect on Plant Growth (σ)

The increase in temperature stimulated the net primary production in 2050 by 5% to 11.1 Pg C yr^{-1}. Total net ecosystem production is relatively small as compared to the other feedbacks, and fluctuates around zero. The overall carbon flux to the atmosphere of (0.3-0.5 Gt yr^{-1}) is therefore strongly related to the direct carbon fluxes resulting from land use changes.

D. Run 4: Land Cover Changes with only Effect of Temperature and Water Availability on Soil Respiration (ϕ)

An increase of temperature usually will enhance the soil respiration. Depending on the availability of water in the soil, this enhancement can be stimulated or inhibited. The model showed that the NPP decreased by 12% to 9.2 Pg C yr^{-1}. This is the result of the conversions of high productive forests to (relative) low productive agricultural lands or grasslands. This feedback effect can enhance decomposition in such a way that, in combination with the decreased NPP, the net ecosystem production is negative for the whole simulation period. This means an extra CO_2 flux from the terrestrial biosphere into the atmosphere. The overall carbon flux to the atmosphere is the largest of all individual feedbacks ranging from 0.5 to 0.9 Pg C yr^{-1}.

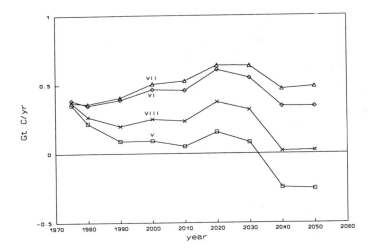

V   = land cover changes with CO_2 fertilization effect + direct temp.effect on plant growth

VI = land cover changes with CO_2 fertilization effect + effect of temperature and water availability on soil respiration

VII = land cover changes with direct temperature effect on plant growth + effect of temperature and water availability on soil respiration.

VIII = land cover changes with CO_2 fertilization effect + direct temperature effect on plant growth + effect of temperature and water availability on soil respiration.

Figure 6. Net carbon flux to the atmosphere from Latin America, as result of land cover changes, combined with multiple feedback mechanisms.

E. Run 5: Land Cover Changes with Combined CO_2 Fertilization Effect and Direct Temperature Effect on Plant Growth

The model computed a relative strong increase in total NPP from 10.8 to 14.2 Pg C yr^{-1} (Figure 6). This increase of 31% is also reflected in the NEP, which is ranging from 0.2 to 0.7 Pg C yr^{-1} during the simulation period. This positive net ecosystem production means an uptake of CO_2 from the atmosphere, so the total net carbon flux to the atmosphere is reduced by these combined feedbacks (ranging from 0.1 to -0.3 Gt yr^{-1}). After 2030 the NEP is even larger than the flux resulting from biomass burning, which is reflected in the total carbon flux becoming negative.

F. Run 6: Land Cover Changes with Combined CO_2 Fertilization Effect and Temperature and Water Availability Effect on Soil Respiration

The relative strong influence of the CO_2 fertilization effect on net primary production greatly outweighs the effect of climate change on soil respiration. These combined effects resulted in a net increase in 2050 in the NPP of 12% from 10.7 to 12.0 Pg C yr^{-1}. The net ecosystem production however, is tempered by these opposite forces and fluctuates around zero during the simulation period. The total carbon flux to the atmosphere ranges from 0.3 to 0.6 Pg C yr^{-1}.

Table 5. Difference in total net C-flux under different NPPI's, compared to a run with reference NPPI

	Test 1	Test 2
No land use changes, no feedback (ref.)	0%	0%
No land use changes, with feedback	-17%	+18%
Land use changes, no feedback	-25%	+25%
Land use changes, with feedback	-36%	+35%

G. Run 7: Land Cover Changes with Direct Temperature Effect on Plant Growth, and Effect of Temperature and Water Availability on Soil Respiration

For this case, the model computed the largest total carbon flux to the atmosphere of all combinations of two or more feedbacks. Total net primary production is hardly stimulated (4% increase to 11.1 Pg C yr^{-1}), and the net ecosystem production is negative during the whole simulation period. The total carbon flux ranges from 0.4 to 0.7 Pg C yr^{-1}.

H. Run 8: Land Cover Changes with CO_2 Fertilization Effect, Direct Temperature Effect on Plant Growth, and Effect of Temperature and Water Availability on Soil Respiration

The total carbon flux to the atmosphere is obviously tempered by the combination of the three feedbacks, when compared to the fluxes with only land use changes involved. The release of CO_2 is enhanced by the temperature effect on soil respiration. However, this is more than compensated by the increased uptake of CO_2 resulting from the CO_2 fertilization effect and the temperature effect on plant growth. Compared with run 1 (only land cover changes), the total carbon flux to the atmosphere dropped by ca. 0.3 Pg C yr^{-1}, ranging from 0.1 to 0.4 Pg C yr^{-1}.

Summarizing the results of all the different runs: CO_2 fertilization is the strongest negative feedback. The temperature effect on plant growth is also negative, but of less importance for this region. The effect of temperature and water availability on soil respiration is a positive feedback, but is does not compensate the other two feedbacks. Combining all feedbacks we found that the CO_2 fertilization effect is somewhat tempered by enhanced soil respiration. This is also emphasized by other authors, e.g. Polglase and Wang (1992), Raich and Schlesinger (1992), and Luxmoore and Baldocchi (1992). Despite all these ecophysiological feedbacks, we found that land cover changes still have the most pronounced effect on the carbon dynamics of the terrestrial biosphere.

Other GCM results show similar patterns: All these climate models predict an increase in global temperature between 1.5 °C and 4.5 °C under doubling of atmospheric CO_2, combined with a global increase in precipitation. However, there are large differences in the predicted temporal and spatial precipitation patterns between the models. Regionally there can thus be great variability in climatic change and this should show up in the regional carbon budgets. However, integrating over a longer period globally should give little differences between GCMs. This difference should become even smaller when also the changes in land cover are taken into account.

I. Sensitivity Analysis

Some extra runs were done with β_i varying plus and minus 50%. The variation caused a difference in the total net carbon flux to the atmosphere of plus and minus 46%. It is clear that the estimate of the value for β_i is very important. It is arguable whether this value can be applied to all the other land cover types, but as we already correct this initial value to the local circumstances it seems to be justifiable.

Changes in NPPI can have a significant effect on the total net carbon flux to the atmosphere. Therefore two extra runs were made with varying NPPI. The results are given in Table 5. The observed and/or measured NPP data are given in Table 1 for different land cover types. When compared to the values used in IMAGE most of the data are close to the average of the different estimates, and all of them are within the range given.

J. Interactions

In the reference situation (no feedbacks), a total of 31.1 Pg C is released by deforestation activities during the simulation period 1970-2050. Another 20.3 Pg C is released by natural processes (uptake minus respiration), so that the total net carbon flux to the atmosphere is 52.4 Pg C. As result of the CO_2 fertilization effect, the release through deforestation is slightly increased to 34.5 Pg C, because more carbon is stored in the converted biomass. On the other hand, no carbon is released in natural processes, but instead a 8.9 Pg C is sequestered. The total C flux to the atmosphere is now 25.6 Pg C, which is a decrease with more than 50% with the reference case. The temperature effect on plant growth released 33.4 Pg C by deforestation. Only 0.6 Pg C is released by natural processes. The total flux in that case is 34 Pg C, a decrease of 35%. The effect of temperature and water availability on soil respiration released a total of 32.6 Pg C through deforestation, and another 38 Pg C is extra released by enhanced decomposition. The total C flux amounts now 60.6 Pg C, which is an increase of 16%. When all feedbacks are included, the total release through land conversions in 2050 are 36.3 Pg C, and 10.9 Pg C is sequestered. The total C flux to the atmosphere during the period 1970-2050 is 25.4 Pg C, this is a decrease of more than 50% compared with the refenerence scenario.

V. Summary and Conclusions

1) Land conversions, such as deforestation, must be taken into account when looking at carbon fluxes between the atmosphere and the biosphere.
2) The CO_2 fertilization effect is the strongest negative feedback of the three feedbacks studied here.
3) The direct temperature effect on plant growth is a relative small negative feedback for Latin America. This may arise because of the relatively small increases in regional temperature given by the climate change scenario used in this paper. This effect could be of more importance at high latitudes because of the larger temperature increase computed there by GCM's.
4) The effect of temperature and water availability on soil respiration is a positive feedback, thus enhancing the greenhouse effect. The carbon fluxes involved however are not as large as those by the CO_2 fertilization effect.
5) In general, soil respiration will probably increase in the future and release more carbon to the atmosphere. On the other hand the increasing CO_2 fertilization effect, in combination with the direct temperature effect on plant growth, will compensate for this additional release.
6) The CO_2 fertilization effect causes an extra uptake of carbon dioxide which results in an increase in biomass. This increase results also in slightly larger deforestation fluxes, but the magnitude of these fluxes are smaller than the increase in biomass.
7) This integrative model approach allows for a precize quantification of the different processes and their linkages. Such process based model should be preferred over more empirical models, because it enhances our understanding of the whole system.

Acknowledgements

We like to thank J. Alcamo and R. Leemans (both at National Institute of Public Health and Environmental Protection, The Netherlands) for stimulating discussions and critical review of the manuscript. Furthermore we like to thank G. Zuidema and E. Kreileman for development and maintenance of the model. This work is supported by the Dutch National Research Programme (NRP) grant no. 851042 and no. 851045.

References

Atjay, G.L., P. Ketner, and P. Duvigneaud (eds.). 1979. Terrestrial primary production and phytomass. Scope Report nr.13. *The global carbon cycle*. John Wiley & Sons, NY.

Bastacow, R. and C.D. Keeling. 1973. Atmospheric carbon dioxide and radiocarbon in the natural carbon cycle. II. Changes from AD 1700 to 2070 as deduced from a geochemical model. In: G.M. Woodwell and E.V. Pecan (eds.). *Carbon and the biosphere.* p. 86-135. U.S. Atomic Energy Commission. Washington D.C.

Bazzaz, F.A. 1990. The response of natural ecosystems to the rising global CO_2 level. *Annu. Rev. Ecol. Syst.* 21:167-196.

Bouwman, A.F. 1989. The role of soils and land use in the greenhouse effect. *Netherlands Journal of Agricultural Science* 37:13-19.

CEC. 1992. (Commission of the European Communities), Directorate General for Environment, Nuclear Safety, and Civil Protection. Development of a framework for the evaluation of policy options to deal with the greenhouse effect.

Cooper, C.F. 1983. Carbon storage in managed forests. *Canadian Journal of Forest Research* 13:155-166.

Dewar, R.C. 1990. A model of carbon storage in forests and forest products. *Tree Physiology* 6:417-428.

Dewar, R.C. 1991. Analytical model of carbon storage in the trees, soils and wood products of managed forests. *Tree Physiology* 8 (3):239-258.

Dewar, R.C. and M.G.R. Cannell. 1992. Carbon sequestration in the trees, products and soils of forest plantations: an analysis using UK examples. *Tree Physiology* 11:49-71.

Esser, G. 1991. Osnabruck Biosphere Model: structure, construction, results. In: G. Esser and D. Overdieck (eds.). *Modern Ecology* p. 679-709. Elsevier, Amsterdam.

FAO. 1987. Report on the agro-ecological zones project. Vol. 3. Methodology and results for South and Central America. Food and Agricultural Organization of the United Nations, Rome. World Soil Report 48/3. 251 pp.

Fitter, A.H. and R.K.M. Hay. 1981. *Environmental physiology of plants.* Academic Press, London.

Goudriaan, J. and H.E. de Ruiter. 1983. Plant growth in response to CO_2 enrichment, at two levels of nitrogen and phosphorus supply. 1. Dry matter, leaf area and development. *Neth. J. Agric. Sci.* 31:157-169.

Goudriaan, J. and P. Ketner. 1984. A simulation study for the global carbon cycle including man's impact on the biosphere. *Climatic Change* 6:167-192.

Houghton, J.T., G.J. Jenkins and J.J. Ephraums (eds.). 1990. *Climate change: the IPCC scientific assessment (I).* Cambridge University Press, Cambridge.

Houghton, R.A. 1991. Tropical deforestation and atmospheric CO_2. *Clim. Change* 19:99-118.

Idso, S.B., B.A. Kimball, M.G. Andserson, and J.R. Mauney. 1987. Effects of CO_2 enrichment on plant growth: The interactive role of air temperature. *Agri. Eco. Environment* 20:1-10.

Kienast, F. and R.J. Luxmoore. 1988. Tree ring analysis and conifer growth responses to increased atmospheric CO_2 levels. *Oecologica* 76:487-495.

King, A.W., W.R. Emanuel, and W.M. Post. 1992. A dynamic model of terrestrial carbon cycling response to land-use change. Pers. comm.

Körner, C. and M. Diemer. 1987. In situ photosynthetic responses to light, temperature and carbon dioxide in herbaceous plants from low and high altitude. *Functional Ecology* 1:179-194.

Kortleven, J. 1963. *Kwantitatieve aspecten van humusopbouw en humusafbraak.* Thesis. Landbouwhogeschool, Wageningen.

Larcher, W. 1980. *Physiological plant ecology.* 2nd Ed. Springer-Verlag, Berlin.

Leemans, R. and W.P. Cramer. 1991. The IIASA database for mean monthly values of temperature, precipitation and cloudiness on a global terrestrial grid. Research Report RR-91-18, International Institute of Applied Systems Analysis, Laxenburg.

Leemans, R. and A. Solomon. 1993. The potential change in yield and distribution of the earth's crops under a warmed climate. Pers. comm.

Luxmoore, R.J. and D.D. Baldocchi. 1993. Modelling interactions of carbon dioxide, forests, and climate. In: Woodwell, G.M. (ed.). *Biotic feedbacks in the global climatic system: will the warming speed the warming.* Oxford University Press, Woods Hole, USA.

Manabe, S. and R.T. Wetherald. 1987. Large scale changes in soil wetness induced by an increase in carbon dioxide. *J. Atm. Sci.* 44:1211-1235.

Melillo, J.M., A.D. McGuire, D.W. Kicklighter, B. Moore III, C.J. Vorosmarty, and A.L. Schloss. 1993. Global climate change and terrestrial net primary production. *Nature* 363:234-239.

Olson, J.S., J.A. Watts, and L.J. Allison. 1983. *Carbon in live vegetation of major world ecosystems.* ESD Pub. Oak Ridge National Laboratory, TN. No. 1997.

Parton, W.J., D.S. Schimel, C.V. Cole, and D.S. Ojima. 1987. Analysis of factors controlling soil organic matter levels in Great Plains grasslands. *Soil Sci. Soc. Am. J.* 51:1173-1179.

Polglase, P.J. and Y.P. Wang. 1992. Potential CO_2-enhanced Carbon Storage by the Terrestrial Biosphere. Pers. comm.

Post, W.M., W.R. Emanuel, P.J. Zinke and A.G. Stangenberger. 1982. Soil carbon pools and world life zones. *Nature* 298:156-159.

Prentice, I.C., W. Cramer, S.P. Harrison, R. Leemans, R.A. Monserud, and A.M. Solomon. 1992. A global biome model based on plant physiology and dominance, soil properties and climate. *Journal of Biogeography* 19:117-134.

Raich, J.W., E.B. Rastetter, J.M. Mellilo, D.W. Kicklighter, P.A. Steudler, B.J. Peterson, A.L. Grace, B. Moore III, and C.J. Vörösmarty. 1991. Potential net primary productivity in South America: application of a global model. *Ecol. Appl.* 1 (4):399-429.

Raich, J.W. and W.H. Schlesinger. 1992. The global carbon dioxide flux in soil respiration and its relationship to vegetation and climate. *Tellus* 44B:81-99.

Rotmans, J. 1990. *IMAGE, an integrated model to assess the greenhouse effect*. Kluwer Academic Publishers, Dordrecht.

Sage, R.F. and T.D. Sharkey. 1987. The effect of temperature on the occurrence of O_2 and CO_2 insensitive photosynthesis in field grown plants. *Plant Physiol.* 84:658-664.

Schlesinger, W.H. 1977. Carbon balance in terrestrial detritus. *Ann. Rev. Ecol. Syst.* 8:51-81.

Vloedbeld, M. and R. Leemans. 1993. Quantifying feedback processes in the response of the terrestrial carbon cycle to global change: The modelling approach of IMAGE-2. *Wat. Air Soil Pol.* 70:615-628.

Waring, R.H. and W.H. Schlesinger. 1985. *Forest ecosystems, concepts and management*. Academic Press, Inc. Harcourt Brace Jovanovich, London.

Whittaker, R.H. and G.E. Likens. 1975. Primary productivity of the biosphere. In: Lieth, H. and R. Whittaker (eds.). *The biosphere and man*. Springer Verlag, Berlin.

World Bank. 1987. World development report 1987. Oxford University Press, NY.

List of Symbols

Symbol	Description	Units
$B_{g,k}$	Carbon in biomass component k of grid cell g	Mg C ha^{-1}
H_g	Carbon in humus in grid cell g	Mg C ha^{-1}
K_g	Carbon in long-living soil carbon in grid cell g	Mg C ha^{-1}
L_g	Carbon in litter in grid cell g	Mg C ha^{-1}
$NPP_{g,t}$	Net primary production of grid cell g, at time t	Mg C ha^{-1}yr^{-1}
$NEP_{g,t}$	Net ecosystem production of grid cell g, at time t	Mg C ha^{-1}yr^{-1}
k	Biomass component number (1 = leaf, 2 = branch, 3 = stem, 4 = root)	--
l	Landcover number (1-17)	--
p_{lk}	Fraction of NPP partitioned to component k in land cover type l	--
$\beta_{g,t}$	CO_2 fertilization factor in grid cell g at time t	--
λ_l	Humification factor in land cover type l	--
δ	Carbonization factor of humus upon decomposition	--
τ	Lifetime	yr
$y_{g,t}$	Fractional increase of NPP in grid cell g, at time t	--
b	Scaling parameter of carbon storage	Mg C ha^{-1}yr^{-1}
b_m	Carbon content of old-growth land cover types	Mg C ha^{-1}
b_o	Carbon content of land cover type after conversion	Mg C ha^{-1}
$\phi_{g,t}$	Effect of temperature and water availability on soil respiration, in grid cell g at time t	--
$\sigma_{g,t}$	Temperature effect on plant growth, in grid cell g at time t	--
γ^{tmp}	Effect of temperature on CO_2 fertilization	--
γ^{wat}	Effect of water availability on CO_2 fertilization	--
γ^{alt}	Effect of altitude on CO_2 fertilization	--
γ^{spe}	Effect of plant species on CO_2 fertilization	--
γ^{nut}	Effect of nutrient availability on CO_2 fertilization	--

CHAPTER **34**

Modeling the Dynamics of Organic Carbon in a Typic Haplorthod

Marcel R. Hoosbeek and Ray B. Bryant

I. Introduction

Carbon dioxide (CO_2) is thought to be the major contributor, an estimated 50%, to greenhouse gas warming because of its abundance, high annual emission, and long atmospheric residence. An understanding of the dynamics of organic carbon in the soil system is an integral part of the study of CO_2 contributions to global change. In his concluding remarks to the conference "Soils and the Greenhouse Effect", held in Wageningen, The Netherlands, Bretherton (1989) stated: "The principal, indeed the only, way that scientists of different disciplines can communicate effectively in a thoroughly multi-disciplinary program like this is through the use of quantitative modelling." Bouwman (1989) provided an overview of models developed for calculating plant residue decomposition rates and soil organic matter levels: 1) Models based on first order decomposition rates constant in time, e.g. Jenny et al. (1949), Greenland and Nye (1959), Kortleven (1963), and Olsen (1963); 2) Models with decomposition rates variable in time, e.g. Janssen (1984), and 3) Models with different fractions of soil organic matter (SOM), each having a different decomposition rate, e.g. Jenkinson and Rayner (1977), Van Veen and Paul (1981), Parton et al. (1987), and Jenkinson et al. (1991).

Van Breemen and Feijtel (1989) presented a case study of forested and reclaimed Ferrasols in Surinam to illustrate the application of simulation models to estimate levels of soil organic carbon (SOC) under steady state conditions and under conditions of changing land use. The model, developed by Feijtel and Meijer (1989), simulates soil water and organic carbon dynamics for 4 soil layers, each characterized by thickness, texture, moisture content permeability, and root percentage. Mean monthly precipitation, temperature, and potential evapotranspiration are the driving input variables.

In a recent paper, Hoosbeek and Bryant (1992) presented a framework for the classification of soil process models based on relative degree of computation, complexity, and level of organization. They described models at the catena and soil region (i + 2 and i + 3) levels (e.g. CENTURY model, Parton et al., 1987), at the pedon (i) level (e.g. Van Breemen and Feijtel, 1989), and at the horizon (i - 1) level (e.g. the ORTHOD model described in this paper). Many models (e.g. the CENTURY model, Parton et al., 1987, and the model used by Van Breemen and Feijtel, 1989) are both quantitative and largely based on functional relationships. The carbon version of the pedodynamic ORTHOD model, used in this paper, is quantitative as well, but is based on a more mechanistic approach.

"Pedodynamics" was defined as: "The quantitative integrated simulation of physical, chemical, and biological soil processes acting over short time increments in response to environmental factors" (Hoosbeek and Bryant, 1994a). The ORTHOD model is a pedodynamic model consisting of newly developed and existing soil process submodels. The objective of the model is to simulate the chemistry and movement of organic carbon, aluminum, silica, and other major chemical components for each horizon in Spodosols. The individual submodels were described elsewhere by Hoosbeek and Bryant (1994a); the submodels used for the carbon version of the ORTHOD model will be described briefly in this paper. The carbon version of the ORTHOD model is not intended to simulate major changes in soil profile development over long periods of time but rather to simulate the production of CO_2 and dissolved organic carbon (DOC), the movement of DOC through the studied profiles, and the changes in SOC for each horizon under modern climate conditions for a limited number of years.

II. Materials and Methods

A detailed study site was installed in the Adirondack Mountains, New York. The site is located 21 km northwest of the village of Tupper Lake at an elevation of 458 m. The combination of climate and vegetation, predominantly Red Spruce (Picea rubens) and Balsam Fir (Abies balsamea), classified as Boreal forest. The parent material consists of deep sandy glacial outwash deposited shortly after the glaciers retreated some 11,000 years ago (New York State Museum/Geological Survey, 1991). The soil is a well expressed Typic Haplorthod (sandy, mixed, frigid) and classifies as an Adams series.

Two pits were dug, about ten meters apart, for soil characterization and collection of minimally disturbed samples for use in laboratory experiments. Three sets of tensiometers, soil solution samplers, thermistors, and platinum electrodes were installed in 1990. Each set consisted of 6 tensiometers (Oa, E, Bhs, Bs, BC, and C horizons), 5 soil solution samplers (Oa, E, Bhs, Bs, and C horizons), 6 thermistors (Oi, Oa, E, Bhs, Bs, and BC horizons), and 3 platinum electrodes (Oi, Oa, and E horizons). One set was installed adjacent to the "north" pit, the two other sets were installed at each side of the "south" pit. Open field and canopy throughfall precipitation collectors were made of heavy PVC pipe (length: 1 m; o.d: 0.22 m) in which an assembly of a 4 L bottle and a funnel (effective catchment diameter: 0.20 m) was placed. A nylon mesh screen in the top of the collection bottle prevented vegetation litter and insects from entering the bottle. One precipitation collector was installed in an open spot in the vegetation approximately 50 m from the monitoring site. Four throughfall collectors were installed on the site adjacent to the monitoring sets.

The soil profiles of each pit were described according to USDA-SCS guidelines. Bulk samples from each horizon were collected for particle size analyses, organic carbon determination, several extractions, and X-ray analyses. Particle size percentages were determined with standard sieves and the pipet method (NSSL, 1991). Organic Carbon percentages were determined with a Leco Corporation induction furnace.

Ring samples were carefully excavated from each horizon to measure bulk densities and establish moisture retention curves. The ring samples were equilibrated on pressure plates (Soil Moisture Corporation) at pressures of 0.1, 0.2, 0.3, 0.5, 1.0, 2.0, 3.0, 5.0, and 15 bar. Measurement of the water content at each pressure resulted in θ-h relations for each horizon.

Precipitation, throughfall, and soil solution samples were collected every 4 weeks during the 1991 and 1992 frost-free seasons. The solution samples were transported and stored at low temperatures (1-4 °C) to minimize microbial processes. All major cations were determined with an ICP. Organic and inorganic carbon was analyzed with an IO Corporation Carbon Analyzer.

The ORTHOD model simulated soil processes per layer with a thickness of 0.025 m. Each layer was assumed to be physically and chemically homogeneous. Each profile was divided into 44 layers (1.1 m depth) in which each soil horizon consisted of two or more layers. Chemical fluxes and budgets were calculated for soil volumes (layers) with a square area of 1.0 m^2 and a thickness of 0.025 m. The submodels of the carbon version of the ORTHOD model are discussed in the following paragraphs.

A. Water Movement

The movement of water in the soil is fundamental to any quantitative mechanistic pedogenetic model since water plays a major role in transfer and transformation processes in soils. Realistic representation of saturated and unsaturated flow in a typically non-isotropic medium like soil is a complex problem. Several deterministic models have been developed that simulate the flow of water in the unsaturated zone based on Darcy's law and the continuity principle:

$$q = -K\frac{\delta H}{\delta z} \; (cm \; day^{-1}) \qquad \frac{\delta \theta}{\delta t} = -\frac{\delta q}{\delta z} \; (day^{-1})$$

Combination yields a partial differential equation in terms of hydraulic head, called the Richard's equation:

$$\frac{\delta \theta}{\delta t} = \frac{\delta}{\delta z}[K\frac{\delta H}{\delta z}]$$

Using the pressure head form of the flow equation and the differential moisture capacity (C), defined as $C = d\Theta/d\psi$, the flow equation for predicting water movement in layered soils is:

$$\frac{\delta \varphi}{\delta t} = \frac{1}{C(\varphi)}\frac{\delta}{\delta z}[K(\varphi)(\frac{\delta \varphi}{\delta z} - 1)]$$

The sandy soil at the study site permits the use of a model based on the Richard's equation. The LEACHM model (Wagenet and Hutson, 1987) uses a numerical solution to the Richard's equation and was selected for use in the ORTHOD model.

Real-time weather data were obtained from the nearby Tupper Lake Sunmount weather station. Undisturbed core samples of each horizon were used to determine bulk densities and water retention curves. The water retention curves were established by equilibrating the undisturbed cores on pressure plates at pressures ranging from 10 kPa to 1500 kPa. Measurement of the water content at each pressure resulted in θ-h relations for each horizon. The RETFIT program (Hutson and Cass, 1987) was used to fit the retention data and to calculate the "a" and "b" parameters of Campbell's equation (Campbell, 1985) for use in LEACHM. With the θ-h relations defined for each horizon, LEACHM was calibrated by adjusting the saturated hydraulic conductivities of each horizon. In the LEACHM and ORTHOD models the soil profile is divided into layers of equal thickness, 25 mm, whose properties are assumed to be concentrated at the nodes (a horizon may have several layers). Simulated values gave a good fit for measured values (Hoosbeek and Bryant, 1994a). Given precipitation input, this submodel gives a relatively good representation of soil moisture conditions and water movement in all horizons at any point in time and over the full range of soil moisture conditions.

B. DOC Movement

The LEACHM model (Wagenet and Hutson, 1987) uses a numerical solution to the convection-dispersion equation to estimate chemical fluxes. Solute movement consists of three components: convective, diffusive, and dispersive transport. Combining the equations of these three processes leads to the following expression (van Genuchten and Wierenga, 1986) for solute flux (J_s):

$$J_s = -\theta D(\delta C/\delta x) + qC$$

where θ is the volumetric water content, C is the volume-averaged solute concentration, x is distance, q is the volumetric fluid flux density, and D is the summation of the molecular diffusion and mechanical dispersion coefficients. Both the diffusion and dispersion coefficients can be assumed to be negligible for a sandy soil with a relative large downward convective flow throughout the year. The solute movement between two layers is then dominantly convective flow, which is the product of the volumetric fluid flux density and the volume averaged solute concentration (Wagenet, 1986):

$$J_s = qC$$

This simplification significantly reduces computation time and does not cause significant error when using relatively thin layers (25 mm) in the simulations.

The submodel calculates daily DOC fluxes for each layer. For downward flow, the DOC in-flux of layer n is calculated by multiplying the DOC concentration of the overlaying layer (n + 1) by the water flux across the boundary of the n + 1 and n layers. DOC out-flux is obtained from the C_{DOC} of layer n and the flux across the n and n - 1 boundary.

C. Soil Temperature

Heat transfer and soil profile temperatures are simulated in LEACHM as described by Tillotsen et al. (1980). The heat flow equation is:

$$\rho C_p \delta T / \delta t = \delta(K_t(\theta)\delta t/\delta z)/\delta z$$

where ρ is bulk density (kg.m^{-3}), C_p is gravimetric heat capacity of the soil (J.m^{-3}.°C^{-1}), T is temperature (°C), t is time (s), z is depth (m), and $K_t(\theta)$ is the thermal conductivity of the soil (J.m^{-1}.s^{-1}.°C^{-1}) at water content θ (m^3.m^{-3}). The volumetric heat capacity is calculated from:

$$\rho C_p = \rho_s C_s + \theta \rho_w C_w$$

where ρ_s and C_s are the bulk density and the gravimetric heat capacity of the solid phase and ρ_w and C_w are the density and the gravimetric heat capacity of the liquid phase.

Mean weekly air temperatures and temperature amplitudes were calculated from daily air temperatures and from daily minimum and maximum air temperatures for input to the submodel. Simulated temperature profiles gave a good fit for the measured soil temperatures. Simulated temperature profiles are used in the microbial decomposition submodel to adjust respiration rates.

D. Microbial Decomposition of Organic Matter

Microbial decomposition is considered to take place in several pools (Bohn, 1985; Parton et al., 1987), e.g: structural and metabolic plant remains, active soil organic carbon (SOC) (decomposing plant residues, live microbes), slow SOC (microbial metabolites), passive SOC (humified material), and dissolved organic carbon (DOC). The general equation for the rate of decomposition per pool within a soil environment is:

$$dC_x/dt = K_x * \theta * T * C_x$$

where
- C_x = Carbon state variable of pool x
- K_x = decomposition rate for pool x
- θ = volumetric water content
- T = soil temperature

The decomposition products are $CO_2 + H_2O$, SOC flowing to other pools, and DOC. The ratio in which these products are produced depends on the volumetric water content.

Field measurements of redox potentials indicated aerobic conditions throughout the year. A laboratory experiment was designed to determine rates of microbial CO_2 and DOC production in horizons of the Typic Haplorthod at our field monitoring site. Soil from the Oi, Oa, E, and Bh/Bhs horizons was air-dried for one day, weighed, and promptly rewet to minimize damage to the microbial population. Per horizon, one series of samples was used to measure the production of CO_2, and an other series was used to measure both the CO_2 and DOC production. Samples were rewet or quickly dried (by flushing with dry compressed air) to $\theta = 0.7$, $\theta = 0.5$, $\theta = 0.3$, and $\theta = 0.1$ moisture. The samples used to measure the 'CO_2 + DOC' production were prepared by leaching extensively with deionized water using a mechanical vacuum extractor (Concept Eng. Inc., Lincoln, NE) until no DOC could be detected. During the experiment, CO_2 produced by microbial respiration was flushed out daily for 5 minutes with CO_2-free compressed air and trapped in a 1.0 N NaOH solution. The CO_2 concentration in these NaOH solutions, trapped as CO_3^{2-}, was measured every 7 days with a OI Corporation, Model 700, carbon analyzer. The 'CO$_2$ + DOC' samples were also flushed with deionized water every 7 days to collect the DOC. Blanks were included in all series.

Initially (day 1 through 14), the respiration rates were low. Microbial populations then grew exponentially from day 15 through 35. Following day 35, the microbial populations reached stable sizes and CO$_2$ and DOC respiration rates were steady.

CO$_2$ respiration rates of the Oi horizon (day 71 through 91) were greatest at a volumetric water content of about 0.3 (m^3·m^{-3}) and were relatively lower at higher and lower water contents (Hoosbeek and Bryant, 1994a). Reduced oxygen availability and reduced physiological activity may explain the reduced rates for the respectively wetter and drier samples. CO$_2$ respiration rates of the Oa, E and Bh/Bhs horizons indicated similar trends, although to a lesser extent. The following polynomials were obtained:

$$dCO_{2,Oi}/dt = 38.2 + 89.8\ \theta - 138.8\ \theta^2 \qquad r^2 = 0.99$$
$$dCO_{2,Oa}/dt = 10.0 + 22.0\ \theta - 9.4\ \theta^2 \qquad r^2 = 0.82$$
$$dCO_{2,E}/dt = 2.9 + 2.1\ \theta - 3.8\ \theta^2 \qquad r^2 = 0.97$$
$$dCO_{2,Bh/Bhs}/dt = 2.1 + 10.7\ \theta - 11.9\ \theta^2 \qquad r^2 = 0.50$$

where $dCO_{2,x}/dt$ is the CO$_2$ respiration rate (μg C.day^{-1}.g^{-1} soil). DOC production rates could be described with linear equations (Hoosbeek and Bryant, 1994a):

$$dDOC_{Oi}/dt = 2.9 + 19.1\ \theta \qquad r^2 = 0.88$$
$$dDOC_{Oa}/dt = 4.0 + 4.1\ \theta \qquad r^2 = 0.97$$
$$dDOC_{E}/dt = 1.6 + 3.2\ \theta \qquad r^2 = 0.92$$
$$dDOC_{Bh/Bhs}/dt = 11.0 + 3.7\ \theta \qquad r^2 = 0.85$$

where $dDOC_x/dt$ is the DOC production rate (μg C·day^{-1}·g^{-1} soil)

These CO$_2$ and DOC production rate equations were established with respiration data measured at 20 °C. For use in the microbial decomposition submodel, these rates need to be adjusted for the particular soil temperature of each simulated layer. A Q_{10}-type temperature response is assumed for the microbial activity, with Q_{10}, the factor by which the rate constant changes over a 10 °C interval, equal to two (Johnsson et al., 1987; Hutson and Wagenet, 1992). The temperature correction factor, $T_{corr.}$, is calculated as:

$$T_{corr.} = Q_{10}^{0.1(Tsoil-Tbase)}$$

where Q_{10} is assumed to be 2, Tsoil is the soil temperature (°C), and Tbase = 20 °C.

E. DOC Adsorption

DOC adsorption is described by the Freundlich equation (Bohn et al., 1985; Sposito, 1989):

$$A = K*C^n$$

where A is the weight of adsorbate per unit weight of adsorbent (kg C/kg soil), C is the equilibrium concentration of adsorbate in solution (kg C/m^3), and K and n are empirical constants. The linear form of the equation, with n = 1, is used in the submodel. The K value for each horizon was obtained through calibration.

III. Results and Discussion

Soil characterization data of the north and south pits are summarized in Tables 1 and 2. Measured dissolved organic carbon (DOC) concentrations of rain and canopy throughfall samples collected during the 1991 and 1992 seasons are presented in Figure 1. All DOC concentrations of canopy throughfall samples are significantly higher than for samples collected in the open field and show seasonal variation. Variations

Table 1. Characterization of north and south soil pits

Horizon	Depth (cm)	Texture class	Sand (%)	Silt (%)	Clay (%)	O.M. (%)
North						
Oi	7-0					
Oa	0-5		71.2	19.0	9.8	39.0
E	5-13	S	88.9	9.7	1.4	1.6
Bh	13-18	CoS	91.5	6.4	2.1	7.7
Bhs	18-23	CoS	91.9	6.6	1.5	5.6
Bs	23-47	S	97.5	1.7	0.8	1.2
2BC	47-63	CoS	96.6	2.8	0.6	1.1
3BC	63-83	CoS	99.7	0.1	0.2	0.1
3C	83-137+	CoS	99.6	0.4	0.0	0.1
South						
Oi	10-0					
Oa	0-9	LCoS	80.4	18.4	1.2	37.5
E	9-19	S	85.8	12.7	1.5	1.2
Bh	19-24	S	86.3	11.7	2.0	8.0
Bhs	24-29	S	88.1	10.4	1.5	8.3
Bs	29-45	S	87.8	10.2	2.0	4.6
BC	45-73	S	98.4	1.4	0.2	1.0
2Bsm	73-79	CoS	97.3	2.5	0.2	1.1
2BC/C	79-107+	CoS	98.7	1.0	0.3	0.2

Table 2. Soil pH of north and south profiles

Horizon	pH (H_2O)	pH (KCl)	pH ($CaCl_2$)
North			
Oa	3.50	2.45	2.87
E	3.81	3.06	3.04
Bh	3.81	3.20	3.04
Bhs	4.02	3.54	3.55
Bs	4.61	4.39	4.34
2BC	4.57	4.30	4.27
3BC	4.52	4.40	4.33
3C	4.40	4.54	4.36
South			
Oa	3.39	2.83	3.36
E	3.79	2.92	3.01
Bh	3.76	3.07	3.11
Bhs	4.09	3.50	3.50
Bs	4.59	4.13	4.08
BC	4.63	4.39	4.32
2Bsm	4.63	4.30	4.24
2BC/C	4.64	4.55	4.38

among the throughfall collectors are due to differences in canopy morphology. These DOC concentrations are used as input to the upper boundary of the DOC movement submodel.

DOC measured in soil solutions of the three sites are presented in Figures 2, 3, and 4. The seasonal variations are well pronounced in the O and E horizons, are dampened in the Bh/Bhs horizons, and are nearly absent in the BC and C horizons.

Modeling the Dynamics of Organic Carbon in a Typic Haplorthod

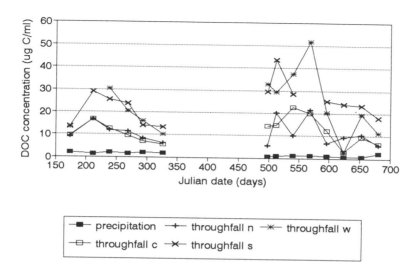

Figure 1. DOC concentrations in precipitation and canopy throughfall collectors (north, west, central, and south).

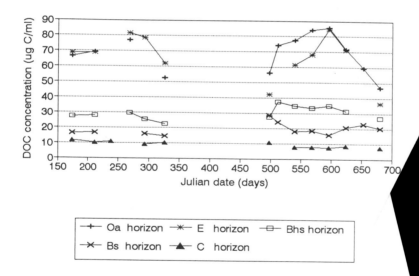

Figure 2. Soil solution DOC concentrations of the north side.

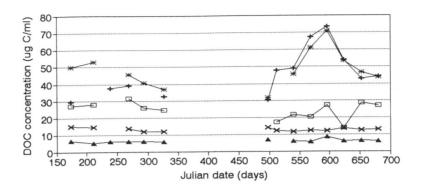

Figure 3. Soil solution DOC concentrations of the central site.

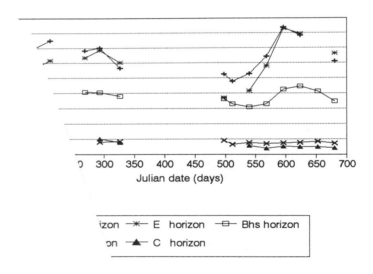

concentrations of the south side.

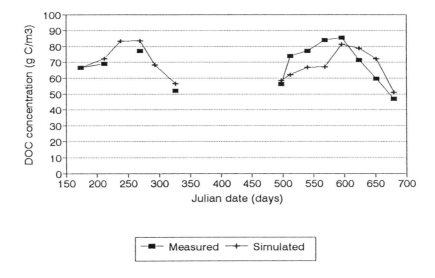

Figure 5. Measured and simulated DOC concentrations of the Oa horizon, north site.

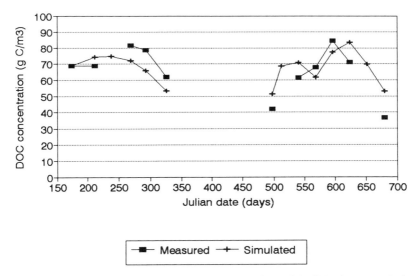

Figure 6. Measured and simulated DOC concentrations of the E horizon, north site.

Results of the simulated DOC concentrations of the Oa, E, and Bh/Bhs horizons of the north site are presented in Figures 5, 6, and 7. Simulated DOC concentrations provide a good fit to the measured data for all horizons of the three monitoring sites.

The adsorption coefficients for the Freundlich adsorption isotherms differ considerably for the Oi/Oa, E, Bh/Bhs, and Bs horizons within each profile, but are in close range across the three sites (Table 3). The O horizons retain a part of the DOC by adsorption. Only a small amount of the DOC passing through the E horizon is adsorbed. Most of the DOC is adsorbed in the Bh/Bhs horizons. Van Breemen and Feijtel (1989) suggested that DOC concentrations in drainage water from Spodosols and Histosols would be high compared to other soil groups. However, the B horizons of the Typic Haplorthod investigated in this study

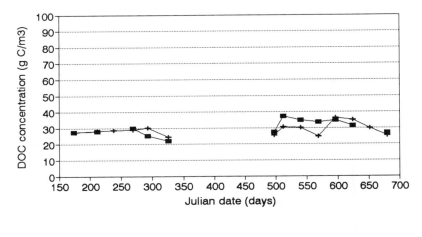

Figure 7. Measured and simulated DOC concentrations of the Bh/Bhs horizons, north site.

Table 3. Adsorption coefficients, K, for the Freundlich equation; $A = K*C_{DOC}$ with A (kg C/kg soil) and C_{DOC} (kg C/m^3 solution)

Horizon	North site	Central site	South site
Oi/Oa	60.0E-06	90.0E-06	70.0E-06
E	13.0E-06	17.0E-06	13.0E-06
Bh/Bhs	140.0E-06	145.0E-06	115.0E-06
Bs	1.5E-06	4.0E-06	10.0E-06

were found to be effective barriers to DOC leaching causing relatively low DOC concentrations in the C horizons. Cumulative DOC drainage (kg C m^{-2}) at 1.10 m depth is presented in Figure 8. Carbon drainage in the form of HCO_3^- was assumed to be negligible because of the low base status and pH of the soils. The north profile loses 2.1 10^{-3} kg C m^{-2} y^{-1} through drainage while the central and south profiles lose 0.4 10^{-3} kg C m^{-2} y^{-1}. These differences may be explained by differences in chemistry and adsorption of the Bs horizons. Organically and inorganically bound aluminum in the Bs of the north pit (north profile) are respectively 18.0 and 15.2 mg per g soil (Hoosbeek and Bryant, 1994b). The Bs of the south pit (central and south profiles) has respectively 108.0 and 33.0 mg $Al_{org.}$ and $Al_{inorg.}$ per g soil. The higher aluminum contents of the central and south Bs horizons resulted in higher adsorption coefficients (Table 3) and more DOC adsorption.

SOM from litter and roots was added to the O horizons throughout the year. SOC percentages of the O horizons remained constant over the years but fluctuated during the seasons due to enhanced microbial activity during the summer. The E horizons of the three sites lost a net average of 0.047% C per year, the Bh and Bhs horizons lost 0.053 and 0.063 % C. The Bs horizons are slowly enriched in SOC, an average of 0.003%, by an in-flux of DOC and subsequent adsorption.

CO_2 is assumed to diffuse into the atmosphere over time. Simulated cumulative CO_2 production rates for the north, central, and south profiles are respectively 0.20, 0.25, and 0.24 kg C m^{-2} y^{-1} (Figure 9). These rates are in agreement with rates compiled by Raich and Schlesinger (1992) for boreal forests. The rate differences for the three profiles can be explained by differences in carbon pool size of each profile. In particular the Oa horizons of the central and south profiles (Table 1; south pit) are thicker than the Oa of the north profile.

Differences between DOC in- and out-flux per layer provide insight into the dynamics of carbon related processes through the seasons (Figures 10-13). Enhanced DOC production and capillary rise resulted in a positive DOC in-flux into the O horizons during the dry summer months of June and July. Late summer

Modeling the Dynamics of Organic Carbon in a Typic Haplorthod

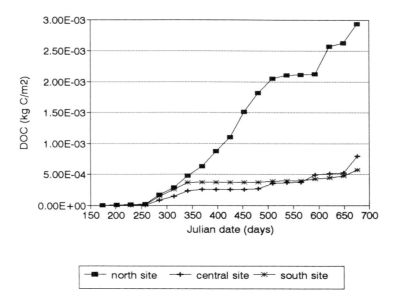

Figure 8. Cumulative DOC drainage from the bottom of 1.0*1.0*1.1 m³ profiles.

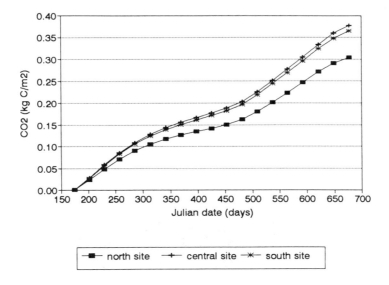

Figure 9. Cumulative CO_2 respiration from 1.0*1.0*1.1 m³ profiles.

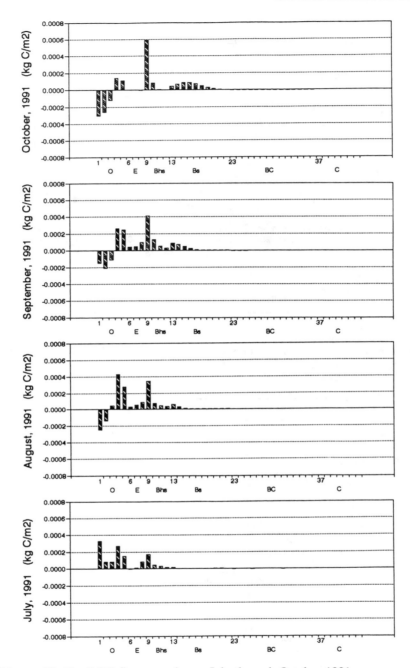

Figure 10. Net DOC fluxes per layer, July through October 1991.

Modeling the Dynamics of Organic Carbon in a Typic Haplorthod

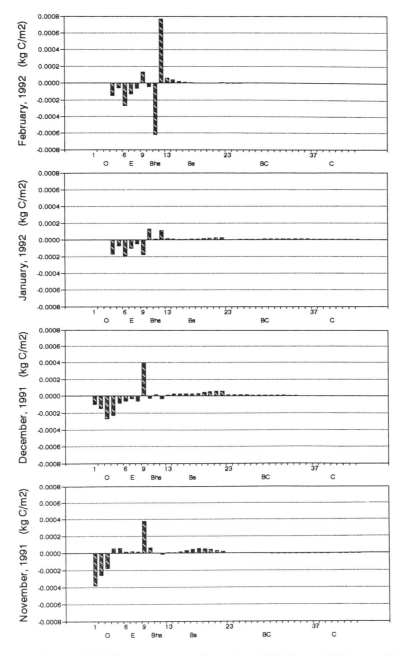

Figure 11. Net DOC fluxes per layer, November 1991 through February 1992.

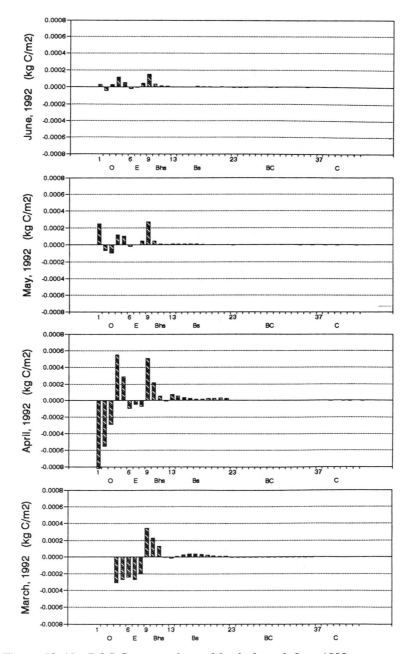

Figure 12. Net DOC fluxes per layer, March through June 1992.

Modeling the Dynamics of Organic Carbon in a Typic Haplorthod

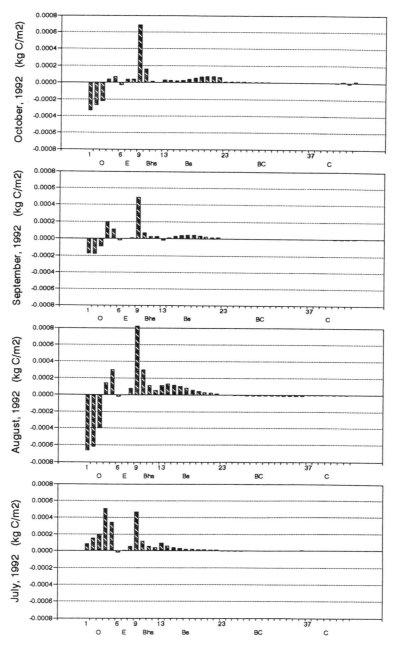

Figure 13. Net DOC fluxes per layer, July through October 1992.

and fall precipitation leached DOC from the Oi into the Oa, E, and B horizons. In particular, the upper layer of the Bhs (Bh) received a relatively large in-flux of DOC coming through the E horizon, a pulse of DOC moved through the layers of the Bs horizon downward. The Oi horizon was frozen during the months of January through March, 1992, but a strong out-flux of DOC occurred after snowmelt in April. The Oa, Bhs, and Bs horizons experienced a corresponding in-flux of DOC after snowmelt. The fall of 1992 showed similar trends compared to 1991, although with slightly larger net fluxes due to a "wetter" year. DOC was leached from the Oi and was adsorbed (and decomposed over time) in the Oa, Bhs, and Bs horizons. Adsorption took place primarily in the upper Bhs (Bh)horizon. DOC influx in the Bs gradually shifted from the upper part to the lower part of the horizon during the period of August through October. DOC movement in the BC and C horizons was very low throughout the simulated period.

Simulation results from the pedodynamic model were consistent with the observed morphology of the soil profile and provided insights into the seasonality of soil forming processes. The simulation approach appears to be a valid method for assessing or predicting changes in C sequestration in response to various global change scenarios, especially when changes in seasonal conditions are not reflected in mean annual climate parameters.

Acknowledgements

This paper is dedicated to Mr. Ralph Friedman who died on July 21, 1992 at the age of 88. We are grateful to his son, Mr. Robert E. Friedman and the members of the Kildare Club for allowing this research on the Kildare Club property. The authors thank Basil and Norma Cheney for their hospitality and friendship and Judson Sanford for his friendship and discussions about nature. The authors acknowledge the help of Dr. Leonard W. Lion of the Department of Civil and Environmental Engineering, Cornell University, and the use of the department's carbon analyzer.

References

Bohn, H.L., B.L. McNeal, and G.A. O'Connor. 1985. *Soil Chemistry*. Wiley-Interscience, New York.
Bouwman, A.F. 1989. Exchange of Greenhouse Gases Between Terrestial Ecosystems and the Atmosphere. In: A.F. Bouwman (ed.). *Soils and the Greenhouse Effect*. John Wiley & Sons, New York.
Bretherton, F. 1989. Concluding Remarks. In: A.F. Bouwman (ed.). *Soils and the Greenhouse Effect*. John Wiley & Sons, New York.
Campbell, G.S. 1985. *Soil physics with BASIC; transport models for soil-plant systems*. Elsevier, Amsterdam.
Feytel, T.C.J. and E.L. Meijer. 1989. Simulation of Soil Forming Processes. Dept. of Soil Science and Geology, Agricultural University Wageningen, The Netherlands.
Greenland, D.J and P.H. Nye. 1959. Increase in the carbon and nitrogen contents of tropical soils under natural fallows. *J Soil Science* 10:284-299.
Hoosbeek, M.R. and R.B. Bryant. 1992. Towards the quantitative modeling of pedogenesis - a review. *Geoderma* 55:183-210.
Hoosbeek, M.R. and R.B. Bryant. 1994a. Developing and Adapting Soil Process Submodels for use in a Pedodynamic Model. In: R.B. Bryant and R.W. Arnold (eds.). *Quantitative modeling of soil forming processes*. SSSA Special Publication, WI. (In press)
Hoosbeek, M.R. and R.B. Bryant. 1994b. A pedodynamic approach to Typic Haplorthod. (In preparation)
Hutson, J.L. and A. Cass. 1987. A retentivity function for use in soil- water simulation models. *J. Soil Sci.* 38:105-113.
Hutson, J.L. and R.J. Wagenet. 1992. LEACHM; Leaching Estimation And Chemistry Model. Dept. of Soil, Crop and Atmospheric Sciences. Research Series No. 92-3. Cornell University.
Janssen, B.H. 1984. A simple method for calculating decomposition and accumulation of "young" soil organic matter. *Plant and Soil* 76:297-304.
Jenkinson, D.S., D.E. Adams, and A. Wild. 1991. Model estimates of CO_2 emissions from soil in response to global warming. *Nature* 351:304-306.
Jenkinson, D.S. and J.H. Rayner. 1977. The turnover of soil organic matter in some of the Rothamsted classical experiments. *Soil Sci.* 123:298-305.

Jenny, H., S.P. Gessel, and F.T. Bingham. 1949. Comparative study of decomposition rates of organic matter in temperate and tropical regions. *Soil Sci.* 68:419-432.

Johnsson, H., L. Bergstrom, P-E. Jansson, and K. Paustian. 1987. Simulated nitrogen dynamics and losses in a layered agricultural soil. *Agriculture. Ecosystems and Environment* 18:333-356.

Kortleven, J. 1963. Kwantitative aspecten van humusopbouw en humusafbraak. Verslagen Landbouwkundige Onderzoekingen 69-1. PUDOC. Wageningen, The Netherlands. 109 pp.

New York State Museum/Geological Survey. 1991. Geology of New York. The State Education Dept., The University of the State of New York. Educational leaflet no. 28.

NSSL. 1991. National Soil Survey Laboratory Methods & Procedures. Lincoln, Nebraska.

Olson, J. 1963. Energy storage and the balance of producers and decomposers in ecological systems. *Ecology* 44:322-331.

Parton, W.J., D.S. Schimel, C.V. Cole, and D.S. Ojima. 1987. Analysis of Factors Controlling Soil Organic Matter Levels in Great Plain Grasslands. *Soil Sci. Soc. Am. J.* 51:1173-1179.

Raich, J.W. and W.H. Schlesinger. 1992. The global carbon dioxide flux in soil respiration and its relationship to vegetation and climate. Tellus 44B:81-99.

Sposito, G. 1989. *The Chemistry of Soils*. Oxford University Press. New York.

Tillotson, W.R., C.W. Robbins, R.J. Wagenet, and R.J. Hanks. 1980. Soil water, solute, and plant growth simulation. Bulletin 502. Utah State Agr. Exp. Stn., Logan, Utah.

Van Breemen, N. and T.C.J. Feijtel. 1989. Soil Processes and Properties Involved in the Production of Greenhouse Gases, with Special Relevance to Soil Taxonomic Systems. In: A.F. Bouwman (ed.), *Soils and the Greenhouse Effect*. John Wiley & Sons, New York.

Van Genuchten and Wierenga. 1986. Solute Dispersion Coefficients and Retardation Factors. p. 1025-1054. In: A. Klute (ed.). Methods of Soil Analysis, Part 1, Physical and Mineralogical Methods. ASA/SSSA, Madison, WI.

Van Veen, J.A. and E.A. Paul. 1981. Organic carbon dynamics in grassland soils. 1. Background information and computer simulation. *Canadian J. Soil Sci.* 61:185-201.

Wagenet, R.J. 1986. Water and Solute Flux. p. 1055-1088. In: A. Klute (ed.). Methods of Soil Analysis, Part 1, Physical and Mineralogical Methods. ASA/SSSA, Madison, WI.

Wagenet, R.J. and J.L. Hutson. 1987. LEACHM: Leaching Estimation and Chemistry Model. Continuum Vol.2. Water Resources Inst., Cornell University. Ithaca, NY.

CHAPTER **35**

Towards Improving the Global Data Base on Soil Carbon

R. Lal, J. Kimble, E. Levine, and C. Whitman

I. Introduction

Radiatively-active or "greenhouse" gases trap longwave radiation which warm the earth and may influence global temperature. Although water vapor (H_2O) is the most predominant greenhouse gas, it is the atmospheric concentrations of carbon dioxide (CO_2), methane (CH_4), nitrous oxide (N_2O), and chlorofluorocarbons (CFCs) that have been increasing steadily since the 1890s and substantially since the 1950s. Present atmospheric concentrations of these gases are 345 ppm (parts per million) for CO_2, 90 ppb (parts per billion) for CH_4, and 1.65 ppm for N_2O. Annual increase in atmospheric concentration of these gases is 0.5% for CO_2, 0.8% for CH_4 and 1.0% for N_2O. Per molecule, CH_4 is 32 times more effective, and N_2O is approximately 150 times more effective at trapping longwave radiation than CO_2. The primary concern is that additional energy trapped in the atmosphere may lead to global warming and concomitantly to a change in global climate.

II. Soils and Biogeochemical Cycles

World soils play a major role as a source or sink for CO_2 and CH_4 and to a lesser extent for N_2O. Soil use and management affect the rate, type, and direction of gaseous flux, and may have an important impact on global climate. World soils contain about twice as much carbon as contained in the atmosphere and about three times that contained in the world's biota. Small changes in carbon pool in soils could have large effects on concentrations of atmospheric CO_2, CH_4, and N_2O. Soil and vegetation are estimated to have contributed 90 to 120 Pg of C to the atmosphere from 1850 to 1980, or an average of about 0.7 to 0.9 Pg of C annually. Burning fossil fuel releases about 6 Pg (Pg = petagram = 1 x 10^{15} g) of carbon annually, of which 3.4 Pg is retained in the atmosphere.

World soils also contain large amounts of nitrogen. Along with fertilizers, organic manures and urban wastes applied to the soil, soil nitrogen pool is the principal source of N_2O and other NO_x gases released into the atmosphere. About 80 percent of nitrogen stored in the world soils is contained in the top 10 cm of the soil surface, the layer of active disturbance for all agricultural land uses.

III. Land Use and Carbon Fluxes

There are several land use activities that accentuate carbon fluxes from soils in solid or gaseous forms. Important among these are deforestation and afforestation, biomass burning, cultivation, crop residue management, application of inorganic fertilizers and organic manures, rice paddy cultivation, and farming systems. Once the land is changed from forest or prairies to agricultural use, there is a rapid change in soil organic carbon (SOC) within first 5 to 25 years with losses ranging from 25 to 75% in the upper most soil horizons and somewhat less in the lower horizons. The loss is usually more from soils containing high than low levels of SOC. In addition to agricultural land use, forest ecosystems and forest soils also play a major role in global carbon budget, and account for about 60% of the terrestrial carbon. Therefore, management of forest resources and forest soils is also important to the global carbon balance.

IV. Soil Processes and Carbon Fluxes

Important processes affecting carbon fluxes from soils to the atmosphere include mineralization of SOC, soil erosion, nitrogenous transformations, and anaerobiosis. The magnitude of fluxes by these processes varies widely among soil types and climates. Rates of mineralization of SOC and accelerated erosion are generally more in the tropics than in temperate zone environments. The current net annual loss of carbon in gaseous form from plants and soils in the tropics is about 10 times that from the temperate region e.g. 2 Pg C/yr vs. 0.2 Pg C/yr. Total carbon displaced annually from world soils by water erosion is 5.7 Pg of which about 0.5 Pg is translocated to oceans and 1.14 Pg may be emitted into the atmosphere as CO_2.

Cold soils in northern ecosystems represent 25 to 33 percent of the total world soil carbon pool. Arctic tundra systems alone are estimated to contain about 12 percent of soil carbon. In the event of global warming, these ecosystems could become a major carbon source. Wetlands, peats, and rice paddies are other major components of the global carbon budget. Wetlands occupy over 6% of the land surface and are a major sink for carbon. Carbon transformation in wetlands mostly occurs under reducing conditions. Carbon addition to wetlands by peat deposition is estimated at 0.06 to 0.08 Pg/yr. However, about 25% of annual carbon stored in wetlands is released as CH_4 amounting to emission rate of 0.09 Pg C/yr. Methane emission from rice paddies is an additional 0.08 Pg C/yr.

V. Global Warming and Gaseous Emission from Soils

One of the major concerns is the impact of possible global warming on gaseous emission from soils. Increase in global temperature may cause a massive shift in carbon now sequestered in soils. A possible increase of 3°C in temperature over the next 50 years may decrease SOC by 10 to 15% in the top 30 cm of soils leading to a total emission of 50 to 100 Pg of C into the atmosphere. Deforestation for expansion of cultivated areas in the tropical regions may lead to an overall reduction in the global soil carbon pool by 20% and a total emission of 150 Pg of carbon. This magnitude of carbon loss from world soils would exacerbate the adverse effects of global warming.

VI. Soil Management for Carbon Sequestration

Judicious management of world soils provides an opportunity to sequester or store carbon in terrestrial ecosystems, reduce greenhouse gas emissions from soils, and increase their sink strength. Currently, the global area of cropland is estimated at 1.5 billion ha. These soils represent a potential source to sequester carbon with proper management. Additionally, some degraded soils now estimated at 2.0 billion ha could be restored and managed to sequester additional carbon.

Important soil management practices to sequester carbon are conservation tillage, mulch farming, water management including irrigation and drainage and in-situ conservation of water in the root zone, soil fertility management including use of organic wastes and rhizobium inoculation, liming, and acidity management, use of improved cultivars, and introduction of new and efficient crop species. With improved land use cultivated and degraded soils can annually sequester 0.1 to 1 Pg of carbon depending on the management.

VII. Policy Considerations

There are several important policy considerations which may encourage adoption of land use practices that minimize gaseous emissions and enhance carbon sequestration in soils. Costs and returns and economics of changing agricultural practices and land use systems play an important role. Farmers must survive within changing world grain market and world prices, and the costs of compensating farmers to change their practices is significant and needs to be properly evaluated. Arctic and forest ecosystems play an important role in global carbon budget. Economics, policy ,and social and cultural factors are important considerations in sustainable management of these ecosystems. Socio-cultural factors must be addressed in judicious management of fragile and ecologically-sensitive ecosystems.

VIII. Researchable Priorities

Considering the importance of world soils in global carbon and nitrogen balances and their potential role as source or sink of greenhouse gases, it is important to remove uncertainties and replace estimates and myths with facts. The following issues are important researchable concerns that need to be addressed.

A. Long-Term Soil Management Experiments

Long term field experiments conducted in widely different environments and agroecosystems are needed to generate data for critical regions. The importance of long term research, much of it developed initially not because of questions related to gaseous emissions, but because of practical concerns of agronomy, or scientific curiosity cannot be overemphasized. Future research should include elements of both problem-directed research and research that is investigator initiated based on new ideas driven by scientific curiosity. Coordination of the former is essential to solving some of the current questions, and may require the redirection of funds from areas with an already reasonable database (North America and western Europe, for example) to critical biomes in the tropics or high latitude regions. Long-term research is needed for understanding of fundamental processes including the following:

(a) enhancing understanding of the turnover and processes in C cycling in agroecosystems and managed ecosystems,

(b) determining the fate of carbon in soil displaced by water, wind and other agents of erosion,

(c) predicting the dynamics of carbon in cold soils, with regards to the magnitude of temperature change, increase in biomass production, nutrient availability and net photosynthesis, soil moisture regime, radiation levels, species composition, and soil properties, etc,

(d) assessing the ability of long-lived forest species to respond to both the abiotic environment and to biotic feedback processes at the ecosystem level,

(e) resolving uncertainties about the long term fate of carbon stored in landfills, and

(f) developing soil maps at local and regional scales for more realistic assessment of terrestrial C status.

B. Soil Dynamic in Fire-Dependent Ecosystems

Measurements of pools and fluxes of carbon and nitrogen from fire-dependent and agricultural ecosystems indicate that fire is a major disturbance that can substantially influence direct contributions of CO_2, CH_4, and N_2O to the atmosphere. The global magnitude of these contributions is not known. Simply counting the numbers of fires is not enough. In depth information on the effects of burning on pools and fluxes of soil carbon and nitrogen is not yet available.

C. Global Fluxes

With certain gases, obtaining global estimates of flux and sources are severely limited by the near absence of data. Measurements are often sparse in areas that are thought to be critical, such as estimates of N_2O from warm, moist biomes such as the savannas, and CH_4 evolution from cold and wet tundra biomes.

D. Methodology Standardization

Standardization of methods is needed to enable data to be compared across space and time. Frequency of measurement is crucial to capture critical short-lived events. Static and dynamic methods for measuring fluxes should be compared, calibrated and, standardized. In all experiments, a minimum data set for management and environment should be collected. This minimum data set should be developed jointly by experimenters and modelers. The time-step and spatial scale for collecting this minimum data set should be specified:

(a) Geographic information systems (GIS) are an important tool for managing large databanks. Each GIS system should be designed with a specific purpose, so that relevant data could be included. Errors in data could be propagated through the various GIS layers and be magnified. Compensating errors can mask the magnitude of error in large-scale estimations. The need for developing information for GIS layers at appropriate scales cannot be overemphasized. In addition to classifying information, GIS can also provide a means for establishing relationships among properties, factors, and processes. GIS can provide weighted averages on properties for more realistic assessment.

(b) Owing to the high spatial and temporal variability of the experimental results, and their highly site specific nature, the need for simulation models that can mimic conditions in the field becomes increasingly evident. This need requires that experimenters and modelers work in a more coordinated fashion. Modelers can also help experimenters by conducting sensitivity analyses to identify parameters that require high degree of accuracy and precision, and advise the experimenters of critical knowledge gaps that prevent modelers from making progress.

E. Policy Issues

High priority should be given to socioeconomic and policy research. The macroeconomic analysis can be used to link climate change scenarios with climate yield models and world agricultural trade models. The results of such analysis are useful in the design of response strategies, possible future economics development in less developed countries, and insights into economic and other social systems even without accurate regional climate prediction. Agricultural models can also be used to evaluate the impact of agronomic practices on SOC sequestration and N_2O emissions.

Microeconomic analyses are needed for evaluating changes in agronomic practices and agricultural land use. Costs and returns of changing practices vary with region. The feasibility of a practice such as a change in the crop rotation by inclusion of a winter cover crop is determined by physical impacts such as soil moisture, costs of planting and plowing, and the returns that a farmer can get at the market place. A farmer is dealing now with a world grain market and world prices. In both the analyses of farm practices and tree planting, the impact of changes to other economic sectors is important. These practices need to be thoroughly evaluated.

Index

adsorption 172, 174, 311, 419, 424, 425, 432
aerosols 266, 268
afforestation 3, 5, 50, 54, 57, 433
agricultural land 16, 17, 46, 47, 50, 131, 222, 399-402, 405, 406, 408, 409, 433, 436
agroecosystems 145, 147, 150, 209, 211, 216, 385, 394, 435
air temperature 120, 122, 161, 179, 180, 183, 196, 203, 238, 364, 369, 378-380
airborne fraction 16
albedo 265, 266, 268, 273
ambient temperature 126
amino acids 71, 172, 317, 322, 324, 347
Andisols 29, 35, 36, 40, 307
arctic soils 125
atmospheric oxidants 221
biomass 1, 3-6, 15, 17, 18, 20, 32, 35, 46, 50-54, 56, 57, 67, 96, 145-147, 199, 205, 207, 210, 211, 213-215, 221, 222, 231, 262, 265, 267, 268, 272, 273, 277, 279, 297-301, 312, 344, 346, 348, 351-359, 385, 397, 404-407, 409, 410, 412, 433, 435
biotic pump 16
bogs 3, 4, 117, 153, 155, 224, 279, 368, 377
boreal 3, 4, 17, 19, 46, 48, 49, 51, 54, 56, 117, 126, 135, 153, 155, 156, 158, 159, 161, 166, 222, 226, 268, 271, 368, 399-401, 405, 408, 416, 425
buffaloes 149, 150
burning 1, 3-5, 19, 41, 46, 50, 54, 67, 143, 145-147, 150, 221, 222, 226, 231, 279, 343, 405, 409, 410, 433, 435
^{13}C 11, 16, 56, 167-169, 174, 279, 298, 301-304, 317, 318, 320-324, 326, 351-359
calcic horizon 87
caliche 2, 17
carbon cycle 2, 4, 9, 11, 14-17, 27, 41, 45, 68, 209, 275-278, 290, 339, 343, 344, 396, 405
carbon dating 298, 301, 304
carbon dioxide 1, 4, 9, 10, 16, 31, 46, 68-70, 72, 74-77, 105, 119, 143, 150, 161, 165, 177, 178, 181, 184, 199, 201, 209, 213, 224, 239, 243, 275-277, 279, 285, 361, 362, 366, 373, 395-397, 404, 412, 415, 433
carbon dioxide flux 119, 177, 181, 184, 362
carbon dynamics 28, 131, 138, 209, 213, 351, 385, 411, 415
carbon sequestration 3, 20, 21, 27, 32, 67, 101, 102, 146, 150, 385, 394, 434
carbon storage 4, 17, 20, 21, 27, 45, 46, 58, 61, 67, 69, 71, 75, 77, 82, 93, 95, 101, 102, 117, 150, 277, 397
carbonate 2, 10-12, 14, 17, 27, 28, 33, 35, 36, 39, 41, 81-86, 88, 89, 209, 330
cattle 47, 145, 149, 150, 386
CH_4 see also methane 1-3, 75, 105, 117, 118, 120, 121, 123-126, 147, 149, 153, 155, 157, 158, 160-162, 189-194, 196, 220, 221, 232, 243, 247-250, 276, 279, 315, 367, 368, 377-379, 381-383, 396, 433-435
chemical fractionation 298, 346, 348
chemodenitrification 223, 225, 237
classical fractionation 346
climate change 1, 3, 18, 20, 27, 81, 117, 118, 120, 125-127, 161, 162, 177, 266, 276, 278, 280, 285, 298, 302, 315, 339, 343, 345, 377, 385, 396, 406, 407, 410, 412, 436
CO_2-fertilization 17
coal 11, 14, 68, 69, 222
comets 9
compost 191, 192
corn 112, 137, 178, 193, 209, 211, 213, 214, 227, 265, 301, 302, 347, 354, 355, 358, 379, 385-387, 390, 394
crop rotations 106, 107, 385
cropland 105, 112, 131, 177, 434
crust 9-12, 185
cryic soils 307
cultivated land 47, 52, 150, 172
decomposition 1, 3, 4, 11, 19-21, 29, 31, 46, 57-59, 68-70, 72-77, 95, 96, 101, 117, 125, 127, 135, 138, 165, 166, 168-174, 177, 178, 183, 185, 209-212, 223, 230, 278, 297-299, 301, 304, 313, 326, 339, 353, 354, 357, 358, 362, 383, 385, 388, 397, 403-406, 409, 412, 415, 418, 419
deforestation 1, 3, 5, 28, 46, 47, 50-54, 56, 57, 59, 61, 145, 146, 150, 222, 231, 278, 405, 406, 409, 412, 433, 434
degradation 5, 47, 50-52, 56, 57, 69, 72-74, 94, 102, 143, 145, 169, 210, 213, 285, 297, 355, 357-359
denitrification 174, 182, 223-228, 230-232, 237, 238, 240, 243, 244, 381
desert soils 35
dolomite 14
drainage 4, 31, 41, 93, 97, 102, 125, 132, 134, 153, 158, 160-162, 230, 258, 260, 265, 269, 272, 279, 425, 427, 434
ecology 149, 189

elemental analyses 317, 326
evapotranspiration 108, 131, 132, 161, 361, 369, 370, 372, 402, 404, 405, 415
extractants 168, 309, 346
extraction efficiency 346
FACE 17, 298, 302, 303
fertilization 17, 21, 57, 105, 106, 125, 171, 189, 199, 203, 226, 231, 329, 339, 387, 388, 390, 394, 397, 402, 409-412
fertilizer 3, 17, 20, 21, 144, 149, 150, 189-193, 196, 244, 297, 329, 386-388, 390, 393, 394
flora 126
forest 3-6, 16, 17, 19-21, 32, 35, 45-54, 56, 57, 61, 67, 89, 96, 101, 105, 117, 126, 137, 138, 143, 145-147, 157, 161, 165, 166, 172, 199, 200, 202, 221, 226, 227, 228, 232, 259, 260, 265, 268, 269, 270, 271, 290, 301, 308, 311, 312, 329-332, 335, 338-340, 343, 347, 351, 361, 363, 364, 366, 368, 370, 372, 373, 399-401, 405, 406, 408, 416, 433-435
fossil fuel 1, 11, 16, 18-20, 45, 56, 61, 105, 221, 222, 275, 343, 395, 433
fractal geometry 257
fulvic acids 71, 307, 309, 311-313, 315-326
geostatistics 257
global warming 1-3, 14, 17, 19, 27, 58, 61, 88, 94, 102, 117, 125, 126, 143, 147, 231, 247, 275-277, 290, 315, 364, 377, 433, 434
goats 145, 150
gypsum 11, 14
haplorthod 271, 415, 416, 418, 425
Henry's Law 9, 16, 72, 182
humic acids 71, 170, 309-313, 315-320, 322-326
humic substances 19, 166, 168, 169, 171, 172, 307, 308, 311, 318
irrigation 105, 106, 143, 224, 228, 386, 434
kinetics 301, 302, 343, 358
land use 1, 5, 6, 17, 32, 34, 41, 45, 46, 50, 52, 54-59, 61, 101, 105, 136, 143, 144, 196, 232, 259-261, 278, 280, 343, 395, 396, 405, 409, 411, 415, 433, 434, 436
leaching 232, 310, 316, 321, 419, 425
livestock 47, 143-145, 149, 150, 396
logging 46, 50, 51, 53, 54, 56, 57, 61, 260, 316, 330
manure 191-193, 213, 214, 237-240, 242-244, 297, 377, 379-381
Mars 10
mathematical models 297
methane see also CH_4 1, 4, 5, 35, 68, 69, 72, 74-77, 105, 118, 119, 143, 147-150, 153-162, 189, 192, 243, 247, 248, 276, 277, 279, 315, 326, 338, 361, 362, 366-368, 373, 377, 380, 381, 383, 433, 434
methane flux 4, 118, 119, 148, 154, 155, 157-159, 161, 162, 366-368, 377, 380, 381, 383
methane production 149, 155, 157-162, 247, 248, 315, 367, 368, 381
microbial biomass 297-301, 344, 346, 348, 351-353, 355, 357-359, 385
microbial decomposition 29, 31, 68, 183, 418, 419
missing sink 17, 21, 277
modeling 6, 17, 105, 106, 108, 113, 229, 231, 260, 265, 268, 273, 275, 276, 280, 290, 347, 415
^{15}N 238-242, 244
net primary production 15, 21, 45, 117, 125, 209, 210, 277, 396, 397, 399, 400, 404, 406, 407, 409-411
nitrification 223-228, 230-232
nitrogen 1-3, 6, 10, 17, 20, 21, 35, 40, 41, 58, 59, 61, 88, 149, 150, 171, 177, 193, 199, 202, 205, 206, 241-243, 277, 317, 329, 330, 335, 339, 347, 348, 351, 362, 385-387, 390, 393, 394, 402, 433, 435
nitrogen fractionation 347
no-till 20, 178, 183, 185, 359, 386-388, 390, 392-394
nonhumic substances 309, 313
nutrient availability 125, 397, 402, 435
nutrient mineralization 117, 126, 277
oceans 1, 4, 9-11, 14, 16, 17, 45, 94, 131-133, 135, 137, 138, 141, 221, 222, 275, 277, 278, 343, 434
organic matter 14, 17, 19-21, 27-29, 31, 32, 46, 51-54, 58, 59, 61, 68, 69, 71, 82, 95, 96, 118, 126, 135, 143, 150, 165, 166, 168-174, 177, 178, 182, 183, 185, 189, 190, 192, 193, 196, 209-211, 213-215, 221, 223, 230, 248, 249, 250, 259, 261, 277-280, 286-288, 290, 297, 298, 300, 307, 316, 318, 330-332, 334, 339, 344-348, 351-354, 356, 358, 359, 362, 381, 385, 386, 388, 394, 415, 418
ORTHOD model 415-417
ozone 1, 27, 219, 237, 243, 266, 362
pasture 20, 47-49, 52, 56, 221, 226, 372, 386, 388
peatlands 4, 19, 33, 41, 153, 155, 158, 161, 162, 165, 367, 368
pedodynamics 415
pedology 257
pedosphere 257-260
petroleum 11, 68, 69, 302

photosynthesis 1, 11, 14, 58, 59, 117, 118, 125, 177, 215, 344, 362-364, 378, 396, 402, 435
photosynthetic capacity 125
population 17, 21, 69, 85, 126, 135, 144, 145, 149, 150, 166, 169, 326, 346, 358, 359, 406, 409, 418
precipitation 14, 19, 32, 61, 73, 82, 87-89, 96, 106-108, 110, 117, 125, 131, 150, 161, 200, 222, 224, 231, 267, 269, 277, 280, 309, 317, 348, 380-383, 386, 397, 405, 406, 411, 415-417, 422, 432
proximate analysis 346
pyrite 11, 14
radioisotope fractionation 347, 348
rainfed conditions 150
redox 189, 190, 192, 237, 247-250, 418
reductant capacity 247, 249, 250
reforestation 21, 46, 47, 50, 51, 54, 57
remote sensing 143, 265, 266, 268, 272, 273, 282
residues 1, 17, 19, 29, 69, 166, 169, 170, 172-174, 191, 210-213, 215, 227, 297, 299, 300, 303, 344, 346, 385, 388, 418
respiration 1, 6, 11, 15, 19, 20, 46, 57-59, 61, 117, 177, 178, 180, 183, 185, 201-203, 205, 209-212, 211, 213, 215, 216, 248, 250, 277, 279, 297, 300, 304, 329, 338, 346, 362, 363, 366, 378, 396, 397, 402, 405, 409, 410, 411, 412, 418, 419, 427
rhizosphere 161, 189
rice 3, 4, 143, 144, 147-150, 189-193, 195, 196, 247, 279, 377, 381, 433, 434
rivers 4, 10, 131, 132, 134-137, 141, 190, 193
satellite 50, 155, 219, 265, 266, 268, 269, 273, 280
sediments 10-14, 19, 27, 29, 67, 68, 85, 94, 95, 97, 102, 132, 135, 137, 141, 224, 258
sheep 145, 149, 150
shifting cultivation 5, 47-50, 52-54, 57, 144-146, 405
simulation models 28, 230, 232, 260, 415, 436
snowfall 118
soil carbon 2, 3, 5, 6, 19, 20, 27, 31, 34, 41, 51, 52, 54, 56, 81, 84-86, 93, 95, 96, 101, 102, 105, 141, 150, 177, 178, 183, 199, 215, 257, 260, 261, 265, 266, 268, 271-273, 275, 277-279, 287, 290, 295, 310, 344, 345, 347, 385, 386, 388, 394, 397, 406, 433-435
soil freezing 118, 241
soil leachates 309

soil moisture 33, 46, 121, 122, 124, 201-203, 205, 215, 265-268, 273, 277, 331, 340, 364, 365, 369, 370, 372, 405, 416, 417, 435, 436
soil organic matter pools 347, 359
soil processes 4, 19, 31, 189, 260, 415, 416, 434
soil properties 96, 105, 120, 160, 238, 249, 257-261, 265, 273, 277, 278, 280, 283-286, 435
soil survey 34, 82-85, 88, 95, 99, 191, 257-262, 269, 270, 279, 283, 285, 288, 307-309, 312, 330, 344
soil temperature 19, 20, 162, 182, 183, 189, 199, 202-204, 206, 215, 225-227, 230, 231, 239, 273, 279, 308, 331, 332, 362, 364, 379, 381, 382, 390, 394, 403, 418, 419
soil texture 32, 106, 112, 226, 259, 386
soil water content 118, 120, 124-126, 180, 201, 225, 226, 228-231
solar radiation 1, 179, 189, 196, 368, 377
soybeans 112, 209-212, 215, 301, 302
spatial data 105-108, 110, 113, 114, 282-284
spodosols 29, 31, 34, 36, 40, 307, 308, 312, 315, 415, 425
statistics 108, 131, 138, 144, 145, 190, 193, 195, 257, 258, 260, 287
sulfate reduction 11
tallgrass prairie 209, 210, 213
terrestrial carbon 3, 34, 45, 46, 54, 56, 60, 61, 67, 81, 102, 277, 343-345, 395, 396, 401, 406, 433
thawing 237, 240, 244
thermistors 416
throughfall 416, 420-422
tillage 5, 28, 101, 105-108, 173, 174, 177-186, 209, 213, 216, 297, 351, 356, 379, 385, 386, 387-391, 393, 394, 434
topography 155, 257, 266, 396
tree rings 16, 61
tropical forest 35, 47, 51, 226-228, 265
tropics 4, 5, 16, 17, 19, 31, 33, 35, 46, 47, 50, 51, 53, 54, 56, 61, 138, 279, 301, 405, 406, 434, 435
tundra 3, 4, 17, 19, 20, 41, 48, 49, 117-120, 123-126, 135, 155, 158, 161, 221, 307, 377, 399-402, 408, 434, 435
urea 149, 150, 330
Venus 10
volcanoes 10, 12
water movement 276, 416, 417
water table 117, 124, 155, 157-162, 363, 367

water vapor flux 361, 368, 378
weathering 10, 11, 13, 14, 89, 307
wetland 4, 95, 96, 102, 154, 155, 158, 159, 161, 162, 247, 248, 250, 290, 367, 368, 408
wheat 112, 178, 179, 183, 185, 193, 209-216, 297, 299, 301, 347, 363, 364, 385